SUPPLY CHAIN GAMES:
OPERATIONS MANAGEMENT AND RISK VALUATION

* A list of the early publications in the series is at the end of the book *

SUPPLY CHAIN GAMES:

OPERATIONS MANAGEMENT AND RISK VALUATION

Konstantin Kogan
Bar-Ilan University, Israel

Charles S. Tapiero
Polytechnic University of New York, US and ESSEC, France

 Springer

Konstantin Kogan
Bar-Ilan University
Israel

Charles S. Tapiero
Polytechnic University of NY, USA
and ESSEC, France

Series Editor:
Fred Hillier
Stanford University
Stanford, CA, USA

Library of Congress Control Number: 2007929861

ISBN 978-0-387-72775-2   e-ISBN 978-0-387-72776-9

Printed on acid-free paper.

springer.com

For Kathy and Carole
For their continuous care and support

PREFACE

Operations and industrial modeling and management have a long history dating back to the first Industrial Revolution. Scheduling, inventory control, production planning, projects management, control charts, statistical records, customer satisfaction questionnaires, rankings and benchmarking. are some of the tools used for the purpose of better managing operations and services. The complexity of operations and logistics problems have increased, however, with the growth of supply chains, rendering traditional operational and risk management issues far more complex and strategic-game-like at the same time. Similarly, we have gained increased experience in defining, measuring, valuing and managing risks that result from the particular environment that supply chains create. Increasingly, there is a felt need for convergence between the traditional tooling of industrial-logistics and the economic realities of supply chains operating on a global scale. This book provides students in logistics, risk engineering and economics as well as business school graduates the means to model and analyze some of the outstanding issues currently faced in managing supply chains.

The growth and realignment of corporate entities into strategic supply chains, global and market sensitive, are altering the conception of operations modeling. Now far more strategic and sensitive to external events and to their externalities, they require new avenues of research. There is a need to rethink and retool traditional approaches to operations logistics and technology management so that these activities will be far more in tune with an era of global, cross-national supply chains.

Today, supply chains are an essential ingredient in the quest for corporate survival and growth. Operations strategy in supply chains have mutated, however, assuming ever-expanding and strategic dimensions and augmenting appreciably the operational complexity and risks that modern enterprises face when they operate in an interdependent supply chain environment.

These operational facets imply a brand new set of operational problems and risks that have not always been understood or managed. Supply chain managers have thus an important role to assume by focusing attention on these operations and risks and in educating corporate managers about what these operation problems and their risks imply.

Our purpose in this book will be to consider these problems in depth and to draw essential conclusions regarding their management in supply chains. For example, traditional operational problems (such as inventory control, quality management and their like) are expressed in a strategic and intertemporal manner that recognizes the complexity and the interdependency of firms in a supply chain environment. Examples that highlight our concerns and how to deal technically with these problems will be extensively used.

The book is directed necessarily towards advanced undergraduate students but will be made accessible to students, including those in operations engineering, who have a basic understanding of mathematical tools such as optimization, differential equations and some elements of game theory. When necessary, the book will utilize appendices to review basic mathematical tools, emphasizing their application rather than the theoretical underpinnings. Similarly, a number of computer programs will be used for calculations, bridging the gap between theory and practice.

The book consists of three areas, each intimately dependent on one another, each emphasizing important facets of supply chains management operations. These include:

- Supply Chains and Operations Modeling and Management
- Intertemporal Supply Chains Management
- Risk and Supply Chain Management

The first area provides both traditional static and discrete-time models and their gradual extension to a supply chain environment, highlighting the new concerns of the supply chain environment. In addition, it emphasizes both one- and two-period problems while in the second area, we address essentially inter-temporal problems as differential games. The differential games are presented as natural continuous-time extensions of the corresponding static models so that the effect of various types of dynamics on supply chains may be assessed and insights gained. The third area deals with risk and supply chains as well as with numerous applications to the management of quality in a supply chain environment and in managing interdependent (both in substance and in decision-making) operations. In this sense, the book highlights and resolves some important problems that address directly the needs and the complexity of supply chain management in a tractable and strategic setting.

CONTENTS

Part III: Risk and Supply Chain Management

PART I
SUPPLY CHAINS AND
OPERATIONS
MODELING AND
MANAGEMENT

1 SUPPLY CHAIN OPERATIONS MANAGEMENT

Operations and industrial modelling and management have a long history dating back to the first industrial revolution. With the growth of supply chains, the complexity of operations and logistics problems has increased rendering traditional operational management issues far more complex and of increased strategic importance. Similarly, we have a growing experience in quantitatively modelling, evaluating and using computer-aided analyses that contributes to our ability to better manage operations, in their inter-temporal as well as their strategic and risk settings. Such experience and knowledge makes feasible the operations management of supply chains.

Simultaneously, the growth and realignment of corporate entities into strategic supply chains, global and market sensitive, have altered our conception of operations, their modeling and performance measurements. rendering them far more strategic and sensitive to external events and to the externalities that beset the operations of supply chains. For this reason, in an era of global supply chains, operations performance and management have evolved, providing new and essential challenges and concerns (see for example, Agrawal and Sheshadri 2000; Bowersox 1990, Cachon 2003, Christopher 1992, 2004; Tsay et al. 1998).

Supply chains are an essential ingredient in the quest for corporate survival and growth. Operations in supply chains have mutated, however, assuming ever-expanding dimensions, providing, on the one hand, greater opportunities for managing these operations and, on the other, augmenting appreciably, the operational complexity that modern enterprises face. Many of these problems are ill-understood and poorly valued. As a result the operations may be poorly managed, augmenting the risks that corporate entities face.

Although these concerns are pre-eminent in corporate strategies, they require an understanding of issues that have not been addressed in traditional approaches to operations management. The supply chain manager has thus an important role to assume by focusing attention on supply chain operations and educating corporate managers about what these operations imply, how to value them and to "internalize" them into the firm's economic analyses so that the supply chain can be managed better.

We begin in this chapter by an overview of the transformation of logistics and operations into a concern for the management of supply chains operations.

1.1 SUPPLY CHAIN OPERATIONS: A METAMORPHOSIS

Operations and logistics are undergoing a metamorphosis originating in major changes in market forces, leading to a global competition, the increased and determinant role of customers and new technologies that have altered traditional operations in firms into cooperative and at time "managed operations" across supply chains. This metamorphosis is being driven by the needs of firms to be "here and there" at all times and to thrive in a global environment where all operations involve multiple and coordinated agents. As a result, operational corporate strategies have changed and become far more sensitive, adaptable to the complex growth that confront operations, such as integrating production plans across independent firms which have a common interest to function in a coordinated manner. A common interest arises from a specialization of functions, economies of scales, greater flexibility and the ability to operate on a global-scale at a lower cost. For these reasons, firms focus on specific functions such as logistics, services, back-office finances, distribution and marketing. This has led to selective outsourcing becoming a crucial factor in attaining a competitive advantage.

At the same time, the restructuring of operations in supply chains has also increased the risk to firms that are unaware of the consequences. There are bi-polar forces at play, upstream and downstream, acting simultaneously and setting new trends and raising new problems in the management of operations in a supply chain environment. Within these trends, production, traditionally separated across function and firms, is becoming more and more integrated. For example, pressures are already exercised within firms to outsource some functions to carefully selected suppliers and to coordinate the planning of operations. By the same token, product design, production and distribution are no longer viewed as being the resulting effort of one firm but rather that of a collective of firms operating in a common underlying purpose (e.g., Lalonde and Cooper 1989; Mc Ivor 2000; Newman 1988; Rao and Young 1996; Van Damme et al. 1996).

Similarly, purchasing of materials, inventory control and the shop-floor management (scheduling, -routing, etc.) are now viewed in a supply chain setting in which synergies are sought with both suppliers upstream, distributors and downstream clients. To implement these changes and to manage and to manipulate these systems, personnel must acquire new skills and knowledge. Competition is increasingly international. Globalization has set in, making it possible to operate simultaneously and instantly in a number of countries, each providing a strategic advantage for a specialized function.

This setting has unsettled the traditional and secured environment of operations and logistics and introduced a far greater awareness that adaptation, strategic issues (arising from competitive postures and uncertainty) are now an essential part of what supply chains must deal with. *In other words, the management of operations in terms of concept, scope, and techniques has been altered by global competition and integration of production outsourcing; technology and its holistic integration in the logistic and manufacturing processes; the emergence of market major forces and market metamorphosis* . As was the case following the first industrial revolution, albeit at a far greater scale, these forces have altered the economics of operations, providing a potential for profit through specialization of function and economies of scale. At the same time, the re-organization of operating supply chains has also augmented a firm's risks and consequently the need to introduce approaches to management that are both adaptable and recognize the potential and the inequities that arise in supply chains. This book will seek to highlight some of these problems and at the same time provide a number of insights arising from the analysis of operation in their strategic game-like environment.

1.2 MOTIVATIONS AND ORGANIZATION

Supply chains arise as determinant organizational forms due to the many motivations and purposes. As a result there are many definitions of what constitute a supply chain. The traditional view of a supply chain is that of a "loosely aligned, fragmented series of paired relationships among different firms, agents and parties, independent or not that function within an agreed set of rules, contracts or contractual agreements". For example, supply chains include a broad variety of collaborative agreements between a manufacturer-wholesaler; a wholesaler-retailer; a retailer-consumer and their integration within a collaborative network. These agreements are meant to coordinate collaboration between these parties and promote long-term strategic cooperation by legally binding agreements or through a shared economic interest among independent enterprises.

In an operational and narrower sense, a supply chain and its management consist of the management of a network of facilities, the exchange of communications, distribution channels and the supply chain entities that procure materials, transform these materials into intermediate and finished products, and distribute the finished products to customers. As a result of these wide range of functions, a supply chain can be viewed as an emerging operational and organizational form integrating all firms and entities that cannot, either by design or by economic interest, pursue by itself all these activities. Due to this inclusiveness, supply chain management is a potent and important alternative to the common use of centralized and authoritarian-based approaches to management.

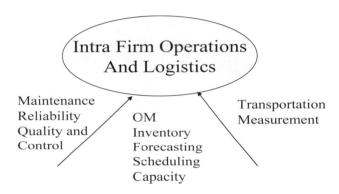

Figure 1.1. Operations Management—Intra Firms Operations and Logistics

In industrial and logistics management, supply chains have become the dominant organizational model, fed by and feeding the important changes in technology and operations management of the last half century. In Figures 1.1, 1.2 and 1.3, for example, we emphasize the growing concerns of supply chain management from intra-industry and self-management to include a far greater complexity based on intrinsically more global approaches and the elements that define a supply chain as stated above. Explicitly, from a concern for operational problems associated to inventory management, capacity planning, transportation, quality control, etc., to problems of supplier selection, collaborative ventures, co-production, contractual agreements and negotiations. Some of these elements are shown in Figure 1.2 and emphasize an upstream sensitivity.

Figure 1.2. Operations Management-Intra Firm and Up-stream Collaboration

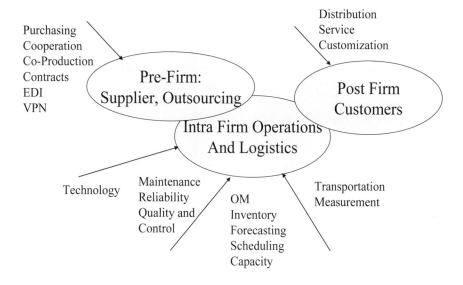

Figure 1.3. Supply Chain: Upstream and Downstream Integration

Technological developments such as electronic data interchange (EDI), Internet and intra-nets as well as virtual private networks (VPN) have, of course, contributed to the growth of supply chains by making it possible for relevant parties to communicate instantly and economically. Subsequently, customer-focused strategies have also led to the development of a similar growth and integration of services in a firm's operations and thereby to the basic supply chain structures commonly observed in medium-sized and large corporations operating nationally and internationally. Such a supply chain is outlined in Figure 1.3. Note that new functions such as distribution and logistics, service-focused customization in both products and services as well as the many activities associated with post-sales, "loyalty management of customers" etc. have now been added to the basic and traditional operational function of firms engaged in operating and production. But, these functions have also contributed to an exponential growth of complexity in the management of operations and to the basic principles of management in such an environment. In turn, managerial challenges have led to important new issues in finance, marketing, and in all facets of business, emphasizing problems of integration and the interface of the supply chain entities so that their economic promises can be realized and sustained. Furthermore, the concept of competition between firms, has also evolved, emphasizing a competition between "global supply chains", with means, markets, capacities and need broadly distributed and at times loosely coordinated. Below we shall consider a few such examples, highlighting in fact the extreme diversity of supply chains.

Examples: The Many Faces of Supply Chains

The common view of supply chains, as stated in Figures 1.1, 1.2 and 1.3 involves supply chains with intra-firms operations focused on production and inventory management and on procurement and the management of logistics and distribution. Often, some of these functions are outsourced as well, leading to considerable diffusion in a firm's operations. Upstream, the supply chain might consist of one or several suppliers operating under collaborative or contractual agreements that emphasize various degrees of commitment and exchange between the firm and its suppliers. Suppliers may be rated according to their reliability, the sustainability of the on-going relationship that has developed as well as their commitment to the firm's activities and their responsibilities. Downstream, marketing channels and selling organizations such as franchises (which we will consider subsequently) and other forms of exchange with retailers and intermediaries might be worked out. In these supply chains, the essential elements to reckon with and which have altered the manner in which we manage operations are:

- procurement and outsourcing
- strategic collaboration, exchange and contractual agreement defining rules of engagement, responsibility and sharing
- channels and organizational structure and infrastructure
- relational management, upstream (with suppliers) and downstream (with customers and distribution and marketing channels)

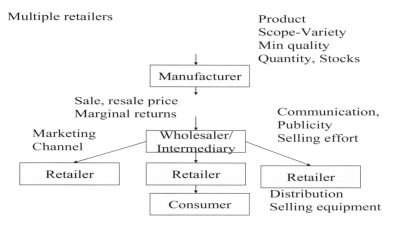

Figure 1.4. Centralized-Decentralized Downstream Supply Chain

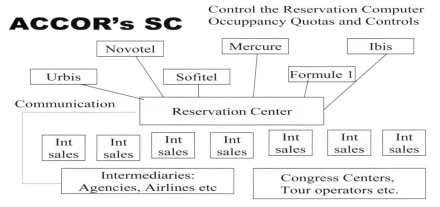

Figure 1.5. The reservation Center at the core of ACCOR

These supply chains arose as an alternative to centralized and vertical "supply chains" where material, manufacturing and products management, logistics and distribution, wholesaling and retailing and marketing channels are integrated as a whole and centrally designed and managed. The vertically integrated chains are based on a collective of agents and firms that are dependent and therefore apply rules of management dictated by a central authority. This is in contrast to "horizontal supply chains" which consist of independent (or nearly independent) agents and firms that pursue their own self-interest yet, at the same time, are dependent and sustained by the successful operations of the "global supply chain". A centralized, downstream supply chain is represented in Figure 1.4 .

Supply chains with many other faces and forms exist, however, their existence justified by their emphasis on economic success or by some other factors that can sustain their existence. An outstanding example is the growth of hotel supply chains (such as Hilton, Marriot, Accor etc.). In the case of Accor, economies of scale in reservation computer centers managing occupancy and, in some cases, the chain cash flow, have provided a strong impetus for individual hotels to affiliate with the Accor chain. This is represented graphically in Figure 1.5, where we highlight the centrality of the reservation center between intermediaries, and the many hotels categories that comprise Accor.

Of course, other examples abound. ISO 9000 and various other certifications seek to develop a supply chain "esprit" Strategic alliances in information technology, in car industries, in airlines etc. are all meant to create a collaborative environment and exchange platform that profits individually and collectively the whole supply chain. Franchises of all sorts (McDonald, Benetton, and so on.) are also based on the principles economic and operational that underlies the management of supply chains.

Recently, the Internet has contributed immensely to the growth of various firms built on supply chain principles. For example, collaborative logistics in cooperation and coordination within a community of shippers and carriers using an Internet service to streamline business relations substantially improved profitability and performance of all the companies involved. Leveraging the power of Internet as a computing platform has grown into a real opportunity for collaborative logistics.

1.3 SUPPLY CHAINS: NEEDS AND RISKS

In an early paper, "OM Factors Explaining the Need for SCM: in Suppliers Positioning" (1984) Hayes and Wheelright presented a number of considerations spanning demand volatility and the interaction effects of delays and uncertainty in supplies; assets intensity favoring upstream suppliers that tend to be more focused and have greater economies of scale; standardization;

profitability; technological change; and scale and balance (economy and scope). In recent years, these issues have been further emphasized and interpreted into strategic and managerial needs including the need to: earn profits in both the short- and long-run; maintain services close to end customers; comply with regulations and government interventions (both nationally and internationally); and to maintain the ability to manage the firm when it grows quantitatively and in complexity and must face the strategic implications of business at a global scale. These needs and risks, underlying the trend toward supply chain management have been the essential engine that leads firms to restructure into "lean and complex" organizational forms, where "what one sees is a lot less than what one has" as it is the case in supply chains.

The number of considerations that justify the growth of supply chains is large. What should a firm do in overcoming barriers in foreign markets? How can a firm adopt a strategy of focusing on its core competence and at the same time maintain its diversity and viability? How can a firm reduce the risks of its non-sustainability by operating alone? Can such a firm augment its market share on its own? Can it acquire, at a reasonable price, all the patents it needs to maintain its inventiveness and its technology savvy? These are among the many considerations that successful firms meet at defining moments, when future growth and oblivion are confronted.

Mini-case: Strategic Alliances in the Airlines Industries

Companies form strategic alliances because each of the parties gains something that they could not get on their own—either at all or at a reasonable price. Globalization, liberalization and privatization have been the factors, among others, that influence the formation of strategic alliances between airlines. It is the relative dominance of demand-side forces for large network carriers that have driven the need to join an alliance (these include fewer connections, higher frequencies, more cities served, lower information costs and frequent flyer programs). In effect if an airline intends to pursue a market strategy of being a full-service, broadly-based carrier, a necessary condition for success is joining an alliance . Which alliance to join also makes a difference. The benefits and costs of an alliance are determined by the underlying demand and supply-side drivers, and differ according to who the alliance partners are and the nature of the alliance. For an airline, the alliance can range from a simple marketing arrangement to a strong equity position. As more formal arrangements are made, the benefits rise but the costs of adjustment and integration increase as well. These factors have combined to create together "airlines supply chains" flying under various names, yet operating as a common whole, exchanging and allocating flights and seats.

For airlines in particular, the environmental factors pulling the growth of strategic alliances are: globalization, liberalization, and privatization, combined with basic industry-specific factors such as bilateral and regulatory restrictions. Demand-side and supply-side economics have also been pushed by the risks that airlines with a global outreach face if they wish to remain national airlines. Some of these risks have included: the presence of barriers to entry; vertical integration to exclude rivals from the market; increased competition costs; and indirect control by the competition's actions (such as control of feeder carriers). To circumvent such difficulties airlines have pursued bottom-up and side strategies including, for example, simple alliances (such as marketing agreements for preferential exchange of traffic, code-sharing, frequent flyer participation) as well as strong carrier alliances (including equity swaps as was the case between Air France, KLM, Alitalia etc.).

1.4 SUPPLY CHAINS AND OPERATIONS MANAGEMENT

Operations in supply chains have evolved in tune with their needs and the risks they imply. Today operations in supply chains are viewed as operating in a coherent rather than a fragmented whole with responsibilities for various segments allocated to functional areas such as purchasing, manufacturing, sales and distribution. In addition, however, supply chain management calls for, and depends on strategic decisions-making. 'Supply', the shared objective of every element in the chain, is of particular strategic importance due to its impact on overall costs and market share.

Figure 1.6. From Operational to Strategic Operations

Various operations problems that arise are motivated (as with all operations problems) by some of the following considerations:

- Time and cycle time management, seeking to reduce time in all the firm does such as using the Internet in communications. Cycle time reduction, "at any price", is a strategic objective of industrial management that contributes to stock cost reduction, to reactivity, to quality, to service and to customization.
- The necessity of reducing geographical distances as a means of responding to the global outreach of supply chains and of achieving savings in resource consumption.
- Customization and flexibility, variability, adaptability, and fixed cost reduction
- Differentiation, quality and standardization

These strategic prerequisites of supply chain management are also changing the manner in which we represent and analyse models. In addition, attention is moving increasingly from intra-firm to strategic inter-firms as Figure 1.6. shows above.

From Traditional to Supply Chain Aggregate Production Planning

A traditional periodic production model consists of determining a production plan responding to the problem of how much to produce and when. For example, weekly, monthly or daily production decisions may be reached on the basis of a demand forecast. These decisions may, however, involve a broad number of concerns such as inventory costs, transport capacity etc. each of which, in a supply chain, may be managed by independent agents. Traditional aggregate planning models have emphasized a centralized approach, applying available resources and their associated information for a common purpose—centralized management. In a supply chain, some of these functions involve at times completely independent agents, generally a number of agents who reach their decisions independently.

Current supply chains exhibit simultaneously an upstream and a downstream sensitivity. For example, supplies and customer delivery needs are often conflicting and have to be reconciled in a manner that rather than just the production needs with only a customer focus. Outsourcing on a global scale has added another aspect: "remote" production management. The effects of this concept have not yet been fully integrated into the consciousness of operations managers. However, concern is gradually increasing for the management of services, of quality "controlled at a distance" and for production that the end firm cannot always control.

Traditional production models are essentially based on "costs", "infrastructure", "quality" and "constraints". Strategic issues were essentially neglected. Essential strategic concerns include:

- Costs of materials and components which are determined by supplier's selection, handling and the use of logistics.
- Costs, determined by the organization, of production, efficiency and productivity-related issues.
- Costs of people, salaries and related costs which are market driven.
- Costs of inventory which are both direct and indirect, determined by the inventory policy, financial costs and related considerations.

In addition, the production environment is presumed constrained by resources such as: capacity; people (both as a function of quantity, quality and working modes); procurement and outsourcing; transportation and logistics. Intelligence gathering and forecasting continue to be important while technology constraints include: process structure; product assembly; organization; customer requirements; multiple plants and products; and warehouses. The integration of all these elements provides an underlying reality that supply chain managers must reckon with.

These factors of cost, quality, infrastructure and constraint have generated complex decision problems which have been extensively studied, both in practice and in theory. They involve short- and long-run decision making problems, both under certainty and uncertainty. For example, problems of inventory management have been treated by using specific models that recognize both the production infrastructure and constraints, and that minimize costs. The advent of MRP systems has augmented the dimensions of inventory management by adding supply coordination, Just-in-Time management and managed outsourcing. These systems have grown into ERP as well as supply chain ERP systems. The trend is to harness technology to handle ever greater and more complex problems, while under-valuing, in some cases, the importance the managerial and motivating facets of the production function which independent managers now co-manage. by . Fig. 1.6. highlights these facets of the production system.

Globalization and the growth of logistic costs, relative to the costs of production and related issues and the ever-increasing costs of competition and customer dissatisfaction, have altered managerial practices, in particular, aggregate production and logistic planning. The timely delivery of schedules, logistic costs and production services are emphasized far more.

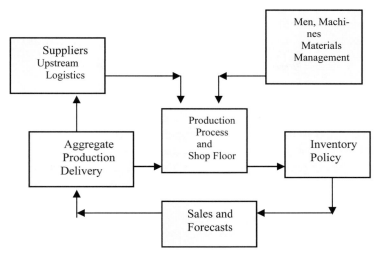

Figure 1.7. Traditional Production and Inventory

As a result, aggregate production planning, is, at the same time, far more upstream –supplier sensitive, and also increasingly logistic and customer-driven. In such an environment, other considerations are introduced that determine production plans and their management, including greater concern for:

- Supply delays, supply reliabilities and their costs
- Information exchange and communication technology
- Incentives management
- Technological, political and social constraints
- Re-scheduling, supply and products delay management and their penalties
- Transportation scheduling and logistic controls
- Service and product quality and their controls
- Customer prioritization, upstream and downstream

The integration of aggregate operations into supply chains, thus indicates a growth of complexity implied in a strategic management of operation. These problems and models that address a number of selective issues will be dealt with in forthcoming chapters.

Production and Supply Chains: A Simplified Model

In the traditional and centralized approach to production management, a firm is presumed to plan aggregate production over a given period of time (say weekly, over a period of a year) based on a data base which it controls. Information regarding its capacity to produce (how much it can produce at most in any one week), cost data pertaining to how much it costs to produce one unit of a product, its inventory costs etc. are then gathered. Finally, even though information regarding the demands for its product is imperfect, the firm uses a forecast, essentially replacing the unknown demand with the presumed forecast. On the basis of these assumptions and any other a production manager may care to make, the firm proceeds to determine a production plan.

The pull to producing in a supply chain environment arises because it provides advantages – economic and otherwise. For example, to assure client loyalty by adopting a collaborative environment, a firm may expect to retain a number of clients who would be both loyal and allow early planning of the production program of the firm, (thereby, reducing the risk in production-demands and perhaps making it possible "to make do more with less"). The firm, which is facing such an eventuality is considering a relationship that will be maintained with other firms which it will supply as well as it can. These firms in turn will contribute to a synchronization of production schedules such that demands will be given ahead of time together with some tolerance regarding delivery dates.

1.5 SUPPLY CHAINS AND INVENTORY MANAGEMENT

Theory and practice in inventory management have produced an extremely large number of inventory models motivated by the need to reduce production and operations costs while managing at the same the business risks associated with either excess or inventory shortages (Lambert 1982; Cachon and Fisher 2000; Tapiero 2005; Ritchken and Tapiero 1986). Traditional models of inventory have thus been based by their location within the production process, such as raw materials, in process products, finished goods inventory, etc. Classified by functional and statistical demand properties, inventories are meant to meet various types of demand: discrete, continuous, deterministic, stochastic, constant, single-period or multi-period (and their combination thereof). Then there are demands by the policy that inventory managers seek to adopt including (Q,T), (s,S)-two bins, multiple-barrier policies etc. Finally, many inventory models, presuming essentially a centralized control, have been classified by the underlying process structure (whether in production, distribution, in retailing, in supplier inventories etc.), namely, single-stage; multiple stages in series and parallel; and assembly, to name some of the systems. Other names that capture this class of models include multi-echelon and demand independent and dependent models. This latter feature, independence, is particularly important, since the presumption that demands are independent of the order policies greatly simplifies the modeling and the management of inventories. In supply chains, this is not the case and consequently we confront the particular complexity of dealing with such models and problems.

The simplest textbook inventory problem consists in minimizing the following costs: ordering, inventory holding and out-of stock inventory stock (the latter being particularly difficult to price since it is often difficult to estimate the costs associated with lost sales). Inventory policies contemplated are then merely defined by a set of parameters on the basis of which the model is formulated and resolved. Typical policy examples include: an order cycle time or T policy (also called periodic review) in which the decision to order or not is reviewed over fixed intervals of time; an order-size (EOQ) or Q policy, consisting of ordering fixed quantities at variable or at fixed instants of time (depending on whether the problem assumes a demand uncertainty or not); a stock security (s,S) "feedback" policies, consisting in launching quantity orders S at specific inventory states (with supplies provided with or without delays). Other policies exist of course and have been extensively treated in the literature.

The following model (see Figure 1.8) represents, for example, a typical EOQ model with a constant demand, which in this special case can be interpreted as a T or Q policy. It also presents a typical (s,S) security policy where s is both a stock held for potential and incoming demands and a trigger to an order launched whenever inventory that is being is felt to be falling below secure stock levels.

Such issues as "permitted shortages", "production inventory models with on-going production", "recuperating part or whole lost sales", "deterministic and stochastic demands", "single or multiple items" etc. are introduced in such models, complicating their analysis but not their structure as is the case when we consider inventory models in supply chains. In such situations, questions and problems pertinent to supply chains arise as we shall see subsequently. In the cases mentioned above, determining the inventory policy requires information regarding: demand (Is it constant? Time variable?

Deterministic? Stochastic?); storage of items (Are they inert? Active? Deteriorating or appreciating?); the selective main contributing costs to inventories–directly and indirectly (fixed costs, variable costs, out of stock costs, holding costs etc.); and, finally, what are the essential decision parameters to reckon with. An extremely simple EOQ model will be considered as a starting point to demonstrate the fact that the analysis of these simple problems assume an added complexity when the supply chain environment is considered.

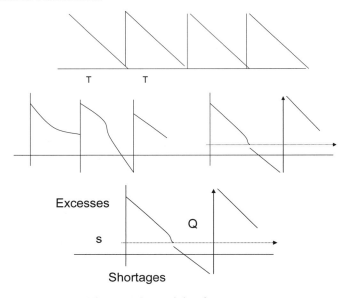

Figure 1.8. Models of Inventory Management

Example: The EOQ Model Revisited

Consider an inventory model with K the set up cost and c_1 the unit holding cost. No shortages are tolerated. In such a case the optimal order policy can be shown to be the EOQ and based on the minimization of the average "inventory system costs". The solution is thus:

$$\text{EOQ: } Q = \sqrt{2xK/c_1} \; ;$$
$$\text{Cycle time: } T = \sqrt{K/c_1x} \; ;$$
$$\text{Average cost } AC = \sqrt{2xKc_1} \; .$$

Note that this model implies that a number of business functions intervene in determining the inventory policy. For example, financial considerations arise about the interest rate paid on money which is tied down; manufacturing can have an effect by seeking investments in a flexible manufacturing technology which reduces the set-up costs marketing has an effect based on the requirements it imposes on demand satisfaction (i.e. the cost of shortages). The relative dominance of each of the business functions will determine the inventory policy. While the EOQ formula is simple to apply, it is most revealing in integrating multiple strategic functions of business.

These functions include
- marketing, through determination of the demand x
- purchasing, through the effects of the fixed order
- finance, through the cost of money, or inventory holding c_1
- Tech SMED through the fixed cost K

For example, if the cost of money goes up, how would that affect inventories? If demand is expected to fall, how would that affect inventories? If purchasing costs decrease, would we order more or less often? If demand increases by 30% how much would average inventory increase? If the price of goods held in inventory increases, would we have on the average more or less inventory? Of course, by altering some of the assumptions made to obtain the EOQ formula, more general models can be derived and solved while other EOQ formulas may be obtained

Some elementary calculations, based on our assuming a departure from the centralized solution (consisting in optimizing the average total inventory cost) would reveal a departure from such a solution and the effects of multiple interests at play in a supply chain environment. Explicitly, consider a demand to be set to kx instead of x, and reflecting the desire of a marketing manager who supplies such information to assure larger inventory levels. In this case, since the EOQ formula resulting from an average cost minimization:

$$AC = \frac{K + c_1 QT/2}{T}, \quad T = \frac{Q}{x}$$

yields

$$Q^* = \sqrt{2xK/c_1} \quad \text{and} \quad AC^* = \frac{Kx}{Q^*} + \frac{c_1 Q^*}{2} = \sqrt{2xKc_1}$$

Thus, by a mis-specification (or weighted inventory cost, by k, we have:

$$Q' = \sqrt{2xK/kc_1}, AC_1 = \sqrt{2xKkc_1} \ .$$

When we compare this to the centralized solution ($k=1$), we obtain the following costs:

$$\frac{AC_1}{AC} = \sqrt{k}$$

and therefore

$$\frac{AC_1 - AC}{AC_1} = 1 - \sqrt{\frac{1}{k}}$$

In other words, the percentage growth in average inventory costs is indeed a function of stating (over or under) the true demand, imposed by the party who has the responsibility and potentially, the power to do so. Generalizing to multiple parties, in other words to additional departures from the centralized solution, it is easy to show that:

$$Q_1^* = \sqrt{\frac{2(k_1 K)(k_3 x)}{k_2 c_1}} \; ,$$

which will alter the calculations of the average costs as seen by each of the parties. Note that if each of these parties is "an independent agent", the resulting inventory policy will be the outcome of a "game" between these parties, each selecting voluntarily the information to be supplied. The game can be stated as follows:

$$\underset{k_1}{Min}\, F_1(k_1, Q_1^*), \quad \underset{k_2}{Min}\, F_2(k_2, Q_1^*), \quad \underset{k_3}{Min}\, F_2(k_3, Q_1^*)$$

$$\text{Subject to:} \quad Q_1^* = \sqrt{\frac{2(k_1 K)(k_3 x)}{k_2 c_1}}$$

The determination of agreed upon models and cost parameters as a supply chain determining the inventory policy is thus a particularly difficult issue. This involves an intra-chain political process where the power of managers can be determinant. Is marketing or production determinant in the firm strategy? If it is the former, it is possible that the out-of-stock cost may be overstated while the holding cost would be understated.

Such an approach recognizes one facet of the management of supply chains inventory problems and some of the specific characteristics we are led to consider in such situations. In other words, inventory management in supply chains necessarily changes.

SCM, as we saw earlier, involves a process of integration and collaboration that optimizes the internal and external activities of the firm involved in delivering a greater perceived value to customers. For inventory management, new technologies based on EDI and Internet-VPNs (Virtual Private

Networks) has provided the means to simplify and augment collaborative efforts and provide an exchange of information. These technologies have simplified the process of communication in general and that of managing orders in particular. Nonetheless, when agents in a SC are independent, and independently reach their decisions, information is unevenly distributed throughout the chain. As a result, uncertainties may be induced endogenously, generating additional costs. For example, car manufacturers often over-supply cars to distributors in order to transfer some of the inventory holding costs and to motivate them to greater sales efforts. Distributors may be aware of such behavior and as a result they may understate the orders in the expectation of over supplies. In this sense, mutual uncertainty is induced, implicit in the behavior of the manufacturers and the distributors. By the same token, manufacturers have the tendency to "load" shelf space in supermarkets to augment the probability of selling. In this sense, they have a great interest in over supplying supermarkets As a result, the conventional wisdom that the less inventory, the better, is not always right. For end-product inventory, close to selling points, there may be an incentive to maintain inventory. This is well known, as noted, in the car industry, in pharmaceuticals as well as in brand facing in supermarkets, where visible inventory is used as a mean to induce sales. In these cases, point-of-sales effort is associated to the size and the quality of the display (and therefore to the inventory investment incurred for a particular brand). Sales campaigns combined with the manufacturer's subscription to inventory carrying charges as well as other expenses are then used as an incentive to improve sales performance at the selling point. Such practices are applied over a broad range of other type of products and in various manners.

Similar observations may be recorded about various types of franchises where end-product inventory is shifted to franchisees. For example, inventory ordering in supply chains considers the individual objectives of firms in the supply chain and their organization (defined by the franchise contract or the organization under which they operate) or their position in the supply chain. The inventory policy must then reflect the objectives and the conflicts-or-collaboration (which can be solved under alternative organizational and informational assumptions) between the supply chain members.

By the same token, modeling the inventory process in outsourcing (compare Figures 1.9 and 1.10) involves many issues that are often neglected in traditional inventory models. We outline below some issues, each of which is important in its own right.

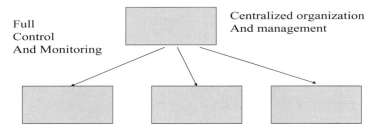

Figure 1.9. Inventory Management in Vertical-Horizontal SC

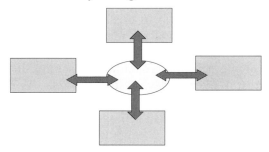

Figure 1.10. Inventory and Outsourcing

- What are the rules of leadership? Who is leading? Who has the information? Who has the power and can exercise it or not?
- What are the supply priorities, guarantees and related issues associated with products and goods transferred from one party to the other?
- What are the information flows? Who gets what and when and by whom?
- What are the objectives of the members of the supply chain?
- What are the principles of equity, distribution and control?
- What are the policy variables that each of the parties can exercise?
- Who controls what?
- What are the sources of uncertainty? Are they internally induced or do they occurr externally?
- What are the constraints on each of the parties? The individual? The collective?
- What are the objectives that each of the parties optimizes?

- What are the relevant models to consider and which are agreed upon (or not) by the supply chain parties?
- How can we solve them?
- How can a solution that has been deemed appropriate be implemented?
- What are the post-implementation monitoring tools and grievance resolution modes (economic and otherwise)?

1.6 QUALITY AND SUPPLY CHAIN MANAGEMENT

Quality and its control are important in all facets of supply chain management—economic and otherwise. For example, a substantial part of warranty claims can be traced to purchased items that are failing and to failed services. Similarly, faulty supply parts invariably lead to extremely large costs for the manufacturer and the retailer. Traditionally, quality and its control have emphasized the use of statistical control techniques which seek to detect deviations from agreed upon quality standards. In a supply chain environment, the dependence of agents and their self-motivation, requires another approach, game sensitive, which recognizes the supply chain structure, its economic exchanges mechanisms and its rules of leadership. Such an approach will, necessarily, have an effect on both the suppliers' propensity to supply quality products and the control procedures implemented. It is for these reasons that industrial supply contracts, special relationships and coordination between producers and suppliers are so important. In TQM (see Tapiero 1994) attempts are made to integrate supplier control procedures into a broad management framework. These attempts, however, often fail to recognize the complex motivations that underlie supplier behavior in a contractual environment and the specific characteristics of supply contracts. Due to the importance of this problem, guarantees of various sorts are sought in practice to assure that quality complies with its promise, as specified by the quality supply contract. Of course, such problems are not specific to industrial producers and suppliers but are quite general, spanning the gamut of business transactions where there is an exchange between parties (e.g. buyer-seller of a product or service, franchises, etc.). In a supply chain, since some of the problems that arise may be due to the multiple, partially conflicting, objectives of supply chain agents, a great deal of effort is invested in developing a collaborative framework to meet both quality standards and objectives (Reyniers and Tapiero 1995a, 1995b; Tapiero 1995, 1996, 2006).

For discussion purposes, we shall consider a number of cases which will be outlined and dealt with in far greater detail in Chapter 8, devoted to quality and its control in supply chains. Consider supplier-producer relationships, franchises and situations where there may be both an opportunity and a reason to collaborate and yet, there might also be reasons for conflict between the parties. The problems we consider are relatively simple and are used to highlight some of the basic considerations we ought to be aware of in such situations. First, industrial quality and business exchanges are often defined in terms of contracts. For example, a product which is sold has a quality responsibility associated with it, defined and protected by contractual agreements. Further, the transaction itself may have as well a service contract associated with it, assuring the buyer that product performance will conform to its advertisement. Warranties of various sorts are then designed to convey both a signal of quality and to manage the risk of product acquisition. These and other mechanisms are increasingly used by buyers who are demanding risk protection clauses to assure that they obtain what they expected at the time the transaction was realized. This "downstream sensitivity" has also an "upstream sensitivity", in maintaining both a trusting and functional relationship with suppliers. There is also an also an economical aspect. The cost of quality control is invariably cheaper upstream than downstream. As a result, in a supply chain, the collective welfare of the chain is improved if quality starts at the source and not discovered at its end-point when the customer is located.

Technically, the traditional approach to the statistical control of quality (based on Neymann-Pearson theory) has ignored both the elements of conflict and measurement costs in constructing statistical tests. As a result, the traditional approach has underestimated the strategic (and game-like) importance of controls when these are tied to contracts that have retaliatory clauses and when the agreed upon quality is not supplied. In such cases, threats, the nature of the contract (whether it is a short- or a long-term contract) the information available to each of the parties and so on have an important effect on the selection of quality control strategies. Although economic theory has studied such problems, the traditional view of statistical quality control has not explicitly considered these effects.

Of course, the TQM approach has increased awareness that industrial cooperation between producers, suppliers, wholesalers and retailers is necessary to guarantee better quality and profitability and supply chain competitiveness. Underlying this belief, are the basic facts that conflict is pervasive and that it is detrimental to productivity. The cases below provide a motivation for dealing with such problems.

Intra-Firm Supplier-Producer Relationships

Increasingly, some firms have reorganized their production and business operations as supplier-producer relationships. It is believed that this decentralizes the firm's operational units which can then be managed through proper incentives and controls as a supply chain and thereby attain the desired performance. In practice, top management invests considerable effort in demonstrating that the supplier-producer relationship is a win-win relationship, leading to improved communication, coordination and synchronization of operations schedules. A leading steel manufacturing firm in Europe has followed such a path and has emphasized management audits of services and operations. In particular, "inter-unit" contracts and agreements are assessed and evaluated in terms of performance, transparency, opportunistic behavior (such as cheating and conniving) and the maintenance over time of intra-firm contractual agreements. Through such a system, the firm has observed that it was possible to remove from inter-unit exchanges conflicts which are not related to the unit profit and cost objectives (such as jealousy, personal conflicts, self-aggrandizement etc.) and to construct a system of procedures where responsibility, participation, self-control and decentralization can be induced.

Marketing Channels and Quality

Marketing channels of various sorts lead to the creation of supplier-producer relationships involving intermediaries which can be complex to manage. There can, of course, be no intermediaries, in which case a direct relationship between a supplier and a producer is established. In both cases, the supply and the control of quality are affected by the management of the relationship and by the contracts which are used to regulate it. Problems such as supplier (producer) liability and the responsibility of the intermediary and how it can be managed, audited and controlled are part of an array of business and operational tools which can be used to manage the quality from its inception to its delivery and consumption by the producer (or the consumer). When there are complex marketing channels consisting of many suppliers, producers, wholesalers, semi-wholesalers, retailers and consumers, the problems of quality management and control become that much greater. This situation leads to problems of managing the intermediaries and their control. In practice, since we encounter such problems, insights regarding the effects of conflict, channel structure etc. on the supply and the control of quality are clearly needed to improve our potential to manage quality in the supply chain.

Quality Management and Franchises

Franchises involve a mutual relationship between a firm (say a manufacturer) and one or more firms (say retailers) in which some contractual rules are established for the conduct of business and the sharing of revenues and costs. For example, a franchiser may solely provide the products to be sold by the franchisee at an agreed upon price and quality. The franchisee, involved directly in the marketing of products, may assume part of the costs as well as (partly or wholly) some of the costs associated with post-sales product failures, repairs and other services.

A franchiser-franchisee agreement is usually bound by contractual agreements which are maintained over time and which guarantee their mutual incentive to operate and cooperate during this period. Quality and its control can thus provide an added incentive to maintain and to sustain the partnership. For a manufacturing franchiser, which is extremely sensitive to its national image and quality, the potential to consistently produce goods of advertised quality are essential to sales and to the franchise growth. In most franchises (such as in fast-food industries, services, etc.) however, the uniformity of the product quality through franchises is an important feature of the franchise business itself and therefore the control of quality and its management are extremely important and is, in most cases, an essential feature of a franchise. Some recurring questions in such relationships include the incentives for the supply of quality by the franchiser and the incentives to perform for the franchisee. Also of interest are the effects of sharing post-sale costs (warranty, post-sales servicing, liability costs etc.) for the franchiser and the franchisee incentives for control.

To assess these questions, the development of "conflict-prone" game models and their analysis can be used. In these models both the franchiser and the franchisee engage in risk-sharing; privately taken actions by any of the parties affects the outcomes of interest and their probabilities. For example, the franchiser may design an incentive scheme for the purpose of inducing the franchisee to act in the franchiser's interest. The franchiser may also agree to contracts that induce the franchisee to expend a greater marketing effort (such as maintaining a high advertising budget) to stimulate sales (from which the franchisee benefits) and to deliver quality products. In this latter case, the franchisee and the franchiser may reach a price-incentive contract which is sensitive to delivered quality. These contracts will, of course, affect the amount of inspection conducted by both the franchiser and the franchised (Bank 1996, see also, Chapters 7-8).

Virtual Supplier Integration

Robert N. Boyce, former CEO of SEMATECH (Austin, Texas), claims that supplier integration is replacing vertical integration. He calls this type of integration "Virtual Supplier Integration". Explicitly, Boyce states that the Japanese have created a competitive edge through vertical integration. We can learn from it by establishing "virtual vertical integration" through Partnering with customers and suppliers. Just like a marriage, we need to give more than we get and believe that it will all work out better in the end. We should take a long term view, understanding suppliers' need for profitability and looking beyond this year's.

Partnering is referred to as a shift from traditional open-market bargaining to cooperative buyer-vendor relationships. Of course, Partnering implies a broad variety of actions taken simultaneously by the buyer and the vendor. It can involve the increased use of long term contractual agreements, reduction of the number of suppliers, negotiation procedures based on management tradeoffs rather than conflict management, strategic coordination and cooperation in product development and market evaluation, integration of computer support systems and internet derived products and most of all developing a relationship based on trust and mutual support.

More on Franchises

Franchises are an important source of supply chain organizations. It involves a mutual relationship between a firm (a manufacturer, for example) and one or more firms (retailers, possibly) in which some contractual rules are established for conducting business and sharing of revenues and costs. There are many definitions for such agreements. Caves and Murphy (1976) define franchises as an agreement lasting for a definite or an indefinite period of time, in which the owner of a protected trademark grants to another person or firm, for some consideration, the right to operate under this trademark for the purpose of producing or distributing a product or service. Thus, franchise contracts involve a sharing of intangible assets (especially trademark, goodwill) between independent firms. Because the value of such assets is defined by their use, these contracts involve difficult contractual relations. Rey and Tyrole, 1986 (see also Rey 1992) define a franchise agreement as usually involving several provisions, which relate to transfers from one side to the other (generally monetary transfers from the franchisee to the franchisor and technological or know-how transfers from the franchisor to the franchisees), and restrictions on the side of both the franchisor and the franchisee.

In practice these definitions are quite accurate, representing many sorts of franchises. For example, in some franchises production may be centralized and the distribution franchised (e.g. car sales, some food and department stores, fast food, Benetton etc.). Then there are franchises based on a branded image with advertising centralized but production decentralized to the franchisee. Some franchises require an appreciable investment by franchisees (due to the very high set-up costs in selling and generally diseconomies of scale in retailing and some logistic systems). In such cases, a franchisor may puts up some, if not all, of the required local investment, while the franchisee may have to self-invest by transferring part of his initial capital to the franchisor. In general, franchises are created due to the possibility they offer of reducing costs or risk to revenue.

The typical franchise consists of a contract between two legally independent firms establishing a long-term relationship giving the franchisee the right to use the franchiser's trademark. In exchange a payment of a lump sum fee and annual royalties at an agreed percentage of sales is signed. There may, of course, be many other provisions including for example: franchisee fee royalties or commission; resale price maintenance; quantity fixing; exclusive territories; exclusive dealing and tie-in. Transfer payment schemes can also be varied. For example, the franchisee's lump sum transfer may be refundable in case of success and if the relationship is maintained (as a way to commit the franchisee to entrepreneurial activity). Royalties on gross sales (in general between 3 to 5 percent) have both positive and negative effects since they can create disincentives but reduce the risk for the franchisor and can be negotiated and depend on a large number of factors. Taxes have also an important role to play as they can be used to transfer liabilities from one agent to another, depending on the pricing of inputs.

To assess these schemes, numerous approaches are used, based on economic and theoretical (games) assessments of franchisee and franchisor relationships and intentions. These approaches consist essentially of the following: resource constraints, in which the franchisee has access to financial capital, to market expertise and to the managerial talent of the franchisor; incentive issues, in which case the franchisor provides incentives to franchisees to perform in the interests of the franchisor and supply chain. These issues are important because of the acute problems that franchises create. Some of these problems cover: conflicting motivations; information asymmetry where information is not distributed evenly among the two partners (non-observable, hidden action); moral hazard in which case conflicting motivations combined with information asymmetry can lead to moral cheating or to opportunistic behavior which is not in the interest of the supply chain. This latter problem is particularly significant as it is often difficult to detect. As a result, supply chain managers will seek to institute audits to test the performance of the supply chain parties and at times threaten retaliation in case non-conforming behavior is detected. Standardization of parts and managerial practices needed to assure uniformity, often an important feature of many franchises, makes it possible to control the quality prevalent in the supply chain and deliver quality-conforming goods to the supply chain clients. And finally, the design of the franchise contracts, carefully balancing the costs, the risks and their consequences and the opportunities the supply chain parties are facing. Of course, these contracts, issues to reckon with and their management provide an ample opportunity for stud, some of which will be considered subsequently.

1.7 GAMES AND SUPPLY CHAIN MANAGEMENT

Supply chain governance and the independence of decision-making agents lead necessarily to a multi-decision- maker framework. For example, outsourcing an industrial activity to an independent supplier implies that the decisions and the policies implemented by the outsourcing firm and its supplier are based on their own self-interest. Collaboration and coordination of their industrial activities in a supply chain framework will hopefully lead to added benefits for both firms even though the benefits of the collaboration will have to be split. For a review of game theory there are numerous texts and papers such as Friedman 1986; Fudenberg and Tirole 1991; Moulin 1995; Nash 1950; Von Neumann and Morgenstern 1944.

To motivate and simplify our presentation, we shall outline below some extremely simple problems that were solved using game theoretic notions in the next chapter and in an intertemporal differential game approach in subsequent chapters. Consider for example, the traditional one-period inventory problem. Such problems are based on the minimization of some centralized objective (usually costs borne by the operations manager) subject to an estimate of future demands. In a supply chain framework, both the decisions reached by the supplier delivering the goods and the order set by the operations manager are reached independently, albeit in a coordinated and collaborated manner. These problems are therefore dynamic problems, although we shall consider first a static version of such problems. This implies that within a given period, the problem's parameters are assumed to be fixed. A decision is reached at the beginning of a period and implemented at the end of the period together with the revelation of the demand subsumed in the inventory model. We shall also assume that products are delivered by the end of the period and then instantly sold to meet the period's demand. Of course, such a problem ignores intermediate inventories (and their associated costs) incurred in pre-season production and inventory build-ups.

With these limitations in mind, we focus our attention on *stock and pricing policies* which will mitigate the costs of shortages as well as excess orders, accounted for at the end of the period. In addition, and for matters of simplicity, we also assume that the information needed for the decisions to be reached by both the supplier and the operations manager are fully known and shared by the supply chain participants and that order lead-time are smaller than the period's length such that all deliveries are provided on time. (The problems we outline next will, of course, be resolved in a subsequent chapter, in their static, one-period game and dynamic (differential) game frameworks.)

The Pricing Game

As a departure point, consider the classical deterministic pricing game due to Bertrand's price competition model . This games involves two vertical participants in a supply chain. For example, a single supplier might sell a product to a single retailer over a single time period . Let the retailer face an endogenous demand, $q(p)$, a downward-sloping function of the retail price p, i.e., $\dfrac{\partial q}{\partial p} < 0$. We assume that the supplier incurs a unit production cost c and sells the unit at the wholesale price w. The retailer's price per unit is $p=w+m$, where m defines the retailer's margin. Thus, the supplier's profit is

$$J_s(w,m)=(w-c)q(w+m)$$

while the retailer's profit is:

$$J_r(w,m)=mq(w+m).$$

In the Bertrand pricing model, both the supplier and the retailer seek to maximize profits—the supplier by choosing the wholesale price and the retailer by selecting the retail price, p, and hence the order quantity $q(p)$. Of course, each of these has dependent profits, depending on the other's (independently reached) decisions. Therefore, a game theoretical approach, expressing the information available to each, the power relationship that co-exists between these players and the business rules at hand, determine a framework which we can use to assess the implications and the decisions that each of these players ought to reach.

The game we shall consider consists in the following: the supplier sets the wholesale price while the retailer selects the retail price and thus the quantity to order. The supplier then delivers the quantity ordered. Since this pricing game is deterministic, all products that the retailer orders will be sold (since there is no point in ordering quantities to be added to an inventory cost). As a result, we face such questions as: What is the effect of the vertical competition between the supplier and the retailer on the retail prices (i.e., customers)? On the wholesale price and the quantities sold by the chain?. Does collaboration in a supply chain framework matter and how much? Further, if collaboration pays, then how are the spoils of such collaboration distributed in a sustainable manner between the supplier and the operations manager.

The Production Game

A similar problem, dealing with the quantity to produce and production competition is defined by the well known Cournot model. Two essential features distinguish this model compared to the Bertrand pricing game described above. In this model, we represent a product price as a function of demand, $p=p(q)$, $\dfrac{\partial p}{\partial q}<0$, or $q=q(p)$, $\dfrac{\partial q}{\partial p}<0$, i.e., using an inverse demand function. A second feature deals with the types of competition presumed by the Cournot model. In the supply chain under consideration, we assume two horizontal participants (two independent firms reaching their own decisions independently), consisting of a manufacturer and a supplier (or both manufacturers and suppliers). Assume two manufacturers, each incurring the unit production cost c when producing the same (or substitutable) product type. Further, assume that both compete in selling to the same retailer.

The retailer employs the so-called vendor-managed inventory policy and thus will not interfere in the manufacturers' competition. This implies that both manufacturers, say "1" and "2" decide on production quantities q_1 and q_2 respectively, supplied in turn to the retailer. (It is assumed that the retailer relies on the manufacturers' decisions and that sales are transparent to both manufacturers-suppliers.) Consequently, the product price is a function of aggregate demand, $p=p(q_1+q_2)$. In other words, a manufacturer selecting a production quantity, will necessarily affect the other's profit. The resulting game is as follows: both manufacturers set the quantities to produce and supply to the retailer. The retailer will then sell the products at a price which is an aggregate function of these quantities. In a supply chain with a collaborative environment, a number of issues can then be raised. For example, how does horizontal competition compare to a collaborative framework and can such a framework be sustainable? Further, if such a collaboration can be maintained, what are the essential factors that define a "successful" supply chain?

The Stocking Game

The simplest single-period inventory problem is a "Newsvendor problem" consisting of a producer or retailer deciding to order to stock a fixed product quantity q, when the period demand is assumed to be uncertain. Such a problem is subject to numerous modifications and extensions, some of which are considered in the next chapter. In the classical newsvendor problem, there is no initial inventory on hand; the demand, d, is exogenous with a known probability distribution function; the setup (or order) cost is negligible; the purchasing price is fixed; and the decision to order q units is made at the beginning of the period. Since the true demand D is known only by the end of the period, it is likely that either we incur a shortage ($D-q>0$) whose unit cost is denoted by h^- or an inventory excess ($q-D>0$) whose unit holding cost is denoted by h^+. A retailer's objective is then a "linear regret" objective, seeking to minimize the least weighted shortage and holding costs (see also Tapiero, 2004, 2005). However, if in addition, we include a supplier who seeks to sell his product to the retailer, a relationship between the supplier and the retailer may, under numerous circumstances, alter the wholesale price, w, (the purchase price for the retailer). In this sense, the newsvendor problem turns out to be a game. Indeed, the retailer's purchase price w will no longer be fixed, in contrast to the classical newsvendor problem which assumes such a price is fixed.

In a game framework, the supplier sells at a wholesale price w, incurs a production cost c, and maximizes the profit $J_s=(w-c)q$. This profit function is deterministic. However, due to the random demand, the retailer's profit will be uncertain. Therefore, either an expected profit is maximized, $E[.]$, (in which case, we assume that the retailer is risk neutral) or a risk sensitive objective is defined, leading to a "robust" ordering policy, which will be insensitive to some of the actions that the suppliers may take. In this framework, the game consists in the following: after the supplier chooses a wholesale price, w, the retailer determines the order quantity, q. The supplier then produces the products and delivers them by the end of the period. Of course, there are also variations and extensions to this game, in both their static and in their inter-temporal frameworks. Some of these will be considered in Chapters 2 and 3 respectively.

The Outsourcing Game

Outsourcing consists in the transfer of previously in-house production or other activities to a third party. This problem has been the subject of considerable analysis due to the current awareness that a large segment of industrial, logistics and service activities are outsourced both nationally to local firms and internationally. A simplistic version of this model, based on the classical make-buy decision problem can be construed as a single-period newsvendor model with a setup cost added. The assumed setup cost C, is a fixed irreversible cost which the manufacturer incurs for each in-house production order. In addition, a variable per unit cost c_m, is assumed. The outsourcing decision is presumed to relieve the industrial firm from the fixed costs it assumes, augmenting thereby reactivity to market demands and reducing its aggregate costs. Of course, such presumptions are simplistic since there are many mitigating factors and strategic considerations implied in the outsourcing decision. In our simplified model, the basic newsvendor assumptions remain unchanged; the demand is random with a known probability distribution; and the selling season is assumed to be short, so that if the order quantity is less than the true demand at the end of the period, then shortage h^- cost per unit of unsatisfied demand is incurred and there is no time for additional orders. Otherwise, if there is a surplus with respect to the quantity that the manufacturer is able to sell, the inventory cost incurred per unit is h^+ for inventory left over at the end of period. Such a framework will provide, nonetheless, a model in which some of the salient factors determining to outsource or not can be defined and discussed.

The resulting outsourcing game is defined as follows: the supplier sets the wholesale price and then the manufacturer decides whether to outsource or to produce in-house. If the decision is to outsource, the manufacture responds with an order quantity, which the supplier delivers by the end of the period. If the decision is to produce in-house, then the manufacturer decides on the quantity to produce and initiates the production. This game, simple in a static framework, becomes elaborate in an inter-temporal framework.

Inventory Game with Buy-Back, Sell-Back and other Options

The profusion of optional features in inventory and production contracts has greatly expanded the number of issues addressed in operations and supply chain management problems. Such features have been considered particularly important since they allow directional risk transfer between parties who possess different information at the time the contracts are negotiated and signed as well as vastly differing risk attitudes. To present one such optional feature, we shall consider a buy-back option in an inventory contract. In classical models, all expenses and risks related to the order that has been made is assumed by the "order taker", i.e., the party who assumes the difference between the quantity, q, stocked and the actual demand, D, revealed at the end of period, which is paid for solely by the retailer. In an optional buy-back of the newsvendor problem, the retailer has the option to return unsold products at the end of period at a price, $b(w)$, which is below the wholesale price, w. This implies that if a buy-back contract is signed, the supplier mitigates the retailer's surplus, x^+, related to the risk it assumes and thus encourages the retailer to buy more. As a result, in addition to the retailer's uncertain profit, the supplier's payoff function $J_s(q,w) = (w-c)q - E[b(w)x^+]$ now involves a random surplus, x^+. This is in contrast to the outsourcing and inventory games, where the supplier's profit was defined at the time of sale.

The game is defined as follows: the supplier sets a wholesale price w and a buy-back price $b(w)$; the retailer orders quantity q, which the supplier delivers, contracting for surplus products (if any) at the end of period, once the optional decision by the "buyer-manufacturer" is reached ex-post when demand is revealed.

The Inventory Game with a Purchasing Option

Similar to the sell-back and buy-back options, a purchasing option provides a supplier with the means to mitigate the retailer's risk associated with uncertain demands. To do so, an agreement regarding inventory shortage costs sharing rather than surplus costs may be assumed. Specifically, the supplier may agree to carry additional inventories by providing the retailer a purchasing option, complementing the regular order at the wholesale price, w. The option allows the retailer to issue an urgent or emergency order at a predetermined price, $u(w) > w$, $\dfrac{\partial u(w)}{\partial w} \geq 0$, close to the end of the selling season and to be shipped immediately. The retailer, of course, will exercise this option only if customer demand exceeds its inventories. The quantity, which is the difference between the retailer's shortage (back-order) and the supplier's inventory position at the end of period, will be delivered then as an emergency order. If the supplier is not able to satisfy such a backorder in full, a supply contract might stipulate that associated retailer's losses will be covered by the supplier (in whole or in part). Under this type of option, the supplier explicitly assures the buyer, providing a customer service level at the retailer's site (assuming as a result, part or the whole retailer's backlog cost).

Similar to the buy-back option, a purchasing option affects the supplier's profit and introduces some uncertainty and thus augments the risk associated with random demands faced by the buyer. Nevertheless, such contract provides both a sale incentive on the one hand and is justified by the economies of scale that the supplier wishes to implement (presuming that the supplier has a diversified group of such buyers).

The game is defined as follows: the supplier chooses the wholesale and purchasing option prices, then the retailer and supplier choose quantities for their regular orders, the supplier then delivers the regular order. When the demand is realized, if there is a shortage, the backlogged units are urgently delivered to the retailer.

1.8 RISK AND SUPPLY CHAIN MANAGEMENT

Risks and its management have traditionally been used as a panacea for the many ills, real, potential or imaginary, that corporate management deals with or sustains, either internally or externally. However, the growth and realignment of corporate entities into strategic supply chains, global and market sensitive, are altering conceptions of corporate risk and as a result the management of supply chains. These concerns are today far more in tune with the operational challenges faced by cross-national supply chains. These topics will be developed further in Chapter 7.

Supply chains are based on exchange and dependence between firms, all drawing financial benefits from the arrangement. These benefits include risks which must be sustained and managed, frequently in as many ways the mind can measure and the imagination suggests. Collaboration for example, is a well-trumpeted mechanism for maximizing profits while at the same time managing the dependence risks between firms engaged in supply chain exchanges. Collaboration is not always possible, however, for agreements may be difficult to self-enforce and as a result dependence risks are strategic and potentially overwhelming. These issues, specific to supply chains, combined with the operational and external risks that supply chains are subject to and create, require that specific attention be directed to their measurement and to their management. Such measurement will require a greater understanding of a firm's motivations in entering supply chain relationship. Here, supply chain managers have an important role in achieving this understanding by focusing attention on these risks and in educating corporate managers about what these risks imply, how to measure, evaluate, and internalize them in the costs and benefits calculations they use to reach decisions.

It should be kept in mind that risk is also a great "motivator", energizing technological innovation, development and growth. Without risk, there can be no profit as well. In other words, in the spirit of financial theory, profits can be realized if supply chain entrepreneurs take risk and these profits are a compensation for the risk they are willing to assume. The reverse might not be true, however—risk taking does not imply profits! For these reasons, risk is a two-edged sword, an inducement to creative change but also bearing the possibility of a negative consequence.

Supply chains have expanded hand-in-hand with the globalization of the economic environment and technological change and the emergence of financial markets entailing corporate objects and business risks that cannot be sustained by individual firms. Risk-sharing through joint ventures, supply chains and other inventive organizational frameworks has both justified the trend to ever-larger supply chains entities but at the same time it has raised a number of issues about risks and their control (such as operational risks; sustainability; political risks; and risk externalities sustained by supply chains). In contrast to the traditional focus on internal and external risks in operations and logistics (for example, dealing only with the risks of supplies or meeting demands as outlined above), supply chains are far more subject to *strategic risks* and to *risk externalities*. Examples that highlight these risks will be considered here; basic models to deal with such problems in a supply chain environment will be considered in Chapters 7 and 8.

Supply chains are an essential ingredient of the quest for corporate survival and growth. Risk in supply chains has assumed, however, added dimensions, providing, on the one hand, greater opportunities to manage these risks and, on the other, augmenting appreciably, the risks that modern enterprises face. Since many of these risks are ill-understood and poorly evaluated, they are poorly managed. As a result they are also poorly measured, augmenting the risks that supply chains entities face.

Risks Galore

In contrast to the traditional focus on internal and external risks, we shall distinguish between:

- Operational risks (such as intra-firms operational risks)
- External risks (such as technology, financial markets, political and market structure risks)
- Strategic risks (such as inter-firms risks)
- Risk externalities

In Figure 1.11, a plethora of such risks are included, emphasizing the fact that risks are both varied and numerous.

Figure 1.11. Risk galore

Operational risks (see Table 1.1) concern the direct and indirect adverse consequences of outcomes and events resulting from operations and services that were not accounted for, that were ill-managed or ill- prepared. They affect individual clients, customer-firms or society at large (their externalities). The risks result from many reasons and may be induced both internally and externally. Internal consequences are the result of failures in operations and service management while in the latter instance they derive from external uncontrollable events we were not ready for or were unable to attend to. Operations risks are thus a measurement of these consequences. By contrast, operation attributes (product and service quality, for example) may be objective and subjective and can be measured in many disparate ways. Quality may be based on a measure of excellence (measured absolutely, or relatively as is the case in benchmarking), or be defined as the ability to meet consumer specifications (as is the case in industry). Or, equivalently, measured in terms of the firm's ability to meet "customers expectations", it can be broadly used to measure service quality. Zeithamal at el. (1990), for example, concluded that "service quality as perceived by consumers results from a comparison of perceived service with expected service". "Unquality" is then defined as the "risk" of deviating from consumer expectations without actually measuring the consequence of such a deviation—which measures risk!. Other approaches have also been suggested, however, based on the socio-psychology of persons (the servers, the serviced and the encounter between the two) involved in the service process (Klaus 1991, pp. 261-263). In other words, in services, quality and risk are highly intertwined concepts, often one expressing the other. In this sense, risk is an essential attribute of quality in services (as well as in industry). Operational risk and quality are therefore intimately related—one is used to measure, to define and to manage the other.

In industrial quality, since the definition of quality risks is based on the management of variations, their measurement and control are far more specific than in the case of services. Such differences arise because service quality may be person-specific i.e. quality may be measured or valued in various ways by different persons and firms and in diverse circumstances. Further, it may depend on both "the service provider—the supplier" and the "serviced–producer", each with their own interacting characteristics and wants, etc. For example, a firm over-emphasizing on-time delivery to a member of the supply chain (because of absolute requirements in synchronization) may neglect some of its intangible attributes (the meeting and the discussion that were on-going between clients and delivery persons) leading thereby to a subsequent loss of customers.

External risks derive from events over which the firms within the supply chain have little control. These events are now assuming a far greater importance, providing a source of concern and worry to supply chains. For example, financial markets in particular have created immense possibilities for corporate risk transfer and risk valuation of corporate enterprises and therefore a more efficient risk management. However, they have also become a two-edged sword—misused in a manner to render certain financial decisions to be non-transparent and generally tending to favor short-term gains over longer ones. Further, based on the presumption that there is no profit without risks, financial markets have been used to exercise and assume exuberant (and irrational) risks. Size and scale in financial markets have also contributed to dwarfing any potential control can be exercised by the firms and the supply chains aggregates upon which they depend .

By the same token, the current supply chain environment –"globalization"– has fostered the growth of many external threats that had previously been kept at bay . Globalization is thus both an opportunity and a threat. It is an opening to markets (with many specificities and risks) while at the same time there is a risk that "global" competition may invade what may have been traditional and protected markets. In other words, globalization means also that the "world" enters freely into an enterprise's traditional markets, threatening its ability to compete.

Approaches to risk measurement, valuation and management might differ due to cultural environments, values and society's traditions. Each emphasizes perspectives often neglected by the other. Such problems often contribute to misunderstandings in business practices and exchanges. Thus risk can be culture- sensitive, as has often been observed in practice in the US, Europe and Asia. In this spirit, the measurement of risk in a supply chain must reflect the many intricacies that local habits and culture imply as well as the many opportunities and threat they open for the supply chain (as is the case in China, for example).

Technology, by the same token, is both an external as well as a strategic risk. It is an external risk because technology innovation is broadly diffused with firms having little control over the process. Further, the "democratization" of innovation has removed the center of gravity in technological innovation from in-grown and managed R&D to innovations appearing in a seemingly spontaneous manner throughout the global chain. For these reasons, some firms have abandoned the in-house process of innovation and inventiveness management in favor of permanently scouting for talent and innovation. While in some cases, this might seem as a mechanism to reduce costs. In other, more likely cases, it is far more a losing battle fought by strengthening the protective walls that supply chains create to augment the control that firms within the chain have over their markets.

Another strategic, external risk that supply chains encounter lies in the fact that firms within the chain have become major consumers of fast-changing technologies, in particular IT. At the same, they are increasingly losing control over these technologies. This in turn amplifies the technological risks that supply chains and enterprises face. IT outsourcing, a current fad, is a revealing signal of helplessness in managing a technology, imbedded in a strategic rationale; it has dire consequences for enterprises in the long run.

Strategic risks (see Table 1.1): Supply chains are based on exchange and collaboration. The former aspect means that for firms to engage in a supply chain, their utility must at least be larger than a "going-it-alone". Therefore risk arises when enterprises exchange with several other firms whose motivations may differ from the enterprise's aims. In this case, collaboration may be impossible to maintain. Information and power asymmetries, "a tyranny of minorities" etc., make it possible then for the few to threaten and control the many through moral hazard and adverse selection risks. This leads to supply chains breaking down. Further, even if firms do collaborate and find it economical to sustain the supply chain, often a randomized strategy is Pareto efficient, which induces an additional strategic risk. In such circumstances, the mere fact that firms engage in a collaborative supply chain relationship induces a risk unlike the risk sustained by firms that "go-it-alone". Strategic risks arise then from an uncertainty with risk derivatives that are no longer consequences of a latent environmental (bad enough) uncertainty but the outcome of strategic (and potentially malevolent) behaviors.

Corporate realignment along supply chains and well integrated business entities is an example of the modern corporate work environment. Traditional and basic functions such as quality (risk) control, inventory (risk) management in a supply chain, etc. can no longer be dealt with in the "risk neutral" context and "conflict free" environment in which such problems are taught in the classroom and applied automatically in industrial management. Rather, a strategic approach to risk assessment is needed. Recently, some papers along these lines have appeared (for example, Akerlof 1970; Barzel 1982; Riordan 1984, and many others as will be outlined in Chapter 7). Underlying strategic risk is an information asymmetry whose risk effects can be summarized by adverse selection and moral hazard. In *"Adverse Selection"*, Akerlof points out that goods of different qualities may be uniformly priced when buyers cannot realize that there are quality differences (Akerlof 1970). For example, one may buy a used car, not knowing its true state, and therefore the risk of such a decision may induce the customer to pay a price which would not truly reflect the value of the car. In other words when there is such an information asymmetry, pricing of quality is ill-defined because of the mutual risks that exists between the buyer and the seller (who has a better information).

"The Moral Hazard Problem" implies that a quality that cannot be observed induces a risk to the customer. For example, there is a possibility that the supplier (or the provider of quality) will use that fact to his advantage and not deliver the right level of quality. Of course, if we contract the delivery of a given level of quality and if the supplier does not knowingly maintain the terms of the contract that would be cheating. We can deal with such problems with various sorts of (risk-statistical) controls combined with incentive contracts which create an incentive not to cheat or lie. In supply chain services, the control of such risks may be treated in many different ways. For example, some restaurants might open their kitchen to their patrons to convey a message of truthfulness in so far as cleanliness is concerned. A supplier would let the buyer visit the manufacturing facilities as well as reveal procedures relating to the control of quality, service record and reputation etc.

Examples of these problems are numerous. We outline a few. A transporter may not feel sufficiently responsible for the goods shipped by a company to a demand point. As a result, it is necessary to manage the transporter relationship and thereby the risks implied in such a relationship. Otherwise, this may lead to a greater probability of transport failure. The "de-responsabilization" of workers also induces a moral hazard. It is for this reason that incentives, performance indexation and "on-the-job-responsibility" are so important and needed to minimize the risks of moral hazard (irrespective of whether these are tangibles or intangibles). For example, decentralization of the work place and getting people involved in their jobs may be a means to make them care a little more about their job and to provide an appropriate performance in everything they do. A supplier who has a long-term contract might not care to supply on time for a buyer who is locked into such a relationship (contract). Within these examples, there are also negative inducements to performance.

The relationship between strategic risks, conflict and control as well as the role of statistical sampling in improving the control of supply chain conflict has to a large measure been neglected. For example, the failure of statistics to reflect conflict arose from the presumption that "uncertainty is not motivated". In other words, randomness is an act of G-D and has no known purpose or is not directed towards any special purpose. Interpreting uncertainty and reducing its effects is then based on the presumption that our measurements and our acts are independent of the origins of such uncertainty. Randomness arises as well due to "moral hazard and adverse selection", as pointed out earlier, because of information asymmetry and conflicting interests that induce a greater need for controls in order to assure that "what is intended will occur". Here again, the use of sampling (because measurements are costly) as a technique to mitigate the effects of information asymmetry on decision efficiency has been ignored. For example, insurance contracts with binding sampling clauses, may be designed not only as a means of exchange but also as a way to induce post-contract behaviour which is compatible with a contract's intentions. Similarly, strategic audits always have a number of messages that they convey: a control, a signal to the auditor on the firm's intentions and, of course, to collect information which is needed to reach an economic decision. For example, in a bilateral monopoly under information asymmetry, there is ample room for opportunistic behaviour! (That is to say, when only two parties are involved in decision-making, information asymmetry can lead to opportunistic behaviour, or, simply said, cheating). The control of exchanges between parties should, therefore, keep in mind the parties' intentionality imbedded in their preferences, the exchange terms, as well as the information each will use in respecting or not the intended terms of their exchange. In some papers (Reyniers and Tapiero, 1995a, 1995b; Tapiero 1996, 2006), these effects have been stressed in producer-supplier relationships and in the design of contracts for controlling quality.

Table 1.1. Risks

Operational Risk	External Risks	Strategic Risks	Risk Externalities
Supply delay risks Synchronization and delays risks Measurement risks Inventory risks Quality risks	Political Regulation Financial markets Macroeconomic Risk Bias (Leptokurtic, chaos etc.) Measurement risks	Dependence Outsourcing Exchange Information asymmetry Moral hazard Adverse selection Non-transparency Measurement risks	Environmental risks Non detection risks Collective risks Ethics-Social Regulation

An externality is a cost or benefit that is experienced by someone who is not a party to the transaction that produced it. A negative externality is a cost experienced by someone who is not a party to the transaction that produced it. A positive externality is a benefit experienced by someone who is not a party to the transaction that produced it. Externalities are important because they can create incentives to engage in too much or too little of an activity, from an efficiency perspective. When all of the costs and benefits of a transaction are internal, meaning that all costs and benefits are experienced by someone directly involved, we expect the transaction to take place only if the benefits are greater than the costs. Say, for example, that a good is produced through a supply chain. A price can then be agreed on if both the clients –the public and the supply chain – can profit. What if, in making the product, the supply chain also contributes significantly to pollution without sharing the costs of cleaning the pollution it has created? In that case, the fact that a product was produced and sold does not necessarily mean that wealth was created because of such an exchange. To know for sure, we'd have to find out the economic value of the pollution damage.

In general, the problem is that externalities create a divergence between private and social costs and that can be very risky because supply chains, due to their size and power, are often competing with public and political institutions. Further, corporate responsibility, ethics, environmental consciousness and other such concepts are used by firms to deal with risk externalities. Currently they are mostly used as popular buzz words in corporate strategies with risk externalities that are mostly misunderstood and not valued. Corporations react in a similar manner to patients that are told that they must diet urgently—which they do with a great appetite giving an indigestion. The problem is that externalities create a risk for corporate firms—the risk that a divergence between private costs and social costs will entail appreciable damage for the firm.

Measurement of risk in supply chains is thus essential and ought to emphasize the specific characteristics and motivations of supply chains in managing their priorities, operational and otherwise. However, if the quest for profit is not without risk, there are no profits without an efficient risk management. Risk can no longer be a consequence of corporate strategy in its quest for an efficient supply chain management but an essential aspect, feeding and fed by this strategy.

APPENDIX: ESSENTIALS OF GAME THEORY

Game theory involves decision-making between two or more parties competing against one another for the purpose of reaching an objective. Each of the parties may depend on the other. These problems are, in general, difficult to analyze, since risks of various sorts and many other factors must be taken into account. To properly describe how to behave in these circumstances involves an appreciation of psychological, sociological as well as economic motives. Behavioral decision-making, which seeks to study how decisions are made in fact, has been used to investigate and understand how and why certain decisions are reached in complex and conflicting situations. "Rational decision-making", that is, the theory of how we rationally behave in competing with others is known as game theory. Areas of application span economic analysis, for example, how agents reach decisions in an economic environment, how market forces operate, and so on.

Game theory was first proposed and defined by the French mathematician Emil Borel in 1921. The famed mathematician John von Neumann provided an analysis of games in 1928. In 1944, von Neumann, assisted by the economist Oskar Morgenstern, published the first thorough and to this day, most complete, work on game theory, "The Theory of Games and Economic Behavior". This book was published about the same time that

Dantzig developed the simplex algorithm in linear programming. A few years later, a relationship between certain types of games (explicitly, zero-sum games) and their solution by linear programming was pointed out. Here we are concerned with two-persons zero-sum games. Situations where there may be more than one player, potential coalitions, cooperation, asymmetry of information (where one player may know something the other does not) etc. are practically important but are not within our scope of study.

Two-Persons Zero-Sum Games

Two-persons zero-sum games involve two players. Each has only one move (decision) to take and both make their moves simultaneously. Each player has a set of alternatives, say A $=(A_1, A_2, A_3,......., A_n)$ for the first player and B$=(B_1, B_2, B_3,......., B_m)$ for the second player. When both players make their moves (i.e. they select a decision alternative) an outcome O_{ij} follows, corresponding to the pair of moves (A_i, B_j) which was selected by each of the players respectively. In two-persons zero-sum games, additional assumptions are made: (1) $A_1, A_2, A_3,......., A_n$ as well as $B_1, B_2, B_3,......., B_m$ and O_{ij} are known to both players. (2) Players do not know with what probabilities the opponent's alternatives will be selected. (3) Each player has a preference that can be ordered in a rational and consistent manner. In strictly competitive games, or zero-sum games, the players have directly opposing preferences, so that a gain by a player is a loss to its opponent. That is;

The Gain to Player 1 = The Loss of Player 2

The concepts of pure and mixed strategies, minimax and maximin strategies, saddle points, dominance etc. are also defined and elaborated. For example, two rival companies, A and B, are the only ones. Company A has three alternatives A_1, A_2, A_3 expressing different strategic while B has four alternatives B_1, B_2, B_3, B_4. The payoff matrix to A (a loss to B) is given by:

	B_1	B_2	B_3	B_4
A_1	.6	-.3	1.5	-1.1
A_2	-7	.1	.9	.5
A_3	-.3	0	-.5	.8

This problem has a solution, called a *saddle-point*, because the least greatest loss to B is equal to the greatest minimum gain to A. When this is the case, the game is said to be stable, and the pay-off table is said to have a saddle-point. This saddle-point is also called the value of the game, which is the least entry in its row, and the greatest entry in the column. Not all games can have a pure, single strategy, saddle-point solution for each player. When a game has no saddle point, a solution to the game can be devised by adopting a mixed strategy. Such strategies result from the com-bination of pure strategies, each selected with some probability. Such a mixed strategy will then result in a solution which is stable, in the sense that player 1's maximin strategy will equal player 2's minimax strategy. Mixed strategies therefore induce another source of uncertainty.

Non-Zero Sum Games

Consider the bimatrix game $(\mathbf{A}, \mathbf{B}) = (a_{ij}, b_{ij})$. Let $\mathbf{x}$ and $\mathbf{y}$ be the vector of mixed strategies with elements x_i and y_j, and such that $\sum_{i=1}^{n} x_i = 1$, $0 \leq x_i \leq 1$, $\sum_{j=1}^{m} y_j = 1$, $0 \leq y_j \leq 1$. The value of the game for each of the players is given by:

$$V_a = \mathbf{x}\mathbf{A}\mathbf{y}^\mathbf{T}, V_b = \mathbf{x}\mathbf{B}\mathbf{y}^\mathbf{T}$$

and an equilibrium is defined for each strategy if the following conditions hold $\mathbf{A}\mathbf{y} \leq V_a, \mathbf{x}\mathbf{B} \leq V_b$. For example, consider the 2*2 bimatrix game. We see that

$$V_a = (a_{11} - a_{12} - a_{21} + a_{22})xy + (a_{12} - a_{22})x + (a_{21} - a_{22})y + a_{22}$$
$$V_b = (b_{11} - b_{12} - b_{21} + b_{22})xy + (b_{12} - b_{22})x + (b_{21} - b_{22})y + b_{22}$$

Then, for an admissible solution for the first player, we require that

$$V_a(1, y) \leq V_a(x, y); V_a(0, y) \leq V_a(x, y),$$

which is equivalent to

$$A(1-x)y - a(1-x) \le 0; \quad Axy - ax \ge 0,$$

where, $A = (a_{11} - a_{12} - a_{21} + a_{22}); a = (a_{22} - a_{12})$. That is when,

$$\begin{cases} x = 0 & then \quad Ay - a \le 0 \\ x = 1 & then \quad Ay - a \ge 0. \\ 0 < x < 1 & then \quad Ay - a = 0 \end{cases}$$

In this sense there can be three solutions (0,y), (x,y) and (1,y). We can similarly obtain a solution for the second player using parameters B and b. Say that $A \ne 0$ and $B \ne 0$, then a solution for x and y satisfies the following conditions:

$$\begin{cases} y \le a/A & if \quad A > 0 \\ y \ge a/A & if \quad A < 0 \\ x \le b/B & if \quad B > 0 \\ x \ge b/B & if \quad B < 0 \end{cases}$$

As a result, a simultaneous solution leads to the following equations for (x,y), which we have used in the text:

$$y^* = \frac{a}{A} = \frac{(a_{22} - a_{12})}{(a_{11} - a_{21} - a_{12} + a_{22})}$$

$$x^* = \frac{a}{A} = \frac{(a_{22} - a_{12})}{(a_{11} - a_{12} - a_{21} + a_{22})}.$$

In this case, the value of the game is:

$$V_a = (a_{11} - a_{12} - a_{21} + a_{22})xy + (a_{12} - a_{22})x + (a_{21} - a_{22})y + a_{22}$$

$$V_b = (b_{11} - b_{12} - b_{21} + b_{22})xy + (b_{12} - b_{22})x + (b_{21} - b_{22})y + b_{22}$$

$$x^* = \frac{a}{A} = \frac{(a_{22} - a_{12})}{(a_{11} - a_{12} - a_{21} + a_{22})}$$

$$y^* = \frac{b}{B} = \frac{(b_{22} - b_{12})}{(b_{11} - b_{12} - b_{21} + b_{22})}$$

For further study of games and related problems we refer to Moulin 1981; Nash 1950; Von Neumann and Morgenstern 1944; Thomas 1986.

REFERENCES

Agrawal V, Seshadri S (2000) Risk intermediation in supply chains. *IIE Transactions* 32: 819-831.

Akerlof G (1970) The Market for Lemons: Quality Uncertainty and the Market Mechanism. *Quarterly Journal of Economics*, 84: 488-500.

Bank D (1996) Middlemen find ways to survive cyberspace shopping. *Wall Street Journal,* 12 December, p. B6.

Barzel Y (1982) Measurement cost and the organization of markets. *Journal of Law and Economics* 25: 27-47

Bowersox DJ (1990) The strategic benefits of logistics alliances. *Harvard Business Review*, July-august 68: 36-45.

Cachon G (2003) Supply chain coordination with contracts. In: De Kok, AG, Graves S (Eds.), Handbooks in Operations Research and Management Science. Elsevier, Amsterdam.

Caves RE, Murphy WE (1976) Franchising firms, markets and intangible assets. *Southern Economic Journal* 42: 572-586.

Christopher M (1992) Logistics and Supply Chain Management, Pitman, London.

Christopher M (2004) Creating resilient supply chains. Logistics Europe, 14-21.

Friedman JW (1986) *Game theory with Applications to Economics.* Oxford University Press.

Fudenberg D, Tirole J (1991) *Game Theory.* MIT Press. Cambridge, Mass.

Klaus P (1991) Die Qualitat von Bedienungsinteraktionen, in M. Bruhn and B. Strauss,(Eds.), Dienstleistungsqualitat: Konzepte-Methoden-Erfahrungen, Wiebaden

Lafontaine F (1992) Contract theory and franchising: some empirical results. *Rand Journal of Economics* 23(2): 263-283.

La Londe B, Cooper M (1989) *Partnership in providing customer service: a third-party perspective.* Council of Logistics Management, Oak Brook, IL.

Lambert D, Stock J (1982) Strategic Physical Distribution Management, R.D. Irwin Inc., Ill., p. 65.

McIvor R (2000) A practical framework for understanding the outsourcing process. *Supply Chain Management: an International Journal* vol.5(1): 22-36.

Moulin H (1995) Cooperative *Microeconomics: A Game-Theoretic Introduction.* Princeton University Press. Princeton ,New Jersey

Nash F (1950) Equilibrium points in N-person games. *Proceedings of the National Academy of Sciences* 36: 48-49.

Newman R (1988) The buyer-supplier relationship under just in time, *Journal of Production and Inventory Management* 4: 45-50.

Rao K, Young RR (1994) Global Supply Chain: factors influencing Outsourcing of logistics functions. *International Journal of Physical Distribution & Logistics Management* 24(6).

Rey P (1992) The economics of franchising, ENSAE Paper, February, Paris.

Rey P, Tirole J (1986) The logic of vertical restraints. *American Economic Review* 76: 921-939

Reyniers DJ, Tapiero CS (1995a) Contract design and the control of quality in a conflictual environment. *Euro J. of Operations Research*, 82(2): 373-382.

Reyniers, Diane J, Tapiero CS (1995b) The supply and the control of quality in supplier-producer contracts. *Management Science* 41: 1581-1589.

Riordan M (1984) Uncertainty, asymmetric information and bilateral contracts. Review of Economic Studies 51: 83-93.

Ritchken P, Tapiero CS (1986) Contingent Claim Contracts and Inventory Control. *Operations Research* 34: 864-870

Rubin PA, Carter JR (1990) Joint optimality in buyer–supplier negotiations. *Journal of Purchasing and Materials Management* 26(1): 54-68.

Tapiero CS (1995) Acceptance sampling in a producer-supplier conflicting environment: Risk neutral case, *Applied Stochastic Models and Data Analysis* 11: 3-12

Tapiero CS (1996) *The Management of Quality and Its Control,* Chapman and Hall, London

Tapiero CS (2005a) Risk Management, John Wiley *Encyclopedia on Actuarial and Risk* Management, Wiley, New York-London

Tapiero CS (2005b) Value at Risk and Inventory Control, *European Journal of Operations Research* 163(3): 769-775.

Tapiero CS (2006) Consumers Risk and Quality Control in a Collaborative Supply Chains. *European Journal of Operations Research,* (available on line, October 18)

Thomas LC (1986) *Game Theory and Application,* Ellis Horwood Ltd, Chichester.

Tsay A, Nahmias S, Agrawal N (1998) Modeling supply chain contracts: A review, in Tayur S, Magazine M, Ganeshan R (eds), *Quantitative Models of Supply Chain Management*, Kluwer International Series

Van Damme DA, Ploos van Amstel MJ (1996) Outsourcing Logistics management Activities. *The International Journal of Logistics management* 7(2): 85-94.

Van Laarhoven P, Berglund M, Peters M (2000) Third-party logistics in Europe – five years later. *International Journal of Physical Distribution & Logistics Management* 30(5): 425-442.

Von Neumann J, Morgenstern O (1944) *Theory of Games and Economic Behavior*. Princeton University Press.

Williamson OE (1985) *The Economic Institutions of Capitalism*, New York, Free Press.

Zeithaml V, Parasuraman A, Berry LL (1990) *Delivering Quality Service*, Free Press, New York.

2 SUPPLY CHAIN GAMES: MODELING IN A STATIC FRAMEWORK

A supply chain can be defined as "a system of suppliers, manufacturers, distributors, retailers, and consumers where materials flow downstream from suppliers to customers and information flows in both directions" (Geneshan et. al. 1998). The system is typically decentralized which implies that its participants are independent firms each with its own frequently conflicting goals spanning production, service, purchasing, inventory, transportation, marketing and other such functions. Due to these conflicting goals a decentralized supply chain is generally much less efficient than the corresponding centralized or integrated chain with a single decision maker. Efficiency suffers from both vertical (e.g., buyer-vendor competition) and horizontal (e.g., a number of vendors competing for the same buyer) conflicts of interest.

How to manage competition in supply chains is a challenging task which comprises a variety of problems. The overall target is to make, to the extent possible, the decentralized chain operate as efficiently as its benchmark, the corresponding centralized chain. This particular aspect of supply chain management is referred to as coordination. This chapter addresses simple static supply chain models, competition between supply chain members and their coordination.

2.1 STATIC GAMES IN SUPPLY CHAINS

In research and management literature where supply chain problems and related game theoretic applications have gained much attention in recent years, we see extensive reviews focusing on such aspects as taxonomy of supply chain management (Geneshan et. al. 1998); integrated inventory models (Goyal and Gupta 1989); game theory in supply chains (Cachon and Netessine 2004); operations management (Li and Whang 2001); price quantity discounts (Wilcox et. al. 1987); and competition and coordination (Leng and Parlar 2005).

In the literature, supply chains are distinguished by various features such as: types of decisions; operations; competition and coordination; incentives; objectives; and game theoretic concepts. In this chapter we deal with three essential features of static supply chains, i.e., the supply chains with decisions independent of time: customer demand, competition and risk. In this sense we distinguish between

- deterministic and random demands; endogenous and exogenous demands
- vertical and horizontal competition within supply chains
- no risk involved, risk incurred by only one of the parties and risk shared between the parties.

In this chapter, supply chain games are combined into three groups. The first group of games represents classical horizontal production and vertical pricing competition under endogenous demands. These games involve decisions about either product prices or quantities with respect to two types of endogenous demands: (i) the quantity demanded for a product as a function of price set for the product and (ii) an inverse demand function with price as a function of the quantity produced or sold. In both cases the demands are deterministic, which implies that all produced/supplied products are sold and thus there is no risk involved.

Random exogenous demand for products characterizes the second group of games which is related to the classical newsvendor problem. The parties vertically compete by deciding on a price to offer and a quantity to order for a particular price. Since the demand is uncertain, the downstream party, which faces the demand, runs the risk of overestimating or underestimating it. The risk involves costs incurred due to choosing the quantity to order and stock before customer demand is realized. We refer to this group of games as stocking / pricing competition with random demand.

The third group of games represents classical risk-sharing interactions between supply chain members. Similar to the second group, the competition is vertical and the demand is exogenous and random. Unlike the second group, however, incentives to mitigate risk may be offered to a party which faces uncertain customer demands. Since the incentives include buyback and urgent purchase options, some of the uncertainty is transferred from one party to another. In such a case, the risk associated with random demand is shared and the inventories of all involved parties are affected when deciding on what quantities to stock.

Motivation

We describe a few production, pricing and inventory-stock related problems which have been found in various service and industry-related supply chains. Most of these problems have been extensively studied and can be found virtually in every survey devoted to supply chain management including those mentioned above. It is worth noting that, in general, the number of basic supply chain problems is significant and selecting just a few of them for an introductory purpose is not a simple matter.

Our selection criterion is based on one of the overall goals of this book–to show how optimal pricing and inventory policies evolve when static operation conditions become dynamic. Under such conditions, we find particularly interesting the static problems which allow for straightforward and, yet natural, dynamic extensions. The problems which we discuss in this chapter will be discussed again in the following chapters to show the effect of production and service dynamics on managerial decisions.

The static feature of the problems we select implies that the period of time that the problems encompass is such that no change in system parameters is observed. Since all products are delivered at once by the end of the period and then instantly sold, these problems ignore the intermediate inventories (and associated costs) before and during the selling season. Due to the focus on *stock and pricing policies*, shortages as well as leftovers are avoided, as much as possible, by the end of the period. In all the problems that we consider, it is assumed that the information needed for decision-making is available and transparent to the supply chain participants and that the overall order lead-time is smaller than the length of the period so that all deliveries are provided on time.

This chapter introduces and discusses basic models of horizontal and vertical competition between supply chain members, the effect of uncertainty and risk sharing as well as basic tools for coping with the competition by coordinating supply chains. The analysis which we employ includes (i) formal statements of problems of each non-cooperative party involved as well as the corresponding centralized formulations where only one decision-maker is responsible for all managerial decisions in the supply chain; (ii) system-wide optimal and equilibria solution for competing parties; (iii) analysis of the effect of competition on supply chain performance and of coordination for improving the performance. In analyzing the problems we use Nash and Stackelberg equilibria which we briefly present next.

Nash and Stackelberg equilibria

Game theory is concerned with situations involving conflicts and cooperation between the players. Our focus is on two important concepts of Nash
and Stackelberg equilibria intended respectively for dealing with simultaneous and sequential non-cooperating decision-making by multiple players. Consider a game, with the strategies y_i, $i=1,..,N$ being feasible actions
which the N players may undertake. All possible strategies of a player, i,
form a strategy set Y_i of the player. A payoff (objective function), $J_i(y_1,$
$y_2,..,y_N,)$, $i=1,..,N$ is evaluated when each player i selects a feasible strategy,
$y_i \in Y_i$. We assume that the games are played on the basis that complete
information is available to all players. Since two-player games can be
straightforwardly extended to multiple players and to simplify the presentation, we further assume that there are only two players A and B.

Each player's goal is to maximize his own payoff. The following definition presents the concept of a Nash equilibrium (Nash 1950)

Definition 2.1

A pair of strategies (y_A*, y_B*) *is said to constitute a Nash equilibrium if
the following pair of inequalities is satisfied for all* $y_A \in Y_A$, *and* $y_B \in Y_A$

$$J_A(y_A*, y_B*) \geq J_A(y_A, y_B*) \text{ and } J_B(y_A*, y_B*) \geq J_B(y_A*, y_B).$$

The definition implies that the Nash solution is

$$y_A* = \arg\max_{y_A \in Y_A}\{J_A(y_A, y_B*)\} \text{ and } y_B* = \arg\max_{y_B \in Y_B}\{J_B(y_A*, y_B)\},$$

and a unilateral deviation from this solution results in a loss. If this problem is static, strategy sets are not constrained and the payoff functions are
continuously differentiable. The first-order (necessary) optimality condition results in the following system of two equations in two unknowns y_A*,
y_B*:

$$\frac{\partial J_A(y_A, y_B*)}{\partial y_A}\bigg|_{y_A=y_A*} = 0 \text{ and } \frac{\partial J_B(y_A*, y_B)}{\partial y_B}\bigg|_{y_B=y_B*} = 0.$$

In addition, the second order (sufficient) optimality condition which
ensures that we maximize the payoffs is

$$\frac{\partial^2 J_A(y_A, y_B*)}{\partial y_A^2}\bigg|_{y_A=y_A*} < 0 \text{ and } \frac{\partial^2 J_B(y_A*, y_B)}{\partial y_B^2}\bigg|_{y_B=y_B*} < 0.$$

Equivalently, one may determine $y_A^R(y_B) = \arg\max_{y_A \in Y_A}\{J_A(y_A, y_B)\}$ for each

$y_B \in Y_B$ to find the best response function, $y_A = y_A^R(y_B)$, of player A and of player B, $y_B = y_B^R(y_A)$ which constitute a system of two equations in two unknowns.

The examples we shall consider here will be elaborated later in this and subsequent chapters.

Example 2.1

Consider a supply chain consisting of one supplier, s, and one retailer r. The supplier offers products at wholesale price w and the retailer buys q product units and sets retail price $p = w + m$. This is the classical pricing game where the two firms want to maximize their profits. Let the supplier and retailer costs be negligible and the demand is linear and downward in price, $d = a - bp = a - b(w+m)$, $a > 0$, $b > 0$. Then the retailer's optimization problem is

$$J_r(m,w) = m(a - b(w+m)) \to \max,$$

$$0 \le m \le \frac{a}{b} - w$$

and the suppliers problem is

$$J_s(m,w) = w(a - b(w+m)) \to \max,$$

$$w \ge 0.$$

First we observe that both objective functions are strictly concave in their decision variables. Thus, the first-order optimality condition is necessary and sufficient. Using the first-order optimality condition we have

$$a - bw - 2bm = 0 \text{ and } a - 2bw - bm = 0.$$

If our constraints are not binding, the two best response functions are

$$m = m^R(w) = \frac{a - bw}{2b} \text{ and } w = w^R(m) = \frac{a - bm}{2b}.$$

Solving these two equations (or equivalently the previous two) we find a unique Nash equilibrium

$$m^n = \frac{a}{3b} \text{ and } w^n = \frac{a}{3b}.$$

The equilibrium is evidently feasible and all constraints are met, as $\frac{a}{3b} > 0$,

hence, $m^* > 0$, $w^* > 0$, and $\frac{a}{3b} < \frac{a}{b} - w^n = \frac{2a}{3b}$, hence, $m^n < \frac{a}{b} - w^n$.

Stackelberg strategy is applied when there is an asymmetry in power or in moves of the players. As a result, the decision-making is sequential rather than simultaneous as is the case with Nash strategy. The player who first announces his strategy is considered to be the Stackelberg leader. The follower then chooses his best response to the leader's move. The leader thus has an advantage because he is able to optimize his objective function subject to the follower's best response. Formally this implies that if, player A, for example, is the leader, then $y_B = y_B^R(y_A)$ is the same best response for player B as determined for the Nash equilibrium. Since the leader is aware of this response, he then optimizes his objective function subject to $y_A = y_A^R(y_B) = y_A^R(y_B^R(y_A))$.

Definition 2.2

In a two-person game with player A as the leader and player B as the follower, the strategy $y_A^* \in Y_A$ is called a Stackelberg equilibrium for the leader if, for all y_A,

$$J_A(y_A^*, y_B^R(y_A^*)) \geq J_A(y_A, y_B^R(y_A)),$$

where $y_B = y_B^R(y_A)$ is the best response function of the follower.

Definition 2.2 implies that the leader's Stackelberg solution is

$$y_A^* = \arg\max_{y_A \in Y_A}\{J_A(y_A, y_B^R(y_A))\}.$$

That is, if the strategy sets are unconstrained and the payoff functions are continuously differentiable, the necessary optimality condition for the leader is

$$\frac{\partial J_A(y_A, y_B^R(y_A))}{\partial y_A}\bigg|_{y_A = y_A^*} = 0.$$

To make sure that the leader maximizes his profits, we check also the second-order sufficient optimality condition

$$\frac{\partial^2 J_A(y_A, y_B^R(y_A))}{\partial y_A^2}\bigg|_{y_A = y_A^*} < 0.$$

Example 2.2

Consider again Example 2.1 but assume that the supplier is the leader. That is, the supplier sets first his wholesale price. In response, the retailer, in setting his retail price, determines the product quantity he orders. Then, to find the Stackelberg solution, we substitute the best retailer's response $m = m^R(w) = \dfrac{a - bw}{2b}$ (see Example 2.1) into the supplier's objective function.

$$\max_w J_s(m, w) = \max_w w(a - b(w + \frac{a - bw}{2b})) = \max_w (\frac{aw}{2} - \frac{bw^2}{2}).$$

The supplier's objective function is evidently strictly concave. Consequently, the first-order optimality condition results in

$$w^s = \frac{a}{2b}, \; m^s = m^R(w^s) = \frac{a}{4b}.$$

The found equilibrium is evidently unique and feasible, as $\dfrac{a}{2b} > 0$,

$\dfrac{a}{4b} > 0$ and $\dfrac{a}{b} - w^s = \dfrac{a}{2b}$ and, thus, $m^s = \dfrac{a}{4b} < \dfrac{a}{b} - w^s = \dfrac{a}{2b}$, i.e., all constraints are met.

For comparative reasons we shall also consider a centralized supply chain with no competition (game) involved. The centralized problem can be viewed as a single-player game.

Example 2.3

Consider again Example 2.1 but assume that there is only one decision-maker in the system. Then the centralized objective function is

$$\max_{m,w} J(m,w) = \max_{m,w} [\, J_r(m,w) + J_s(m,w)] = \max_{m,w} (w+m)(a-b(w+m)).$$

Applying the first-order optimality condition we get two identical equations for m and n. This implies that there is only one decision variable p, so that the system-wide optimal solution is, $m^* + w^* = p^* = \dfrac{a}{2b}$.

2.2 PRODUCTION/PRICING COMPETITION

We discuss here two classical problems arising in supply chains characterized by deterministic demands and either vertical supplier-retailer or horizontal supplier-supplier competition. The competition is represented by games. We first analyze pricing equilibrium based on Bertrand's competition model and then production equilibrium according to Cournot's competition model. Since the problems are deterministic, they can be viewed as both single-period and continuous review models.

2.2.1 THE PRICING GAME

Consider a two-echelon supply chain consisting of a single supplier selling a product type to a single retailer over a period of time. The supplier has ample capacity and the period is longer than the supplier's leadtime which implies that the supplier is able to deliver on time any quantity q ordered by the retailer. The retailer faces a concave endogenous demand, $q=q(p)$, which decreases as product price p increases, i.e., $\dfrac{\partial q}{\partial p} < 0$ and $\dfrac{\partial^2 q(p)}{\partial p^2} \le 0$.

The supplier incurs unit production cost c and sells at unit wholesale price w, i.e., the supplier's margin is $w-c$. Note that this formulation is an extension of that employed in Example 2.1, where a specific, linear in price, demand was considered.

Let the retailer's price per unit be $p=w+m$, where m is the retailer's margin. Both players, the supplier and the retailer, want to maximize their profits – margin times demand which are expressed as $J_s(w)=(w-c)q(w+m)$ and $J_r(p)=mq(w+m)$ respectively (see Figure 2.1). This leads us to the following problems.

The supplier's problem

$$\max_w J_s(w,m)= \max_w (w-c)q(w+m) \tag{2.1}$$

s.t.

$$w \ge c. \tag{2.2}$$

The retailer's problem

$$\max_m J_r(w,m)= \max_m mq(w+m) \tag{2.3}$$

s.t.

$$m \ge 0, \tag{2.4}$$

$$q(w+m) \ge 0. \tag{2.5}$$

Note that from $w \ge c$ and $m \ge 0$, it immediately follows that $p=w+m \ge c$. In contrast to the vertical competition between the two decision-makers as determined by (2.1)-(2.5), the supply chain may be vertically integrated or centralized. Such a chain is characterized by a single decision-maker who is in charge of all managerial aspects of the supply chain. We then have the following single problem as a benchmark.

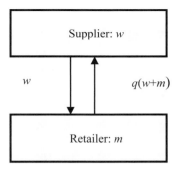

Figure 2.1. Vertical pricing competition

The centralized problem

$$\max_{m,w} J(m,w) = \max_{m,w} [\, J_r(m,w) + J_s(m,w)] = \max_{m,w} (w+m-c)q(w+m) \qquad (2.6)$$

s.t.

$$m \geq 0,\, q(w+m) \geq 0.$$

To distinguish between different optimal strategies, we will use below superscript n for Nash solutions, s for Stackelberg solutions and $*$ for centralized solutions.

System-wide optimal solution

We first study the centralized problem by employing the first-order optimality conditions

$$\frac{\partial J(m,w)}{\partial m} = q(w+m) + (w+m-c)\frac{\partial q(p)}{\partial p} = 0,$$

$$\frac{\partial J(m,w)}{\partial w} = q(w+m) + (w+m-c)\frac{\partial q(p)}{\partial p} = 0.$$

Since both equations are identical, only the optimal price matters in the centralized problem, p^*, while the wholesale price $w \geq 0$ and the retailer's margin $m \geq 0$ can be chosen arbitrarily so that $p^*=w+m$. This is because w and m represent internal transfers of the supply chain. Thus, the proper notation for the payoff function is $J(p)$ rather than $J(m,w)$ and the only optimality condition is

$$q(p^*) + (p^* - c)\frac{\partial q(p^*)}{\partial p} = 0. \qquad (2.7)$$

Let $q(P)=0$, $P>c$. Then it is easy to verify that,

$$\frac{\partial^2 J(p)}{\partial p^2} = \frac{\partial q(p)}{\partial p} + \frac{\partial q(p)}{\partial p} + (p - c)\frac{\partial^2 q(p)}{\partial p^2} < 0,$$

that is, the centralized objective function (2.6) is strictly concave in price for $p \in [c, P]$. This implies that equation (2.7) has a unique solution which maximizes (2.6).

Game Analysis

We consider now a decentralized supply chain characterized by non-cooperative or competing firms and assume first that both players make their decisions simultaneously. The supplier chooses the wholesale price w and the retailer selects his price, p, or equivalently his margin, m, and hence buys $q(p)$ products. The supplier then delivers the products. Since this pricing game is deterministic, all products that the retailer buys will be sold.

Using the first-order optimality conditions for the retailer's problem, we find that the retailer's best response is determined by the following expression

$$\frac{\partial J_r(m, w)}{\partial m} = q(w + m) + m\frac{\partial q(p)}{\partial p} = 0. \tag{2.8}$$

It is easy to verify that the retailer's objective function is strictly concave in m and, thus, (2.8) has a unique solution, or, in other words, the retailer's best response function is unique. Comparing (2.8) and (2.7) and taking into account that $w > c$ (otherwise the supplier has no profit), we conclude with the following result:

Proposition 2.1. *In vertical competition of the pricing game, if the supplier makes a profit, i.e., $w > c$, the retail price will be greater and the retailer's order less than the system-wide optimal (centralized) price and order quantity respectively.*

Proof: Substituting $p = w + m$ into (28) we have

$$q(p) + (p - w)\frac{\partial q(p)}{\partial p} = 0. \tag{2.9}$$

Comparing (2.7) and (2.9) we observe that

$$q(p) + (p - w)\frac{\partial q(p)}{\partial p} = q(p^*) + (p^* - c)\frac{\partial q(p^*)}{\partial p} = 0, \tag{2.10}$$

while taking into account that $w > c$ and $\frac{\partial q}{\partial p} < 0$,

$$q(p^*)+(p^*-w)\frac{\partial q(p^*)}{\partial p} > q(p^*)+(p^*-c)\frac{\partial q(p^*)}{\partial p}=0. \quad (2.11)$$

Next, by denoting $f(p) = q(p)+(p-w)\dfrac{\partial q(p)}{\partial p}$, and recalling $\dfrac{\partial q}{\partial p}<0$

and $\dfrac{\partial^2 q(p)}{\partial p^2} \le 0$, we find that

$$\frac{\partial f(p)}{\partial p}=\frac{\partial q(p)}{\partial p}+\frac{\partial q(p)}{\partial p}+(p-w)\frac{\partial^2 q(p)}{\partial p^2}<0$$

Thus, to have (2.10) we need $f(p)<f(p^*)$, which, with respect to the last inequality, requires, $p>p^*$ and, hence, $q(p)<q(p^*)$, as stated in Proposition 1.

Note, that our conclusion that vertical pricing competition (2.1)-(2.5) increases retail price and decreases the retailer's order quantity does not depend on whether both players make a simultaneous decision or whether the supplier first sets the wholesale price and plays the role of the Stackelberg leader, as is often the case in practice. In either of the two cases, the overall efficiency of the supply chain deteriorates under vertical competition.

Equilibrium

To determine the Nash pricing equilibrium, which corresponds to simultaneous moves of the supplier and retailer, we next consider the optimality conditions for the supplier's objective function,

$$\frac{\partial J_s(m,w)}{\partial w}=q(w+m)+(w-c)\frac{\partial q(w+m)}{\partial p}=0. \quad (2.12)$$

One can readily verify that the supplier's objective function is strictly concave in w, $\dfrac{\partial^2 J_s(m,w)}{\partial w^2}<0$ and, thus, the supplier's best response (2.12) is unique as well. As a result, the Nash equilibrium, (w^n,m^n) is found by solving simultaneously the following system of equations

$$q(w+m)+m\frac{\partial q(w+m)}{\partial p}=0, \quad (2.13)$$

$$q(w+m)+(w-c)\frac{\partial q(w+m)}{\partial p}=0. \quad (2.14)$$

Solving (2.13) and (2.14) results in

$$w-c-m=0 \text{ and } q(c+2m)+m\frac{\partial q(c+2m)}{\partial p}=0.$$

Assuming that the solution $w+m=P$, $q(P)=0$ cannot be optimal since it leads to zero profit for all supply chain members, we conclude with the following result.

Proposition 2.2. *The pair* (w^n, m^n), *where* m^n *satisfies the following equation*

$$q(c+2m^n)+m^n\frac{\partial q(c+2m^n)}{\partial p}=0. \tag{2.15}$$

and $w^n=m^n+c$ *constitutes a unique Nash equilibrium of the pricing game with* $0<m^n<(P-c)/2$.

Proof: To see that a solution of equation (2.15) always exists and that it is unique, assume $m^n=0$. Then, since $P>c$ and $q(P)=0$, $q(c+2m^n)>0$, while

the second term in (2.15) is zero. Thus, $f(m^n)=q(m^n)+m^n\frac{\partial q(m^n)}{\partial p}>0$

when $m^n=0$. On the other hand, let $c+2m^n=P$, since $q(P)=0$, while the second term in (2.15) is strictly negative as $m^n=(P-c)/2>0$, we have

$f(m^n)=q(m^n)+m^n\frac{\partial q(m^n)}{\partial p}<0$. Finally, taking into account that

$\frac{\partial f(m^n)}{\partial m^n}<0$, we conclude that the solution of $f(m^n)=0$ is unique and

$0<m^n<(P-c)/2$.

Next, we assume that the supplier makes the first move by setting the wholesale price. The retailer then decides on what price to set and, hence, the quantity to order. To find the Stackelberg equilibrium, we need to maximize the supplier's objective with m subject to the best retailer's response $m=m^R(w)$ determined by (2.8),

$$J_s(m,w)=(w-c)q(w+m^R(w)).$$

Differentiating the supplier's objective function we have

$$\frac{\partial J_s(m,w)}{\partial w}=q(w+m^R(w))+(w-c)\frac{\partial q(w+m)}{\partial p}\frac{\partial m^R(w)}{\partial w}=0,$$

where $\frac{\partial m^R(w)}{\partial w}$ is determined by differentiating (2.8) with m set equal to

$m^R(w)$.

$$\frac{\partial q(w+m)}{\partial p}(1+\frac{\partial m^R(w)}{\partial w})+\frac{\partial m^R(w)}{\partial w}\frac{\partial q(p)}{\partial p}+m\frac{\partial^2 q(p)}{\partial p^2}(1+\frac{\partial m^R(w)}{\partial w})=0.$$

Thus

$$\frac{\partial m^R(w)}{\partial w}=-\left(\frac{\partial q(w+m)}{\partial p}+m\frac{\partial^2 q(w+m)}{\partial p^2}\right)\Bigg/\left(\frac{\partial q(w+m)}{\partial p}+\frac{\partial q(w+m)}{\partial p}+m\frac{\partial^2 q(w+m)}{\partial p^2}\right). \tag{2.16}$$

Equation (2.16) naturally implies

the greater the supplier's wholesale price w, the lower the retailer's margin m.

Based on (2.16) and (2.8) we conclude that a pair (w^s, m^s) constitutes a Stackelberg equilibrium of the pricing game if there exists a joint solution in w and m of the following equations

$$q(w+m) + (w-c)\frac{\partial q(w+m)}{\partial p}\frac{\partial m}{\partial w} = 0 \,,$$

$$q(w+m) + m\frac{\partial q(w+m)}{\partial p} = 0 \,,$$

where

$$\frac{\partial m}{\partial w} = -\left(\frac{\partial q(w+m)}{\partial p} + m\frac{\partial^2 q(w+m)}{\partial p^2}\right) \Big/ \left(\frac{\partial q(w+m)}{\partial p} + \frac{\partial q(w+m)}{\partial p} + m\frac{\partial^2 q(w+m)}{\partial p^2}\right)$$

We do not study here the existence and uniqueness of the Stackelberg solution. Instead we revisit Examples 2.1 and 2.2, which determine both Stackelberg and Nash solutions for a special case of the pricing game.

Example 2.4

Let the demand be linear in price, $q(p)=a-bp$ and the supplier's cost negligible, $c=0$. Thus we obtain the problem solved in Example 2.1. Note that the demand requirements, $\frac{\partial q}{\partial p} = -b < 0$ and $\frac{\partial^2 q}{\partial p^2} \le 0$ are met for the selected function. Using Proposition 2.2. we solve (2.15),

$$q(2m^n) + m^n\frac{\partial q(2m^n)}{\partial p} = a - b2m^n + m^n(-b) = 0 \,, \ w^n = m^n$$

to find Nash equilibrium $w^n = m^n = \dfrac{a}{3b}$, hence, $p^n = w^n + m^n = \dfrac{2a}{3b}$ and

$q(p^n) = \dfrac{a}{3}$, as is also the case in Example 2.1. The payoff for the equilibrium

is identical for both players, $J_r(m^n, w^n) = J_s(m^n, w^n) = \dfrac{a^2}{9b}$. Similarly, one can

verify that the Stackelberg solution is the same as in Example 2.2,

$$w^s = \frac{a}{2b} \,, \ m^s = \frac{a}{4b} \,, \ p^s = w^s + m^s = \frac{3a}{4b} \,, \ q(p^s) = \frac{a}{4} \,,$$

$$J_s(m^s, w^s) = \frac{a^2}{8b} \ \text{and} \ J_r(m^s, w^s) = \frac{a^2}{16b} \,.$$

Finally, the centralized solution (2.7) (see also Example 2.3) is

$$q(p^*) + (p^*-c)\frac{\partial q(p^*)}{\partial p} = a-bp^*+p^*(-b)=0,$$

that is,

$$m^*+w^*=p^*=\frac{a}{2b}, \; q(p^*)=\frac{a}{2} \text{ and } J(p^*)=\frac{a^2}{4b}.$$

Comparing these results we find that the system-wide optimal order is greater than that of the Nash or Stackelberg strategy

$$q(p^s)=\frac{a}{4}<q(p^n)=\frac{a}{3}<q(p^*)=\frac{a}{2},$$

which agrees with Proposition 2.1. Correspondingly, the retail prices increase under vertical competition

$$p^s=\frac{3a}{4b}>p^n=\frac{2a}{3b}>p^*=\frac{a}{2b}.$$

and the overall chain payoff deteriorates

$$J_s(m^s,w^s)+J_r(m^s,w^s)=\frac{3a^2}{16b}<J_r(m^n,w^n)+J_s(m^n,w^n)=\frac{2a^2}{9b}<J(p^*)=\frac{a^2}{4b}.$$

Example 2.5

The goal of this example is twofold. First of all, it is rarely possible to find an equilibrium analytically. This example illustrates how to conduct the analysis numerically with Maple. Secondly, the condition imposed on the second derivative of demand is sufficient for the equilibrium to be unique, but it is not necessary, as the example demonstrates.

Let the demand be non-liner in price, $q(p)=a-bp^\alpha$. Assuming that $0<\alpha<1$, we observe that the demand requirements with respect to the first derivative are met, $\frac{\partial q}{\partial p}=-b\alpha p^{\alpha-1}<0$, while with respect to the second $\frac{\partial^2 q}{\partial p^2}=b\alpha(1-\alpha)p^{\alpha-2}>0$ is not. Using Proposition 2.2., we employ (2.13) and (2.14) to obtain numerically the retailer's and supplier's best response respectively, $m=m^R(w)$ and $w=w^R(m)$. Specifically, we first set the left-hand side of (2.13) as L1

```
>L1:=a-b*(w+m)^alpha-m*alpha*(w+m)^(alpha-1);
```

$$L1 := a - b\,(w+m)^\alpha - m\,\alpha\,(w+m)^{(\alpha-1)}$$

and the left-hand side of (2.14) as L2.

```
> L2:=a-b*(w+m)^alpha-(w-c)*alpha*(w+m)^(alpha-1);
```

$$L2 := a - b\,(w+m)^{\alpha} - (w-c)\,\alpha\,(w+m)^{(\alpha-1)}$$

Next we substitute specific parameters of the example $\alpha=0.5$, $a=15$, $b=2$, $c=1$ to have numeric left-hand sides L11 and L12 respectively

```
>L11:=subs(alpha=0.5, a=15, b=2, c=1, L1);
```

$$L11 := 15 - 2\,(w+m)^{0.5} - \frac{0.5\,m}{(w+m)^{0.5}}$$

```
> L12:=subs(alpha=0.5, a=15, b=2, c=1, L2);
```

$$L12 := 15 - 2\,(w+m)^{0.5} - \frac{0.5\,(w-1)}{(w+m)^{0.5}}.$$

Next we find the equilibrium by solving the system of equations L11=0 and L12=0

```
>solve({L11=0, L12=0}, {m,w});
```

$$\{m = 21.83319513,\ w = 22.83319513\,\}$$

To verify that the equilibrium is unique, we find the best retailer's response $m^R(w)$ numerically as mR

```
> mR:=solve(L11=0,m);
```

$$mR := 18. + 1.200000000\ \sqrt{225. + 5.\,w} - 0.8000000000\ w,$$

$$18. - 1.200000000\ \sqrt{225. + 5.\,w} - 0.8000000000\ w$$

and the inverse function $mRinv$ of the best supplier's response $w^R(m)$

```
>mRinv:=solve(L12=0,m);
```

$$mRinv := 28.37500000\ + 1.875000000\ \sqrt{229. - 4.\,w} - 1.250000000\ w,$$

$$28.37500000\ - 1.875000000\ \sqrt{229. - 4.\,w} - 1.250000000\ w$$

Both responses have two solutions, positive and negative. Since the margin is non-negative, we select only positive solutions $mR[1]$ and $mRinv[2]$ and plot them on the same graph.

```
>plot([mR[1],mRinv[1]],w=1..45,legend=["Retailer",
"Supplier"]);
```

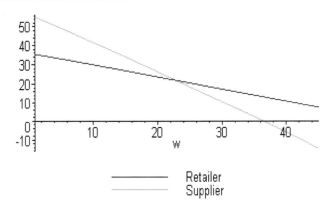

Figure 2.2. The pricing equilibrium

From Figure 2.2 we observe that there is only one point where the responses intersect. This is the Nash equilibrium point which we found numerically as $m^n = 21.833$ and $w^n = 22.833$.

The centralized solution (2.7) is found similarly with Maple

```
> L:=a-b*p^alpha-(p-c)*alpha*p^(alpha-1);
```

$$L := a - b\,p^{\alpha} - (p - c)\,\alpha\,p^{(\alpha - 1)}$$

```
> L11:=subs(alpha=0.5, a=15, b=2, c=1, L);
```

$$L11 := 15 - 2\,p^{0.5} - \frac{0.5\,(p - 1)}{p^{0.5}}$$

```
> popt:=solve(L11=0,p);
```

$$popt := 36.39890107$$

Comparing the system-wide optimal price with the equilibrium Nash price, we find that $p^* = 36.398 < p^n = m^n + w^n = 21.833 + 22.833 = 44.666$.

Coordination

According to Proposition 2.1, vertical competition has a negative effect on the supply chain. The retailer orders less, the retail price goes up and profits shrink. Moreover, although the supplier's leadership allows the supplier to increase his profit, in the specific case of linear price demand (see Example 2.4), the leadership is also destructive as it further reduces the total profit in the supply chain. The negative effect of the vertical competition is due to the well-known double marginalization effect. This effect takes place if the retailer ignores the supplier's profit margin, $w-c$, when ordering as shown in Proposition 2.1. Specifically, when recalling that $p=w+m$, the retailer's best response (2.9)

$$q(p)+(p-w)\frac{\partial q(p)}{\partial p}=0 ,$$

can be written as

$$q(p)+m\frac{\partial q(p)}{\partial p}=0 ,$$

which implies that though the demand depends on price $p=w+m$, the retailer accounts only for his margin m instead of ordering as indicated by the centralized approach (2.7)

$$q(p)+(p-c)\frac{\partial q(p)}{\partial p}=q(p)+(w-c+m)\frac{\partial q(p)}{\partial p}=0$$

and thus adding the supplier's margin, w-c, to m. Equivalently, from equation (2.14)

$$q(p)+(w-c)\frac{\partial q(p)}{\partial p}=0$$

we observe that the supplier ignores the retailer's margin m when setting the wholesale price. The remaining question is how to induce the retailer to order more, or the supplier to reduce the wholesale price, i.e., how to coordinate the supply chain and thus increase its total profit. Of course, the supplier may set the wholesale price at his marginal cost, $w=c$, or the retailer may set his margin at zero. Equation (2.7) then becomes identical to (2.9) and the supply chain is perfectly coordinated. However, the supply chain member who gives up his margin gets no profit at all. The most popular way of dealing with such a problem is by discounting or by collaboration for profit sharing.

One approach to discounting is a simple two-part tariff. If the supplier is the leader, he can set $w=c$, but charge the retailer a fixed fee. In this way, the supplier can regulate his share in the total supply chain profit without a special contract. Moreover, if the supplier sets the fixed fee very close to the centralized supply chain profit, $J(p^*)$, then the retailer gets almost no profit and still orders the system-wide optimal quantity $q(p^*)$ as well as sets system-wide optimal price p^*.

Regardless of whether there is a leader or not, signing a profit-sharing contract is an alternative way to mitigate the double marginalization. In such a contact, the parties would explicitly set their shares of the total supply chain profit, $J(p^*)$ with η, $0\le \eta \le 1$, so that the retailer gets $\eta J(p^*)$ and the supplier $(1-\eta)J(p^*)$. This, however, is already cooperative rather than competitive behavior. To illustrate one possibility for coordination with cooperation, we briefly consider an example of bargaining over the wholesale price and retailer's margin in terms of the Nash bargain, which solves

$$\max_{m,w} [J_r(w,m)\text{-}j_r][J_s(w,m)\text{-}j_s],$$

where j_r and j_s represent the outside options to each party. Employing the demand function of this section and assuming that all outside options are normalized to zero, i.e., j_r =0 and j_s =0, we have the following bargaining problem:

$$\max_{m,w} J^B(m,w) = \max_{m,w} mw[q(w+m)]^2.$$

If $q(w+m)$ is such that $J^B(m,w)$ is concave, then applying the first-order optimality conditions we obtain the following two equations

$$q(m+w) + 2m\frac{\partial q(m+w)}{\partial p} = 0,$$

$$q(m+w) + 2(w-c)\frac{\partial q(m+w)}{\partial p} = 0.$$

From these equations we immediately find that $m=w\text{-}c$ and thereby the two equations result in a single condition:

$$q(m+w) + (m+w-c)\frac{\partial q(m+w)}{\partial p} = 0.$$

Taking into account that $p=m+w$, we observe that the derived condition is identical to the system-wide optimality condition (2.7). Thus, if $J^B(m,w)$ is concave, the Nash bargain perfectly coordinates the supply chain for the case of the pricing game. The only difference is that the system-wide optimal solution specifies only the optimal price p^* (since the transfer costs are not important for a centralized system), while the Nash bargain solution of the pricing problem results in equal margins, $m=w\text{-}c$, and shares, $J_r(w,m)=J_s(w,m)$, for both parties.

The multi-echelon effect

It is intuitively clear that the greater the number of the upstream suppliers involved, the more margins are added to the supply chain and thereby the greater the deterioration of the expected system performance. Specifically, let an upstream distributor have a marginal cost c_d per product and let him sell his products to the supplier at a price w_d. Then the retail price would be $p= w+m$, $w \geq c+w_d$ and the resulting problems of the three-echelon supply chain are defined as follows.

The distributor's problem

$$\max_{w_d} J_d(w_d,w,m) = \max_{w_d} (w_d\text{-}c_d)q(w+m)$$

s.t.

$$w_d \geq c_d.$$

The supplier's problem

$$\max_{w} J_s(w_d, w, m) = \max_{w} (w - c - w_d) q(w + m)$$

s.t.

$$w \geq c + w_d.$$

The retailer's problem

$$\max_{m} J_r(w_d, w, m) = \max_{m} mq(w + m)$$

s.t.

$$m \geq 0, \; q(w + m) \geq 0.$$

The centralized problem

$$\max_{m,w} J(m, w) = \max_{m,w} (\, m + w - c - c_d)q(w + m)$$

s.t.

$$m \geq 0, \; q(w + m) \geq 0, \; w \geq c + w_d.$$

Consequently the system-wide optimal retail margin is determined by

$$\frac{\partial J(m, w)}{\partial m} = q(p) + (m + w - c - c_d)\frac{\partial q(p)}{\partial p} = 0,$$

while the equation for an optimal margin when the parties are non-cooperative remains the same

$$\frac{\partial J_r(m, w)}{\partial m} = q(p) + m\frac{\partial q(p)}{\partial p} = 0 .$$

We thus observe that the retailer when ordering, accounts for his margin m and ignores both the supplier's margin $w - c - w_d$ and the distributor's margin $w_d - c_d$, which is, $w - c - c_d$ in total. Again, by employing the two-part tariff, the supply chain becomes perfectly coordinated. This is accomplished if the distributor and the supplier set the wholesale prices equal to their marginal costs, i.e., $w_d = c_d$ and $w = c + c_d$, respectively and charge a fixed cost per transaction.

2.2.2 THE PRODUCTION GAME

Previously we were concerned with vertical competition. Now we shall study the effect of horizontal production competition (see Figure 2.3). Consider two manufacturers producing the same or substitutable types of product over a period of time and thus competing horizontally for the same customers, possibly for the same retailer. Accordingly, the manufacturers are suppliers with ample capacity and the order period is longer than the suppliers' lead-time. This means that both suppliers are able to deliver on time any quantity q_1 and q_2 to the retailer. The retailer, on the other hand, adopts the so-called vendor managed inventory (VMI) policy, in which the suppliers decide on the quantities to deliver while the retailer simply charges a fixed percentage from sales. Since the retailer has no part in the competition, he does not affect the system-wide optimal solution, equilibrium order quantities, or prices.

Further, in the previous section we assumed that the retailer demand is a function of product price which is referred to as Bertrand's model of competition pricing. In this section we assume that the retail price is a function of customer demand which is referred to as Cournot's model of production competition. Specifically, the product is characterized by an endogenous price function of total demand $Q=q_1+q_2$, $p=p(Q)$, which, since the products are fully substitutable, is symmetric in q_1 and q_2. We assume that this symmetric function is down-sloping (concave) in the total quantity of the products, i.e., $\dfrac{\partial p}{\partial q_1}=\dfrac{\partial p}{\partial q_2}<0$ and concave, $\dfrac{\partial^2 p}{\partial Q^2}\leq 0$, i.e.,

$\dfrac{\partial^2 p}{\partial q_1{}^2}=\dfrac{\partial^2 p}{\partial q_2{}^2}=\dfrac{\partial^2 p}{\partial q_1\partial q_2}\leq 0$. The suppliers incur identical unit production cost c, $c<p(0)$, and seek to maximize profits, i.e., they maximize their margins, $p(Q)-c$, times the demand, q_1 or q_2.

The problem of supplier 1

$$\max_{q_1} J_1(q_1,q_2)=\max_{q_1}\ q_1[p(q_1+q_2)-c] \qquad (2.17)$$

s.t.

$$q_1\geq 0,\, p(q_1+q_2)\geq c.$$

The problem of Supplier 2

$$\max_{q_2} J_2(q_1,q_2)=\max_{q_2}\ q_2[p(q_1+q_2)-c] \qquad (2.18)$$

s.t.

$$q_2\geq 0,\, p(q_1+q_2)\geq c,$$

where $p(Q)$ is the price at which the retailer can sell Q product units; q_1 and q_2 are the quantities produced by suppliers (manufacturers) 1 and 2 respectively and sold by the retailer; $Q=q_1+q_2$ is the total quantity sold by the retailer; and c is the unit production cost for both suppliers.

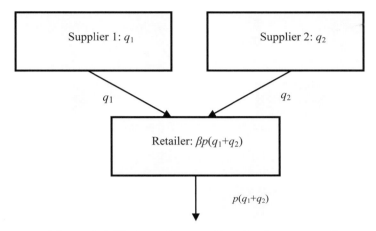

Figure 2.3. Horizontal competition for the same retailer

Exactly, (2.17) and (2.18) can be presented as

$$\max_{q_1} J_1(q_1,q_2)=\max_{q_1} \beta q_1[p(q_1+q_2)-c];$$

$$\max_{q_2} J_2(q_1,q_2)=\max_{q_2} \beta q_2[p(q_1+q_2)-c],$$

where β is percentage paid to the retailer by each manufacturer. Since coefficient β does not affect the optimality conditions, it is omitted. Moreover, since the retailer's profit is

$$J_r(q_1,q_2)= (1-\beta)q_1[p(q_1+q_2)-c] + (1-\beta)q_2[p(q_1+q_2)-c],$$

the centralized objective function does not involve β at all since it represents internal supply chain transfers. Thus, if the supply chain is horizontally integrated, that is, if a single decision maker is in charge, then we have the following single problem as a benchmark.

The centralized problem

$$\max_{q_1,q_2} J(q_1,q_2) =\max_{q_1,q_2} [J_1(q_1,q_2)+J_2(q_1,q_2)]=$$

$$\max_{q_1,q_2} q_1[p(q_1+q_2)-c]+ q_2[p(q_1+q_2)-c] \qquad (2.19)$$

s.t.

$$q_1 \geq 0, q_2 \geq 0, p(q_1+q_2) \geq c.$$

System-wide optimal solution

We first study the centralized problem by employing the first-order optimality conditions

$$\frac{\partial J(q_1,q_2)}{\partial q_1} = p(q_1 + q_2) - c + q_1 \frac{\partial p(Q)}{\partial Q} \frac{\partial Q}{\partial q_1} + q_2 \frac{\partial p(Q)}{\partial Q} \frac{\partial Q}{\partial q_1} = 0,$$

$$\frac{\partial J(q_1,q_2)}{\partial q_2} = p(q_1 + q_2) - c + q_2 \frac{\partial p(Q)}{\partial Q} \frac{\partial Q}{\partial q_2} + q_1 \frac{\partial p(Q)}{\partial Q} \frac{\partial Q}{\partial q_2} = 0.$$

Since the two problems are symmetric, $Q=q_1+q_2$, $\frac{\partial p}{\partial q_1} = \frac{\partial p}{\partial q_2} = \frac{\partial p}{\partial Q}$, only total order Q matters in terms of optimality. Considering the symmetric solution to the above system of equations as well, $q^* = q_1^* = q_2^*$, we obtain the following equation

$$p(2q^*) - c + 2q * \frac{\partial p(2q^*)}{\partial Q} = 0. \tag{2.20}$$

Define Q' so that $p(Q')=c$. Then it is easy to verify that,

$$\frac{\partial^2 J}{\partial q_1^{\,2}} = \frac{\partial^2 J}{\partial q_2^{\,2}} = \frac{\partial^2 J}{\partial q_1 \partial q_2} = 2 \frac{\partial p}{\partial Q} + q_1 \frac{\partial^2 p}{\partial Q^2} + q_2 \frac{\partial p^2}{\partial Q^2} < 0.$$

This implies that the Hessian of $J(q_1,q_2)$ is semi-definite negative and thus the function $J(q_1,q_2)$ is jointly concave in production quantities q_1 and q_2 for $q_1 + q_2 \in [0,Q']$. Though this does not ensure the uniqueness of the optimal solution, by differentiating the left-hand side of equation (2.20) in $q=q^*$ we obtain for the symmetric solution

$$\frac{\partial^2 J}{\partial q^2} = 4 \frac{\partial p}{\partial Q} + 4q \frac{\partial^2 p}{\partial Q^2} < 0,$$

that is, the left-hand side of (2.20) is strictly monotone in q. Thus, equation (2.20) has a unique solution as formalized in the following proposition.

Proposition 2.3. *The pair* (q_1^*,q_2^*)*, where* $q_1^* = q_2^* = q^*$ *satisfy equation (2.20) constitutes a unique symmetric system-wide optimal order with* $0<q^*<Q'/2$.

Proof: Since the left-hand side of equation (2.20) is strictly decreasing in q, if there is a feasible solution to (2.20), it is unique. To see that a solution of (2.20) always exists, assume $q=0$, then, since $p(0)>c$, the left-hand side of (2.20) is positive. On the other hand, if $2q=Q'$, since $p(Q')=c$, while the last term of (2.20) is strictly negative as $q=Q/2>0$, we find that the left-hand side of (2.20) is negative. Thus a feasible solution always exists and $0<q<Q'/2$.

Game analysis

Consider now a decentralized supply chain characterized by non-cooperative firms and assume that both players simultaneously decide how many products to produce and supply to the retailer. Using the first-order optimality conditions for the suppliers' problems we find

$$\frac{\partial J(q_1,q_2)}{\partial q_1} = p(q_1 + q_2) - c + q_1 \frac{\partial p(q_1 + q_2)}{\partial q_1} = 0,$$

$$\frac{\partial J(q_1,q_2)}{\partial q_2} = p(q_1 + q_2) - c + q_2 \frac{\partial p(q_1 + q_2)}{\partial q_2} = 0.$$

Again, since the two problems are symmetric, the competition is symmetric. That is, the solution to this system of equations is $q = q_1 = q_2$, which satisfies the following equation

$$p(2q) - c + q \frac{\partial p(2q)}{\partial Q} = 0. \tag{2.21}$$

Comparing (2.21) and (2.20), we conclude with the result highlighting the differences between the centralized and (Nash) game solution.

Proposition 2.4. *In horizontal competition of the production game with equal power players, the retail price will be lower and the quantities produced by the manufacturers higher than the system-wide optimal price and production quantity respectively.*

Proof: Comparing (2.21) and (2.20) we observe that if $q = q^*$, then

$$p(2q) - c + q \frac{\partial p(2q)}{\partial Q} > p(2q^*) - c + 2q * \frac{\partial p(2q^*)}{\partial Q} = 0,$$

while the derivative of the left-hand side of this inequality with respect to q is negative. Thus, $q > q^*$, which, in regard to the down-sloping price function $p(2q)$, means that $p(2q) < p(2q^*)$.

Nash solution

Since it is easy to verify that the suppliers' objective functions are strictly concave in their production quantities, each supplier has a unique, best-response function. In addition, since the derivative of the left-hand side of (2.21) is strictly negative, (2.21) has a unique solution.

Proposition 2.5. *The pair (q_1^n, q_2^n), which satisfies $q_1^n = q_2^n = q^n$ and*

$$p(2q^n) - c + q^n \frac{\partial p(2q^n)}{\partial Q} = 0 \tag{2.22}$$

constitutes a unique Nash equilibrium of the production game with $0 < q^n < Q'/2$.

Proof: The proof is identical to that for proposition (2.3).

The uniqueness of the Nash solution implies that both parties will tend to attain the equilibrium when pursuing their own profits.

The effect of partial product substitutability

Let the product that the second supplier produces partially substitute for the brand of the first supplier. This is expressed by the ratio $0 \le \lambda \le 1$, so that $p = p(Q) = p(q_1 + \lambda q_2)$. Then, the Nash optimality conditions take the following form

$$\frac{\partial J(q_1, q_2)}{\partial q_1} = p(Q) - c + q_1 \frac{\partial p(Q)}{\partial Q} = 0,$$

$$\frac{\partial J(q_1, q_2)}{\partial q_2} = p(Q) - c + q_2 \frac{\partial p(Q)}{\partial Q} \lambda = 0.$$

Though these conditions are no longer symmetric, subtracting one equation from the other we find

$$q_1{}^n = \lambda q_2{}^n.$$

Thus, $Q = q_1{}^n + \lambda q_2{}^n = 2 \lambda q_2{}^n$ and $q_2{}^n$ is determined by

$$p(2\lambda q_2{}^n) - c + q_2{}^n \frac{\partial p(2\lambda q_2{}^n)}{\partial Q} \lambda = 0.$$

In other words, the equilibrium exists, but the production quantities are now proportional rather than identical.

Stackelberg solution

Next we assume that one of the suppliers is the leader, say supplier-one. To find the Stackelberg equilibrium, we need to maximize supplier-one's objective with q_1, subject to the best supplier-two's response $q_2 = q_2{}^R(q_1)$. Let $q_2 = q_2{}^R(q_1)$ satisfy the following equation

$$p(q_1 + q_2) - c + q_2 \frac{\partial p(q_1 + q_2)}{\partial q_2} = 0. \tag{2.23}$$

The Stackelberg equilibrium is determined by maximizing the following function

$$\max_{q_1} J_1(q_1) = \max_{q_1} q_1[p(q_1 + q_2{}^R(q_1)) - c].$$

Differentiating this function we find

$$\frac{\partial J_1(q_1)}{\partial q_1} = p(q_1 + q_2{}^R(q_1)) - c + q_1 \frac{\partial p(q_1 + q_2{}^R(q_1))}{\partial Q} (1 + \frac{\partial q_2{}^R(q_1)}{\partial q_1}) = 0, \tag{2.24}$$

where $\dfrac{\partial q_2{}^R(q_1)}{\partial q_1}$ is determined by differentiating (2.23) with q_2 set equal to $q_2{}^R(q_1)$

$$\frac{\partial p(Q)}{\partial Q}\left(1+\frac{\partial q_2{}^R(q_1)}{\partial q_1}\right)+\frac{\partial q_2{}^R(q_1)}{\partial q_1}\frac{\partial p(Q)}{\partial Q}+q_2{}^R(q_1)\frac{\partial p^2(Q)}{\partial Q^2}\left(1+\frac{\partial q_2{}^R(q_1)}{\partial q_1}\right)=0.$$

Thus

$$\frac{\partial q_2{}^R(q_1)}{\partial q_1}=-\left(\frac{\partial p(Q)}{\partial Q}+q_2{}^R(q_1)\frac{\partial p^2(Q)}{\partial Q^2}\right)\Bigg/\left(2\frac{\partial p(Q)}{\partial Q}+q_2{}^R(q_1)\frac{\partial p^2(Q)}{\partial Q^2}\right). \quad (2.25)$$

Equation (2.25) implies, $\dfrac{\partial q_2{}^R(q_1)}{\partial q_1}<0$,

the greater the production of the first supplier, q_1, the lower the production of the second supplier, $q_2{}^R(q_1)$.

Based on (2.23), (2.24) and (2.25) we conclude that the pair $(q_1{}^s,q_2{}^s)$ constitutes the Stackelberg equilibrium of the production game if there exists a joint solution in q_1 and q_2 of the following equations:

$$p(q_1+q_2)-c+q_1\frac{\partial p(q_1+q_2)}{\partial Q}\left(1+\frac{\partial q_2}{\partial q_1}\right)=0,$$

$$p(q_1+q_2)-c+q_2\frac{\partial p(q_1+q_2)}{\partial Q}=0,$$

where

$$\frac{\partial q_2}{\partial q_1}=-\left(\frac{\partial p(Q)}{\partial Q}+q_2\frac{\partial p^2(Q)}{\partial Q^2}\right)\Bigg/\left(2\frac{\partial p(Q)}{\partial Q}+q_2\frac{\partial p^2(Q)}{\partial Q^2}\right),\ Q=q_1+q_2.$$

We illustrate this with the following example:

Example 2.6

Let the price be linear in production quantity, $p=a-bQ$, $Q=q_1+q_2$, $p(0)=a>c$.

Note that the price requirements, $\dfrac{\partial p}{\partial q_1}=\dfrac{\partial p}{\partial q_2}=-b<0$ and $\dfrac{\partial^2 p}{\partial q_1{}^2}=\dfrac{\partial^2 p}{\partial q_2{}^2}=$

$\dfrac{\partial^2 p}{\partial q_1\partial q_2}=0$ are met for the selected function. Using Proposition 2.5 we

solve (2.22),

$$p(2q^n)-c+q^n\frac{\partial p(2q^n)}{\partial Q}=a-2bq^n-c+q^n(-b)=0$$

and find that $q_1^n = q_2^n = \dfrac{a-c}{3b}$, hence, $p^n = \dfrac{1}{3}a + \dfrac{2}{3}c$. The payoffs for the equilibrium are thus identical for both players, $J_1(q_1^n, q_2^n) = J_2(q_1^n, q_2^n) = \dfrac{(a-c)^2}{9b}$.

Based on (2.23) we can identify the best response function of the second supplier

$$p(q_1 + q_2) - c + q_2 \frac{\partial p(q_1 + q_2)}{\partial q_2} = a - b(q_1 + q_2) - c + q_2(-b) = 0,$$

and thus

$$q_2 = q_2^r(q_1) = \frac{a - bq_1 - c}{2b}.$$

This response is then employed in (2.24) and (2.25) to find the Stackelberg equilibrium. Equivalently, by substituting this response into the first supplier objective function

$$\max_{q_1} \; q_1[p(q_1 + q_2^R(q_1)) - c] = \max_{q_1} \; q_1\left[\frac{a}{2} - \frac{bq_1}{2} - \frac{c}{2}\right].$$

and using the first-order optimality conditions, we obtain an explicit resolution of equation (2.24) for our example,

$$\frac{\partial J_1}{\partial q_1} = \left[\frac{a}{2} - \frac{bq_1}{2} - \frac{c}{2}\right] + q_1\left[-\frac{b}{2}\right] = 0.$$

Accordingly, $q_1^s = \dfrac{a-c}{2b}$, $q_2^s = \dfrac{a-c}{4b}$, $p^s = \dfrac{a+3c}{4}$, $J_1(q_1^s, q_2^s) = \dfrac{(a-c)^2}{8b}$ and $J_2(q_1^s, q_2^s) = \dfrac{(a-c)^2}{16b}$. Note that instead of equal payoff under a simultaneous Nash strategy, the first supplier, who is the leader, gains a profit which is twice as much as the follower's profit under a sequential Stackelberg strategy.

Finally, the centralized solution (2.20) is

$$p(2q^*) - c + 2q^* \frac{\partial p(2q^*)}{\partial Q} = a - 2bq^* - c + 2q^*(-b) = 0.$$

Or, $q_1^* = q_2^* = \dfrac{a-c}{4b}$, hence, $p^* = \dfrac{1}{2}a + \dfrac{1}{2}c$ and the system-wide optimal supply chain profit is $J(q_1^*, q_2^*) = \dfrac{(a-c)^2}{4b}$.

Comparing these results, we find for the first supplier, that his production quantity under the centralized approach is smaller than both that of the Nash strategy and that obtained when the supplier is the Stackelberg leader

$$q_1^s = \frac{a-c}{2b} > q_1^n = \frac{a-c}{3b} > q_1^* = \frac{a-c}{4b}.$$

For the second supplier, the production level is the same under the Stackelberg follower strategy and the system-wide policy, but higher for the Nash strategy.

$$q_2^n = \frac{a-c}{3b} > q_2^s = q_2^* = \frac{a-c}{4b}.$$

Both results agree with Proposition 2.4 which compares Nash and system-wide strategies. Correspondingly, given $p(0)=a>c$, the retail prices decrease

$$p^s = \frac{a+3c}{4} < p^n = \frac{1}{3}a + \frac{2}{3}c < p^* = \frac{1}{2}a + \frac{1}{2}c$$

and the overall supply chain payoff deteriorates under horizontal competition,

$$J_1(q_1^s,q_2^s) + J_2(q_1^s,q_2^s) = \frac{3(a-c)^2}{16b} < J_1(q_1^n,q_2^n) + J_2(q_1^n,q_2^n) =$$

$$\frac{2(a-c)^2}{9b} < J(q_1^*,q_2^*) = \frac{(a-c)^2}{4b}.$$

Example 2.7

This example illustrates how the equilibrium can be analyzed numerically. Let the price be exponential in the production quantity, $p=ae^{-bQ}$, $Q=q_1+q_2$, $p(0)=a>c$. Note that, $\dfrac{\partial p}{\partial q_1} = \dfrac{\partial p}{\partial q_2} = -abe^{-bQ} < 0$, while for the second order

condition $\dfrac{\partial^2 p}{\partial q_1^2} = \dfrac{\partial^2 p}{\partial q_2^2} = \dfrac{\partial^2 p}{\partial q_1 \partial q_2} = ab^2 e^{-bQ} > 0$ implying that the equilibrium

is not necessarily unique. The Nash equilibrium is determined by (2.22)

$$ae^{-b2q^n} - c - q^n abe^{-b2q^n} = 0.$$

Setting the left-hand side of this equation as L in Maple
```
>L:=a*exp(-b*2*q)-c-q*a*b*exp(-b*2*q);
```

$$L := a\,e^{(-2\,b\,q)} - c - q\,a\,b\,e^{(-2\,b\,q)}$$

and substituting specific parameters of the problem a=15,b=0.1,c=1, we have

```
> L1:=subs(a=15, b=0.1, c=1, L);
```

$$L1 := 15\,\mathbf{e}^{(-0.2\,q)} - 1 - 1.5\,q\,\mathbf{e}^{(-0.2\,q)}$$

The solution to this transcendental equation is found with Maple's SOLVE

```
> solve(L1=0, q);
```

$$7.191168444$$

To verify that the Nash equilibrium is unique, we construct a plot of the left-hand side $Y=L1$

```
> plot(L1, q=0..10);
```

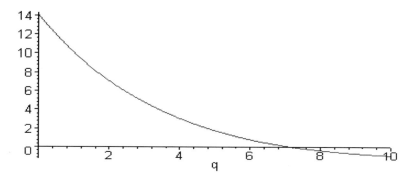

Figure 2.4. The Nash equilibrium

From this plot (see Figure 2.4) we observe that for feasible orders $q_1{}^n = q_2{}^n \geq 0$, there is only one intersection of $Y=L1$ with line, $Y=0$, which is the Nash equilibrium, $q_1{}^n = q_2{}^n = 7.191168444$.

Similarly, employing equation (2.20) to find the system-wide optimal solution with Maple:

```
> LL:=a*exp(-b*2*q)-c-2*q*a*b*exp(-b*2*q);
```

$$LL := a\,\mathbf{e}^{(-2\,b\,q)} - c - 2\,q\,a\,b\,\mathbf{e}^{(-2\,b\,q)}$$

```
> LL1:=subs(a=15, b=0.1, c=1, LL);
```

$$LL1 := 15\,\mathbf{e}^{(-0.2\,q)} - 1 - 3.0\,q\,\mathbf{e}^{(-0.2\,q)}$$

```
> solve(LL1=0, q);
```

$$4.224140740$$

Comparing the system wide optimal production quantity with the Nash quantity we find $q^* = 4.224 < q_1{}^n = q_2{}^n = 7.191$.

Coordination

According to Proposition 2.4, although retailers and consumers may benefit from non-cooperating suppliers leading to a fall in retail prices and an increase in production as well as consumption of products, the horizontal competition has a negative effect on the supply chain's profits. Thus, just as with the double marginalization effect, the deterioration in the supply chain performance arises because each manufacturer, when deciding on the quantity to produce, ignores the quantity which the other manufacturer is producing. This can be termed a "double quantification". Indeed, in vertical competition the supplier sells the retailer products which are then resold to the customers. Two margins are being imposed then on the same product quantity. On the other hand, in horizontal competition, each supplier produces a number of products, but sells them at the same price. The price is due to the two quantities being produced. Ignoring one of the quantities, such as ignoring one of the margins, yields results that are different from the system-wide optimal solution.

The essential means to coordinate horizontal competition is thus to cooperate. By simply agreeing to simultaneously set the production quantities equal to the system-wide optimal quantity, rather than to the non-cooperative equilibrium quantities, the suppliers, will be perfectly coordinating the supply chain and increasing their profits equally without any internal supply chain transfers.

The multi-echelon effect

Recalling the effect of vertical competition on the supply chain discussed in the previous section, it is apparent that the more upstream suppliers that are involved, the more margins are added to the supply chain. This results in a decrease in the quantity produced and an increase in prices. This is to say, double marginalization may coordinate the supply chain if its effect is not stronger than that of the horizontal competition. Specifically, let an upstream distributor who has a marginal cost c_d per product play a supply part or sell products to both suppliers at price w_d. (Of course, if the suppliers are not symmetric, then the wholesale price that they can get from the distributor may be different). The corresponding problems of the three-echelon supply chain with two horizontally competing suppliers are as follows (as aforementioned in this section, we consider the case when the retailer does not compete and therefore his problem is not accounted for):

The problem of supplier 1

$$\max_{q_1} J_1(q_1,q_2)=\max_{q_1} \ q_1[p(q_1+q_2)-c-w_d]$$

s.t.

$$q_1 \geq 0, p(q_1+q_2) \geq c+w_d.$$

The problem of supplier 2

$$\max_{q_2} J_2(q_1,q_2) = \max_{q_2} \; q_2[p(q_1+q_2)-c-w_d]$$

s.t.

$$q_2 \geq 0, \, p(q_1+q_2) \geq c+w_d$$

The distributor's problem

$$\max_{w_d} J_d(w_d,w,m) = \max_{w_d} \; (w_d-c_d)(q_1+q_1)$$

s.t.

$$w_d \geq c_d.$$

The centralized problem

$$\max_{q_1,q_2} J(q_1,q_2) = \max_{q_1,q_2} \; q_1[p(q_1+q_2)-c-c_d] + q_2[p(q_1+q_2)-c-c_d]$$

s.t.

$$q_1 \geq 0, \, q_2 \geq 0, \, p(q_1+q_2) \geq c+c_d.$$

Assuming that the suppliers are at a Nash equilibrium, the equation for an optimal order quantity $q=q_1=q_2$ for the symmetric suppliers is similar to (2.21). The only difference could be that w_d is subtracted

$$p(2q) - c - w_d + q \frac{\partial p(2q)}{\partial Q} = 0.$$

A system-wide optimal solution, on the other hand, is similar to (2.20) but corrected by c_d,

$$p(2q^*) - c - c_d + 2q * \frac{\partial p(2q^*)}{\partial Q} = 0 .$$

Comparing these two equations, we find that both suppliers account for their margins, $p(2q)-c-w_d$, and ignore the distributor's margin w_d-c_d, which, if added, as in the centralized solution, results in a total of $p(2q)-c-c_d$. Since $w_d > c_d$ and the derivatives of the left hand sides of these equations are negative, the Nash production quantity q decreases compared to the system-wide optimal solution. On the other hand, when the quantity which the other party produces is ignored (as discussed in this section), the (Nash) production quantity q decreases compared to the system-wide optimal solution. Thus, if for $q=q^*$ the following holds

$$p(2q) - c - w_d + q \frac{\partial p(2q)}{\partial Q} > p(2q^*) - c - c_d + 2q * \frac{\partial p(2q^*)}{\partial Q},$$

or, equivalently,

$$-q*\frac{\partial p(2q*)}{\partial Q} > w_d - c_d,$$

Then, the effect of horizontal competition between the two suppliers is stronger than that of the vertical competition between the suppliers and additional upstream parties coordinate the supply chain. More precisely, the quantity produced and sold by the three-echelon supply chain will be lower than that of the corresponding two-echelon chain which does not involve an additional upstream distributor.

Finally, it is worth noting that horizontal competition in multi-echelon supply chains opens up a whole spectrum of collaboration activities. For example, horizontally competing producers may coordinate the quantities they order from an upstream supplier to bargain lower wholesale prices. Interested readers are referred to Davidson (1988), Horn and Wolinsky (1988) and Viehoff (1987) who have addressed the benefits of various bargaining schemes.

2.3 STOCKING COMPETITION WITH RANDOM DEMAND

In contrast to the previous section, we now assume that the retailer demand is random and proceed to adapt two classic newsvendor models into two stocking/pricing games. In one game the supplier sets the wholesale price to sell some of his stock while the retailer decides on the quantity to purchase in order to replenish his stock. The retailer incurs no fixed order cost. We refer to this game as the stocking game.

The other game is related to a manufacturer who pays a setup cost for each production order. To avoid this irreversible cost, the manufacturer has the alternative of outsourcing current in-house production to a supplier. Similar to the stocking game, the supplier decides on the wholesale price and does not charge a fixed order cost. Unlike the stocking game, the manufacturer determines first whether to outsource the production at this wholesale price or to produce in-house and then determining the proper quantity to order. We refer to this game as the outsourcing game.

2.3.1 THE STOCKING GAME

The classical, single-period, newsboy or newsvendor problem formulation assumes random exogenous demand, d, in contrast to previously discussed pricing and production problems with deterministic but endogenous demands. The selling season is short and there is no time for additional orders so if the retailer orders less than the demand at the end of period, then shortage h^- cost per unit of unsatisfied demand is incurred. The shortage cost normally includes lost sales and a loss of customer goodwill. On the other hand, if the retailer orders more than he is able to sell, unit inventory cost h^+ (mitigated by salvage cost) is incurred for units left over at the end of period. The fixed-order cost is assumed to be negligible. The retailer's goal is to find order quantity, q, to maximize expected overall profits. The described newsvendor problem assumes that the product purchasing cost is fixed and given. However, if we take into account a supplier who independently maximizes his profit and thus impacts the retailer's optimal solution by choosing a wholesale price, w, the newsvendor problem is reduced to a game.

Let retailer's margin, m, be fixed, $f(D)$ and $F(a) = \int_0^a f(D)dD$ be the demand probability density and cumulative distribution functions respectively. Then, the retailer's problem is formulated as follows.

The retailer's problem

$$\max_q J_r(q,w) = \max_q \{E[ym - h^+x^+ - h^-x^-] - wq\}, \tag{2.26}$$

s.t.

$$x = q - d, \tag{2.27}$$

$$q \geq 0, \tag{2.28}$$

where $x^+ = \max\{0, x\}$ and $x^- = \max\{0, -x\}$ are inventory surplus and shortage at the end of selling season respectively, and $y = \min\{q,d\}$ is the number of products sold.

Applying conditional expectation to (2.26), the objective function transforms into the following form

$$\max_q J_r(q,w) = \max_q \{$$

$$\int_0^q mDf(D)dD + \int_q^\infty mqf(D)dD - \int_0^q h^+(q-D)f(D)dD - \int_q^\infty h^-(D-q)f(D)dDwq\}. \tag{2.29}$$

The first term in the objective function $E[ym] = \int_0^q mDf(D)dD + \int_q^\infty mqf(D)dD$ represents income from selling y product units; the second and the third terms, $E[h^+x^+] = \int_0^q h^+(q-D)f(D)dD$, $E[h^-x^-] = \int_q^\infty h^-(D-q)f(D)dD$ represent losses due the inventory surplus and shortage respectively; and the last term, wq, is the amount paid to the supplier.

Note, that the retailer orders products from the supplier if he expects non-negative profit. In other words, there is a maximum wholesale price, w^M, that the supplier can charge. Taking this into account, as well as the unit production cost, c, of the supplier, we formulate the supplier's problem.

The supplier's problem

$$\max_{w} J_s(q,w) = (w-c)q \qquad (2.30)$$

s.t.

$$c \le w \le w^M. \qquad (2.31)$$

The corresponding centralized problem is based on the sum of two objective functions (2.30) and (2.26), which results in a function independent of the wholesale price, w, representing a transfer within the supply chain.

The centralized problem

$$\max_{q} J(q) = \max_{q} \{E[ym - h^+x^+ - h^-x^-] - cq\} \qquad (2.32)$$

s.t.

$$x = q - d, \ q \ge 0.$$

System-wide optimal solution

We first study the centralized problem. Similar to (2.29), by determining the expectation of (2.32), we obtain

$$\max_{q} J(q) = \max_{q} \{$$

$$\int_0^q mDf(D)dD + \int_q^\infty mqf(D)dD - \int_0^q h^+(q-D)f(D)dD - \int_q^\infty h^-(D-q)f(D)dD - cq\}.$$

By employing the first-order optimality condition to this function, we have

$$\frac{\partial J(q)}{\partial q} = mqf(q) - mqf(q) + \int_q^\infty mf(D)dD - \int_0^q h^+f(D)dD + \int_q^\infty h^-f(D)dD - c = 0,$$

which, after simple manipulations, results in

$$m(1 - F(q)) - h^+F(q) + h^-(1 - F(q)) - c = 0 \ .$$

Thus we find that the traditional newsvendor expression for the optimal order quantity q^*, which is feasible if $m + h^- > c$,

$$F(q^*)= \frac{m + h^- - c}{m + h^- + h^+}. \tag{2.33}$$

We can also verify the sufficient condition, i.e., that the objective function (2.30) is concave,

$$\frac{\partial^2 J(q)}{\partial q^2} = -(m + h^+ + h^-)f(q) \le 0. \tag{2.34}$$

Let $f(D)>0$ for $d^{min} \le D \le d^{max}$. Then, since ordering less than the minimum demand, d^{min}, as well as more than the maximum demand, d^{max}, does not make any sense, the centralized objective function is strictly concave and thus we find a unique solution.

The effect of initial inventory

Note that if the retailer has an initial inventory, x^0, that is, $x = x^0 + q - d$, then by using the same arguments we observe that the only change in (2.33) is in the argument of $F(.)$:

$$F(x^0 + q)= \frac{m + h^- - c}{m + h^- + h^+}. \tag{2.35}$$

Let s satisfy the equation,

$$F(s)= \frac{m + h^- - c}{m + h^- + h^+}, \tag{2.36}$$

then s is the base stock, and the optimal order quantity is interpreted as the well-known order-up-to policy,

$$q^* = \begin{cases} s - x^0, \text{if } s > x^0 \\ 0, \text{otherwise.} \end{cases}$$

Service level

For the risk of shortage, we have the probability $P[x<0]=1- \alpha$, where α is referred to as the service level. From (2.32) it follows that the service level in the centralized supply chain is $P[x \ge 0]= F(q^*)$, or, equivalently,

$$\alpha = \frac{m + h^- - c}{m + h^- + h^+}. \tag{2.37}$$

When $x^0>s$, the service level is higher than the specified level α.

Game analysis

We consider now a decentralized supply chain characterized by non-cooperative firms and assume first that both players make their decisions simultaneously. The supplier chooses the wholesale price w and the retailer selects the order quantity, q. The supplier then produces q units at unit cost c and delivers them to the retailer.

Using the first-order optimality conditions for the retailer's problem, we have

$$\frac{\partial J(q,w)}{\partial q} = mqf(q) - mqf(q) + \int\limits_q^\infty mf(D)dD - \int\limits_0^q h^+ f(D)dD + \int\limits_q^\infty h^- f(D)dD - w = 0$$

Thus, we find that the maximum wholesale price, $w^M = m + h^-$, so that if $w \le w^M$, the best retailer's response is determined by

$$F(q) = \frac{m + h^- - w}{m + h^- + h^+} . \tag{2.38}$$

From (2.38) we observe, that if $w = w^M$, the retailer does not order at all, while if $w < w^M$, then comparing (2.33) and (2.38) and taking into account $w \ge c$ and $\dfrac{\partial F(q)}{\partial q} > 0$, we conclude with results similar to those found for the pricing game with endogenous demand.

Proposition 2.6. *In vertical competition of the stocking game, if the supplier makes a profit, i.e., $w > c$, the retailer's order quantity and the customer service level are lower than the system-wide optimal order quantity and service level.*

Note that if the retailer would account for the supplier's margin, $w - c$, by including it into the numerator of (2.38), equation (2.38) would transform into (3.33). We thus find the double marginalization effect discussed in the pricing game. In addition, this effect decreases the customer service level unless the supplier does not want to profit from the sale and sets $w = c$. On the other hand, since the supplier's objective function (2.30) is linear in w, we conclude that the supplier would set the wholesale price as high as possible, i.e., $w = w^M$ under the Nash strategy. In such a case, the retailer makes no profit and orders nothing. As a result of the Nash strategy, there is neither business nor customer service between the supplier and the retailer.

Similar to the pricing game of the previous section, the statement of Proposition 2.6 that vertical competition causes the supply chain performance to deteriorate does not depend on whether the players make a simultaneous decision or if the supplier first sets wholesale price, as is often the case in practice. In what follows, we show that under the supplier's leadership, the Stackelberg equilibrium's wholesale price does not equal the maximum purchasing price w^M.

Equilibrium

Assume that the supplier is a leader in the Stackelberg game. The supplier's objective function with q subject to the optimal retailer's response $q=q^R(w)$ is determined by (2.38),

$$J_s(q,w)= (w-c)\, q^R(w).$$

Differentiating the supplier's objective function, we have

$$\frac{\partial J_s(q,w)}{\partial w} = q^R(w)+(w-c)\frac{\partial q^R(w)}{\partial w} = 0. \qquad (2.39)$$

The value of $\dfrac{\partial q^R(w)}{\partial w}$ is determined by differentiating (2.38) with q set equal to $q^R(w)$,

$$f(q^R(w))\frac{\partial q^R(w)}{\partial w}=-\frac{1}{m+h^-+h^+}.$$

As a result:

The greater the wholesale price, the lower the quantity that the retailer orders and by substituting

$$\frac{\partial q^R(w)}{\partial w}=-\frac{1}{\left(m+h^-+h^+\right)f(q^R(w))}$$

into (2.38), we have

$$\frac{\partial J_s(q,w)}{\partial w} = q^R(w)-\frac{w-c}{(m+h^-+h^+)f(q^R(w))} = 0, \qquad (2.40)$$

where

$$F(q^R(w))= \frac{m+h^--w}{m+h^-+h^+}. \qquad (2.41)$$

We conclude with the following proposition.

Proposition 2.7. *Let $f(D)>0$ for $D\geq0$, otherwise $f(D)=0$. The pair (w^s,q^s), where w^s and $q^s= q^R(w^s)$ satisfy*

$$q^R(w^s)-\frac{w^s-c}{(m+h^-+h^+)f(q^R(w^s))} = 0,\; F(q^R(w^s))=\frac{m+h^--w^s}{m+h^-+h^+},$$

constitutes a Stackelberg equilibrium of the stocking game with $c<w^s< m+h^-=w^M$.

Proof: First we consider equation (2.40) and verify that

$$\frac{\partial J_s(q,c)}{\partial w} = q^R(c) > 0, \; \frac{\partial J_s(w^M)}{\partial w} = -\frac{w^M-c}{(m+h^-+h^+)f(0)} < 0.$$

Since $f(D)>0$ for $D\geq0$ we observe that

$$\frac{\partial J_s(q,w)}{\partial w} = q^R(w)-\frac{w-c}{(m+h^-+h^+)f(q^R(w))}$$

is a continuous function for $c \leq w \leq w^M$. We conclude that there is at least one root, $\dfrac{\partial J_s(q, w^s)}{\partial w} = 0$, $c < w^s < w^M$, as stated in Proposition 2.7.

To have a unique Stackelberg wholesale price, however, we require that the supplier's objective function be strictly concave, $\dfrac{\partial^2 J_s(q, w)}{\partial w^2} < 0$, that is,

$$\frac{\partial q^R(w)}{\partial w} - \frac{1}{(m+h^- +h^+)f(q^R(w))} + \frac{w-c}{(m+h^- +h^+)[f(q^R(w))]^2} \frac{\partial f(q^R)}{\partial q^R} \frac{\partial q^R(w)}{\partial w} < 0, \quad (2.42)$$

which apparently does not hold for every distribution.

Example 2.8

Let the demand be characterized by the uniform distribution,

$$f(D) = \begin{cases} \dfrac{1}{A}, & \text{for } 0 \leq D \leq A; \\ 0, & \text{otherwise} \end{cases} \quad \text{and } F(a) = \frac{a}{A}, \ 0 \leq a \leq A.$$

Then the supplier objective function is strictly concave, as (2.42) holds. Using (2.40) - (2.41) we find

$$q^R(w^s) - \frac{w^s - c}{(m+h^- +h^+)} A = 0 \ \text{ and } \ F(q^R(w^s)) = \frac{q^R(w^s)}{A} = \frac{m+h^- - w^s}{m+h^- +h^+}.$$

Thus,

$$\frac{m+h^- - w^s}{m+h^- +h^+} A - \frac{w^s - c}{(m+h^- +h^+)} A = 0,$$

which results in

$$w^s = \frac{m+h^- +c}{2}, \quad q^s = q^R(w^s) = \frac{m+h^- -c}{m+h^- +h^+} \frac{A}{2}, \quad (2.43)$$

while the system-wide optimal order quantity is twice as large,

$$q^* = \frac{m+h^- -c}{m+h^- +h^+} A. \quad (2.44)$$

Recalling our assumption that $w^M = m+h^- > c$, we observe that $c < w^s < w^M$ and $0 < q^s < A/2$. Thus, this problem has always a unique Stackelberg equilibrium.

Example 2.9

Let the demand be characterized by an exponential distribution, i.e.,

$$f(D) = \begin{cases} \lambda e^{-\lambda D}, & \text{for } D \geq 0; \\ 0, \text{ otherwise} \end{cases} \quad \text{and } F(a) = 1 - e^{-\lambda a}, a \geq 0.$$

Then according to (2.40), we have the equation for the Stackelberg wholesale price

$$q^R(w) - \frac{w-c}{(m+h^- +h^+)\lambda e^{-\lambda q^R(w)}} = 0,$$

where according to (2.41)

$$1 - e^{-\lambda q^R(w)} = \frac{m+h^- -w}{m+h^- +h^+}$$

and thus

$$q^R(w) = \frac{1}{\lambda} \ln \frac{m+h^- +h^+}{w+h^+}.$$

Substituting this into the equation of the Stackelberg wholesale price, we obtain the following expression

$$\frac{1}{\lambda} \ln \frac{m+h^- +h^+}{w+h^+} - \frac{w-c}{(w+h^+)\lambda} = 0.$$

We solve this equation with Maple by first setting the left hand side as L

```
>L:=ln((m+hplus+hminus)/(w+hplus)-(w-c)/(w+hplus);
```

$$L := \ln\left(\frac{m+hplus+hminus}{w+hplus}\right) - \frac{w-c}{w+hplus}$$

Then substituting specific values for m=15, hplus=1, hminus=10, c=2

```
>L1:=subs(m=15, hplus=1, hminus=10, c=2, L);
```

$$L1 := \ln\left(\frac{26}{w+1}\right) - \frac{w-2}{w+1}$$

we verify with a plot Y= L1 that it crosses line Y=0 only once and thus the Stackelberg wholesale price is unique.

```
>plot(L1, w=2..15);
```

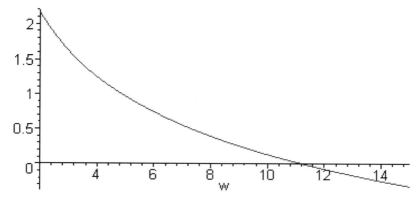

Figure 2.5. The Stackelberg wholesale price

Next we solve equation L1=0 in a general form
```
> ws:=solve(L1=0, w);
```

$$ws := -\frac{\text{LambertW}\left(\dfrac{3}{26}\,\mathbf{e}\right) - 3}{\text{LambertW}\left(\dfrac{3}{26}\,\mathbf{e}\right)}$$

and evaluate the result numerically
```
> evalf(ws);
```
$$11.22512050$$

Finally we calculate the equilibrium order quantity by using the best retailer's response function $q^R(w) = \dfrac{1}{\lambda}\ln\dfrac{m+h^- +h^+}{w+h^+}$.
```
> q:=1/lambda*ln((m+hplus+hminus)/(w+hplus));
```

$$q := \frac{\ln\left(\dfrac{m + hplus + hminus}{w + hplus}\right)}{\lambda}$$

and substituting the specific parameters of the problem
```
>qR:=subs(m=15, hplus=1, hminus=10, lambda=0.1,
w=evalf(ws), c=2, q);
```
$$qR := 10.\ln(2.126768403)\,.$$

Evaluating numerically the last result leads to
```
> qs=evalf(qR);
```
$$qs = 7.546036459\,.$$

Thus $w^s=11.225$ and $q^s= 7.546$. The system-wide optimal order quantity is determined by (2.33)

$$q^* = \frac{1}{\lambda}\ln\frac{m+h^- +h^+}{c+h^+}\,,$$

which with Maple results in
```
> qopt:=1/lambda*ln((m+hplus+hminus)/(c+hplus));
```

$$qopt := \frac{\ln\left(\dfrac{m + hplus + hminus}{c + hplus}\right)}{\lambda}$$

```
>qswopt:=subs(m=15,   hplus=1,   hminus=10,   lamb-
da=0.1,  c=2,  qopt);
```

$$qswopt := 10.\ \ln\left(\frac{26}{3}\right)$$

```
>evalf(qswopt);
```

$$21.59484249$$

Comparing the system-wide optimal solution with the equilibrium solution we find that the system-wide optimal order is almost three-times as large.

$$q^s = 7.546 < q^* = 21.594.$$

Coordination

According to Proposition 2.6, vertical competition under exogenous random demand has a negative effect on the supply chain: the retailer orders less and the service level decreases. This is similar to the pricing competition considered in the previous section and again the negative effect is due to the double marginalization. As opposed to the pricing game, there is no Nash equilibrium in the stocking game while the supplier's leadership has a positive effect on the chain. More precisely, there is an equilibrium if the supplier assumes leadership.

Due to the same double marginalization effect, the coordination in this game is similar to that discussed for the pricing game: discounting and profit sharing. We present here a straightforward approach for developing a coordinating quantity discounting scheme.

First we generalize the supplier's objective function $J_s(q,w)=(w-c)q$ to make the wholesale price dependent on the order quantity, q,

$$J_s(q,w) = w(q) - cq.$$

Then the retailer's best-response (2.38) takes the following form

$$F(q) = \frac{m + h^- - \partial w/\partial q}{m + h^- + h^+}. \tag{2.45}$$

We do not specify any specific requirement for wholesale price $w(q)$ but impose conditions on the rate of change of $w(q)$

$$\frac{\partial w(q)}{\partial q} < c, \quad \frac{\partial^2 w(q)}{\partial q^2} \geq 0, \text{ if } q < q^* \text{ and } \frac{\partial w(q)}{\partial q} \geq c, \text{ if } q > q^*.$$

These conditions imply that the function $w(q)$ may have various discounting schemes for $0 \leq q \leq q^*$. Next we show that if the conditions are met, the supplier can select any value for $w(q)$, $w(q^*) < w^M$, and still have the retailer ordering the system-wide optimal quantity.

Proposition 2.8. *Let $w(q^*) < w^M$, and the discounting scheme be such that if $w(q)$ is a continuous function of q, $\dfrac{\partial w(q)}{\partial q} < c$ and $\dfrac{\partial^2 w(q)}{\partial q^2} \leq 0$ for $q < q^*$, and $\dfrac{\partial w(q)}{\partial q} \geq c$ for $q > q^*$, then the supplier orders the system-wide optimal quantity q^*.*

Proof: Since $w(q)$ is continuous, $\dfrac{\partial^2 w(q)}{\partial q^2} \geq 0$ for $q < q^*$ and $\dfrac{\partial w(q)}{\partial q} \geq c > 0$ for $q > q^*$, the wholesale price $w(q)$ is a convex function, a solution which satisfies (2.45). Note that derivative of $w(q)$ at $q = q^*$ is not required to exist. We thus represent it by the sub-gradient, $\dfrac{\partial w(q^*)}{\partial q} = e$, $a \leq e \leq c$ where $a = \lim\limits_{q \to q^*, q < q^*} \dfrac{w(q) - w(q^*)}{q - q^*} < c$. There can be three possible solutions to (2.45). Assume there exists an optimal solution q', $q' < q^*$, such that $\dfrac{\partial w(q)}{\partial q} \leq a < c$ and (2.45) is met. Recalling that $F(q^*) = \dfrac{m + h^- - c}{m + h^- + h^+}$, we find that if (2.45) is met and $\dfrac{\partial w(q)}{\partial q} < c$, then $q' > q^*$, which contradicts our initial assumption. Similarly, we observe that another solution, say q'', $q'' > q^*$ and thus $\dfrac{\partial w(q)}{\partial q} \geq c$ contradicts (2.45). The only solution left is $q''' = q^*$, $\dfrac{\partial w(q^*)}{\partial q} = e$. Substituting this into (2.45) we find

$$F(q''') = \frac{m + h^- - e}{m + h^- + h^+},$$

which is satisfied for $e = c$ as $a \leq e \leq c$ and $q'' = q^*$.

A trivial example of linear discounting that satisfies Proposition 2.8 is

$$w(q) = \begin{cases} A - aq, & 0 \leq q \leq q^*; \\ A - aq^* + c(q - q^*), & \text{otherwise}, \end{cases}$$

where $A - aq^* < w^M$.

2.3.2 THE OUTSOURCING GAME

In this section, the classical, single-period newsvendor model with a setup cost is turned into an outsourcing game. We consider a single manufacturer with two potential situations. He either incurs a fixed cost per each production order or the product produced is characterized by frequently changing characteristics and/or technology. These changes may be due to new product features and/or technological developments so that each change induces a non-negligible fixed cost. The basic assumptions remain unchanged: the demand is random with known density, $f(D)$ and cumulative $F(a)$ distribution function. In addition we assume a short selling season. If the manufacturer's production or supply order is less than the demand realized at the end of period, then a shortage cost h^- per unit of unsatisfied demand is incurred and there is no time for additional orders. Otherwise, if there is a surplus, the unit inventory cost h^+ is incurred at the end of period.

Accordingly, the manufacturer has two options. One is to order the production in-house, which incurs an irreversible fixed cost C as well as variable cost c_m per unit product. This is in contrast to the newsvendor model considered in the previous section, where the retailer's fixed-order cost was assumed to be negligible. The other option involves outsourcing the production to a single supplier. Then the manufacturer incurs only the variable purchasing cost w per product unit and the supplier incurs a unit production cost c. We assume that $c>c_m$, no initial inventory, and a profitable in-house production (at least when there is no initial inventory at the manufacturer's plant). Otherwise outsourcing is always advantageous. Both the manufacturer and the supplier are profit maximizers.

The manufacturer's problem

$$\max_q J_m(q,w)=$$

$$\max\{ \max_q \{E[ym -h^+x^+- h^-x^-]-wq\}, \max_q \{E[ym -h^+x^+- h^-x^-]-c_mq-C\}\}, \quad (2.46)$$

s.t.

$$x=q-d, \quad (2.47)$$

$$q\geq 0, \quad (2.48)$$

where $x^+=\max\{0, x\}$ and $x^-=\max\{0, -x\}$ are respectively inventory surplus and shortage at the end of a period, and $y=\min\{q,d\}$ is the number of products sold.

The manufacturer's objective function (2.46) consists of two parts. The first part $\max_q \{E[ym - h^+x^+ - h^-x^-]-wq\}$ represents the profit which the manufacturer can gain if he decides to outsource the production. The other part is the profit from in-house production (assuming that the production is profitable). Since the first part is identical to that studied in the previous section, application of conditional expectation to the first part of (2.46) results into (2.29). Thus, the optimal manufacturer's outsourcing order q' for (2.29) is given by (2.38),

$$F(q')= \frac{m+h^- - w}{m+h^- +h^+}.$$

If we assume that $C=0$, then the second part of (2.46) differs from the first part by c_m only, replaced with w. Consequently, if $C=0$, then the optimal response for the second part of (2.46), q'', is

$$F(q'')= \frac{m+h^- - c_m}{m+h^- +h^+}.$$

Introduce a cost function, $\pi(q)$, such that

$$\pi(q)=E[ym - h^+ x^+ - h^- x^-]. \tag{2.49}$$

Then,

$$\pi(q')-wq'=$$

$$\int_0^{q'} mDf(D)dD + \int_{q'}^{\infty} mq'f(D)dD - \int_0^{q'} h^+(q'-D)f(D)dD - \int_{q'}^{\infty} h^-(D-q')f(D)dD\, wq'$$

is the maximum profit if outsourcing is selected (the first part of (2.46)). The maximum profit when in-house production is selected (the second part of (2.46)) is

$$\pi(q'')-c_m q''-C=$$

$$\int_0^{q''} mDf(D)dD + \int_{q''}^{\infty} mq''f(D)dD - \int_0^{q''} h^+(q''-D)f(D)dD - \int_{q''}^{\infty} h^-(D-q'')f(D)dD - c_m q''-C.$$

Thus, the optimal manufacturer's choice for a given wholesale price is summarized by

$$q = \begin{cases} q', \text{if } \pi(q') - wq' \geq \pi(q'') - c_m q''-C \\ q'', \text{otherwise}, \end{cases} \tag{2.50}$$

where q' is the outsourcing order, while q'' is the in-house production (according to our assumption that in-house production is at least worthwhile, $\pi(q'')-c_m q''-C>0$). Furthermore, condition (2.50) assumes that outsourcing is a dominating strategy when profits from in-house production and outsourcing are identical.

Let outsourcing at supplier's marginal cost be advantageous compared to in-house production profit,

$$\pi(q'')-c_m q''-C \le \pi(q')-cq', \quad F(q')=\frac{m+h^- -c}{m+h^- +h^+}.$$

This, along with (2.50) and the fact that outsourcing profit decreases when the wholesale price increases, implies that the maximum purchase price $w^o \ge c$ always exists such that

$$\pi(q'')-c_m q''-C=\pi(q') - w^o q', \quad F(q')=\frac{m+h^- -w^o}{m+h^- +h^+}.$$

Using (2.49), w^o is the smallest root of the expression below

$$\int_0^{q''} mD f(D)dD+ \int_{q''}^{\infty} mq''f(D)dD- \int_0^{q''} h^+(q''-D)f(D)dD- \int_{q''}^{\infty} h^-(D-q'')f(D)dD-c_m q''-C=$$

$$\int_0^{q'} mD f(D)dD + \int_{q'}^{\infty} mq'f(D)dD - \int_0^{q'} h^+(q'-D)f(D)dD - \int_{q'}^{\infty} h^-(D-q')f(D)dD -$$

$$w^o q', \tag{2.51}$$

where $F(q'')=\dfrac{m+h^- -c_m}{m+h^- +h^+}$ and $F(q')=\dfrac{m+h^- -w^o}{m+h^- +h^+}$.

On the other hand, if outsourcing is not advantageous, then $\pi(q'')-c_m q''-$
$C>\pi(q')-cq'$, $F(q')=\dfrac{m+h^- -c}{m+h^- +h^+}$ and $c_m<w^o<c$. Thus condition (2.50) can be reformulated as follows

$$q = \begin{cases} q', \text{if } c \le w \le w^o, \\ q'', \text{if } w^o < c, \end{cases} \tag{2.52}$$

where $F(q'')=\dfrac{m+h^- -c_m}{m+h^- +h^+}$ and $F(q')=\dfrac{m+h^- -w^o}{m+h^- +h^+}$.

The interpretation of (2.52) is straightforward. If purchasing at the marginal cost of the supplier is not beneficial compared to the in-house production, then there is no wholesale price, $w>c$, to encourage outsourcing.

The supplier's problem is similar to that of the previous section.

The supplier's problem

$$\max_{w} J_s(q,w)= (w-c)q \tag{2.53}$$

s.t.

$$c \le w \le w^o. \tag{2.54}$$

Note that if $\pi(q'')-c_m q''-C \le \pi(q')-cq'$, then the supplier's problem has a feasible solution. Otherwise, $c_m<w^o<c$, and the supplier's problem has no

feasible solution since, in order to compete with in-house production, the supplier has to set the wholesale price below his marginal cost, $w < c$.

Correspondingly, the centralized problem is split into two cases. If $\pi(q'') - c_m q'' - C \leq \pi(q') - cq'$, or equivalently, $w^o \geq c$, the centralized problem is reduced to that considered in the previous section. Indeed, if the supply chain is integrated, then wholesale-related costs represent a transfer within the chain which does not affect the system-wide optimal solution. Then the supplier will deliver products at his marginal cost c and no fixed irreversible cost will be paid since in-house production is not implemented.

The centralized problem

$$\max_q J(q) = \max_q \{E[ym - h^+ x^+ - h^- x^-] - cq\} \tag{2.55}$$

s.t.

$$x = q - d, \; q \geq 0.$$

If $w^o < c$, then the centralized objective function is identical to the second part of (2.46), which is the classical newsvendor problem with a setup cost

$$\max_q \{E[ym - h^+ x^+ - h^- x^-] - c_m q - C . \tag{2.56}$$

In other words, the manufacturer's problem and the centralized problem become identical in such a case.

System-wide optimal solution

The centralized problem (2.55) was studied in the previous section. If outsourcing is selected, i.e., $w^o > c$, the system-wide optimal order quantity $q^{*\prime}$ is unique and defined by (2.33).

$$F(q^{*\prime}) = \frac{m + h^- - c}{m + h^- + h^+}.$$

Note that if the supply chain is centralized, then it simply has two options to produce the product (at the manufacturer and at the supplier). Therefore, it is the production at the supplier option (if chosen) rather than outsourcing.

Similarly, if production at the manufacturer is selected, $w^o < c$, the optimal solution is the newsvendor solution

$$F(q^{*\prime\prime}) = \frac{m + h^- - c_m}{m + h^- + h^+}. \tag{2.57}$$

The effect of initial inventory

Since the supplier does not impose any fixed-order cost, the effect of initial inventories on outsourcing is identical to that for the centralized system as discussed in the previous section,

$$F(x^0+q^*)=\frac{m+h^- - c}{m+h^- + h^+}.$$

To study the effect of initial inventories on production at the manufacturer's plant, let $x^0 < S$, (otherwise it is not optimal to produce at all) and $x = x^0+q-d$. Then the profit from not ordering anything is

$$\pi(x^0)=$$

$$\int_0^{x^0} mDf(D)dD + \int_{x^0}^{\infty} mx^0 f(D)dD - \int_0^{x^0} h^+(x^0 - D)f(D)dD - \int_{x^0}^{\infty} h^-(D-x^0)f(D)dD.$$

On the other hand, if the manufacturer produces $q>0$ products, the profit is

$$\pi(q+x^0)-c_m q-C.$$

The optimal solution for this objective function is determined by (2.57)

$$F(q^{*\prime\prime}+x^0)=\frac{m+h^- - c_m}{m+h^- + h^+}.$$

Denote $S = q^{*\prime\prime}+x^0$, then the optimal in-house profit for a given x^0 is

$$\pi^0(S)-c_m(S- x^0)-C.$$

Note that if $x^0=0$, then assuming that in-house production is profitable under conditions of no initial inventory, we have, $\pi(S)-c_m(S- x^0)-C>0$, while $\pi(x^0)<0$ since we do not sell anything when $x^0=0$. That is,

$$\pi(S)-c_m(S- x^0) -C> \pi(x^0),$$

or equivalently,

$$\pi(S)-c_m S -C> \pi(x^0)-c_m x^0,$$

which implies that it is optimal to produce in-house when $x^0=0$. When initial inventories increase $x^0>0$, then the left-hand part of the inequality remains unchanged while the right-hand part increases towards its maximum which is attained at $x^0=S$. Thus, when $x^0=S$, $C>0$, we have

$$\pi(S)-c_m S - C< \pi(x^0)-c_m x^0,$$

which implies that it is optimal not to produce when $x^0=S$. The right-hand side of the inequality represents the traditional newsvendor objective function, $\pi(x^0)-c_m x^0$, which monotonically increases when x^0 increases towards S. We conclude that there exists $x^0=s<S$, such that,

$$\pi(S)-c_m S - C= \pi(s)-c_m s.$$

Thus, if $x^0<s$, then $\pi(S)-c_m S -C> \pi(x^0)-c_m x^0$ and it is profitable to produce so that $S= q^{*\prime\prime}+x^0$. On the other hand, if $x^0>s$, then $\pi(S)-c_m S - C< \pi(x^0)-c_m x^0$ and it is not profitable to produce. Consequently, in contrast to the optimal order-up-to policy when no fixed order cost is incurred, we obtain the so-called security stock (s, S) policy which is widely used in industry as well,

$$q^{*''} = \begin{cases} S - x^0, \text{if } x^0 < s \\ 0, \text{otherwise}, \end{cases}$$

where s is the smallest value that satisfies $\pi(S) - c_m S - C = \pi(s) - c_m s$.

Game analysis

To simplify the presentation, we assume $x^0 = 0$ and consider now a decentralized supply chain characterized by non-cooperating firms. Let the supplier first set the wholesale price. If $w^o < c$, then regardless of the wholesale price, an in-house production for q" is chosen. Otherwise, the manufacturer decides to outsource and issues an order, q', which the supplier delivers.

Since in-house (2.57) and the centralized in-house solutions are identical, we further focus on outsourcing, i.e., $w^o \geq c$. Let us first assume that $w^o = c$, then the supplier has zero profit by setting $w = c$, and simply sustains himself since the manufacturer's dominating policy is to outsource (2.50) when the profit from in-house production is equal to the outsourcing profit.

Let $w^o > c$. Using the results from the previous section, the optimal order is determined by (2.38)

$$F(q') = \frac{m + h^- - w}{m + h^- + h^+}.$$

This, similar to Proposition 2.6, implies the double marginalization effect.

Proposition 2.9. In the outsourcing game, *if $w^o > c$ and the supplier makes a profit, i.e., $w > c$, the manufacturer's order quantity and the customer service level are lower than the system-wide centralized order quantity and service level.*

Again, similar to the observation from the previous section, since the supplier's objective function is linear in w, the supplier would want to set the wholesale price as high as possible, i.e., $w = w^o$ under the Nash strategy. This causes supply chain performance to deteriorate. In contrast to the inventory game of the previous section, if the manufacturer's dominating policy is to outsource when the profit from in-house production is equal to the profit from outsourcing, then the manufacturer will still outsource at $w = w^o$.

Equilibrium

Given $w^o > c$, Proposition 2.7 proves that there is a Stackelberg equilibrium price $c < w^s < m + h^-$. However, since $q' > 0$ and $\pi(q') - w^o q' = \pi(q'') - c_m(q'') - C > 0$, then $w^o < w^M = m + h^-$. This implies that the Stackelberg wholesale price found with respect to Proposition 2.7 may be greater than w^o. In such a case it is set to $w^s = w^o$.

Based on Proposition 2.7 and the manufacturer's optimal response (2.52), we summarize our results.

If $w^o < c$, then produce q'' products in-house, where

$$F(q'') = \frac{m + h^- - c_m}{m + h^- + h^+}.$$

If $w^o = c$, then outsource; the equilibrium wholesale price is $w^s = c$, and the outsourcing quantity q' is such that

$$F(q') = \frac{m + h^- - c}{m + h^- + h^+}.$$

If $w^o > c$, then outsource; find w' and $q' = q^R(w')$ (according to Proposition 2.7), i.e.,

$$q^R(w') - \frac{w' - c}{(m + h^- + h^+)f(q^R(w'))} = 0, \ F(q^R(w')) = \frac{m + h^- - w'}{m + h^- + h^+}.$$

If $w' < w^o$, then the equilibrium wholesale price is $w^s = w'$ and the outsourcing order is q', otherwise $w^s = w^o$ and the outsourcing order q' is such that $F(q') = \dfrac{m + h^- - w^o}{m + h^- + h^+}$.

Example 2.10

Let the demand be characterized by the uniform distribution,

$$f(D) = \begin{cases} \dfrac{1}{A}, & \text{for } 0 \le D \le A; \\ 0, \text{otherwise} \end{cases} \quad \text{and } F(a) = \frac{a}{A}, \ 0 \le a \le A.$$

Then using the results of Example 2.8, we have a unique solution for each case.

If $w^o < c$, then produce $q'' = \dfrac{m + h^- - c_m}{m + h^- + h^+} A$ products in-house, which is equivalent to the system-wide optimal solution.

If $w^o=c$, then we outsource; the equilibrium wholesale price is $w^s=c$ and the outsourcing quantity is $q^s=\dfrac{m+h^- -c}{m+h^- +h^+}A$ products, which is equivalent to the system-wide optimal order.

If $\dfrac{m+h^- +c}{2} \leq w^o$ (and thus $w^o>c$), then we outsource; the equilibrium wholesale price is $w^s = \dfrac{m+h^- +c}{2}$ and the outsourcing order is

$q^s = q' = \dfrac{m+h^- -c}{m+h^- +h^+}\dfrac{A}{2}.$

If $\dfrac{m+h^- +c}{2} >w^o>c$, then we outsource; the equilibrium wholesale price is $w^s = w^o$ and outsourcing order quantity is $q^s = q' = \dfrac{m+h^- -w^o}{m+h^- +h^+}\dfrac{A}{2}$ products,

where w^o satisfies the expression

$$\int_0^{q''} m\frac{D}{A}dD + \int_{q''}^{\infty} m\frac{q''}{A}dD - \int_0^{q''}\frac{h^+}{A}(q''-D)dD - \int_{q''}^{\infty}\frac{h^-}{A}(D-q'')dD - c_m q'' - C\}$$

$$= \int_0^{q'} m\frac{D}{A}dD + \int_{q'}^{\infty} m\frac{q'}{A}dD - \int_0^{q'}\frac{h^+}{A}(q'-D)dD - \int_{q'}^{\infty}\frac{h^-}{A}(D-q')dD -$$

$$w^o q'\}, \ q''=\frac{m+h^- -c_m}{m+h^- +h^+}A \text{ and } q'=\frac{m+h^- -w^o}{m+h^- +h^+}A .$$

Example 2.11

Let the demand be characterized by an exponential distribution, i.e.,

$$f(D) = \begin{cases} \lambda e^{-\lambda D}, & \text{for } D \geq 0; \\ 0, \text{otherwise} \end{cases} \text{ and } F(a) = 1 - e^{-\lambda a}, a \geq 0.$$

We first formalize equation (2.51) for w^o which, for the exponential distribution yields,

$$\int_0^{q''} [mD - h^+(q''-D)]\lambda e^{-\lambda D}dD + \int_{q''}^{\infty} [mq'' - h^-(D-q'')]\lambda e^{-\lambda D}dD - c_m q'' - C =$$

$$= \int_0^{q'} [mD - h^+(q'-D)]\lambda e^{-\lambda D} dD + \int_{q'}^{\infty} [mq' - h^-(D-q')]\lambda e^{-\lambda D} dD - w^0 q',$$

where $q'' = \dfrac{1}{\lambda} \ln \dfrac{m + h^- + h^+}{h^+ + c_m}$ and $q' = \dfrac{1}{\lambda} \ln \dfrac{m + h^- + h^+}{h^+ + w^0}$.

We calculate this expression with Maple. Specifically, we set the order quantities q'' and q' as q2 and q1 respectively,

```
> q2:=1/lambda*ln((m+hplus+hminus)/(cm+hplus));
```

$$q2 := \frac{\ln\left(\dfrac{m + hplus + hminus}{cm + hplus}\right)}{\lambda}$$

```
> q1:=1/lambda*ln((m+hplus+hminus)/(w0+hplus));
```

$$q1 := \frac{\ln\left(\dfrac{m + hplus + hminus}{w0 + hplus}\right)}{\lambda}$$

Next we define the left-hand side and right-hand side of (2.51) as LHS and RHS

```
>LHS:=int((m*D-hplus*(q2-D))*lambda*exp(-lambda*
D),D=0..q2)+int((m*q2-hminus*(D-q2))*lambda*exp(-
lambda*D),  D=q2..infinity)-cm*q2-C:
>RHS:=int((m*D-hplus*(q1-D))*lambda*exp(-lambda*
D),D=0..q1)+int((m*q1-hminus*(D-q1))*lambda*exp(-
lambda*D),  D=q1..infinity)-w0*q1:
```

Then to see how fixed cost, C, effects the solution, specific values are substituted for the parameters of the problem except for C.

```
> LHSC:=subs(m=15,hplus=1,hminus=10,cm=2,lambda=0.1,
LHS);
> RHS1:=subs(m=15,hplus=1,hminus=10,cm=2,lambda=0.1,
RHS);
```

After evaluating the left-hand side and the right-hand side

```
> LHSCe:=evalf(LHSC);
```

$$LHSCe := 65.2154725 - 1.\, C$$

```
> RHSe:=evalf(RHS1);
```

$$RHSe := -15.76923077 \; \ln\left(\frac{26.}{w0 + 1.}\right) + 168.7967107 + 8.796710786 \; w0$$

$$- 15.76923077 \; \ln\left(\frac{26.}{w0 + 1.}\right) w0 + 5.769230769 \; \ln\left(\frac{1}{w0 + 1.}\right)$$

$$+ 5.769230769 \; w0 \ln\left(\frac{1}{w0 + 1.}\right)$$

we solve (2.51) in w^0

```
> solutionw0:=solve(LHSCe=RHSe, w0);
```
and plot the solution as a function of the fixed cost
```
>plot(solutionw0, C=0..200);
```

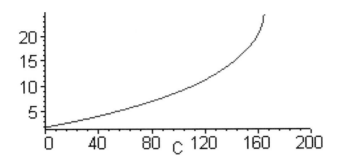

Figure 2.6. The effect of the fixed cost C on the maximum wholesale price w_0

The plot (Figure 2.6) implies that the higher the fixed cost, C, the greater w^0 and thus the smaller the chance that in-house production is beneficial compared to the outsourcing. For example, if $C=120$
```
> LHSes:=subs(C=120, LHSCe);
```

$$LHSes := -54.7845275$$

then
```
>solve(LHSes=RHSe, w0);
```

$$11.26258264$$

$w^0=11.2625$ and thus if supplier's cost $c>11.2625$, the in-house production is advantageous (and is system-wide optimal) at quantity $q''{*}=q2opt=21.594$
```
>q2opt:=evalf(subs(m=15, hplus=1, hminus=10, cm=2,
lambda=0.1, q2));
```

$$q2opt := 21.59484249$$

Otherwise, if $c \leq 11.2625$, then outsourcing is advantageous and the Stackelberg equilibrium wholesale price w' and order quantity q' are calculated as described in the previous section. Note that in case of $w'>w^o$, the Stackelberg wholesale price equals w^o and the order quantity is computed correspondingly.

Coordination

If $w^o>c$, then outsourcing has a negative impact compared to the corresponding centralized supply chain, the manufacturer orders less and the service level decreases. This is similar to the vertical inventory game without a setup cost considered in the previous section. In contrast to that game, this effect is reduced when $c \leq w^o < w^s$, where w^s is calculated under an assumption of no constraints, i.e., according to Proposition (2.7). In addition, there can be a special case when $w^o = c$, and thus the supplier is forced to set the wholesale price equal to its marginal cost, $w = c$. This eliminates double marginalization, the manufacturer outsources the system-wide optimal quantity and the supply chain becomes perfectly coordinated regardless of whether the supplier is leader in a Stackelberg game or the firms make decisions simultaneously using a Nash strategy. On the other hand, since the case when the manufacturer prefers in-house production is identical to the corresponding centralized problem, no coordination is needed. Consequently, the case which requires coordination is when $w^o > c$. This case coincides with that derived for the inventory game with no setup cost. Thus, the coordinating measures discussed in the previous section are readily applied to an outsourcing-based supply chain.

An alternative way of improving the supply chain performance is to develop a risk-sharing contract which would make it possible to coordinate the chain in an efficient manner as discussed in the following section.

2.4 INVENTORY COMPETITION WITH RISK SHARING

In competitive conditions discussed so far, the retailer incurs the overall risk associated with uncertain demands. The fact that expected profit is the criterion for decision-making implies that the retailer does not have an assured profit. The supplier, on the other hand, profits by the quantity he sells. If the supplier is sensitive to the retailer's service level, he may agree to mitigate demand uncertainty by buying back left-over products at the end of selling season or offer an option for additional urgent deliveries to cover cases of higher than expected demand. These well-known types of risk-sharing contracts make it possible to improve the service level as well as to coordinate the supply chain as discussed in the following sections. (See also Ritchken and Tapiero 1986).

2.4.1 THE INVENTORY GAME WITH A BUYBACK OPTION

A modification of the traditional newsvendor problem considered here arises when the supplier agrees to buy back leftovers at the end of selling season at a price, $b(w)$, $\dfrac{\partial b(w)}{\partial w} \geq 0$ and $\dfrac{\partial^2 b(w)}{\partial w^2} \geq 0$. This means that the

uncertainty associated with random demand may result in inventory associated costs, $b(w)x^+$ at the supplier's site while at the retailer's site it is an
income $b(w)x^+$ rather than a cost. Thus the supplier mitigates the retailer's
risk associated with demand overestimation or, in other words, the supplier
shares costs associated with demand uncertainty. The other parameters of
the problem remain the same as those of the stocking game.

The retailer's problem

$$\max_q J_r(q,w)= \max_q \{E[ym + b(w)x^+ - h^-x^-]-wq\}, \tag{2.58}$$

s.t.

$$x=q-d,$$
$$q\geq 0,$$

where $x^+=\max\{0, x\}$, $x^-=\max\{0, -x\}$ and $y=\min\{q,d\}$.

Applying conditional expectation to (2.58) the objective function transforms into

$$\max_q J_r(q,w)=\max_q \{$$

$$\int_0^q mD f(D)dD+ \int_q^\infty mqf(D)dD+ \int_0^q b(w)(q-D)f(D)dD- \int_q^\infty h^-(D-q)f(D)dD-wq\}. \tag{2.59}$$

The first term in the objective function, $E[ym]= \int_0^q mDf(D)dD + \int_q^\infty mqf(D)dD$,

represents income from selling y product units; the second, $E[b(w)x^+]=$

$\int_0^q b(w)(q - D)f(D)dD$, represents income from selling leftover goods at

the end of the period; the third, $E[h^-x^-]= \int_q^\infty h^-(D - q)f(D)dD$, represents

losses due to an inventory shortage; while the last term, wq, is the amount
paid to the supplier for purchasing q units of product. As discussed earlier,
there is a maximum wholesale price, w^M, that the supplier can charge so
that the retailer will still continue to buy products. Taking this into account
we formulate the supplier's problem.

The supplier's problem

$$\max_w J_s(q,w)= \max_w (w-c)q-E[b(w)x^+] \tag{2.60}$$

s.t.

$$c \leq w \leq w^M.$$

The first term $(w-c)q$ in (2.60) represents the supplier's income from selling q products at margin $w-c$, while the second, $E[b(w)x^+]$ is the payment for the returned leftovers to the supplier. To simplify the problem, we here assume that leftovers are salvaged at a negligible price rather than stored at the supplier's site. The centralized problem is then based on the sum of two objective functions (2.59) and (2.60) which results in a function independent of the wholesale price, w.

The centralized problem

$$\max_q J(q) = \max_q \{E[ym - h^-x] - cq\} \tag{2.61}$$

s.t.

$$x = q - d, \ q \geq 0.$$

Note that since w and b represent transfers within the supply chain, system-wide profit does not depend on them.

System-wide optimal solution

Applying conditional expectation to (2.61) and the first-order optimality condition, we find that

$$\frac{\partial J(q)}{\partial q} = mqf(q) - mqf(q) + \int_q^\infty mf(D)dD - \int_q^\infty h^- f(D)dD - c = 0,$$

which results in

$$F(q^*) = \frac{m + h^- - c}{m + h^-}. \tag{2.62}$$

Since this result differs from (2.33) by only h^+ set at zero, the objective function in (2.61) is strictly concave under the same assumptions. Similarly, the service level in the centralized supply chain with a buyback contract is

$$\alpha = \frac{m + h^- - c}{m + h^-}, \tag{2.63}$$

This is different from $\alpha = \dfrac{m + h^- - c}{m + h^- + h^+}$ of the traditional newsvendor problem only because of our assumption that surplus products are salvaged at a negligible price rather than stored at the supplier's site.

Game analysis

Consider now a decentralized supply chain characterized by non-cooperative firms and assume that both players make their decisions simultaneously. The supplier chooses the wholesale price w and thereby buyback $b(w)$ price while the retailer selects the order quantity, q. The supplier then delivers the products and buys back leftovers.

Using the first-order optimality conditions for the retailer's problem, we find $w^M = m + h^-$, so that if $w \le w^M$, then

$$F(q) = \frac{m + h^- - w}{m + h^- - b(w)}. \tag{2.64}$$

Since the retailer's objective function is strictly concave, we conclude from (2.64), the following result.

Proposition 2.10. *In vertical competition, if the supplier makes a profit, i.e., $w > c$, a buyback contract induces increased retail orders and an improved customer service level compared to that obtained in the corresponding stocking game.*

Proof: To prove this proposition, compare the optimal orders with the non-cooperative buyback option

$$F(q) = \frac{m + h^- - w}{m + h^- - b(w)},$$

and without the buyback option

$$F(q) = \frac{m + h^- - w}{m + h^- + h^+}.$$

From Proposition 2.10 we conclude that the buyback contract has a coordinating effect on the supply chain. Moreover, comparing (2.62) and (2.64), we observe that in contrast to the stocking game, with buyback contracts, i.e., $b(c) > 0$, when setting $w = c$, the retailer orders even more than the system-wide optimal quantity since there is less risk of overestimating demands. In such a case, the supplier has only losses due to buying back leftover products. Thus, the supplier can select $w > c$ so that the retailer's non-cooperative order will be equal to the system-wide optimal order quantity. This coordinating choice will be discussed below after analyzing possible equilibria.

Equilibrium

Let us first consider the case of $\dfrac{\partial b(w)}{\partial w} > 0$, $\dfrac{\partial^2 b(w)}{\partial w^2} > 0$ and assume that

$b(w)$ is chosen such that $\lim\limits_{w} J_s(q,w) = -\infty$, i.e., the solution set is compact.

Then the Nash equilibrium can be found by differentiating the supplier's

objective function $J_s(q,w) = (w-c)q - E[b(w)x^+] = (w-c)q - \int\limits_0^q b(w)(q-D)f(D)dD$,

$$\frac{\partial J_s(q,w)}{\partial w} = q - \frac{\partial b(w)}{\partial w} \int\limits_0^q (q-D)f(D)dD = 0. \qquad (2.65)$$

Verifying the second-order optimality condition, we also find

$$\frac{\partial^2 J_s(q,w)}{\partial w^2} = -\frac{\partial^2 b(w)}{\partial w^2} \int\limits_0^q (q-D)f(D)dD < 0. \qquad (2.66)$$

Since the functions of both supplier and retailer are strictly concave and the solution space is compact, we readily conclude that a Nash equilibrium exists (see, for example, Basar and Olsder 1999).

Proposition 2.11. *The pair (w^n, q^n), such that*

$$q^n - \frac{\partial b(w^n)}{\partial w} \int\limits_0^{q^n} (q^n - D)f(D)dD = 0, \ F(q^n) = \frac{m + h^- - w^n}{m + h^- - b(w^n)}$$

constitutes a Nash equilibrium of the inventory game under a buyback option.

An interesting case arises when $b(w)$ is a linear function of w. In such a case, similar to the traditional stocking game, $J_s(q,w)$ depends linearly on w, i.e., the supplier would set the wholesale price as high as possible. Unlike the stocking game, this situation does not lead to no-business under a buyback contract. Indeed, by setting w close to but less than w^M, the supplier may still be able to induce the retailer to order the desired quantity by properly choosing a function $b^* = b^*(w)$. In fact, this strategy leads to perfect coordination regardless of the fact whether the supplier is the Stackelberg leader or the decision is made simultaneously. This is because under any wholesale price w, $b^* = b^*(w)$ would ensure the same response from the retailer by increasing w the supplier increases his profit. Thus, this time we find *the greater the wholesale price, the greater the supplier's profit while the order quantity remains the same.*

Example 2.12

Let $\dfrac{\partial b(w)}{\partial w} > 0$, $\dfrac{\partial^2 b(w)}{\partial w^2} > 0$ and the demand be characterized by the uniform distribution,

$$f(D) = \begin{cases} \dfrac{1}{A}, & \text{for } 0 \le D \le A; \\ 0, & \text{otherwise} \end{cases} \quad \text{and } F(a) = \dfrac{a}{A}, \, 0 \le a \le A.$$

Then using (2.64), we find

$$q^n = \dfrac{m + h^- - w^n}{(m + h^- - b(w^n))} A.$$

Substituting into (2.65) we have

$$\dfrac{m + h^- - w^n}{m + h^- - b(w^n)} A - \dfrac{\partial b(w^n)}{\partial w}\left(\dfrac{m + h^- - w^n}{m + h^- - b(w^n)}\right)^2 \dfrac{A}{2} = 0.$$

Rearranging this last equation we obtain

$$\dfrac{m + h^- - w^n}{m + h^- - b(w^n)} A\left(1 - \dfrac{\partial b(w^n)}{\partial w}\dfrac{m + h^- - w^n}{m + h^- - b(w^n)}\dfrac{1}{2}\right) = 0.$$

Since $w^n = w^M = m + h^-$ results in no order at all, the Nash equilibrium is found by

$$1 - \dfrac{\partial b(w^n)}{\partial w}\dfrac{m + h^- - w^n}{m + h^- - b(w^n)}\dfrac{1}{2} = 0.$$

If for example, $b(w) = \alpha + \beta w^2$, and the buyback price does not exceed the maximum price, $\alpha + \beta[w^M]^2 < m + h^-$, then we have a unique Nash equilibrium

$$w^n = \dfrac{1}{\beta}(1 - \dfrac{\alpha}{m + h^-}), \ q^n = \dfrac{m + h^- - w^n}{m + h^- - \alpha - \beta[w^n]^2} A.$$

On the other hand, the system-wide optimal order is

$$q^* = \dfrac{m + h^- - c}{m + h^-} A.$$

Coordination

As discussed in previous sections, discounting, for example, a two-part tariff is one tool which provides coordination by inducing a non-cooperative solution to tend to the system-wide optimum.

In this section we show that buyback contacts provide an efficient means for coordinating vertically competing supply chain participants. Specifically, when $b(w)$ is a linear function of w, the supplier's objective function depends linearly on w. This implies that it is optimal for the supplier to set the wholesale price as high as possible. However, unlike the traditional stocking game, this situation does not lead to no orders if the supplier chooses $b^*=b^*(w)$ as described below.

Let the best retailer's response q defined by (2.64) be identical to the system-wide optimal solution q^* defined by (2.62),

$$\frac{m+h^- -c}{m+h^-} = \frac{m+h^- -w}{m+h^- -b^*(w)}. \tag{2.67}$$

From (2.67) we conclude that if

$$b^*(w) = (m+h^-)\frac{w-c}{m+h^- -c}, \tag{2.68}$$

then $q=q^*$ for any $w<w^M$. Thus, if $b^*(w)$ is set according to (2.68), the supplier can maximize his profit by choosing w very close to w^M. This would leave the retailer still ordering a system-wide optimal quantity which would perfectly coordinate the supply chain. This result is independent of the fact whether the supplier first sets w and $b^*(w)$ (as Stackelberg leader) or whether decisions on w and q are made simultaneously (Nash strategy) if function $b^*(w)$ is known to the retailer.

Example 2.13

Let the demand be characterized by an exponential distribution, i.e.,

$$f(D) = \begin{cases} \lambda e^{-\lambda D}, & \text{for } D \geq 0; \\ 0, \text{ otherwise} \end{cases} \quad \text{and } F(a) = 1 - e^{-\lambda a}, a \geq 0$$

and $b^*=b^*(w)$ be chosen by the supplier so that the best retailer's response q is identical to the system-wide optimal solution q^*, that is, $b^*(w)$ is determined by (2.68). Then the equilibrium wholesale and buyback prices are

$$w=w^M-\varepsilon=m+h^- -\varepsilon \text{ and } b^*(w) = (m+h^-)(1-\frac{\varepsilon}{m+h^- -c}),$$

where ε is a small number and the equilibrium order quantity is

$$q= \frac{1}{\lambda}\ln\frac{m+h^-}{c}.$$

Note that the smaller the ε, the greater the supplier's share of the risk associated with uncertain demands and the greater the share of the overall supply chain profit that the supplier gains on account of the retailer. When ε is very small, the retailer returns all unsold products at almost the same wholesale price he purchased them. He therefore has no risk at all in case the demand realization will be lower than the quantity stocked.

2.4.2 THE INVENTORY GAME WITH A PURCHASING OPTION

Similar to the buyback option, this modification of the stocking game arises when the supplier is willing to mitigate the risk the retailer incurs with respect to the uncertainty of customer demands. Specifically, similar to a buyback contract, the supplier may agree to have an inventory surplus at the end of the selling season. In contrast to the buyback contract, this surplus is due to an option which is offered to the retailer. The option allows the retailer to issue an urgent or fast order, to be shipped immediately, at a predetermined option price, $m > u(w) > w$, $\dfrac{\partial u(w)}{\partial w} \geq 0$, close to the end of the selling season. The retailer will exercise this option only if customer demand exceeds his inventories. It is this difference between the retailer's backorder and the supplier's inventory level which the retailer's option purchase covers. If the supplier is unable to satisfy such a backorder, he will compensate the retailer for his loss. Thus, under this type of contract, the supplier assumes the customer service level at the retailer's site by mitigating the retailer's backlog costs. We assume that the system parameters are such that the supplier's order q_s exceeds the retailer's order q_r, $q_r < q_s$ (an exact requirement for this to hold is stated in Proposition 2.13) which ensures an inventory game between the retailer and supplier. Furthermore, we assume that the wholesale price and the retailer's margin are fixed and the supplier cost is negligible unless it is an urgent order. This enables us to focus solely on the inventory game where the supplier and retailer have to choose a quantity to order. To draw an analogy with our previous analysis, we allow the wholesales price to change when coordination aspects are discussed.

The retailer's problem

$$\max_{q_r} J_r(q_r, q_s) = \max_{q_r} \{ E[my + (m - u(w))x_r^- - h_r^+ x_r^+ - h_r^- x_s^-] - wq_r \}, \quad (2.69)$$

s.t.

$$x_r = q_r d,$$
$$x_s = q_s - q_r - x_r^-,$$
$$q_r \geq 0,$$

where $x_r^+ = \max\{0, x_r\}$, $x_r^- = \max\{0, -x_r\}$ and $y = \min\{d, q_r\}$,

In this single-period formulation, x_r is the retailer's inventory level by the end of a period prior to an urgent order when realization, D, of random demand d is already known; x_r^+ is the retailer's inventory surplus at the end of the period; x_r^- is the retailer's inventory shortage prior to an urgent order; the urgent quantity ordered by the retailer for immediate shipment,

h_r^+, h_r are the retailer's inventory holding and shortage costs respectively; and q_r is the quantity ordered by the retailer at the beginning of the period and shipped by the end of the period. If the supplier does not have enough products to ship, then a purchase option implies that the supplier covers the difference between the retailer's margin and the option price m-$u(w)$ for unsold product.

Applying conditional expectation to (2.69), the objective function transforms into

$$\max_{q_r} J_r(q_r,q_s)= \max_{q_r} \{$$

$$\int_0^{q_r} mDf(D)dD + \int_{q_r}^{\infty} mq_r f(D)dD + \int_{q_r}^{\infty}(m-u(w))(D-q_r)f(D)dD - \int_0^{q_r} h_r^+(q_r-D)f(D)dD$$

$$- \int_{q_s}^{\infty} h_r^-(D-q_s)f(D)dD - wq_r \}. \qquad (2.70)$$

The first term in the objective function, $E[ym]= \int_0^{q_r} mDf(D)dD + \int_{q_r}^{\infty} mq_r f(D)dD$,

represents the income from selling y=min$\{d,q_r\}$ product units; the second,

$E[(m-u(w))x_r^-] = \int_{q_r}^{\infty}(m-u(w))(D-q_r)f(D)dD$, represents the income from

backlog at the end of the period; the third and the fourth, $E[h_r^+x_r^+]=$

$\int_0^{q_r} h_r^+(q_r-D)f(D)dD$, $E[h_r x_s]= \int_{q_s}^{\infty} h_r^-(D-q_s)f(D)dD$, are the surplus and

shortage costs; and the last term, wq_r, is the amount paid to the supplier for a regular order.

The supplier's problem

$$\max_{q_s} J_s(q_r,q_s)=$$

$$\max_{q_s} \{wq_r + E[(u(w)-c)(x_r^- - x_s^-) - (m- u(w)) x_s^- - h_s^+ x_s^+]\}, \qquad (2.71)$$

s.t.

$$x_s = q_s - q_r - x_r,$$
$$x_r = q_r - d,$$
$$q_s \geq 0,$$
$$x_s^+ = \max\{0, x_s\}, x_s = \max\{0, -x_s\},$$

where x_s is the supplier's inventory level by the end of period after an urgent order; q_s is the quantity ordered by the supplier at the beginning of the period and shipped in time for reshipment from the supplier to the retailer by the end of the period; $u(w)$ is the option price; h_s^+ is the supplier's inventory holding cost; and c is the cost of processing the urgent order.

After simple manipulations with (2.71)

$$J_s(q_r,q_s) = wq_r + E[(u(w)-c)x_r - (m-c)x_s^- - h_s^+x_s^+]$$

and determining expectation, we have

$$J_s(q_r,q_s) = wq_r + \int_{q_r}^{\infty}(u(w)-c)(D-q_r)f(D)dD - \int_{q_s}^{\infty}(m-c)(D-q_s)f(D)dD -$$

$$\int_{q_r}^{q_s}h_s^+(q_s-D)f(D)dD - \int_0^{q_r}h_s^+(q_s-q_r)f(D)dD\}. \tag{2.72}$$

The first term in the objective function, wq_r, is the income from selling q_r products; the second, $E[(u(w)-c)x_r] = \int_{q_r}^{\infty}(u(w)-c)(D-q_r)f(D)dD$, represents income from the optional order; the third, $E[(m-c)x_s] = \int_{q_s}^{\infty}(m-c)(D-q_s)f(D)dD$, represents the compensation paid by the supplier for the part of the optional order which the supplier is unable to deliver (i.e., this is the supplier's shortage cost); and the last term, $E[h_s^+x_s^+] = \int_{q_r}^{q_s}h_s^+(q_s-D)f(D)dD$

$+ \int_0^{q_r}h_s^+(q_s-q_r)f(D)dD$, is the inventory surplus cost incurred by the supplier.

The centralized problem is based on the sum of two of the objective functions (2.69) and (2.71).

The centralized problem

$$\max_{q_r,q_s} J(q_r,q_s) = \max_{q_r,q_s}\{E[my+(m-c)(x_r^--x_s^-) - h_r^+x_r^+ - h_s^+x_s^+ - h_r^-x_s^-]\} \tag{2.73}$$

s.t.

$$x_s = q_s - q_r - x_r,$$
$$x_r = q_r - d,$$
$$q_r \geq 0, q_s \geq 0.$$

Note that since w, $u(w)$ and $(m-c)x_s^-$ represent transfers within the supply chain, the system-wide profit does not depend on w, $u(w)$ and is reduced by $(m-c)x_s^-$ to account only for the satisfied part $(x_r^- - x_s^-)$ of the optional (urgent) order. Applying conditional expectation to (2.73) we have explicitly,

$$J(q_r,q_s)= \int_0^{q_r} mDf(D)dD + \int_{q_r}^{\infty} mq_r f(D)dD + \int_{q_r}^{\infty}(m-c)(D-q_r)f(D)dD -$$

$$\int_{q_s}^{\infty}(m-c)(D-q_s)f(D)dD - \int_{q_s}^{\infty} h_r^-(D-q_s)f(D)dD - \int_0^{q_r} h_r^+(q_r-D)f(D)dD$$

$$- \int_{q_r}^{q_s} h_s^+(q_s-D)f(D)dD - \int_0^{q_r} h_s^+(q_s-q_r)f(D)dD = mE[D] - \int_{q_r}^{\infty} c(D-q_r)f(D)dD$$

$$- \int_{q_s}^{\infty}(m-c)(D-q_s)f(D)dD - \int_{q_s}^{\infty} h_r^-(D-q_s)f(D)dD$$

$$- \int_0^{q_r} h_r^+(q_r-D)f(D)dD - \int_{q_r}^{q_s} h_s^+(q_s-D)f(D)dD - \int_0^{q_r} h_s^+(q_s-q_r)f(D)dD.$$

System-wide optimal solution

The first-order optimality condition with respect to q_r results in

$$\frac{\partial J(q_r,q_s)}{\partial q_r} = \int_{q_r}^{\infty} cf(D)dD - \int_0^{q_r} h_r^+ f(D)dD + h_s^+(q_s-q_r)f(q_r)$$

$$- h_s^+(q_s-q_r)f(q_r)=0.$$

Thus, the system-wide unique optimal order quantity of the supplier is

$$F(q_r^*) = \frac{c}{c + h_r^+}. \tag{2.74}$$

Similarly, the first-order optimality condition with respect to q_s yields,

$$\frac{\partial J(q_r,q_s)}{\partial q_s} = (m-c)(1 - F(q_s)) - h_s^+(F(q_s) - F(q_r)) - h_s^+ F(q_r) +$$

$$h_r^-(1 - F(q_s))=0.$$

Thus, the system-wide unique optimal supplier's order is

$$F(q_s^*) = \frac{m - c + h_r^-}{m - c + h_s^+ + h_r^-}. \tag{2.75}$$

Furthermore, since the first derivative in one of the variables is independent of the other variable, the corresponding Hessian is negative definite and this newsvendor type of the objective function is strictly concave in both decision variables.

Game analysis

Consider now a decentralized supply chain characterized by non-cooperative firms and assume that both players make their decisions simultaneously. After the retailer and supplier choose their orders q_r and q_s , the supplier delivers q_r units as a regular order and $(x_r^- - x_s^-)$ as an urgent order as well as covers the retailer for losses if the urgent order does saturate the demand, x_s^-.

Applying the first-order optimality condition to the retailer's objective function (2.70) we find

$$\frac{\partial J(q_r,q_s)}{\partial q_r} =$$

$$mq_r f(q_r) - mq_r f(q_r) + \int_{q_r}^{\infty} mf(D)dD - \int_{q_r}^{\infty}(m-u(w))f(D)dD - \int_{0}^{q_r} h_r^+ f(D)dD - w =$$

$$= m(1-F(q_r)) - (m-u(w))(1-F(q_r)) - h_r^+ F(q_r) - w = 0 ,$$

that is,

$$F(q_r) = \frac{u(w) - w}{u(w) + h_r^+} . \tag{2.76}$$

Equation (2.76) represents a unique, newsvendor-type, optimal solution. As long as our assumption $u(w)<m$ holds, the regular order is independent of the retailer's margin. Shortage cost h_r^- is not a part of this equation since the purchasing option causes a shortage which depends on the supplier's order quantity rather than on the retailer's decision.

To determine the Nash equilibrium, we next differentiate the supplier's objective function (2.72),

$$\frac{\partial J(q_r,q_s)}{\partial q_s} = \int_{q_s}^{\infty}(m-c)f(D)dD - \int_{q_r}^{q_s} h_s^+ f(D)dD - \int_{0}^{q_r} h_s^+ f(D)dD =$$

$$= (m-c)(1-F(q_s)) - h_s^+(F(q_s)-F(q_r)) - h_s^+ F(q_r) = 0$$

that is,

$$F(q_s) = \frac{m-c}{m-c+h_s^+} . \tag{2.77}$$

This solution is unique and identical to (2.75) if $h_r^-=0$, that is, *the supplier's equilibrium order is system-wide optimal if h_r^- is negligible.*

However, if $h_r^->0$, then $q_s^*>q_s$.

Equilibrium

It is easy to verify that the second derivative with respect to the supplier's order quantity is negative and the supplier's objective function is also strictly concave. Thus, imposing our assumption, $q_r \leq q_s$, we readily conclude with the following statement.

Proposition 2.12. *Let* $\dfrac{m-c}{m-c+h_s^+} \geq \dfrac{u(w)-w}{u(w)+h_r^+}$. *The pair* (q_r^n, q_s^n), *such that*

$$F(q_r^n) = \frac{u(w)-w}{u(w)+h_r^+} \text{ and } F(q_s^n) = \frac{m-c}{m-c+h_s^+}$$

constitutes a unique Nash equilibrium of the inventory game under a purchasing option.

Since $c < u(w) < m$, then we can assume that $u(w)-w \leq c$. If this condition holds, then $F(q_r^*) = \dfrac{c}{c+h_r^+} > F(q_r^n) = \dfrac{u(w)-w}{u(w)+h_r^+}$ which, of course, is not a new discovery. In contrast to previous results, the total order also includes urgent order, $x_r^- - x_s^-$, while the supplier's inventory level, $F(q_s) = \dfrac{m-c}{m-c+h_s^+}$

determines the service level in the supply chain with a purchasing option. We thus conclude with the following property:

Proposition 2.13. *Let* $\dfrac{m-c}{m-c+h_s^+} \geq \dfrac{u(w)-w}{u(w)+h_r^+}$. *In vertical competition, if* $u(w)-w \leq c$, *a contract with a purchasing option induces lower order quantities from the retailer and supplier as well as a lower service level than the system-wide optimal solution.*

Next, comparing the retailer's order with (q_r^n) and without (q_r) purchasing option (see the stocking game in Section 2.3.2), we conclude that

$$F(q_r^n) = \frac{u(w)-w}{u(w)+h_r^+} < F(q_r) = \frac{m+h_r^- - w}{m+h_r^- + h_r^+},$$

as $u(w) < m$.

Proposition 2.14. *Let* $\dfrac{m-c}{m-c+h_s^+} \geq \dfrac{u(w)-w}{u(w)+h_r^+}$. *In vertical competition, a contract with a purchasing option induces a lower regular order quantity by the retailer compared to the contract without a purchasing option, while the service level depends on* h_s^+.

From Proposition 2.14, it follows that unless the supplier's inventory holding cost is too high, a contract with a purchasing option improves the service level, but the regular order quantity decreases. This is expected, since, given the possibility of an urgent order, it is beneficial for the retailer to reduce the regular order and wait for demand to realize and only then increase profit by an urgent purchase if the demand exceeds the regular order stock. Note that since the urgent order is random, $x_r^- - x_s^-$, and always non-negative, it means that

$$E[x_r^- - x_s^-] = \int_{q_r}^{\infty}(D - q_r)f(D)dD - \int_{q_s}^{\infty}(D - q_s)f(D)dD, \qquad (2.78)$$

is not zero and thus the overall quantity ordered by the retailer is greater than that of a regular order. Moreover, the regular order quantity can be increased since a contract with a purchasing option allows efficient coordination by the proper choice of the option price, $u(w)$. These results are demonstrated in the following example.

Example 2.14

Let the demand be characterized by the uniform distribution,

$$f(D) = \begin{cases} \dfrac{1}{A}, & \text{for } 0 \le D \le A; \\ 0, \text{otherwise} \end{cases} \quad \text{and } F(a) = \dfrac{a}{A}, \; 0 \le a \le A.$$

Then using Proposition 2.12, we find the Nash equilibrium

$$q_r^{\,n} = \frac{u(w) - w}{u(w) + h_r^+} A \text{ and } q_s^{\,n} = \frac{m - c}{m - c + h_s^+} A.$$

The centralized solution is

$$q_r^* = \frac{c}{c + h_r^+} A \text{ and } q_s^* = \frac{m - c + h_r^-}{m - c + h_s^+ + h_r^-} A \; .$$

The average urgent order is thus,

$$E[x_r^- - x_s^-] = \int_{q_r}^{\infty}(D - q_r)f(D)dD -$$

$$\int_{q_s}^{\infty}(D - q_s)f(D)dD = (q_s^{\,n} - q_r^{\,n})[1 - \frac{1}{2A}(q_s^{\,n} + q_r^{\,n})] > 0,$$

while the total average retailer's order is

$$q_r^{\,n} + (q_s^{\,n} - q_r^{\,n})[1 - \frac{1}{2A}(q_s^{\,n} + q_r^{\,n})].$$

Coordination

Coordination under a purchasing option is similar to buyback contacts where a proper choice of the buyback price, $b(w)$, induces the retailer to choose a system-wide optimal order quantity. Specifically, if the supplier chooses the option price $u(w)$ as a linear function of w, $u*(w)$, so that

$$\frac{u*(w) - w}{u*(w) + h_r^+} = \frac{c}{c + h_r^+},$$

and thus

$$u*(w) = \frac{w + \dfrac{ch_r^+}{c + h_r^+}}{1 - \dfrac{c}{c + h_r^+}}, \tag{2.79}$$

then $q_r^n = q_r*$. Moreover, since $u*(w)$ is chosen as a linear function of w, the supplier, as is the case with the buyback contacts, can increase the wholesale price very close to its maximum level and thus gain most of the supply chain profit while still having the retailer order the system-wide optimal quantity. The overall game will, however, become perfectly coordinated only if the retailer's shortage cost is negligible. If it is not negligible, sharing inventory-related costs may have a positive effect on the supply chain's performance.

REFERENCES

Basar T, Olsder GJ (1999) *Dynamic Noncooperative Game Theory*, SEAM.

Bertrand J (1983) Theorie mathematique de la richesses sociale, *Journal des Savants*, September pp499-509, Paris.

Cachon G, Netessine S (2004) Game theory in Supply Chain Analysis in *Handbook of Quantitative Supply Chain Analysis: Modeling in the eBusiness Era*. edited by Simchi-Levi D, Wu SD, Shen Z-J, Kluwer.

Cournot AA (1987) *Research into the mathematical principles of the theory of wealth*, Mcmillan, New York.

Davidson C (1988) Multiunit Bargaining in Oligopolistic Industries. Journal of Labor Economics 6: 397-422.

Horn H, Wolinsky A (1988), Worker substitutability and patterns of unionisation, Economic Journal 98: 484-97.

Ganeshan RE, Magazine MJ, Stephens P (1998) A taxonomic review of supply chain management research, in *Quantitative models for supply chain management*, edited by Tayur SR.

Goyal SK, Gupta YP (1989) Integrated inventory models: the buyer-vendor coordination, European journal of operational research 41(33): 261-269.

Leng M, Parlar M (2005) Game Theoretic Applications in Supply Chain Management: A Review, *INFOR* 43(3): 187220.

Li L, Whang S (2001) Game theory models in operations management and information systems. In *Game theory and business applications*, K. Chatterjee and W.F. Samuelson, editors, Kluwer.

Ritchken P, Tapiero CS (1986) Contingent Claim Contracts and Inventory Control. *Operations Research* 34: 864-870.

Viehoff I (1987) Bargaining between a Monopoly and an Oligopoly. Discussion Papers in Economics 14, Nuffield College, Oxford University.

Wilcox J, Howell R, Kuzdrall P, Britney R (1987) Price quantity discounts: some implications for buyers and sellers. *Journal of Marketing* 51(3): 60-70.

3 SUPPLY CHAIN GAMES: MODELING IN A MULTI-PERIOD FRAMEWORK

In this chapter, we extend the single-period newsvendor-type model discussed in Chapter 2 to a multi-period setting. This implies that the supply chain operates in dynamic conditions and that customer demand has a different realization at each period (see, for example, Sethi et al. 2005). In such multi-period cases, the newsvendor problem is turned into a stochastic game. We address here two such games. One is a straightforward extension of the stocking game considered in Chapter 2. The other is a replenishment game, where the decisions are concerned not only with the quantities to order for stock but also with the frequency of orders or, equivalently, with the length of the replenishment period. The meaning of such an extension is not only technical. It is conceptually important for setting the grounds of the management of supply chains in inter-temporal frameworks to be dealt with in forthcoming chapters.

3.1 STOCKING GAME

The multi-period stocking game which we consider in this section presumes that the supply chain operates during a number of production periods. At the beginning of each period, current inventories and demands are observed; the supplier sets a unit wholesale price for the period; and the retailer orders (stocks) a quantity at this price to cope with the demand which will be observed only at the end of the period when it is no longer possible to adjust the quantity ordered. Therefore, any unsold quantities will be stored and any backlogged shortages will be dealt with in the next period.

3.1.1 THE STOCKING GAME IN A MULTI-PERIOD FORMULATION

Let the supply chain consist of a single supplier and a single retailer and consider the straightforward extension of the single-period stocking game studied in Chapter 2. Specifically, assume that there are multiple periods and that at the end of each period, inventories can be reviewed and a decision made by both the supplier and retailer. At each period, the supplier selects a wholesale price at which to sell his stock) while the retailer orders a certain quantity to satisfy customer demands (see Figure 3.1). The supplier has ample capacity and his lead-time is assumed to be shorter than the period length, T. We assume stationary states, i.e., all parameters remain unchanged over the periods and demands at each period are independent and identically distributed variables with $f(.)$ and $F(.)$ denoting the known density and cumulative probability functions res-pectively. Both the supplier and retailer intend to maximize expected profits per period. Unlike the previous chapter, remaining inventories from one period are stored for use in subsequent periods. Sales are not lost. If the demand exceeds the stock, the shortage is backlogged.

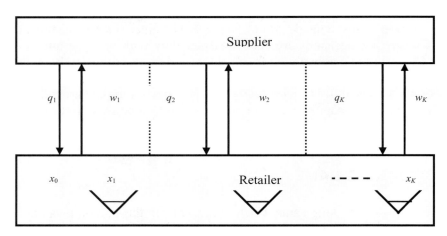

Figure 3.1. The multi-period stocking game

In this context, the general K-period retailer's problem is formulated as follows.

The retailer's problem

$$\max_{q} J_r(q,w) = \max_{q} E[\sum_{t=1}^{K}(y_t m - h_r^+ x^+_t - h_r^- x^-_t - w_t q_t)], \qquad (3.1)$$

s.t.

$$x_{t+1} = x_t + q_{t+1} - d_{t+1}, \; x_0\text{-fixed}, \; t=0,1,..,K\text{-}1 \qquad (3.2)$$

$$q_t \geq 0, \; t=1,..,K$$

$$q=(q_1,q_2,..., q_K), \; w=(w_1,w_2,..., w_K),$$

where, as indicated earlier, $x_t^+ = \max\{0, x_t\}$ and $x_t^- = \max\{0, -x_t\}$ are the inventory surplus and shortage at the end of period t, respectively; $y_t = \min\{q_t + x_{t-1}, d_t\}$ is the quantity of products sold at the end of period t; m is the retailer's margin; h_r^+ and h_r^- are the unit inventory holding and backlog costs respectively; and q_t is the quantity ordered by the retailer and delivered by the supplier at period t.

The corresponding single-period stocking game and the effect of initial inventories have been discussed in Chapter 2. Specifically, applying conditional expectation to (3.1), the multi-period objective function transforms into the following single-period form when there is only one period to go, i.e., $t=K$

$$\max_{q_t} J_r(q_t, w_t) =$$

$$\max_{q_t} \{ \int_0^{q_t + x_{t-1}} mDf(D)dD + \int_{q_t + x_{t-1}}^{\infty} m(q_t + x_{t-1})f(D)dD -$$

$$\int_0^{q_t + x_{t-1}} h_r^+ (q_t + x_{t-1} - D)f(D)dD - \int_{q_t + x_{t-1}}^{\infty} h_r^- (D - q_t - x_{t-1})f(D)dD - w_t q_t \}$$

$$(3.4)$$

Differentiating (3.4) we find the traditional newsvendor solution under initial inventories and $t=K$

$$F(x_{t-1} + q_t) = \frac{m + h_r^- - w_t}{m + h_r^- + h^+}. \qquad (3.5)$$

Thus, the analysis of the problem may change only when there is more than one period left to go, $t<K$.

Similar to the retailer's problem, we formulate the supplier's multi-period problem:

The supplier's problem

$$\max_{\mathbf{w}} J_s(\boldsymbol{q}, \boldsymbol{w}) = \max_{\mathbf{w}} E[\sum_{t=1}^{K} (w_t q_t - ci_t^- - h_s^+ i_t^+)] \qquad (3.6)$$

s.t.

$$i_{t+1} = i_t + o_{t+1} - q_{t+1}, \ i_0\text{-fixed}, \ t=0,1,..,K-1 \qquad (3.7)$$

$$c \leq w_t \leq w_t^M, \ t=1,..,K \qquad (3.8)$$

where w_t is the supplier's wholesale price at period t; c is the supplier's production or processing cost; h_s^+ is the supplier's unit stock holding cost and $i_t^+ = \max\{0, i_t\} = \max\{0, i_{t-1} - q_t\}$ is the supplier's stock at the end of period t; o_t is the quantity produced by the supplier at period t, $o_t = 0$ if $q_t < i_{t-1}$ and $o_t = q_t - i_{t-1}$ if $q_t \geq i_{t-1}$, i.e., $o_t = i_t^- = \max\{0, q_t - i_{t-1}\}$.

The objective function (3.6) involves the stock holding cost $h_s^+ i_t^+$ and no backlog cost (since it is assumed that the supplier has ample capacity and is able to produce as much as needed). This, however, does not prevent the supplier from surplus costs if the initial stock level i_0 is too high so that $i_t > 0$ for $t > 0$.

The corresponding centralized problem is based on the sum of two objective functions (3.6) and (3.1), which results in a function independent on the wholesale (transfer) price.

The centralized problem

$$\max_q J(q) = \max_q E[\sum_{t=1}^{K} (y_t m - h_r^+ x_t^+ - h_r^- x_t^- - h_s^+ i_t^+ - c i_t^-)] \qquad (3.9)$$

s.t.

$$x_{t+1} = x_t + q_{t+1} - d_{t+1}, \ x_0\text{-fixed}, \ t=0,1,..,K-1;$$
$$q_t \geq 0, \ t=1,..,K;$$
$$i_{t+1} = i_t + o_{t+1} - q_{t+1}, \ i_0\text{-fixed}, \ t=0,1,..,K-1;$$
$$c \leq w_t \leq w_t^M, \ t=1,..,K.$$

In what follows we limit our attention to the case of $K=2$.

3.1.2 THE TWO-PERIOD SYSTEM-WIDE OPTIMAL SOLUTION

One-period-to-go system-wide optimal solution

It is shown in the previous chapter that the centralized problem for a single-period case results in the optimality condition (2.35). Given x_1, $K=2$ (only one period to go), this condition remains unchanged if $i_1=0$, that is, for the last review period,

$$F(x_1+q_2) = F(s_2^c) \ \frac{m + h_r^- - c}{m + h_r^- + h_r^+}.$$

Otherwise, the supplier's stock affects the optimal solution.

The effect of the supplier's stock

To facilitate the presentation, let us introduce a function $\gamma(z,a,b)$

$$\gamma(z,a,b) = \begin{cases} z(a-b), \text{if } b < a \\ 0, \text{otherwise} \end{cases},$$

such that

$$\frac{\partial \gamma}{\partial a} = \begin{cases} z, \text{if } b < a, \\ e, \text{if } b = a, \\ 0, \text{if } b > a, \end{cases} \text{ and } -\frac{\partial \gamma}{\partial b} = \begin{cases} z, \text{if } b < a, \\ e, \text{if } b = a, \\ 0, \text{if } b > a, \end{cases} \text{ where } 0 \le e \le z.$$

Then expectation of (3.9) for the second period, $t=2$, is

$$\max_{q_2} J(q_2) = \max_{q_2} \{ \int_0^{q_2+x_1} mDf(D)dD + \int_{q_2+x_1}^{\infty} m(q_2 + x_1)f(D)dD$$

$$- \int_0^{q_2+x_1} h_r^+(q_2 + x_1 - D)f(D)dD - \int_{q_2+x_1}^{\infty} h_r^-(D - q_2 - x_1)f(D)dD$$

$$- \gamma(c, q_2, i_1) - \gamma(h_s^+, i_1, q_2).$$

Differentiating this expression with respect to q_2 we find that,

$$m(1 - F(q_2 + x_1)) - h_r^+ F(q_2 + x_1)) + h_r^-(1 - F(q_2 + x_1)))$$

$$- \frac{\partial \gamma(c, q_2, i_1)}{\partial q_2} - \frac{\partial \gamma(h_s^+, i_1, q_2)}{\partial q_2} = 0 \qquad .$$

We thus conclude

$$F(x_1+q_2) = \begin{cases} \dfrac{m + h_r^- - c}{m + h_r^- + h_r^+}, \text{if } q_2 > i_1; \\[3mm] \dfrac{m + h_r^- + \hat{e}}{m + h_r^- + h_r^+}, \text{if } q_2 = i_1; \\[3mm] \dfrac{m + h_r^- + h_s^+}{m + h_r^- + h_r^+}, \text{if } q_2 < i_1, \end{cases} \qquad (3.10)$$

where $-c \le \hat{e} \le h_s^+$.

From equation (3.10) we observe that the notion of the base-stock derived from the classical newsvendor problem leads to two base-stock levels, s_2' and s_2^c, $s_2^c < s_2'$, given by

$$F(s_2^c) = \frac{m + h_r^- - c}{m + h_r^- + h_r^+} \text{ and } F(s_2') = \frac{m + h_r^- + h_s^+}{m + h_r^- + h_r^+}.$$

Therefore condition (3.10) implies a policy combining order-up-to and order-exactly, or:

$$\text{if } x_1 \ge s_2', \text{ then } q_2 = 0;$$
$$\text{if } s_2' - i_1 \le x_1 < s_2', \text{ then } q_2 = s_2' - x_1;$$
$$\text{if } s_2^c - i_1 \le x_1 < s_2' - i_1, \text{ then } q_2 = i_1; \qquad (3.11)$$
$$\text{if } x_1 < s_2^c - i_1, \text{ then } q_2 = s_2^c - x_1.$$

The centralized objective function is evidently concave when $t=2$ and the found single-period solution is unique under the assumptions identical to those of the single-period stocking problem from the previous chapter.

Two-periods-to-go system-wide optimal solution

To solve the two-period centralized problem, we use the dynamic programming formalism intuitively presented in the appendix to this book. Let the multi-period, profit-to-go function at time t be defined by

$$B_t(x_{t-1}, i_{t-1}) = \max_{q_k, k=t,..K} E[\sum_{k=t}^{K} (y_k m - h_r^+ x_k^+ - h_r^- x_k^- - h_s^+ i_k^+ - ci_k^-)], \ t=1,..,K. \quad (3.12)$$

If there is no initial stock at the supplier's side, $i_1=0$, and thus the supplier's stock does not affect the optimization, then i_t can be omitted in (3.12). Therefore, $i_t^+=0$, $i_t^-=q_t$ and the profit-to-go function (3.12) transforms into

$$B_t(x_{t-1}) = \max_{q_k, k=t,..K} E[\sum_{k=t}^{K} (y_k m - h_r^+ x_k^+ - h_r^- x_k^- - cq_k)], \ t=1,..,K.$$

Introducing the following function

$$\pi(z) = \int_0^z mD f(D)dD + \int_z^\infty mz f(D)dD - \int_0^z h_r^+ (z-D) f(D)dD - \int_z^\infty h_r^- (D-z) f(D)dD$$

and recalling the optimal order $q_2{}^* = \begin{cases} s_2^c - x_1, \text{if } x_1 < s_2^c \\ 0, \text{otherwise} \end{cases}$ when there is only

one period to go, we have for the case $i_1=0$,

$$B_2(x_1) = \begin{cases} \pi(s_2^c) - c(s_2^c - x_1), & \text{if } x_1 < s_2^c \\ \pi(x_1), \text{if } x_1 \geq s_2^c. \end{cases} \quad (3.13)$$

$B_2(x_1)$ is a deterministic function if x_1 is fixed, i.e., when only one period is left to go. However, when there are two periods to go, x_1 is not known and thus $B_2(x_1)$ is random. Specifically, by noting that $x_1= x_0+q_1-d_1$, for the case $i_1=0$, (3.13) transforms into:

$$B_2(x_0 + q_1 - d_1) = \begin{cases} \pi(s_2^c) - c(s_2^c - x_0 - q_1 + d_1), & \text{if } x_0 + q_1 - d_1 < s_2^c \\ \pi(x_0 + q_1 - d_1), \text{if } x_0 + q_1 - d_1 \geq s_2^c. \end{cases} \quad (3.14)$$

On the other hand, if there is a stock, $i_1>0$, for example, $s_2^c - x_1 \leq i_1 < s_2' - x_1$ when $s_2'>x_1$, then we have the following one-period-to-go profit function,

$$B_2(x_1, i_1) = \begin{cases} \pi(x_1 + i_1), & \text{if } x_1 < s_2' \\ \pi(x_1) - h_s^+ i_1, \text{if } x_1 \geq s_2'. \end{cases} \quad (3.15)$$

Then applying Bellman's principle of optimality, the two-period expected profit is

$$B_1(x_0, i_0) = \max_{q_1 \geq 0} \{E[my_1 - h_r^+ x_1^+ - h_r^- x_1^- - h_s^+ i_1^+ - ci_1^- + B_2(x_1)]\}.$$

Assume next that $i_0=0$, which results in

$$B_1(x_0) = \max_{q_1 \geq 0} \{\pi(x_0 + q_1) - cq_1 + E[B_2(x_1)]\}. \tag{3.16}$$

By taking into account (3.14), we find that

$$E[B_2(x_1)] =$$

$$\int_0^{x_0+q_1-s_2^c} \pi(x_0 + q_1 - D)f(D)dD + \int_{x_0+q_1-s_2^c}^{\infty} [\pi(s_2^c) - c(s_2^c - x_0 - q_1 + D)]f(D)dD,$$

which, when substituting into (3.16), leads to

$$B_1(x_0) = \max_{q_1 \geq 0} \{\pi(x_0 + q_1) - cq_1 +$$

$$\int_0^{x_0+q_1-s_2^c} \pi(x_0 + q_1 - D)f(D)dD + \int_{x_0+q_1-s_2^c}^{\infty} [\pi(s_2^c) - c(s_2^c - x_0 - q_1 + D)]f(D)dD\}. \tag{3.17}$$

Differentiating (3.17), we obtain the optimality condition

$$\frac{\partial \pi(x_0 + q_1)}{\partial q_1} - c +$$

$$\pi(s_2^c)f(x_0 + q_1 - s_2^c) + \int_0^{x_0+q_1-s_2^c} \frac{\partial \pi(x_0 + q_1 - D)}{\partial q_1} f(D)dD - \pi(s_2^c)f(x_0 + q_1 - s_2^c)$$

$$+ \int_{x_0+q_1-s_2^c}^{\infty} cf(D)dD = 0.$$

Since

$$\frac{\partial \pi(z)}{\partial z} = m(1 - F(z)) - h_r^+ F(z) + h_r^- (1 - F(z))$$

we obtain

$$m + h_r^- - (m + h_r^- + h_r^+)F(x_0 + q_1) +$$

$$+ \int_0^{x_0+q_1-s_2^c} [m + h_r^- - (m + h_r^- + h_r^+)F(x_0 + q_1 - D)]f(D)dD - cF(x_0 + q_1 - s_2^c) = 0.$$

Consequently we have

$$m + h_r^- - (m + h_r^- + h_r^+)F(x_0 + q_1) + (m + h_r^- - c)F(x_0 + q_1 - s_2^c)$$

$$- \int_0^{x_0+q_1-s_2^c} (m + h_r^- + h_r^+)F(x_0 + q_1 - D)f(D)dD = 0. \tag{3.18}$$

Equivalently, denoting the base-stock level for the first period as $s_1^c = x_0 + q_1$, we observe that $s_1^c \geq s_2^c$ and the optimality equation (3.18) takes the following form

$$m + h_r^- - (m + h_r^- + h_r^+)F(s_1^c) + (m + h_r^- - c)F(s_1^c - s_2^c)$$

$$- \int_0^{s_1^c - s_2^c} (m + h_r^- + h_r^+)F(s_1^c - D)f(D)dD = 0. \qquad (3.19)$$

Differentiating the left-hand side of (3.19) (or of (3.18)), one can verify that the derivative is negative with respect to s_1^c (or q_1). Therefore the centralized objective function is concave and the solution is unique in terms of both s_1^c and q_1. Note that for a found base-stock , s_1^c, the order quantity q_1 is not always feasible. If the initial inventory is too large, $x_0 > s_1^c \geq s_2^c$, then we have, $q_1 = s_1^c - x_0 < 0$, that is, it is optimal not to order at all,

$$q_1^* = \begin{cases} s_1^c - x_0, \text{if } x_0 < s_1^c \\ 0, \text{otherwise.} \end{cases} \qquad (3.20)$$

3.1.3 GAME ANALYSIS

We consider now a decentralized supply chain characterized by non-cooperative parties and $i_0 = 0$. The game is as follows: at the beginning of the first period, the supplier chooses a wholesale price w_1 while the retailer observes his inventory level x_0 and selects an order quantity, q_1. The supplier then produces q_1 units at unit cost c and delivers them to the retailer. At the beginning of the second period, the retailer observes his inventory level x_1 and the game is repeated.

To analyze this dynamic game, we introduce two Bellman functions, one for the retailer, $B_t^r(.)$ and one for the supplier $B_t^s(.)$.

Single-period solution

The solution for the case when only one period is left to go ($t=2$) is derived in the previous chapter and is determined by equation (2.38), where the maximum wholesale price is $w_2^M = m + h_r^-$. Therefore, taking into account initial inventory for the second period, x_1, we have a scenario that if $w_2 \leq w_2^M$, the best retailer's response is determined by

$$F(x_1 + q_2) = \frac{m + h_r^- - w_2}{m + h_r^- + h_r^+}. \qquad (3.21)$$

Comparing (3.21) with the single-period system-wide optimal solution, we confirm the same negative effect of vertical competition on the supply chain performance (see Proposition 2.6).

Again, since the Nash strategy leads to no business and no customer service at all between the supplier and the retailer, we next assume the supplier's leading role. The single-period, Stackelberg price is less than the maximum purchasing price w_2^M which was determined and presented in Chapter 2.)

One- period- to-go equilibrium

The single-period, Stackelberg equilibrium price is determined by Proposition 2.7 from equation (2.40) for the case of zero inventories at the beginning of the period. With respect to the initial inventory level, x_1, this equation takes the following form

$$\frac{\partial J_s(w_2)}{\partial w_2} = q_2^R(w_2) - \frac{w_2 - c}{(m + h_r^- + h_r^+)f(x_1 + q_2^R(w_2))} = 0. \quad (3.22)$$

This result obviously holds if the initial inventory level does not exceed the base-stock level s_2^s induced by the Stackelberg wholesale price, w_2^s, i.e., if $x_1 \le s_2^s$, so that

$$F(s_2^s) = \frac{m + h_r^- - w_2^s}{m + h_r^- + h_r^+}. \quad (3.23)$$

We next introduce the maximum possible wholesale price w_2^* for a given $x_1 < s_2^c$. Price w_2^* satisfies the following equation

$$F(x_1) = \frac{m + h_r^- - w_2^*}{m + h_r^- + h_r^+}, \quad (3.24)$$

If $x_1 \ge s_2^c$, then the retailer orders nothing as the supplier is unable to offer a price which is below his marginal cost. On the other hand, if $x_1 < s_2^c$ or the same $w_2^* > c$, then there exists a Stackelberg equilibrium price w_2^s, $w^* > w_2^s > c$ so that $q_2^s > 0$, as shown in the following proposition.

Proposition 3.1. Let $f(D) > 0$ for $D \ge 0$, otherwise $f(D) = 0$,

$$F(x_1) = \frac{m + h_r^- - w_2^*}{m + h_r^- + h_r^+} \text{ and } F(s_2^c) = \frac{m + h_r^- - c}{m + h_r^- + h_r^+}.$$

If $x_1 \ge s_2^c$, then the retailer orders nothing $q_2^R(w_2) = 0$.
If $x_1 < s_2^c$, then the pair (w_2^s, q_2^s), such that w_2^s and $q_2^s = q_2^R(w_2^s)$ satisfy the equations

$$q_2^R(w_2^s) - \frac{w_2^s - c}{(m + h_r^- + h_r^+)f(x_1 + q_2^R(w_2^s))} = 0,$$

$$F(x_1 + q_2{}^R(w_2{}^s)) = \frac{m + h_r{}^- - w_2{}^s}{m + h_r{}^- + h_r{}^+},$$

constitutes a Stackelberg equilibrium for the second period of the two-period stocking game with $c < w_2{}^s < w_2{}^*$.

Proof: If $w_2{}^* \leq c$, then $x_1 \geq s_2{}^c$ and we order nothing. Let $w_2{}^* > c$ (the same $x_1 < s_2{}^c$), then it is easy to show that $w_2{}^* > w_2{}^s > c$. Specifically, by considering equation (3.22) we straightforwardly verify that

$$\frac{\partial J_s(c)}{\partial w_2} = q_2{}^R(c) > 0, \quad \frac{\partial J_s(w_2{}^*)}{\partial w_2} = -\frac{w_2{}^* - c}{(m + h_r{}^- + h_r{}^+)f(0)} < 0.$$

Taking into account that $f(D) > 0$ *for* $D \geq 0$ we observe that

$$\frac{\partial J_s(w_2)}{\partial w_2} = q_2{}^R(w_2) - \frac{w_2 - c}{(m + h_r{}^- + h_r{}^+)f(x_1 + q_2{}^R(w_2))}$$

is a continuous function for $c \leq w_2{}^s \leq w_2{}^*$. Thus, we conclude that there is at least one root of the equation, $\dfrac{\partial J_s(w_2)}{\partial w_2} = 0$, $c < w_2{}^s < w_2{}^*$, as stated in

Proposition 3.1.

The second part of the proof is by contradiction. Let $x_1 < s_2{}^c$ and there exist $s_2{}^s \leq x_1$. Then $F(s_2{}^s) \leq F(x_1)$ and comparing (3.23) and (3.24) we find that $w_2{}^* \leq w_2{}^s$ which contradicts the fact proven in the first part, that $w_2{}^* > w_2{}^s$. Thus, if $x_1 < s_2{}^c$, then $x_1 < s_2{}^s$ and $q_2{}^s > 0$.

To have a unique Stackelberg wholesale price, we need the supplier's objective function to be strictly concave, $\dfrac{\partial J_s{}^2(w_2)}{\partial w_2{}^2} < 0$, which apparently

does not hold for every distribution. In other words, the uniqueness of the equilibrium depends on the type of demand that the supply chain faces.

Two-periods-to-go – the best retailer's response

To solve the two-period decentralized problem, we use dynamic programming. Consider first the retailer's problem (3.1)-(3.2). Let the profit-to-go from period t be defined as

$$B_t{}^r(x_{t-1}) = \max_{q_k, k=t,...K} E\left[\sum_{k=t}^{K} (y_k m - h_k{}^+ x^+{}_k - h_r{}^- x^-{}_k - w_t q_k) \right], \quad t=1,..,K \quad (3.25)$$

Employing function $\pi(z)$, Proposition 3.1 and recalling that $K=2$, we have for $t=2$ and no supplier's stock, $i_1=0$,

$$B_2{}^r(x_1) = \begin{cases} \pi(s_2{}^s) - w_2{}^s(s_2{}^s - x_1), & \text{if } x_1 < s_2{}^c \\ \pi(x_1), & \text{if } x_1 \geq s_2{}^c. \end{cases} \quad (3.26)$$

Note that according to Proposition 3.1, s_2^s and w_2^s depend on x_1, i.e., $s_2^s = s_2^s(x_1)$ and $w_2^s = w_2^s(x_1)$. When there are two periods to go, x_1 is unknown and thus $B_2^r(x_1)$ is a function of a random variable. Specifically, by taking into account that $x_1 = x_0 + q_1 - d_1 = s_1 - d_1$, (3.26) transforms into

$$B_2^r(x_1) = \begin{cases} \pi(s_2^s(s_1 - d_1)) - w_2^s(s_1 - d_1)(s_2^s(s_1 - d_1) - s_1 + d_1), & \text{if } s_1 - d_1 < s_2^c \\ \pi(s_1 - d_1), & \text{if } s_1 - d_1 \geq s_2^c. \end{cases} \quad (3.27)$$

Then applying the principle of optimality, the expected two-period retailer's profit is

$$B_1^r(x_0) = \max_{q_1 \geq 0} \{ E[my_1 - h_r^+ x_1^+ - h_r^- x_1^- - w_1 q_1 + B_2^r(x_1)] \},$$

which, when accounting for the expectation and $q_1 = s_1 - x_0$, results in

$$B_1^r(x_0) = \max_{s_1 \geq x_0} \{ \pi(s_1) - w_1(s_1 - x_0) + E[B_2^r(x_1)] \}. \quad (3.28)$$

By taking into account (3.27), we find that

$$E[B_2^r(x_1)] =$$

$$\int_0^{s_1 - s_2^c} \pi(s_1 - D) f(D) dD + \int_{s_1 - s_2^c}^{\infty} [\pi(s_2^s(s_1 - D)) - w_2^s(s_1 - D)(s_2^s(s_1 - D) - s_1 + D)] f(D) dD.$$

Substituting into (3.28), results in the profit for the two-periods-to-go case,

$$B_1^r(x_0) = \max_{s_1 \geq x_0} \{ \pi(s_1) - w_1(s_1 - x_0) + \int_0^{s_1 - s_2^c} \pi(s_1 - D) f(D) dD$$

$$+ \int_{s_1 - s_2^c}^{\infty} [\pi(s_2^s(s_1 - D)) - w_2^s(s_1 - D)(s_2^s(s_1 - D) - s_1 + D)] f(D) dD \}. \quad (3.29)$$

Differentiating (3.29). we obtain the first-order optimality condition

$$\frac{\partial \pi(s_1)}{\partial s_1} - w_1 + \pi(s_2^c) f(s_1 - s_2^c) +$$

$$+ \int_0^{s_1 - s_2^c} \frac{\partial \pi(s_1 - D)}{\partial s_1} f(D) dD - \pi(s_2^s(s_2^c)) f(s_1 - s_2^c) + \int_{s_1 - s_2^c}^{\infty} \frac{\partial \pi(s_2^s(s_1 - D))}{\partial s_1} f(D) dD$$

$$+ w_2^s(s_2^c)(s_2^s(s_2^c) - s_2^c) f(s_1 - s_2^c) - \int_{s_1 - s_2^c}^{\infty} \frac{\partial w_2^s(s_1 - D)}{\partial s_1} (s_2^s(s_1 - D) - s_1 + D) f(D) dD -$$

$$- \int_{s_1 - s_2^c}^{\infty} w_2^s(s_1 - D)(\frac{\partial s_2^s(s_1 - D)}{\partial s_1} - 1) f(D) dD = 0.$$

Consequently, taking into account $s_2^s(s_2^c) = s_2^c$, $w_2^s(s_2^c) = c$, and

$$\frac{\partial \pi(z)}{\partial z} = m(1 - F(z)) - h_r^+ F(z) + h_r^- (1 - F(z))$$

we obtain

$$-w_1 + m + h_r^- - (m + h_r^- + h_r^+)F(s_1) +$$

$$\int_0^{s_1 - s_2^c} [m + h_r^- - (m + h_r^- + h_r^+)F(s_1 - D)]f(D)dD +$$

$$\int_{s_1 - s_2^c}^{\infty} [m + h_r^- - (m + h_r^- + h_r^+)F(s_2^s(s_1 - D))]\frac{\partial s_2^s(s_1 - D)}{\partial s_1} f(D)dD$$

$$- \int_{s_1 - s_2^c}^{\infty} \frac{\partial w_2^s(s_1 - D)}{\partial s_1}(s_2^s(s_1 - D) - s_1 + D)f(D)dD -$$

$$\int_{s_1 - s_2^c}^{\infty} w_2^s(s_1 - D)(\frac{\partial s_2^s(s_1 - D)}{\partial s_1} - 1)f(D)dD = 0. \qquad (3.30)$$

Differentiating the left-hand side of (3.30), one can verify whether the derivative is negative with respect to s_1. If this is the case, the objective function is concave and the solution is unique in terms of both s_1 and q_1. Furthermore, $s_1 \geq s^2{}_c$ and the retailer's optimal order is

$$q_1 = \begin{cases} s_1 - x_0, \text{if } x_0 < s_1 \\ 0, \text{otherwise.} \end{cases} \qquad (3.31)$$

Comparing (3.30) with the corresponding system-wide optimal solution (3.19), we observe that unless special conditions are maintained, the base-stocks and thereby the order quantities are different. As a result, the supply chain performance deteriorates not only at the last period, $t=2$, but also when there are two periods to go, $t=1$.

Considering the supplier's objective function (3.6) for the case of no initial stock at the supplier's side

$$\max_w J_s(q,w) = \max_w E[\sum_{t=1}^{K} (w_t-c)q_t],$$

we observe that since the Nash strategy at $t=2$ results in the maximum wholesale price and no business regardless of the value of x_1, the supplier's Nash strategy at $t=1$ is a single-period solution. That is, the supplier's Nash strategy remains the same and there is no business at all at both periods. The Stackelberg equilibrium's wholesale price, however, does not necessarily equal the maximum purchasing price as discussed below.

Two- periods-to-go equilibrium

According to Proposition 3.1, if $x_1 \geq s_2^c$, then $q_2 = 0$ regardless of the wholesale price w_2, otherwise w_2^s and $q_2^s = s_2^s - x_1$ satisfy

$$s_2^s - x_1 - \frac{w_2^s - c}{(m + h_r^- + h_r^+) f(s_2^s)} = 0 , F(s_2^s) = \frac{m + h_r^- - w_2^s}{m + h_r^- + h_r^+}. \quad (3.32)$$

Thus, for the second period, the supplier's payoff function is

$$\max_{w_2} J_s(q_2, w_2) = \max_{w_2} E[(w_2 - c)q_2] =$$

$$B_2^s(x_1) = \begin{cases} 0, & \text{if } x_1 \geq s_2^c; \\ (w_2^s(x_1) - c)(s_2^s(x_1) - x_1), & \text{otherwise.} \end{cases}$$

Function $B_2^s(x_1)$ is deterministic if x_1 is fixed when only one period is left. However, when there are two periods to go, x_1 is unknown and thus $B_2^s(x_1)$ is random. Specifically, by taking into account that $x_1 = x_0 + q_1 - d_1$, , we have

$$B_2^s(x_1) = \begin{cases} 0, & \text{if } s_1 - d_1 \geq s_2^c \\ (w_2^s(s_1 - d_1) - c)(s_2^s(s_1 - d_1) - (s_1 - d_1)), & \text{if } s_1 - d_1 < s_2^c. \end{cases} \quad (3.33)$$

Applying the principle of optimality, the supplier's expected two-periods-to-go profit is

$$B_1^s(x_0) = \max_{w_1 \geq c} \{ E[(w_1 - c)q_1 + B_2^s(x_1)] \}.$$

Accounting for the expectation and for the follower's response $q_1 = q_1^R(w_1)$, results in

$$B_1^s(x_0) = \max_{w_1 \geq c} \{ (w_1 - c)q_1^R(w_1) + E[B_2^s(x_1)] \}, \quad (3.34)$$

where $E[B_2^s(x_1)]$ is calculated with respect to (3.33), as follows

$$E[B_2^s(x_1)] = \int_{s_1 - s_2^c}^{\infty} (w_2^s(s_1 - D) - c)(s_2^s(s_1 - D) - s_1 + D_1) f(D) dD \cdot$$

When substituting into (3.34) and setting $s_1(w_1) = q_1^R(w_1) + x_0$, we have

$$B_1^s(x_0) = \max_{w_1 \geq c} \{ (w_1 - c)(s_1(w_1) - x_0) +$$

$$\int_{s_1(w_1) - s_2^c}^{\infty} (w_2^s(s_1(w_1) - D) - c)(s_2^s(s_1(w_1) - D) - s_1(w_1) + D) f(D) dD \}. \quad (3.35)$$

Finally, differentiating (3.35), we obtain the optimality condition

$$s_1(w_1) - x_0 + (w_1 - c) \frac{\partial s_1(w_1)}{\partial w_1} +$$

$$\int\limits_{s_1(w_1)-s_2^c}^{\infty} (w_2^{\ s}(s_1(w_1)-D)-c)[\frac{\partial s_2^s(s_1(w_1)-D)}{\partial w_1}-\frac{\partial s_1(w_1)}{\partial w_1}]f(D)dD+$$

$$\int\limits_{s_1(w_1)-s_2^c}^{\infty} \frac{\partial w_2^{\ s}(s_1(w_1)-D)}{\partial w_1}[s_2^s(s_1(w_1)-D)-s_1(w_1)+D]f(D)dD=0, \text{ (3.36)}$$

where according to (3.31) if $x_0 > s_1$, then $q_1=0$ regardless of the wholesale price w_1. Otherwise $\dfrac{\partial s_1(w_1)}{\partial w_1}$ is found by differentiating (3.30) with respect

to w_1 and the other two partial derivatives by differentiating both equations of (3.32) in w_1.

We thus derived a system of two equations (3.36) and (3.30). A solution of this system with two unknowns, s_1 and w_1, provides a Stackelberg equilibrium s_1^s and w_1^s for the first period, $t=1$. We summarize the result as follows.

Proposition 3.2. *Let $f(D)>0$ for $D \geq 0$, w_1^s and s_1^s be simultaneous solutions of (3.30) and (3.36) in w_1 and s_1 respectively.*

If $x_0 > s_1^s$, then the retailer orders nothing $q_1^R(w_1)=0$ regardless of the wholesale price w_1,

Otherwise, if $x_0 \leq s_1^s$, then the pair $(w_1^s, q_1^s= s_1^s - x_0)$ constitutes a Stackelberg equilibrium for the first period of the two-period stocking game with $w_1^s > c$.

Example 3.1

Let the demand be characterized by the uniform distribution,

$$f(D)=\begin{cases} \dfrac{1}{A}, & \text{for } 0 \leq D \leq A; \\ 0, \text{otherwise} \end{cases} \quad \text{and } F(a)=\frac{a}{A}, 0 \leq a \leq A.$$

The result of the single-period game with initial inventory is determined by (3.32). Using the uniform distribution, we have

$$s_2^{\ s}-x_1-\frac{w_2^{\ s}-c}{(m+h_r^-+h_r^+)}A=0, s_2^s=\frac{m+h_r^--w_2^{\ s}}{m+h_r^-+h_r^+}A.$$

Thus,

$$w_2^{\ s}=\frac{m+h_r^-+c}{2}-\frac{x_1}{2A}(m+h_r^-+h_r^+), s_2^s=\frac{m+h_r^--c}{m+h_r^-+h_r^+}\frac{A}{2}+\frac{x_1}{2}, \text{ (3.37)}$$

while the system-wide optimal base-stock level is (see (2.44)),

$$s_2^c=\frac{m+h_r^--c}{m+h_r^-+h_r^+}A.$$

Based on (3.37), we find

$$\frac{\partial w_2^{\ s}(x_1)}{\partial x_1} = -\frac{1}{2A}(m + h_r^- + h_r^+) \text{ and } \frac{\partial s_2^{\ s}(x_1)}{\partial x_1} = \frac{1}{2} \qquad (3.38)$$

Next, we consider the retailer's best response (3.30) when there are two periods to go. Substituting (3.37)-(3.38) into (3.30) and taking into account the uniform distribution we have

$$-w_1 + m + h_r^- - (m + h_r^- + h_r^+)\frac{s_1}{A} + \int_0^{s_1 - s_2^c} [m + h_r^- - (m + h_r^- + h_r^+)\frac{1}{A}(s_1 - D)]\frac{1}{A}dD +$$

$$\int_{s_1 - s_2^c}^{\infty} [m + h_r^- - (m + h_r^- + h_r^+)\frac{1}{A}(\frac{m + h_r^- - c}{m + h_r^- + h_r^+}\frac{A}{2} + \frac{s_1 - D}{2})]\frac{1}{2A}dD$$

$$-\int_{s_1 - s_2^c}^{\infty} \frac{1}{2A}(m + h_r^- + h_r^+)(\frac{m + h_r^- - c}{m + h_r^- + h_r^+}\frac{A}{2} + \frac{s_1 - D}{2} - s_1 + D)\frac{1}{A}dD$$

$$-\int_{s_1 - s_2^c}^{\infty} [\frac{m + h_r^- + c}{2} - \frac{s_1 - D}{2A}(m + h_r^- + h_r^+)](-\frac{1}{2A})dD = 0 \qquad (3.39)$$

Consequently, we consider the supplier's optimality condition (3.36), which, with respect to the uniform distribution, takes the following form

$$\int_{s_1(w_1) - s_2^c}^{\infty} (\frac{m + h_r^- + c}{2} - \frac{s_1(w_1) - D}{2A}(m + h_r^- + h_r^+) - c)[-\frac{1}{2}\frac{\partial s_1(w_1)}{\partial w_1}]\frac{1}{A}dD +$$

$$\int_{s_1(w_1) - s_2^c}^{\infty} \frac{m + h_r^- + h_r^+}{2A}\frac{\partial s_1(w_1)}{\partial w_1}[\frac{m + h_r^- - c}{m + h_r^- + h_r^+}\frac{A}{2} + \frac{s_1(w_1) - D}{2} - s_1(w_1) + D]\frac{1}{A}dD$$

$$+ s_1(w_1) - x_0 + (w_1 - c)\frac{\partial s_1(w_1)}{\partial w_1} = 0. \qquad (3.40)$$

To find $\dfrac{\partial s_1(w_1)}{\partial w_1}$, we differentiate equation (3.39) with respect to w_1

$$-1 - (m + h_r^- + h_r^+)\frac{1}{A}\frac{\partial s_1}{\partial w_1} - \int_0^{s_1 - s_2^c} (m + h_r^- + h_r^+)\frac{1}{A^2}\frac{\partial s_1}{\partial w_1}dD +$$

$$+ [m + h_r^- - (m + h_r^- + h_r^+)\frac{1}{A^2}s_2^c]\frac{\partial s_1}{\partial w_1} - \int_{s_1 - s_2^c}^{\infty} (m + h_r^- + h_r^+)\frac{1}{2A}\frac{\partial s_1}{\partial w_1}\frac{1}{2A}dD +$$

$$+ [m + h_r^- - (m + h_r^- + h_r^+)\frac{1}{A}(\frac{m + h_r^- - c}{m + h_r^- + h_r^+}\frac{A}{2} + \frac{s_2^c}{2})]\frac{1}{2A}\frac{\partial s_1}{\partial w_1} +$$

$$+ \int_{s_1-s_2^c}^{\infty} \frac{1}{2A}(m+h_r^- +h_r^+)\frac{\partial s_1}{2\partial w_1}\frac{1}{A}dD + \frac{1}{2A}(m+h_r^- +h_r^+)(\frac{m+h_r^- -c}{m+h_r^- +h_r^+}\frac{A}{2}\frac{s_2^c}{2}-\frac{s_2^c}{2})\frac{1}{A}$$

$$- \int_{s_1-s_2^c}^{\infty} \frac{1}{4A^2}\frac{\partial s_1}{\partial w_1}(m+h_r^- +h_r^+)dD - [\frac{m+h_r^- +c}{2} - \frac{s_2^c}{2A}(m+h_r^- +h_r^+)]\frac{1}{2A}\frac{\partial s_1}{\partial w_1} = 0. \; (3.41)$$

Substituting (3.41) into (3.40), we obtain a single equation in one unknown w_1. Given x_0, a solution to this equation in w_1 provides an equilibrium wholesale price w_1^s for the current inventory, x_0. Finally, substituting w_1 with w_1^s in equation (3.39) results in a single equation with one unknown s_1. A solution to this equation in s_1 provides equilibrium base-stock level s_1^s.

The system-wide optimal base-stock level for the first period is determined by (3.19), i.e.,

$$m + h_r^- - (m + h_r^- + h_r^+)\frac{s_1^c}{A} + (m + h_r^- - c)\frac{1}{A}(s_1^c - s_2^c) -$$

$$- \int_0^{s_1^c-s_2^c} (m+h_r^- +h_r^+)\frac{1}{A^2}(s_1^c -D)dD = 0.$$

Example 3.2

This example illustrates the effect of initial inventories when there is only one period to go. Let the demand be characterized by an exponential distribution, i.e.,

$$f(D) = \begin{cases} \lambda e^{-\lambda D}, & \text{for } D \ge 0; \\ 0, \text{otherwise} \end{cases} \quad \text{and } F(a) = 1 - e^{-\lambda a}, \; a \ge 0.$$

According to Proposition 3.1, the Stackelberg wholesale price and base-stock level are found from

$$s_2^s - x_1 - \frac{w_2^s - c}{(m + h_r^- + h_r^+)\lambda e^{-\lambda s_2^s}} = 0, \; 1 - e^{-\lambda s_2^s} = \frac{m + h_r^- - w_2^s}{m + h_r^- + h_r^+},$$

when $x_1 < s_2^c$, otherwise the retailer orders nothing. Thus,

$$s_2^s = \frac{1}{\lambda}\ln\frac{m + h_r^+ + h_r^-}{w_2^s + h_r^+}$$

and w_2^s is found from

$$\frac{1}{\lambda}\ln\frac{m + h_r^+ + h_r^-}{w_2^s + h_r^+} - x_1 - \frac{w_2^s - c}{(w_2^s + h_r^+)\lambda} = 0. \quad (3.42)$$

We solve these equations with Maple. The base-stock s_2 equation is

```
> s2:=1/lambda*ln((m+hplus+hminus)/(w2s+hplus));
```

$$s2 := \frac{\ln\left(\dfrac{m + hplus + hminus}{w2s + hplus}\right)}{\lambda}$$

The left-hand side of the Stackelberg wholesale price equation (3.42), LHS, is

```
>LHS:=1/lambda*ln((m+hplus+hminus)/(w2+hplus))-x1-
(w2-c)/((w2+hplus)*lambda);
```

$$LHS := \frac{\ln\left(\dfrac{m + hplus + hminus}{w2 + hplus}\right)}{\lambda} - x1 - \frac{w2 - c}{(w2 + hplus)\,\lambda}$$

Next, we substitute specific values into the left-hand side and define the result as LHS1

```
>LHS1:=subs(m=15, hplus=1, hminus=10, c=2, lambda=0.1,
LHS);
```

$$LHS1 := 10.\ \ln\left(\frac{26}{w2 + 1}\right) - x1 - \frac{10.\ (w2 - 2)}{w2 + 1}$$

Similarly, we find the base-stock level s_2 which is defined as s2s,

```
>s2s:=subs(m=15, hplus=1, hminus=10, c=2, lambda=0.1,
s2);
```

$$s2s := 10.\ \ln\left(\frac{26}{w2s + 1}\right)$$

Then equation (3.42) is solved in w_2^s with x_1 unknown

```
>w2s:=solve(LHS1=0, w2);
```

$$w2s := -\frac{1.\,(\text{LambertW}(0.1153846154\ \mathbf{e}^{(0.1000000000\,x1 + 1.)}) - 3.)}{\text{LambertW}(0.1153846154\ \mathbf{e}^{(0.1000000000\,x1 + 1.)})}$$

The system-wide optimal order quantity is determined by the newsvendor type of equation $F(s_2^c) = \dfrac{m + h_r^- - c}{m + h_r^- + h_r^+}$, which with respect to the exponential distribution leads to

$s_2^c = \dfrac{1}{\lambda} \ln \dfrac{m + h_r^+ + h_r^-}{c + h_r^+}$. We calculate it as a benchmark, s2c.

```
>s2c:=1/lambda*ln((m+hplus+hminus)/(c+hplus));
```

$$s2c := \frac{\ln\left(\dfrac{m + hplus + hminus}{c + hplus}\right)}{\lambda}$$

```
>s2ce:=evalf(subs(m=15, hplus=1, hminus=10, c=2,
lambda=0.1, s2c));
```

$$s2ce := 21.59484249$$

To see how the initial inventory level affects the Stackelberg wholesale price and the base-stock level, we produce a graph of w_2^s and s_2^s as functions of inventory x_1 and, for comparison, include the system-wide optimal base-stock level.

```
>plot([w2s,s2s,s2ce],x1=0..30,legend=["wholesale
price", "base-stock", "system-wide base-stock"]);
```

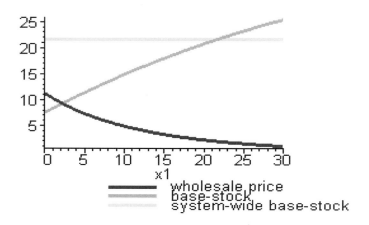

Figure 3.2. The effect of the initial inventory level x on the Stackelberg wholesale price and the base-stock level

From the graph we observe that *as the initial inventory level, x_1, increases, the wholesale price decreases towards the supplier's marginal cost (to encourage the retailer to order even if he has inventories in stock) and the base-stock level increases towards the system-wide optimal level, i.e., the supply chain coordinates when initial stock increases.* In addition, as x_1 exceeds 21.5948, the base-stock level becomes lower than the initial inventory level and thus the optimal order is zero as shown in Proposition 3.1. Thus given $x_1 < 21.5948$, the base-stock level s_2^s and order quantity $q_2^s = s_2^s - x_1$ are found from the graph along with w_2^s for this x_1.

Coordination

Vertical competition under exogenous random demand has a negative effect on the supply chain at each period. Comparing equations (3.31) and (3.19), we observe that the retailer's order can be equal to the system-wide optimal order only under very special conditions. In general, the coordination methods discussed in the previous chapter for a single-period stocking game are applicable to the corresponding multi-period dynamic game.

There are two important coordinating factors related to our dynamic, multi-period stocking game. The equilibrium base-stock level decreases with the period number, $s_1{}^s \geq s_2{}^c \geq s_2{}^s$. Thus if there is more than one period to go, the retailer orders more than at the last period. In fact, he orders even more than the system wide-optimal order under the corresponding static (single-period) game. This improves the efficiency of the supply chain compared to the single-period stocking game and thus multi-period contracts are advantageous. The other factor is due to the initial inventories. The greater the inventory levels at the beginning of a period, the lower the wholesale price. Consequently, the double marginalization effect incurred by the supply chain is smaller unless, of course, the current inventory level is too high so that the retailer does not need to order at all.

3.2 REPLENISHMENT GAME: CASE STUDIES

To illustrate the practical implications of our results, we address here a problem encountered by a large-scale health service supply chain operating in a periodic review mode. Due to the vital nature of the products and services it provides, the number and timing of urgent orders are not limited. As a result, increasingly high transportation costs are incurred and the problem is to select an inventory replenishment (review) period that minimizes transportation costs. Moreover, the supply chain involves multiple retailers who inevitably and independently respond to any change in replenishment policy since it may affect their inventory costs. Such a relationship results in a game between a distribution center and retailers. Since the problem is intractable due to its scale and stochastic nature, we combine a game theoretic approach with empirical analysis.

Many authors have addressed various replenishment policies intended for either continuous or periodic inventory review. The choice of which review policy to use is due to the corresponding costs as well as to practical and organizational considerations (see e.g. Chaing and Gutierrez 1996; Teunter and Vlachos 2001; Rao 2003; Feng and Rao 2006).

Traditionally, and in contrast to the problem we consider in this section, most periodic review inventory systems operating under regular and urgent orders assume that the review period is predetermined. Veinott (1966) and Whittmore et al. (1977) examine an optimal ordering policy only when regular and emergency lead times differ by one time unit. They focus on a situation in which supply lead times are a multiple of a review period. Chaing and Gutierrez (1998) and Chi Chiang (2003) assume a relatively large predetermined review period so that the lead times can be shorter than the review period. (This assumption is similar to ours. In our study, since the retailers are located relatively close to the distributor, the lead time is short.) They develop optimal policies for regular and urgent orders at a periodic review. Teunter and Vlachos (2001) also presume that the lead times can be shorter than the review period in which duration is predetermined. Bylka (2005) in a recent paper assumes that emergency orders arrive immediately (so that the lead time of a regular order is equal to one). The measure of effectiveness is the total (or average per period) expected cost, which includes holding, shortages and both types of order costs. The typical feature of these studies is that they allow only a very restricted number of urgent orders (normally one or two) per review period.

Only a few works (see Flynn and Garstka 1997; Rao 2003, for details) analyze the optimal review or replenishment period for the single supply mode. Flynn and Garstka (1997) develop a model where every T period a retailer observes the current stock level and places orders for the next T periods. They assume that the retailer orders a sequence of deliveries and distinguish between review and delivery intervals. The review period T that they find minimizes the average cost per period. Flynn and Garstka note that T should increase as order setup cost increases; it decreases as the holding and shortage costs as well as the variance in demand increase. Rao (2003) compares two control policies: the periodic review (R, T) policy and the continuous review, reorder point (Q, r). He shows that an economic order interval from a deterministic analysis can provide a good approximation for the optimal T.

A vast body of literature is devoted to inventory coordination or stock-related games (for literature reviews, see, for example, Cachon and Netessine 2004, and Leng and Parlar 2005. However, there are only a relatively small number of papers that focus on supply chains comprising multiple retailers (see, for example, Cachon 2001b, and Wang et al. 2004). Specifically, Cachon (2001b) studies the competitive and cooperative selection of inventory policies and assumes that each location implements a continuous review policy; that demand for the product is Poisson distributed; and that the supplier serves the retailers on a first-come-first-serve basis. He shows that while a Nash equilibrium for a set of reorder points exists, it does not necessarily lead to supply chain efficiency. Thus a competitive solution need not coincide with the global optimum. Wang et. al. (2004) study a system with one supplier and multiple retailers, each with his own lead time and holding cost. Each echelon uses a base-stock policy. Because the players are not cooperative and care only for their own profit, supply chain performance deteriorates. Several contracts for the system-wide optimal cooperation are introduced.

A different setting is studied in Cachon (2001a) where one retailer sells N products with stochastic demands and trucks with finite capacity are dispatched from a warehouse. There is a constant lead time from the warehouse to the retailer. Three policies for dispatching are considered: full-service periodic review; minimum quantity periodic review; and continuous review. Cachon shows that continuous review is less costly if the warehouse is close to the retailers. When the lead time is long, the advantage is small.

In this section we focus on the replenishment period rather than dispatching policies. The warehouse is relatively close-by and the retailers are able to issue an urgent order at any time and as many times as needed. As a result, they are continuously disrupting the periodic replenishment strategy of the distributor, which can be viewed as a constant replenishment period policy with continuous supply adjustments (Kogan et al. 2007).

3.2.1 THE REPLENISHMENT GAME IN A MULTI-PERIOD FORMULATION

We consider a supply chain which comprises a single distributor and multiple retailers. Two supply modes, regular and urgent, characterize the system. A retailer places an order from the distribution center at regular time intervals imposed by the distributor. Thus, the inventories are reviewed and replenished periodically. In a stock-out case, the retailer can place urgent orders. The distributor (the supplier) has ample capacity and his warehouse induces a constant inventory cost which is independent of the level of inventory handled. At the same time, transportation costs incurred by the supplier depend on the retailer's inventory policies. Thus, in this supply chain, the supplier seeks to find a replenishment period which minimizes the transportation cost related to both regular and urgent orders. The retailers, on the other hand, seek to minimize their inventory-related costs, thereby affecting the supplier's goal. This situation is naturally described by the game theoretic framework where the distributor competes with the retailers.

Case studies

Clalit Health Services (Clalit), with an annual budget of $3.3 billion, is the largest healthcare organization in Israel. More than 30,000 employees are engaged in providing highly advanced medical care to 55% of the Israeli population. With healthcare providers issuing five million prescriptions per year, Clalit's logistic operations deliver approximately 5000 types of items to the organization's 14 hospitals, 1380 primary and specialized clinics and 400 pharmacies. Annual operating costs amount to $17.2 million while transportation costs stand at $2.3 million.

Clalit successfully maintains steady operating costs but transportation costs have been increasing. Relative to operating costs, transportation costs reached 11.8% in 2003, 12.2% in 2004, and 13.3% in 2005. This increase is attributed to Clalit's willingness to dispatch urgent supplies when shortages arise. The transportation costs induced by urgent orders have reached 40% of the regular supply transportation costs ($1.4 million per year) and have become a major management concern.

There are a number of causes for the increase. First of all, Clalit distribution centers handle regular supplies of products under a periodic review mode which allows the pharmacies to place as many urgent orders as needed. Second, the competition between healthcare providers in general and pharmacies specifically has resulted in increased frequency of deliveries – sometimes once a day or even twice a day. The average value of the deliveries in these circumstances is about $220 - $280 and includes only 5-7 different items. Thirdly, since the pharmacies were being charged according to total shipments per month regardless of the transportation frequency, they were being encouraged to keep their inventories as low as possible.

In 2001, the distribution companies (which Clalit employed) changed this operating mode since their profits were being erased and the distribution charges became frequency dependent. More specifically, the distributors began to distinguish between regular deliveries, i.e., those planned in advance and which are normally provided once a week and urgent supplies, which must be carried out within a day or two. Naturally, higher charges were imposed on urgent supply deliveries.

The change in transportation charges had a decentralizing effect on the two-echelon supply chain. On the one hand, Clalit is interested in decreeasing supply frequency since it is still responsible for about 60% of transporttation costs. On the other hand, the competition as well as high inventory holding costs induces pharmacies to increase urgent order frequencies in response to less frequent regular supplies. Since urgent orders are more expensive, this significantly affects the overall transportation costs the healthcare company incurs. As a result, even though the pharmacies are part of Clalit, the new transportation charges (imposed on Clalit by its transportation subcontractors) along with the privilege of urgent orders, increase the impact the pharmacies have on the supply chain. This is to say, driven by their goals of minimizing inventory costs, the pharmacies reduce the way that Clalit dominates the chain.

Today, Clalit's distribution center directly supplies 350 pharmacies. About 75% of the pharmacies are supplied every 14 days; 15% - once a month (every 4 weeks); and about 10% - every 7 days. The pharmacies and the distribution center are connected exclusively on an ongoing basis and orders are issued electronically. The major challenge of Clalit's distribution center, which is a leader in the supply chain, is to determine and implement an optimal review or replenishment period that will lead to minimized overall transportation costs with respect to the best response of the pharmacies (i.e., the followers) in terms of regular and urgent orders. It is also important to understand the current position of the players in terms of leadership in the supply chain: Does the distribution center really dominate (thus implementing a Stackelberg strategy), or do the pharmacies succeed in imposing an independent (Nash) strategy on the chain by means of urgent orders? What are the losses and implications associated with the current position? These are the questions which motivated the research presented in this section.

Problem formulation

Consider a distribution center of ample capacity which supplies products to N retailers at each replenishment period t of length T. The distributor has a large automated warehouse. The warehouse is never completely filled up while inventory handling operations incur negligible variable costs compared to the fixed cost of maintaining the warehouse. The cost of transportation to the retailers during period T, $C(T,Q_t)$, on the other hand, is significant and is incurred only by the distributor. The transportation cost depends on the period length T and total order quantity, $Q_t = \sum_{n=1}^{N} q_t^n$, where q_t^n is the regular order of retailer n, $n=1,..,N$ at period t. We assume that urgent orders depend on both T and Q_t and thus affect $C(T,Q_t)$, which will be studied empirically.

Since the overall number of products each retailer orders is overwhelming, similar to supply contracts which specify the total purchase when dealing with multiple products (see, for example, Anupinidi and Bossok 1998), we consider an aggregate order over all items of retailer n, q_t^n, measured in monetary units. Various researchers report that aggregating data in about 150-200 points normally results in less than one percent error in estimating the total transportation costs (Ballou 1992; Hause and Jamie 1981). In addition to the regular orders, the distributor allows for special orders in case of emergency. Since these contingent orders involve small quantities, they do not affect the retailers' inventory costs. However, as mentioned above, they do affect the transportation cost of the distributor, $C(T,Q_t)$, since special, smaller capacity vehicles are employed to carry out urgent orders. This is to say, by increasing the length of period T, or by decreasing the frequency of supplies - both are the same - the distributor diminishes the transportation cost of regular orders. This, however, causes the retailers to boost urgent orders required for the entire period T, thereby inducing additional costly transportation costs for the distributor. As a result, although the supply chain is formally centralized, the situation reflects a classical non-cooperative game in which the distributor is a leader, who sets first the length of the regular review period T and the retailers respond with regular and urgent order quantities. The optimal strategy in this game is referred to as the Stackelberg solution.

The distributor's problem.

The distributor's problem is to minimize his expected transportation cost per period

$$\min_{T} J_d = \lim_{K \to \infty} \frac{1}{KT} E \left[\sum_{t=1}^{K} C(T,Q_t) \right] \qquad (3.43)$$

s.t,

$$T \geq 0 .$$

Note that although order Q_t is the total result of retailer decisions at period t, the length of the period T is independent of t, as the distributor adopts a constant-period review policy.

The retailer's problem

Let d_{it}^{n} be the customer demand rate for retailer n at ith time unit of period t. The demand is random and characterized at each time unit i by probability density $f_n(.)$ and cumulative distribution $F_n(.)$ with mean μ_n and standard deviation σ_n. Denote the demand for T time units at period t as

$$d_t^{Tn} = \sum_{i=1}^{T} d_{it}^{n} , \qquad (3.44)$$

and its density and cumulative functions as $f_{nT}(.)$ and $F_{nT}(.)$ respectively with mean $T\mu_n$ and standard deviation $\sqrt{T}\, \sigma_n$.

The retailer's n problem is to minimize his expected inventory costs per period

$$\min_{\{q_t^n\}} J_r^{n} = \lim_{K \to \infty} \frac{1}{KT} E \left[\sum_{t=1}^{K} h_n^+ \left(X_t^n \right)^+ + h_n^- \left(X_t^n \right)^- \right] \qquad (3.45)$$

where

X_t^n is the retailer n inventory level at the beginning of period t;

$(X_t^n)^+ = \max\{0, X_t^n\}$ and $(X_t^n)^- = \max\{0, -X_t^n\}$;

h_n^+ and h_n^- are the unit surplus and backlog costs per time unit respectively.

The inventory dynamics are described by the following balance equation

$$X_{t+1}^n = X_t^n + q_t^n - d_t^{nT}, q_t^n \geq 0, t=1,2,\ldots \tag{3.46}$$

In what follows we derive the Stackelberg strategy by first solving the retailer's problem and then substituting the solution into the distributor's problem to find an equilibrium review period.

3.2.2 GAME ANALYSIS

According to Stackelberg strategy, the best retailer n response is sought for a given T. This is accomplished by calculating expectation in (3.45):

$$J_r^n = \lim_{K \to \infty} \frac{1}{KT} \sum_{t=1}^{K} [\int_{-\infty}^{X_t^n + q_t^n} h_n^+ (X_t^n + q_t^n - D) f_{nT}(D) dD - \int_{X_t^n + q_t^n}^{\infty} h_n^- (X_t^n + q_t^n - D) f_{nT}(D) dD]. \tag{3.47}$$

Next, applying the first order optimality condition to a single period term of (3.47), we obtain the newsboy-type optimal policy:

$$F_{nT}(X_t^n + q_t^n) = \frac{h_n^-}{h_n^- + h_n^+} . \tag{3.48}$$

Denote the base-stock value by s_{nT},

$$X_t^n + q_t^n = s_{nT} , \tag{3.49}$$

such that $F_{nT}(s_{nT}) = \dfrac{h_n^-}{h_n^- + h_n^+}$. Retailer n then orders up to this stock level s_{nT} if the current level of inventory is less than s_{nT}, otherwise he doesn't order at all. Assuming that initial inventory is less or equal to s_{nT} , we observe that if this single period a myopic solution is applied at each period, then $X_{t+1}^n \leq s_{nT}$ for each t. This argument is then used in the following theorem (detailed proof can be found in, for example, Zipkin 1995).

Theorem 3.1. *The myopic stationary base-stock policy with base-stock level s_{nT} is optimal for multi-period problem (3.45)-(3.46).*

We thus determined the best retailer's n response, $q_t^n = s_{nT} - X_t^n$, to any replenishment period T set by the distributor. In practice, retailers are not always able to calculate their unit backlog costs. However, the service level $\dfrac{h_n^-}{h_n^- + h_n^+}$ (i.e., $1 - \dfrac{h_n^-}{h_n^- + h_n^+}$ is the probability of backlog $P(d_t^{Tn} > s_{nT})$)

is frequently used by management as a goal to be met. The higher the goal (the service level), the greater the base-stock level and thus the lower the risk of backlogs induced by uncertain demands.

Stackelberg equilibrium

Given initial inventory levels, the retailers' orders are deterministic at the first period. Therefore, calculating expectation in (3.45) we find

$$J_d = \lim_{K \to \infty} \frac{1}{KT} E\left[\sum_{t=1}^{K} C(T, Q_t) \right] = \lim_{K \to \infty} \frac{1}{KT} \left\{ C(T, Q_1) + \sum_{t=2}^{K} E[C(T, Q_t)] \right\}. \quad (3.50)$$

Let $f_Q(.)$ and $F_Q(.)$ be the total order Q_t density and cumulative distributions respectively. Then (3.50) can be presented as

$$J_d = \lim_{K \to \infty} \frac{1}{KT} \left\{ C(T, Q_1) + \sum_{t=2}^{K} \int_{-\infty}^{\infty} C(T, \xi_t) f_Q(\xi_t) d\xi_t \right\}. \quad (3.51)$$

The following proposition determines the distribution of the total order, $f_Q(\xi_t)$.

Proposition 3.3. *Distribution of Q_t is identical to the demand distribution,* $\sum_{n=1}^{N} d_t^{nT}$ *for $t \ge 2$, that is, $f_Q(\xi_t)$ is stationary with mean $T\sum_{n=1}^{N} \mu_n$ and standard deviation* $\sqrt{T\sum_{n=1}^{N} \sigma_n^2}$.

Proof: From (3.46) and (3.49) we have

$$X_{t+1}^n = s_{nT} - d_t^{nT}, \quad X_{t+1}^n = s_{nT} - q_{t+1}^n$$

and thus

$$q_{t+1}^n = d_t^{nT}, \, t=1,2,..$$

Consequently, the distribution of the optimal orders q_t^n, $t \ge 2$ (for a replenishment period of length T) is identical to the demand d_t^{nT} distribution. Since the demand distribution per time unit is assumed to be independent of time, i.e., d_t^{nT} is stationary, the distribution of the total order

$$Q_t = \sum_{n=1}^{N} q_t^n = \sum_{n=1}^{N} d_t^{nT}, \, t \ge 2$$

is stationary as well with mean $T \sum_{n=1}^{N} \mu_n$ and standard deviation

$\sqrt{T \sum_{n=1}^{N} \sigma_n^2}$.

Since the problem data are stationary, equation (3.51) simplifies to

$$J_d = \lim_{K \to \infty} \left\{ \frac{C(T, Q_1)}{KT} + \frac{1}{KT} \sum_{t=2}^{K} \int_{-\infty}^{\infty} C(T, \xi_t) f_Q(\xi_t) d\xi_t \right\} =$$

$$\lim_{K \to \infty} \left\{ \frac{C(T, Q_1) - E[C(T, Q)]}{KT} + \frac{1}{T} \int_{-\infty}^{\infty} C(T, \xi) f_Q(\xi) d\xi \right\}.$$

Assume that the probability of extremely high demands is negligible and a much-stretched replenishment period results in enormous transportation costs due to urgent orders. This, along with constraints $T \geq 0$ and $Q_t \geq 0$, implies that the solution sets for T and Q_t are compact. Consequently, the limit in the last expression results in

$$J_d = \frac{1}{T} \int_{-\infty}^{\infty} C(T, \xi) f_Q(\xi) d\xi . \tag{3.52}$$

The equilibrium is then obtained by assuming that $f_Q(.)$ depends on T (see Proposition 3.3) and by applying the first-order optimality condition with respect to T:

$$\int_{-\infty}^{\infty} \left(\frac{\partial [C(T, \xi) f_Q(\xi)]}{T \partial T} - \frac{C(T, \xi) f_Q(\xi)}{T^2} \right) d\xi = 0 . \tag{3.53}$$

Denoting $A = \int_{-\infty}^{\infty} \left(\frac{\partial [C(T, \xi) f_Q(\xi)]}{T \partial T} - \frac{C(T, \xi) f_Q(\xi)}{T^2} \right) d\xi$ and a solution of

(3.53) in T by α, we conclude that if $\frac{\partial A}{\partial T} > 0$ (i.e., $\frac{\partial^2 J_d}{\partial T^2} > 0$) and $\alpha \geq 0$,

then the replenishment period $T^s = \alpha$ and the base-stock level $s_{nT}^s = s_{n\alpha}$ with order quantity $q_t^n = s_{n\alpha} - X_t^n$ for $t = 1, 2, ..; n = 1, .., N$ constitute a unique Stackelberg equilibrium in the replenishment game.

Nash equilibrium

To compare the effect of leadership on the game between the supply chain parties, we next assume that there is no leader in the chain. This implies that the distributor and the retailers make their decisions simultaneously so that in contrast to the Stackelberg strategy, the distributor's objective function

(3.43) is minimized as though the total order quantity Q_t (and, hence, $f_Q(.)$) does not depend on T. Then, applying the first-order optimality condition to the objective function (3.52) we obtain

$$\int_{-\infty}^{\infty}(\frac{\partial C(T,\xi)}{T\partial T} - \frac{C(T,\xi)}{T^2})f_Q(\xi)d\xi = 0. \qquad (3.54)$$

Introducing $B = \int_{-\infty}^{\infty}(\frac{\partial C(T,\xi)}{T\partial T} - \frac{C(T,\xi)}{T^2})f_Q(\xi)d\xi$, and denoting a solution

of (3.54) in T by β, we conclude with the following theorem.

Theorem 3.2. *If $\beta \geq 0$ and $\dfrac{\partial B}{\partial T} > 0$, then the replenishment period $T''=\beta$ and the base-stock level $s_{nT}^n = s_{n\alpha}$ with order quantity $q_t^n = s_{n\beta} - X_t^n$ for t=1,2,...; n=1,..,N constitute a Nash equilibrium in the replenishment game.*

Proof: First note that $\dfrac{\partial B}{\partial T} > 0$, ensures convexity of the distributor's cost, while the newsvendor type of objective (3.47) is evidently convex in s_{nT} and q_t^n as well. Then there must exist at least one simultaneous solution of the first-order optimality conditions (3.54) and (3.48) (see, for example, Debreu 1952), which, if positive, constitutes a Nash equilibrium for the replenishment game.

Results for a normal distribution of the demand

As discussed above, independent optimization of the retailers' responses results in the distribution of the optimal orders q_t^n (for a replenishment period of length T) identical to the demand d_t^{nT} distribution. Assuming that the demand distribution is normal, we observe that $f_{nT}(.)$ and $F_{nT}(.)$ are normal density and cumulative functions with mean $T\mu_n$ and standard deviation $\sqrt{T}\,\sigma_n$. Then the total optimal order, Q_t, is characterized by the normal distribution as well, with the mean and standard deviation as determined in Proposition 3.3. Note that if demand is independent at each time unit (i.e., stationary), then according to the central limit theorem, summation of independent demands over T time units tends to the normal distribution even if demand at each time unit is not normal. In other words, the normality assumption of this section is not very restricting.

   Our first observation with respect to the normal distribution is related to the retailers' objective functions. When $T=1$, the standard deviation of demand d_n per time unit is σ_n. When T increases, the standard deviation reduces, $\dfrac{\sqrt{T}}{T}\sigma_n = \dfrac{\sigma_n}{\sqrt{T}}$. Therefore, similar to the pooling demand effect widely employed in supply chains, the retailer's expected inventory costs per time unit is a monotonically decreasing function of T (see Figure 3.4). This is shown in the following proposition by utilizing the standard normal density function, $\Phi(.)$.

Proposition 3.4. *Let $f_{nT}(.)$ be the normal density function with mean $T\mu_n$ and standard deviation $\sqrt{T}\,\sigma_n$. Then the greater the replenishment period T, the lower the retailer's expected inventory cost per period, J_r^n so that*

$$\frac{\partial J_r^n}{\partial T} < 0 \ and \ \frac{\partial^2 J_r^n}{\partial T^2} > 0 \ .$$

Proof: With respect to Theorem 3.1, retailer n expected cost is

$$J_r^n = \frac{1}{T}[\int_{-\infty}^{s_{nT}} h_n^+(s_{nT} - D)f_{nT}(D)dD - \int_{s_{nT}}^{\infty} h_n^-(s_{nT} - D)f_{nT}(D)dD].$$

Using the fact that $f_{nT}(D) = \dfrac{1}{\sqrt{T}\sigma_n}\Phi\!\left(\dfrac{D - T\mu_n}{\sqrt{T}\sigma_n}\right)$ and introducing a new

variable z, $z = \dfrac{D - T\mu_n}{\sqrt{T}\sigma_n}$ as well as a standardized base-stock level

$s_n^* = \dfrac{s_{nT} - T\mu_n}{\sqrt{T}\sigma_n}$, the expected cost takes the following form

$$J_r^n = \frac{\sigma_n}{\sqrt{T}}[\int_{-\infty}^{s_n^*} h_n^+(s_n^* - z)\Phi(z)dz - \int_{s_n^*}^{\infty} h_n^-(s_n^* - z)\Phi(z)dz]. \tag{4.55}$$

Differentiating this expression with respect to T, we immediately observe that $\dfrac{\partial J_r^n}{\partial T} < 0$ and $\dfrac{\partial^2 J_r^n}{\partial T^2} > 0$, as stated in this proposition.

The following observation is related to the Nash solution. Considering now the distributor's cost J_d, the best responses of the distributor, $T = T^R(s_{nT})$ as well as of retailer n, $s_{nT} = s_{nT}^R(T)$, we obtain the following properties.

Proposition 3.5. *Let $f_{nT}(.)$ be the normal density function with mean $T\mu_n$ and standard deviation $\sqrt{T}\,\sigma_n$. Then the distributor's cost and, hence, the distributor's best response do not depend on retailer n base- stock level,*

$\dfrac{\partial T^R}{\partial s_{nT}} = 0$. *On the other hand, the best retailer n response does depend on* *the replenishment period: the greater the replenishment period T, the larger the base-stock level,* $\dfrac{\partial s_{nT}^{R}(T)}{\partial T} > 0$.

Proof: First note that neither J_d nor its derivative, which is the left-hand side of equation (3.54), denoted by B, explicitly depends on s_{nT}. Furthermore, according to Proposition 3.3, no matter what base-stock level s_{nT} we choose, the quantity that retailer n orders has the same distribution, which depends only on demand. Thus, given replenishment period T, $f_Q(.)$ does not depend on the base-stock policy s_{nT} employed. This is to say that J_d and B are independent on s_{nT}. However, if B does not depend on s_{nT} then the distributor's best response $T=T^R(s_{nT})$ does not depend on s_{nT}, i.e, $\dfrac{\partial T^{R}}{\partial s_{nT}} = 0$.

The retailer's best response is determined with the standardized base-stock level, $s_{nT} = s_{nT}^{R}(T) = T\mu_n + \sqrt{T}\sigma_n s_n *$ (see Proposition 3.4), and thus,

$$\frac{\partial s_{nT}^{R}(T)}{\partial T} = \mu_n + \frac{1}{2\sqrt{T}}\sigma_n s_n * > 0,$$

as stated in the proposition.

There are two important conclusions related to Proposition 3.5. The first conclusion is concerned with the supply chain's performance and thereby the corresponding centralized supply chain. If the supply chain is vertically integrated with one decision-maker responsible for setting both a replenishment period and base-stock level for each retailer, then the centralized objective function is a summation of all costs involved:

$$J(T) = J_d + \sum_n J_r^n .$$

The distributor's cost J_d is independent of the base-stock level, as shown in Proposition 3.5. Therefore, applying the first-order optimality condition to $J(T)$ with respect to either q_t^n or s_{nT}, we obtain equation (3.48). This implies that the condition for the Nash base-stock level is identical to the system-wide optimality condition. Next, to find the system-wide optimality condition for the replenishment period, we differentiate $J(T)$ with respect to T, which, when taking into account (3.54) and (3.55), results in

$$\frac{\partial J(T)}{\partial T} = \int_{-\infty}^{\infty}(\frac{\partial C(T,\xi)}{T\partial T} - \frac{C(T,\xi)}{T^2})f_Q(\xi)d\xi -$$

$$-\frac{\sigma_n}{2\sqrt{T^3}}[\int_{-\infty}^{s_n^{*}}h_n^{+}(s_n^{*}-z)\Phi(z)dz - \int_{s_n^{*}}^{\infty}h_n^{-}(s_n^{*}-z)\Phi(z)dz]=0. \qquad (3.56)$$

Comparing equations (3.56) and (3.54) we find the following property.

Proposition 3.6. *Let* $\dfrac{\partial B}{\partial T} > 0$ *and* $f_{nT}(.)$ *be the normal density function with*

mean $T\mu_n$ *and standard deviation* $\sqrt{T}\,\sigma_n$. *The system-wide optimal replenishment period and base-stock level are greater than the Nash replenishment period and base-stock level respectively.*

Proof: Let us substitute T in equation (3.56) with the Nash period $T^n = \beta$. Then the first term in (3.56) vanishes as it is identical to B from (3.54), while the second term is negative, i.e., $\dfrac{\partial J(\beta)}{\partial T} < 0$. Since both $\dfrac{\partial B}{\partial T} > 0$ and

$-\dfrac{\partial^2 J_r^n}{\partial T^2} > 0$ (see Proposition 3.4), then $\dfrac{\partial J(T)}{\partial T}$ increases if T increases and

thus (3.56) holds only if the system-wide optimal period $T^* > \beta$.

Finally, it is shown in Proposition 3.5, that $\dfrac{\partial s_{nT}}{\partial T} > 0$, i.e., if $T > \beta$, then

$s_{nT} > s_{n\beta}$.

Proposition 3.6 sustains the fact that vertical competition causes the supply chain performance to deteriorate as discussed in Chapter 2. Similar to the double marginalization effect, this happens because the retailers ignore the distributor's transportation cost by keeping lower, base-stock inventory levels. The distributor, on the other hand, ignores the retailers' inventory costs when choosing the replenishment period. Figure 3.4 illustrates the effect of vertical competition on the supply chain.

The second property, which is readily derived from Proposition 3.5, is related to the uniqueness of the Nash solution.

Proposition 3.7. *Let* $f_{nT}(.)$ *be the normal density function with mean* $T\mu_n$ *and standard deviation* $\sqrt{T}\,\sigma_n$. *The Nash equilibrium* (T^n, s_{nT}^n) *determined by Theorem 3.2 is unique.*

Proof: The proof immediately follows from Proposition 3.5 and Theorem 3.2. Indeed the two best response curves $T = T^R(s_{nT})$ and $s_{nT} = s_{nT}^R(T)$ can

intersect only once if $\dfrac{\partial T^R}{\partial s_{nT}} = 0$ and $\dfrac{\partial s_{nT}^R(T)}{\partial T} > 0$, i.e, a solution determined

by Theorem 3.2 is unique.

3.2.3 EMPIRICAL RESULTS AND NUMERICAL ANALYSIS

Empirical Results

The transportation costs were obtained from a sample of 16 pharmacies which are being exclusively supplied every 14 days on a regular basis by Clalit's primary distribution center. The base-stock policy was determined according to service level definition and demand forecasts. Pharmacists place their orders using software that computes replenishment quantities for every item with respect to the base-stock level. The pharmacist electronically sends the completed order to the distribution center for packing and dispatching. If there is a shortage or expected shortage before the next planned delivery, the pharmacist can send an urgent order to be delivered not later than two working days from the time of the order.

An external subcontractor (according to outsourcing agreement) delivers the orders to the pharmacies. The contractor schedules the appropriate vehicle (trucks in case of regular orders and mini-trucks for urgent orders) according to the supply plans for the following day. Delivery costs depend on the type of the vehicle used (track or mini-track) and the number of pharmacies to be supplied with the specific transport.

To estimate the influence of a periodic review cycle on the transportation costs (planned and urgent deliveries) the replenishment period for the 16 pharmacies was changed from the original two weeks to three and four weeks. This resulted in a total of 18 replenishment cycles representing 34 working weeks. Monthly sales of the selected pharmacies varied from \$50,000 to \$136,000. Each order that was sent from a pharmacy was reported, and each transport, with every delivery on it, including invoices that were paid to the vehicle contractor, was reported. The data, processed with SPSS non-linear regression analysis, indicate that the resultant parameters of the transportation cost function are $a=4463$, $b=0.0000163$ while the average estimation error is less than 5%.

Numerical Analysis

The goal of our numerical analysis is to check whether this supply chain is predictable using equilibria and how it is affected by the distributor's leadership. In other words, we compare the objective functions (3.43) and (3.45), as well as the effect on the overall supply chain (the sum of (3.43) and (3.45)). Specifically, with distributor leadership, its expected cost equation

(3.52),   is $J_{d1} = \dfrac{1}{\alpha} \displaystyle\int_{-\infty}^{\infty} C(\alpha,\xi) f_{\varrho}(\xi) d\xi$,   while   without   leadership   it is $J_{d2} = \dfrac{1}{\beta} \displaystyle\int_{-\infty}^{\infty} C(\beta,\xi) f_Q(\xi)\,d\xi$. Since α is found by minimizing the entire objective function J_{d1}, while β assumes the normal probability function independent on the period T, the distributor obviously is better off if he is the leader and therefore decides first rather than when the decision is made simultaneously (no leaders).

Similarly, retailer n expected cost under the distributor leadership is

$$J_{r1}^n = \frac{1}{\alpha}[\int_{-\infty}^{s_{n\alpha}} h_n^+(s_{n\alpha}-D)f_{n\alpha}(D)dD - \int_{s_{n\alpha}}^{\infty} h_n^-(s_{n\alpha}-D)f_{n\alpha}(D)dD],$$

while under no leadership it is

$$J_{r2}^n = \frac{1}{\beta}[\int_{-\infty}^{s_{n\beta}} h_n^+(s_{n\beta}-D)f_{n\beta}(D)dD - \int_{s_{n\beta}}^{\infty} h_n^-(s_{n\beta}-D)f_{n\beta}(D)dD].$$

The numerical results of our empirical studies show that the current equilibrium of Clalit's supply chain, which is an outcome of many adjustments it has undergone during many years of operations, is close to and positioned in between both the Stackelberg and Nash equilibria. This is in contrast to the skepticism of many practitioners who believe that a theoretical equilibrium is hardly attainable in real life. Specifically, the equilibrium replenishment period under equal competition is about 16 days; the current replenishment period is 14 days; and the equilibrium under the distributor's leadership is 11 days. Figure 3.3 presents the equilibria over the distributor's transportation cost function.

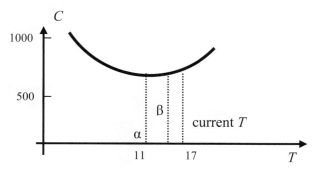

Figure 3.3. The transportation cost as a function of T along with the Stackelberg, Nash and current equilibrium replenishment periods

The Stackelberg equilibrium demonstrates the power the distributor can harness as a leader. The economic implication of harnessing the distributor's power is about 20 NIS per day ($ 4 per day) for the sampled supply volumes. The annual significance, in terms of the overall supply chain, is 1.4 million NIS, or 14% of the total delivery costs. Interestingly enough, the current equilibrium is closer to the Nash replenishment period rather than to the Stackelberg which sustains Clalit's managerial intuition that its distribution centers do not succeed in taking full advantage of their power over the pharmacies.

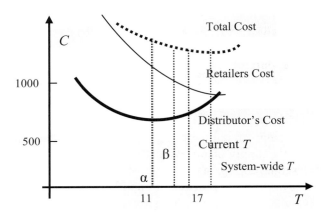

Figure 3.4. Overall supply chain cost, total retailers cost, and distributor's cost

Figure 3.4 presents the results of the calculation for the supply chain as a whole, i.e., including the retailers' inventory management costs and the distributor's transportation costs. In Figure 3.4, the total costs for the Stackelberg, current and Nash strategies as well as the system-wide optimal (global) solution appear as dots on the total cost curve. From this diagram it is easy to observe the effect of the total inventory-related cost on the entire system performance. Specifically, we can see that if the supply chain is vertically integrated or fully centralized and thus has a single decision-maker who is in charge of all managerial aspects, the system-wide optimal replenishment period is 18 days versus the current equilibrium of 14 days. The significance of this gap (which agrees with Proposition 3.6) is that more than 3 million NIS could be saved if the system were vertically integrated. If the distributor attempts to locally optimize (the Stackelberg strategy) this would lead to annual savings in transportation costs of only 1.4 million NIS. However, the significance of such an optimization for the supply chain as a whole is a *loss* of 8 million NIS. This is the price to be paid if the supply chain is either decentralized or operates as a decentralized system.

Coordination

This case study was motivated by increasingly high transportation costs incurred by a large health service provider which is part of a supply chain consisting of multiple retailers (pharmacies) and a distribution center. The costs are attributed to unlimited urgent orders that the retailers could place in the system. Management's approach to handling this problem was to reduce the replenishment period or even transform the policy from periodic to continuous-time review. The latter option in the current conditions would simply imply daily (regular) product deliveries. As shown in Proposition 3.6, such an approach would only lead to further deterioration in supply chain performance due to the double marginalization effect inherent in vertical supply chains. This is also sustained by a numerical analysis of the equilibrium solutions for the case of a normal demand distribution. The analysis shows that if a distributor imposes his leadership on the supply chain, i.e., acts as the Stackelberg leader, then the replenishment equilibrium period is reduced. This makes it possible to cut high transportation costs. However, if instead of an imposed leadership on the supply chain, it is vertically integrated or the parties cooperate, then the potential savings in overall costs are much greater. In such a case, the system-wide optimal replenishment period must increase rather than decrease or transform into a continuous-review policy. Thus, in the short run, imposing leadership by reducing the replenishment period may cut high transportation costs. However, in the long-run, greater savings are possible if, for example, the vendor-managed inventory (VMI) approach is adopted by the retailers or imposed on the retailers by the health provider. In such a case, a distribution center will decide when and how to replenish inventories and the system will become vertically integrated with respect to transportation and inventory considerations. This illustrates the economic potential in cooperation and a total view of the whole supply chain.

REFERENCES

Anupinidi R, Bassok Y (1998) Approximations for multiproduct contracts with stochastic demands and business volume discounts: Single-supplier case. *IIE Transactions* 30: 723-734.

Ballou RH (1992) *Business Logistics Management*, Englewood Cliffs, NJ, Prentice Hall.

Bylka S (2005) Turnpike policies for periodic review inventory model with emergency orders. *International Journal of Production Economics*, 93: 357-373.

Cachon GP (2001a) Managing a retailer's shelf space, inventory, and transportation. *Manufacturing and Service Operations Management* 3: 211-229.

Cachon GP (2001b) Stock wars: inventory competition in a two-echelon supply chain with multiple retailers. *Operations Research* 49: 658-674.

Cachon G, Netessine S (2004) Game theory in Supply Chain Analysis in *Handbook of Quantitative Supply Chain Analysis: Modeling in the eBusiness Era*. edited by Simchi-Levi D, Wu SD, Shen Z-J, Kluwer.

Chiang C (2003) Optimal replenishment for a periodic review inventory system with two supply modes. *European journal of Operational Research* 149: 229-244.

Chiang C, Gutierrez GJ (1996) A periodic review inventory system with two supply modes. *European journal of Operational Research* 94: 389-403.

Debreu D (1952) A social equilibrium existence theorem, *Proceedings of the National Academy of Science* 38: 886-893.

Flynn J, Garstka S (1997) The optimal review period in a dynamic inventory model. *Operations Research* 45: 736-750.

Hause RG, Jamie KD (1981) Measuring the impact of alternative market classification systems in distribution planning, *Journal of Business Logistics* 2: 1-31.

Kogan K, Hovav S, Perlman Y (2007) Equilibrium Replenishment in a Supply Chain with a Single Distributor and Multiple Retailers. *Working paper*, Bar-Ilan University.

Leng M, Parlar M (2005) Game theoretic applications in supply chain management: a review. *INFOR* 43: 187-220.

Simchi-Levi SD, Wu, Shen Z (2004) Handbook of quantitative Supply chain analysis: Modeling in the E-Business era, pp. 13-66.

Sethi, SP, Yan H, Zhang H (2005) Inventory and Supply Chain Management with Forecast Updates. International Series in Operations Research & Management Science, Vol. 81, Springer

Rao US (2003) Properties of the Periodic Review (R, T) Inventory Control Policy for Stationary, Stochastic Demand. *MSOMS* 5: 37-53.

Teunter R, Vlachos D (2001) An inventory system with periodic regular review and flexible emergency review. *IIE Transactions* 33: 625-635.

Veinott Jr.AF (1966) The status of mathematical inventory theory. *Management Science* 12: 745-777.

Wang H, Guo M, Efstathiou J (2004) A game theoretical cooperative mecha-
    nism design for two-echelon decentralized supply chain. *European journal
    of operational research* 157: 372-388.
Whittmore AS, Saunders S (1977) Optimal inventory under stochastic
    demand with two supply options. *SIAM Journal of Applied Mathematics*,
    32: 293–305.

PART II
INTERTEMPORAL SUPPLY
CHAIN MANAGEMENT

4 SUPPLY CHAIN GAMES: MODELING IN AN INTERTEMPORAL FRAMEWORK

So far we have considered discrete-time, single- and multi-period models of competition and coordination in supply chains. In this chapter, we consider continuous-time, intertemporal supply chain models operating in a dynamic environment arising from rapidly changing market conditions including such factors as so-called "word of mouth" and "customer fatigue"; economies of scale; seasonal, fashion and holiday demand patterns; and uncertainty. Since dynamic changes may occur at any point in time, control actions can be exercised continuously. As a result, intertemporal competition between non-cooperative supply chain agents leads to differential games. In some cases, intertemporal relations can be handled by straightforwardly adjusting decision variables as though there is no long-term effect on the supply chain, i.e., by static (myopic) optimization, at each time point independently. However, in most cases, there is a long-term dynamic effect and thus the results obtained for the corresponding static models are no longer valid.

Our goal in this chapter is to illustrate the effect of dynamic conditions on supply chain performance when decisions can be taken at any time point rather than at the beginning (or end) of a certain review period as was the case with the models studied in Chapters 2 and 3. Both periodic and continuous operational review modes are discussed. When inventories and demands are not observable within a review period, continuous in-time decisions are derived based on expected values and thereby known probability distributions.

4.1 DIFFERENTIAL GAMES IN SUPPLY CHAINS

A retailer's ability to collect detailed information about customer purchasing behavior and the ease of changing prices due to new technologies (including Internet and IT) has engendered extensive research into dynamic pricing in general and continuous-time pricing strategies in particular. Increasing attention has been paid to dynamic pricing in the presence of inventory considerations (see, for example, the survey by Elmaghraby and Keskinocak 2003) and to coordinated pricing and production/procurement decisions (see surveys by Chan et. al. 2003; Yano and Gilbert 2002; Cachon 2003). However, despite this range of research interests, relatively few studies are devoted to the continuous interaction between dynamic retail prices, inventory-related costs and wholesale prices in supply chains, i.e., to a dynamic, continuous-time game between supply chain members.

Due to mathematical difficulties inherent in differential games, i.e., games involving decisions that have to be made continuously, the supply chain management literature has been primarily concerned only with the application of deterministic differential models (Cachon and Netessine 2004). Two types of solution approaches have been addressed with respect to the supply chain decision $u(t)$ and state $X(t)$ variables. One is an open-loop solution $u^*=u^*(t)$, which is determined as a function of time. The other, an optimal solution found as a function of state history, $u^*=u^*(t, x(\tau)|0 \leq \tau \leq t$), is referred to as a closed-loop solution. In a special, memory-less case of $u^*=u^*(x(t), t)$, the solution is referred to as a feedback control (for further details, see the appendix to the book). Jorgenson (1986) derives an open-loop Nash equilibrium under static deterministic demand, $d(t)= a(t)-b(t)p(t)$, with demand potential $a(t)$ and customer sensitivity $b(t)$ being constant and thereby not affecting the supply chain dynamics. Eliashberg and Steinberg (1987) use the open-loop Stackelberg solution concept in a game with a manufacturer and a distributor (both with unlimited capacity) involving quadratic seasonal demand potential $a(t)$ and constant sensitivity $b(t)$. Assuming that the wholesale price the manufacturer charges the distributor is constant and that no backlogs are allowed, they investigate the impact of the quadratic seasonal pattern upon the various policies of the distribution channel. They acknowledge that demand uncertainty, together with stock-out costs, may change the results and suggest supplementing the proposed procedure with a sensitivity analysis. Desai (1992) allows demand potential to change with an additional decision variable. To address seasonal demands, he later suggests a numerical analysis for a general case of the open-loop Stackelberg equilibrium under sine form of $a(t)$, constant customer sensitivity $b(t)$ and unlimited manufacturer and retailer capacities (Desai 1996). For more applications of differential games in management science and operations research, we refer the interested reader to a review by Feichtinger and Jorgenson (1983).

In this chapter, we extend the static games considered in Chapter 2 to study various dynamic effects on the supply chain by

- comparing system-wide and equilibrium solutions of dynamic problems with the corresponding solutions of their static prototypes, which we now refer to as myopic solutions that ignore dynamics;
- investigating the effect of system dynamics on vertical and horizontal competition in supply chains under simple demand patterns, demand uncertainty and economy of scale;
- examining the effect of standard (static) as well as dynamic coordinating tools on the performance of dynamic supply chains.

In particular, we find that even though the myopic attitude of a firm is troublesome in many cases, sometimes it remains optimal, as if the problem is static, and sometimes it may even coordinate supply chains. Similarly, standard static coordinating tools in some dynamic conditions result in a perfectly coordinated supply chain. In other cases they are not efficient enough.

We start by considering the effect of learning, with production experience, on vertical pricing and horizontal production competition (Section 4.2). Both static pricing and production games of Chapter 2 are extended with dynamic equations which model production cost reduction as a result of accumulated production experience in economy of scale. In addition to endogenous change in demand, accounted for in the corresponding static games, we assume that the demand for products may evolve gradually with time in an exogenous way as a result of "word of mouth", "customer fatigue", or changes in fashion or the season.

Section 4.3 focuses on inventory competition. In this part of the chapter we discuss two differential games. One game is a straightforward extension of the static pricing game involving the retailer's inventory dynamics. A single supplier and retailer make up a supply chain operating over a production horizon. The supplier sets a wholesale price which is not necessarily constant along the production horizon. In response, the retailer chooses dynamic pricing, production and inventory policies. The need for a dynamic response is due to interaction between a limited processing capacity that features the retailer and exogenous demand peaks which may exceed the capacity. In contrast to the production/pricing games of the first part of this chapter, the exogenous change in demand is instantaneous rather than gradual and is due to special business or high demand periods such as, for example, national holidays and weekends. Such periods are typically affected by the so-called "customer price anticipation" which induces increased price sensitivity. We show that increased price sensitivity, limited processing capacity and available inventory storage lead the retailer to develop sophisticated inventory policies which involve both back-ordering and forward buying. Compared to the static pricing game, these dynamic policies impact the vertical price competition.

As an alternative to the pricing competition with one-side (the retailer's) inventory considerations, the other game discussed in Section 4.3 focuses solely on inventory competition. In this differential inventory game, since the demand is exogenous, pricing has no impact on production. The system consists of one supplier and one retailer. We assume that both the retailer and the supplier have limited capacity. This restriction, along with seasonal demand peaks, induces the supplier and retailer to accumulate inventories and balance production between backlog and surplus inventory costs. Thus, inventory considerations by both sides are involved and the dynamic production policies that the firms employ cause inventory competition which affects the supply chain performance.

Section 4.4 is devoted to two differential games which are extensions of static stocking and outsourcing games. We assume that the demand is random and discuss different forms of subcontracting. One game addresses the question of balancing limited production capacity with an unlimited advance order of end-products. We assume the demand has no peaks; the selling season is short (as in the classical newsvendor problem); and the supply lead-time is long. Therefore, once the season starts, it is too late to outsource the production while in-house capacity can only respond to limited demand fluctuations.

The supply chain involves a single manufacturer and a single supplier (subcontractor) contracting before the selling season starts. The subcontractor sets a wholesale price. In response, the manufacturer selects an order quantity (referred to as advance order) to be delivered by the beginning of the selling season and chooses his production/inventory policy during the season. This description implies that the intertemporal production balancing game is just one of the possible extensions of both the static stocking game and the static outsourcing game (with zero setup cost) considered in Chapter 2. A further extension to these static games as well as to the differential balancing game would be to relax the requirement of only a single advance order contracted out. Such an extension is treated as the differential outsourcing game. In this final intertemporal game of the chapter, production outsourcing is possible at any time point of a production horizon. There are multiple suppliers of limited capacity which determine wholesale prices and a random peak of demand is expected by the end of the production horizon. The manufacturer's goal is to increase capacity to cope with the peak by selecting in-house production, suppliers for outscoring and inventory policies.

The last section of this chapter is devoted to horizontal investment competition in supply chains. The main focus of this section is on feedback equilibrium and cooperation strategies of multiple firms, which co-invest in a supply chain infrastructure.

4.2 INTERTEMPORAL PRODUCTION/PRICING COMPETITION

In this section we consider non-cooperative intertemporal pricing and production games which underlie vertical and horizontal competition in supply chains involved with production experience dynamics.

4.2.1 THE DIFFERENTIAL PRICING GAME

Consider a two-echelon supply chain consisting of a single supplier (manufacturer) selling a product type to a single retailer over a period of time, T. The supplier has ample capacity and can deliver any quantity q at any time t. In contrast to the static model, we assume that the period during which the parties interact is long enough so that the customer demand, which is endogenous in the product price, evolves also over time exogenously. This is to say, we adopt Bertrand's model of pricing competition with the quantity sold per time unit, q, depending not only on product price, p, $\dfrac{\partial q}{\partial p} < 0$ and $\dfrac{\partial^2 q}{\partial p^2} \leq 0$, but also on time t elapsed, $q=q(p,t)$. Therefore, $\dfrac{\partial q(p,t)}{\partial t}$ is not necessarily equal to zero. The exogenous change in demand is due to the interaction of various factors including seasonal fluctuations, fashion trends, holidays, customer fatigue and word of mouth. When the cumulative sales, $\displaystyle\int_0^t q(p(s),s)ds$, i.e., the experience, have little effect on these factors, the dynamic changes can be straightforwardly dealt with by the corresponding price adjustment as in traditional static supply chain models. On the other hand, if production (sales) of large quantities (economy of scale) results in the so-called learning effect, which makes it possible to reduce the unit production cost, $c(t)$, then there is a long-term impact of experience that cannot be studied in the framework of static models.

Let the retailer's price per product unit be $p(t)=w(t)+m(t)$, where $m(t)$ is the retailer's margin at time t and $w(t)$ is the supplier's wholesale price. Then, if both parties, the supplier and the retailer, do not cooperate to maximize the overall profit of the supply chain along period T, their decisions, $w(t)$ and $m(t)$, affect each other's revenues at every point of time, resulting in a differential game. In such a game, the supplier chooses a wholesale price, $w(t)$, at each time point t and the retailer selects a margin, $m(t)$, and thus determines the quantity $q(p,t)$ he will order at price $w(t)$ in order to sell it to his customers at price $p(t)= w(t)+m(t)$. Consequently, the retailer orders $q(p,t)$ products at each time t and the supplier accumulates experience by producing these quantities over time, $\int_0^t q(p(s),s)ds$. As a result, the production cost, $c(t)$, is reduced. We thus have the following problems.

The supplier's problem

$$\max_w J_s(w,m)= \max_w \int_0^T (w(t)-c(t))q(w(t)+m(t),t)dt \qquad (4.1)$$

s.t.

$$\dot{c}(t) = -\gamma q(w(t)+m(t),t), \; c(0)=C \qquad (4.2)$$
$$w(t)\geq c(t), \qquad (4.3)$$

where γ is the learning factor, i.e., the decrease in unit production cost per one more product produced.

The retailer's problem

$$\max_m J_r(w,m)= \max_m \int_0^T m(t)q(w(t)+m(t),t)dt \qquad (4.4)$$

s.t.

$$m(t)\geq 0, \qquad (4.5)$$
$$q(w(t)+m(t),t)\geq 0. \qquad (4.6)$$

Formulations (4.1)-(4.6) assume non-cooperative behavior of the supply chain members which affects the overall supply chain performance. On the other hand, if the supply chain is vertically integrated or centralized, so that a single decision-maker is in charge of all managerial aspects of the supply chain, then we have the following single problem as a benchmark of the best supply chain performance.

The centralized problem

$$\max_{m,w} J(w,m)= \max_{m,w} [J_r(w,m)+J_s(w,m)]=$$

$$\max_{m,w} \int_0^T (w(t)+m(t)-c(t))q(w(t)+m(t),t)dt \qquad (4.7)$$

s.t.

$$(4.2)\text{-}(4.3) \text{ and } (4.5)\text{-}(4.6).$$

We henceforth omit independent variable t wherever the dependence on time is obvious.

System-wide optimal solution

To evaluate the best possible performance of the supply chain, we first study the centralized problem by employing the maximum principle. Specifically, the Hamiltonian for the problem (4.2)-(4.3), (4.5)-(4.6) and (4.7) is

$$H(t) = (w(t)+m(t)-c(t))q(w(t)+m(t),t) - \psi(t)\gamma q(w(t)+m(t),t), \qquad (4.8)$$

where the co-state variable $\psi(t)$ is determined by the co-state differential equation

$$\dot{\psi}(t) = -\frac{\partial H(t)}{\partial c(t)} = q(w(t)+m(t),t), \ \psi(T)=0. \qquad (4.9)$$

Note that since function (4.7) is strictly concave, while all constraints are linear, the maximum principle presents not only necessary but also sufficient optimality conditions and the optimal solution which satisfies these conditions is unique.

The Hamiltonian (4.8) can be interpreted as the instantaneous profit rate, which includes the value $\psi\dot{c}$ of the negative increment in unit production cost created by the economy of scale. The co-state variable ψ is the shadow price, i.e., the net benefit from reducing production cost by one more monetary unit at time t. The differential equation (4.9) states that the marginal profit from reducing the production cost at time t is equal to the demand rate at this point.

From (4.9) we have

$$\psi(t) = -\int_t^T q(w(s)+m(s),s)ds \qquad (4.10)$$

According to the maximum principle, the Hamiltonian is maximized by admissible controls at each point of time. That is, by differentiating (4.8) with respect to $m(t)$ and $w(t)$ and taking into account that $p(t)=w(t)+m(t)$, we have two identical optimality conditions defined by the following equation

$$q(w(t)+m(t),t)+(w(t)+m(t)-c(t)-\psi(t)\gamma)\frac{\partial q(w(t)+m(t),t)}{\partial p(t)}=0,$$

where the shadow price (co-state variable) $\psi(t)$ is determined by (4.10) and the production cost (state variable) $c(t)$ is found from (4.2)

$$c(t) = C - \gamma \int_0^t q(p,s)ds \qquad (4.11)$$

Therefore, as with the static pricing model, only optimal price matters in the centralized problem, $p^* \geq c$, while the wholesale price, $w \geq c$, and the retailer's margin, $m \geq 0$, can be chosen arbitrarily so that $p^* = w + m$. This is due to the fact that w and m represent internal transfers of the supply chain. Thus, the proper notation for the payoff function is $J(p)$ rather than $J(m,w)$ and the only optimality condition is,

$$q(p^*,t) + (p^* - c - \psi\gamma)\frac{\partial q(p^*,t)}{\partial p} = 0. \qquad (4.12)$$

More exactly, p^* is the unique optimal price if it satisfies equation (4.12) and $p^*(t) \geq c(t)$, where c and ψ are determined by (4.11) and (4.10) respectively. Otherwise $p^*(t) = c(t)$ and the supply chain is not profitable at time t.

Let us introduce the maximum price, $P(t)$, at time t, $q(P(t)) = 0$. Naturally assume that $P > c$, then, since, $\psi \leq 0$ (see equation (4.10)), $P > c + \psi\gamma$. Next it is easy to verify that if $p - c - \psi\gamma \geq 0$, then

$$\frac{\partial^2 H}{\partial p^2} = 2\frac{\partial q(p,t)}{\partial p} + (p - c - \psi\gamma)\frac{\partial^2 q(p,t)}{\partial p^2} < 0, \qquad (4.13)$$

and equation (4.12) has an interior solution such that $P > p^* \geq c + \psi\gamma$. This implies that $p^*(t) > c(t)$ does not necessarily hold at each point of time. In such time points the boundary solution $p^*(t) = c(t)$ will be optimal. Comparing the system-wide dynamic optimality condition (4.12) with the optimality condition (2.7) for the corresponding static formulation, we observe that the only difference is due to the product of the shadow price ψ and learning factor γ present in the dynamic formulation. Referring to the static optimal solution at time point t as myopic, since it ignores the future learning effect (the long-run effect γ set at zero) and taking into account that $\psi(t) \leq 0$ for $0 \leq t \leq T$, we find that the myopic attitude leads to overpricing.

Note, that henceforth in the book we distinguish between cases when all supply chain parties have profits at any point of time, $J > 0$ (profitable supply chain) and those when the $J \geq 0$ and thereby the supply chain is sustainable but not necessarily profitable. Similarly, one can characterize separately each party as either profitable or sustainable or as neither of the two.

Proposition 4.1. *In intertemporal centralized pricing* (4.2)-(4.3), (4.5)-(4.6) *and* (4.7), *if the supply chain is profitable, i.e., P>p>c, the myopic retail price will be greater and the myopic retailer's order less than the system-wide optimal (centralized) price and order quantity respectively for* $0 \leq t < T$.

Proof: Comparing (2.7) and (4.12) and employing superscript M for myopic solution we observe that

$$q(p^*,t)+(p^*-c-\psi\gamma)\frac{\partial q(p^*,t)}{\partial p}= q(p^M,t)+(p^M-c)\frac{\partial q(p^M,t)}{\partial p}=0, \text{ (4.14)}$$

while taking into account that $p>c$, $\psi <0$ for $0 \leq t<T$, and $\frac{\partial q}{\partial p}<0$,

$$q(p^M,t)+(p^M-c-\psi\gamma)\frac{\partial q(p^M,t)}{\partial p}<q(p^M,t)+(p^M-c)\frac{\partial q(p^M,t)}{\partial p}=0. \text{ (4.15)}$$

Next, by denoting $f(p)=q(p,t)+(p-c-\psi\gamma)\frac{\partial q(p,t)}{\partial p}$, one can verify

that $\frac{\partial f(p)}{\partial p}<0$.

Thus, from conditions (4.14) and (4.15) we have $f(p^M)<f(p^*)$, which with respect to the last inequality requires that $p^M>p^*$ and, hence, $q(p^M)<q(p^*)$, as stated in Proposition 4.1.

According to Proposition 4.1, myopic pricing derived from static optimization is not optimal. This, however, does not mean that dynamic optimization necessarily leads to time-dependent prices. In other words, an important question is whether the long-term effect of the economy of scale causes the optimal price to evolve with time. It turns out that if the demand does not explicitly depend on time, $q(p,t)=q(p)$, the optimal centralized pricing strategy is independent of time. Otherwise, for example, an exogenous increase in demand monopolistically results in a price increase. This property is stated in the following proposition under the assumption that if $\frac{\partial q(p,t)}{\partial t}<0$, then $\frac{\partial^2 q(p,t)}{\partial p \partial t}\leq 0$ and if $\frac{\partial q(p,t)}{\partial t}>0$, then $\frac{\partial^2 q(p,t)}{\partial p \partial t}\geq 0$.

Proposition 4.2. *In intertemporal centralized pricing* (4.2)-(4.3),(4.5)-(4.6) *and* (4.7), *if the supply chain is profitable, i.e., P>p>c, and there is a demand time pattern q(p,t) such that* $\frac{\partial q(p,t)}{\partial t}$ *exists, then the system-wide optimal price monotonically increases as long as* $\frac{\partial q(p,t)}{\partial t}>0$, *and vice* *versa as long as* $\dfrac{\partial q(p,t)}{\partial t} < 0$. *Otherwise, if* $\dfrac{\partial q(p,t)}{\partial t} = 0$ *at an interval of time, then the system-wide optimal price and order quantity are constant at the interval.*

Proof: Differentiating (4.12), we have

$$\frac{\partial q(p^*,t)}{\partial t} + \frac{\partial q(p^*,t)}{\partial p}\dot{p}^* + (p^*-c-\psi\gamma)[\frac{\partial^2 q(p^*,t)}{\partial p^2}\dot{p}^* + \frac{\partial^2 q(p^*,t)}{\partial p\partial t}] + \dot{p}^*\frac{\partial q(p^*,t)}{\partial p}$$

and thus

$$\dot{p}^*[2\frac{\partial q(p^*,t)}{\partial p} + (p^*-c-\psi\gamma)\frac{\partial^2 q(p^*,t)}{\partial p^2}] = -\frac{\partial q(p^*,t)}{\partial t}$$

$$- (p^*-c-\psi\gamma)\frac{\partial^2 q(p^*,t)}{\partial p\partial t}.$$

Recalling the assumption and (4.13) we readily observe that $\dot{p}^* > 0$ if $\dfrac{\partial q(p^*,t)}{\partial t} > 0$, otherwise, $\dot{p}^* \le 0$.

Game Analysis

We consider now a decentralized supply chain characterized by non-cooperative or competing firms and assume that both players make their decisions simultaneously. The supplier chooses a wholesale price w and the retailer selects a price, p, or equivalently a margin, m, and hence orders $q(p,t)$ products at each $t, 0 \le t \le T$. Since this differential pricing game is deterministic, the retailer sells all the products that he has ordered.

Using the maximum principle for the retailer's problem, we have

$$H(t) = m(t)q(w(t)+m(t),t) - \psi_r(t)\gamma q(w(t)+m(t),t),$$

where the co-state variable $\psi_r(t)$ is determined by

$$\dot{\psi}_r(t) = -\frac{\partial H(t)}{\partial c(t)} = 0, \ \psi_r(T) = 0.$$

Thus, $\psi_r(t) = 0$ for $0 \le t \le T$ and the supplier's production experience does not affect the retailer. This is to say, the myopic pricing is optimal for the non-cooperative retailer and the retailer can simply use the first-order optimality condition to derive pricing strategy for each time point:

$$\frac{\partial J_r(m,w)}{\partial m} = q(w+m,t) + m\frac{\partial q(p,t)}{\partial p} = 0. \tag{4.16}$$

It is easy to verify that since the retailer's objective function is strictly concave in m, (4.16) has a unique solution. Or, by the same token, the retailer's best response function is unique. Comparing (4.12) and (4.16), we conclude that the long-term dynamic effect of production experience causes the supply chain performance to deteriorate even more than in the corresponding static case with no learning.

Proposition 4.3. *In vertical competition of the differential pricing game, myopic pricing is optimal for the retailer. If the retailer and supplier profit at each t, the retail price will be greater and the retailer's order less than the system-wide optimal (centralized) price and order quantity respectively. Moreover, these gaps are even greater than those induced by the corresponding static pricing game.*

Proof: The first statement is due to the fact that $\psi_r = 0$. Employing the fact that $\psi(t) < 0$ for $0 \leq t < T$, the proof of the second statement is similar to that of Proposition 2.1. The last statement of Proposition 4.3 readily results from Proposition 4.1.

Note, that our conclusion that vertical intertemporal pricing competition increases retail prices and decreases order quantities compared to the system-wide optimal solution does not depend on the type of game played. Specifically, it does not depend on whether both players make a simultaneous decision or the supplier first sets the wholesale price and thus plays the role of the Stackelberg leader. As a result, similar to the static pricing game discussed in Chapter 2, the overall efficiency of the supply chain deteriorates under intertemporal vertical competition. Moreover, in addition to the traditional double marginalization effect, we observe the consequence of the learning effect. That is, comparing (4.12) and (4.16), we find that the deterioration of supply chain performance is due to the fact that the retailer myopically ignores not only the supplier's margin, $w-c$, from sales at each time point but also the supplier's profit margin from production cost reduction, $\psi\gamma$. It is because of the latter that the deterioration under dynamic experience in intertemporal supply chain competition is even greater than that which occurs in the static pricing game, as stated in Proposition 4.3. The difference, however, shrinks with time as the shadow price tends to zero by the end of the product production period T.

Equilibrium

To determine the Nash equilibrium which corresponds to the simultaneous moves of the supplier and retailer, we next apply the maximum principle to the supplier's problem. Specifically, we construct the Hamiltonian

$$H(t) = (w(t) - c(t))q(w(t) + m(t), t) - \psi_s(t)\gamma q(w(t) + m(t), t), \quad (4.17)$$

where the co-state variable $\psi_s(t)$ is determined by the co-state differential equation

$$\dot{\psi}_s(t) = q(w(t) + m(t), t), \; \psi_s(T) = 0 . \tag{4.18}$$

Differentiating the Hamiltonian with respect to wholesale price w we have

$$q(p,t) + (w - c - \psi_s \gamma) \frac{\partial q(p,t)}{\partial p} = 0 , \tag{4.19}$$

which implies that an interior optimal solution determined by (4.19) is such that $w - c - \psi_s \gamma > 0$. Next, verifying the second derivative of the Hamiltonian, we find that if $w - c - \psi_s \gamma > 0$, then

$$2 \frac{\partial q(p,t)}{\partial p} + (w - c - \psi_s \gamma) \frac{\partial^2 q(p,t)}{\partial p^2} < 0$$

From equation (4.19) and the last inequality, we observe that (i) although the supplier naturally accounts for his margin from cost reduction with experience, the severe problem of double marginalization persists since the supplier ignores the retailer's margin m; (ii) the intertemporal wholesale price is lower than the myopic wholesale price which is obtained by setting the learning effect γ at zero. The latter implies that the performance of the supply chain further degrades if the supplier adopts a myopic attitude.

It is easy to verify that the supplier's objective function is strictly concave in w and, thus, the supplier's best response (4.19) is unique as well. Thus, the Nash equilibrium (w^n, m^n) is found by solving simultaneously (4.19) and (4.16), which results in

$$w - c - m - \psi_s \gamma = 0 \text{ and } q(c + 2m + \psi_s \gamma, t) + m \frac{\partial q(c + 2m + \psi_s \gamma, t)}{\partial p} = 0 . \tag{4.20}$$

Note that if the second equation of (4.20) has a solution in m, then this solution is such that

$$p = c + 2m + \psi_s \gamma > 0, \; w - c - \psi_s \gamma > 0, \tag{4.21}$$

which however does not ensure that $w = c + m + \psi_s \gamma \geq c$. We conclude with the following result.

Proposition 4.4. *Let ψ_s be determined by (4.18), c by (4,11) and dynamic pair (λ, η) be a solution of system (4.20) in w and m respectively. If $\min\{P - c, \eta\} \geq -\psi_s \gamma$, then the pair $(w^n = \lambda, m^n = \eta)$ constitutes a unique open-loop Nash equilibrium of the differential pricing game with $0 \leq -\psi_s \gamma < m^n < (P - c - \psi_s \gamma)/2 = P - \lambda$.*

Proof: To see that a solution of (4.20) always exists and that it is unique, assume $m^n = 0$ at a point t. Then, since $P(t) > c(t) + \psi_s(t)\gamma$ and $q(P) = 0$,

$q(c+2m^n+\psi_s\gamma,t)>0$, while the second term in the second equation of (4.20) is zero.

Using notation of $f(m^n)$ for the left-hand side of the second equation of (4.20), we find that

$$f(m^n) = q(c+m^n+\psi_s\gamma,t) + m^n \frac{\partial q(c+m^n+\psi_s\gamma,t)}{\partial p} > 0,$$

when $m^n=0$. On the other hand, by letting $c+2m^n+\psi_s\gamma=P$ and accounting for the fact that $q(P,t)=0$, $m^n=(P-c-\psi_s\gamma)/2>0$ and that as a result, the second term of the second equation of (4.20) is strictly negative, we observe that

$f(m^n)<0$. Consequently, taking into account that $\dfrac{\partial f(m^n)}{\partial m^n}<0$, we conclude that the solution of $f(m^n)=0$ is unique and meets the following condition

$$0<m^n<(P-c-\psi_s\gamma)/2.$$

Finally, requiring $m^n \geq -\psi_s\gamma$ and $(P-c-\psi_s\gamma)/2>-\psi_s\gamma$, i.e., $min\{P-c,\eta\} \geq -\psi_s\gamma$, we readily verify that the first equation of (4.20), $w=c+m+\psi_s\gamma$, always has a unique feasible solution as well.

Although, the condition $min\{P-c,\eta\} \geq -\psi_s\gamma$ for the Nash equilibrium is stated in terms of the co-state variable, a sufficient condition can be obtained by assuming the maximum value for the demand $q(c, t)$, i.e.,

$$min\{P(t)-c(t),\eta(t)\} \geq \gamma \int_t^T q(c,s)ds .$$

Note that if c is not replaced with its expression (4.11), then the solution of system (4.20) at time t becomes a function of state variable c, and accordingly can be viewed as closed loop Nash equilibrium.

We next show that similar to the centralized supply chain, a pricing trajectory with respect to the wholesale price and retailer's margin under intertemporal competition is monotonous if the demand time pattern is monotonous. In contrast to the centralized system, where the price p^* barely matters and the only requirement for w and m is $w+m=p^*$, the competition induces not only higher pricing, but also the same rate of change of the margins, $\dot{w}=\dot{m}$. This is shown in the following proposition assuming that conditions of Propositions 4.2 and 4.4. hold.

Proposition 4.5. *In the differential pricing game, if the supply chain is profitable, and there is a demand time pattern $q(p,t)$ such that $\dfrac{\partial q(p,t)}{\partial t}$ exists, then the supplier's wholesale price and the retailer's margin* *monotonically increase at the same rate as long as* $\dfrac{\partial q(p,t)}{\partial t} > 0$, *and they*

decrease as long as $\dfrac{\partial q(p,t)}{\partial t} < 0$. *If* $\dfrac{\partial q(p,t)}{\partial t} = 0$ *at an interval of time, then the Nash equilibrium does not depend on time at the interval.*

Proof: Differentiating both equations of (4.20), we have

$$\dot{w} = \dot{m},$$

$$\frac{\partial q(p,t)}{\partial t} + \frac{\partial q(p,t)}{\partial p}2\dot{m} + m[\frac{\partial^2 q(p,t)}{\partial p^2}2\dot{m} + \frac{\partial^2 q(p,t)}{\partial p \partial t}] + \dot{m}\frac{\partial q(p,t)}{\partial p} = 0$$

and thus

$$\dot{m}[3\frac{\partial q(p,t)}{\partial p} + 2m\frac{\partial^2 q(p,t)}{\partial p^2}] = -\frac{\partial q(p,t)}{\partial t} - m\frac{\partial^2 q(p,t)}{\partial p \partial t}. \qquad (4.22)$$

Taking into account $\dfrac{\partial q}{\partial p} < 0$, $\dfrac{\partial^2 q}{\partial p^2} \le 0$ and $\dot{w} = \dot{m}$, we observe mono-

tonous evolution similar to that obtained for centralized pricing, but with respect to the wholesale price and the retailer's margin.

We next illustrate the results with linear in price demand, $q(p,t)=a(t)-bp$, and the demand potential $a(t)$ first being an arbitrary function of time. Then we plot the solutions for specific supply chain parameters.

Example 4.1.

Let the demand be linear in price with time-dependent customer demand potential $a(t)$, $q(p,t)=a(t)-bp$, $a>bC$. Since the demand requirements, $\dfrac{\partial q}{\partial p} = -b < 0$ and $\dfrac{\partial^2 q}{\partial p^2} = 0$ are met for the selected function, we employ Proposition 4.4 to solve system (4.20), which, for the linear demand, takes the following form:

$$a - b(c + 2m^n + \psi\gamma) - bm^n = 0, \qquad (4.23)$$

$$v^n = c + m^n + \psi\gamma. \qquad (4.24)$$

Using equation (4.22) or, equivalently, by differentiating (4.23) and (4.24) we have

$$\dot{w}^n = \dot{m}^n = \frac{\dot{a}}{3b}$$

and

$$m^n(t) = m^n(T) - \frac{a(T)}{3b} + \frac{a(t)}{3b}, \quad w^n(t) = w^n(T) - \frac{a(T)}{3b} + \frac{a(t)}{3b}.$$

In addition from (4.23) we obtain, $a(T) - bc''(T) - 3bm''(T) = 0$. Thus,

$$m''(T) = \frac{a(T)}{3b} - \frac{c''(T)}{3}.$$

According to (4.24) $w''(T) = c''(T) + m''(T)$, that is,

$$w''(T) = \frac{a(T)}{3b} + \frac{2c''(T)}{3}.$$

Substituting found m'' and w'' into (4.2) we have

$$c''(T) = C - \gamma \int_0^T \{a(t) - b[\frac{2a(t)}{3b} + \frac{c''(T)}{3}]\} dt, \tag{4.25}$$

which results in

$$c''(T) = \frac{3C - \gamma A(T)}{3 - \gamma bT}, \tag{4.26}$$

where $A(T) = \int_0^T a(t) dt$.

Assume that the system parameters are such that the terminal production cost, $c''(T)$, is positive, no matter how experienced the manufacturer becomes, i.e., $\gamma bT < 3$ and $3C > \gamma A(T)$. Consequently, if $\dfrac{a(t)}{3b} \geq \dfrac{3C - \gamma A(T)}{3(3 - \gamma bT)}$, then the Nash equilibrium of the differential pricing game is

$$w''(t) = \frac{a(t)}{3b} + \frac{2(3C - \gamma A(T))}{3(3 - \gamma bT)} \text{ and } m''(t) = \frac{a(t)}{3b} - \frac{3C - \gamma A(T)}{3(3 - \gamma bT)}, \tag{4.27}$$

otherwise at least one of the parties is not always profitable and the equilibrium involves boundary solutions at some intervals of time. Next, the overall price, $m'' + w''$, that the retailer charges and the quantity he orders are

$$p''(t) = \frac{2a(t)}{3b} + \frac{3C - \gamma A(T)}{3(3 - \gamma bT)} \text{ and } q''(t) = \frac{a(t)}{3} - \frac{3C - \gamma A(T)}{3(3 - \gamma bT)} b, \tag{4.28}$$

respectively.

To find the system-wide optimal solution (4.12), which for the linear demand function is determined by the equation

$$a - bp* - (p* - c - \psi\gamma)b = 0, \tag{4.29}$$

we first differentiate it to obtain $\dot{p}* = \dfrac{\dot{a}}{2b}$. Then from (4.29) we have the terminal boundary condition

$$a(T) + bc(T) - 2bp*(T) = 0,$$

that is, $\dfrac{a(T)}{2b} + \dfrac{c(T)}{2} = p*(T)$. Thus, $p* = \dfrac{a}{2b} + \dfrac{c(T)}{2}$. Substituting found centralized solution into (4.2) we have

$$c(T) = C - \gamma \int_0^T a(t) - b(\dfrac{a(t)}{2b} + \dfrac{c(T)}{2})dt , \qquad (4.30)$$

which results in

$$c(T) = \dfrac{2C - \gamma A(T)}{2 - \gamma bT}. \qquad (4.31)$$

Comparing (4.25) and (4.3) and taking into account that $a > bC$, we observe that even if the terminal production costs in the right-hand side of these equations are identical $c(T)=c^n(T)$, the Nash cost $c^n(T)$ in the left-hand side of equation 4.25 is greater than $c(T)$ for the centralized case (equation (4.30)). Consequently, assuming that $\gamma bT<2$ implies $\gamma b<1$ and we have, when comparing (4.26) and (4.31),

$$\dfrac{2C - \gamma A(T)}{2 - \gamma bT} < \dfrac{3C - \gamma A(T)}{3 - \gamma bT}. \qquad (4.32)$$

Then the system-wide optimal price that the retailer charges his customers and the quantity he orders are

$$p*(t) = \dfrac{a(t)}{2b} + \dfrac{2C - \gamma A(T)}{2(2 - \gamma bT)} \text{ and } q*(t) = \dfrac{a(t)}{2} - \dfrac{2C - \gamma A(T)}{2(2 - \gamma bT)} b. \qquad (4.33)$$

Using inequality (4.32), one can immediately observe that both terms of the price-defining equation of (4.33), $\dfrac{a(t)}{2b}$ and $\dfrac{2C - \gamma A(T)}{2(2 - \gamma bT)}$, are smaller than the corresponding terms of the Nash price in (4.28), as stated in Proposition 4.3.

In what follows, we illustrate with Maple the Nash solution (4.28) for specific parameters of the differential pricing game.

Let the demand potential $a(t)$ be exponentially decreasing over time, $a(t)=10e^{-01t}$. The other system parameters are: $b=0.1$, $C=11$, $T=8$, $\gamma=0.05$. We first define the potential $a(t)$ and its cumulative value $A(T)$ with Maple.

```
> a:=10*exp(-0.1*t);
```
$$a := 10\,e^{(-0.1\,t)}$$
```
> A:=int(a, t=0..T);
```
$$A := 100. - 100.\,e^{(-0.1000000000\,T)}$$

Next we determine the Nash wholesale price w; margin m; price p; system-wide optimal price pop; quantity q; shadow price (co-state variable) ψ; and production cost (state variable) c:

```
> w:=a/(3*b)+(2*(3*C-gamma*A))/(3*(3-gamma*b*T));
```

$$w := \frac{10 \, \mathbf{e}^{(-0.1\,t)}}{3 \quad b} + \frac{2\,(3\,C - \gamma\,(100. - 100.\,\mathbf{e}^{(-0.1000000000\,T)}))}{9 - 3\,\gamma\,b\,T}$$

```
> m:=a/(3*b)-(3*C-gamma*A)/(3*(3-gamma*b*T));
```

$$m := \frac{10 \, \mathbf{e}^{(-0.1\,t)}}{3 \quad b} - \frac{3\,C - \gamma\,(100. - 100.\,\mathbf{e}^{(-0.1000000000\,T)})}{9 - 3\,\gamma\,b\,T}$$

```
> p:=2*a/(3*b)+(3*C-gamma*A)/(3*(3-gamma*b*T));
```

$$p := \frac{20 \, \mathbf{e}^{(-0.1\,t)}}{3 \quad b} + \frac{3\,C - \gamma\,(100. - 100.\,\mathbf{e}^{(-0.1000000000\,T)})}{9 - 3\,\gamma\,b\,T}$$

```
> q:=a/3-b*(3*C-gamma*A)/(3*(3-gamma*b*T));
```

$$q := \frac{10}{3}\,\mathbf{e}^{(-0.1\,t)} - \frac{b\,(3\,C - \gamma\,(100. - 100.\,\mathbf{e}^{(-0.1000000000\,T)}))}{9 - 3\,\gamma\,b\,T}$$

```
> psi:=-int(q, t=t..T);
> c:=C-int((gamma*q), t=0..t);
> pop:=a/(2*b)+(2*C-gamma*A)/(2*(2-gamma*b*T));
```

$$pop := \frac{5 \, \mathbf{e}^{(-0.1\,t)}}{b} + \frac{2\,C - \gamma\,(100. - 100.\,\mathbf{e}^{(-0.1000000000\,T)})}{4 - 2\,\gamma\,b\,T}$$

Finally, we substitute the chosen system parameters into the Nash equations and plot the results.

```
> ct:=subs(T=8, C=11, b=0.1, gamma=0.05, c);
> psit:=subs(T=8, C=11, b=0.1, gamma=0.05, psi);
> qt:=subs(T=8, C=11, b=0.1, gamma=0.05, q);
> pt:=subs(T=8, C=11, b=0.1, gamma=0.05, p);
> mt:=subs(T=8, C=11, b=0.1, gamma=0.05, m);
> wt:=subs(T=8, C=11, b=0.1, gamma=0.05, w);
> popt:=subs(T=8, C=11, b=0.1, gamma=0.05, pop);
>plot([pt,mt,wt,popt],   t=0..8,   legend=["Retail
price,  p","Retailer's  margin,  m",  "Supplier's
wholesale price, w", "System-wide opt price p*"]);
```

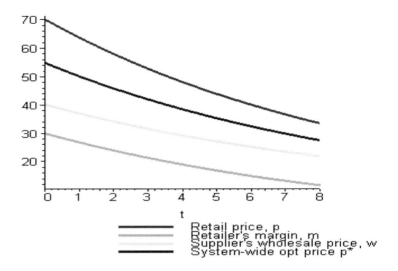

Figure 4.1. Evolution of the retail, wholesale and system-wide optimal prices

```
> plot([psit, ct], t=0..8, legend=["Co-state, psi",
"production cost, c"]);
```

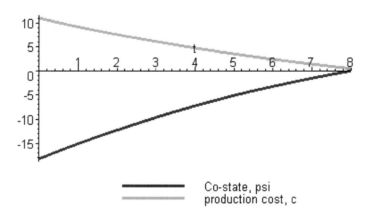

Figure 4.2. Evolution of the co-state and production cost

```
> plot(qt, t=0..8, legend="Demand, q");
```

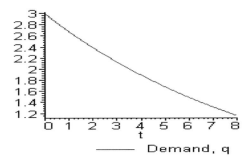

Figure 4.3. Evolution of the order quantity

Coordination

As mentioned above, the negative effect of the intertemporal vertical competition is due to the double marginalization persistent at each time point as in static models and to a dynamic learning effect. It is the learning effect which induces a new margin compared to the corresponding static pricing game. In contrast to the margins from sales, this new margin is gained from reducing production costs. Thus, the deterioration takes place if the retailer ignores both the supplier's profit margin, w-c, and the supplier's margin from cost reduction, $-\psi_s \gamma$. Specifically, recalling that $p=w+m$, the retailer's best response is

$$q(p,t) + m \frac{\partial q(p,t)}{\partial p} = 0,$$

which implies that although the demand depends on two margins, $p=w+m$, and the supplier has a margin from cost reduction, $-\psi_s \gamma$, the retailer takes into account only his margin m rather than ordering with respect to the centralized approach (4.12)

$$q(p^*,t) + (p^*-c-\psi\gamma)\frac{\partial q(p^*,t)}{\partial p} = q(p^*,t) + (w^*+m^*-c-\psi\gamma)\frac{\partial q(p^*,t)}{\partial p} = 0$$

and thus adding the supplier's margins w-c and $-\psi_s\gamma$ to m. At the same time, from equation (4.19)

$$q(p,t) + (w-c-\psi_s\gamma)\frac{\partial q(p,t)}{\partial p} = 0$$

we observe that the supplier ignores the only margin, m, that the retailer has when setting the wholesale price. The question is how to induce the retailer to order more products, or the supplier to reduce the wholesale price, i.e., how to coordinate the supply chain and thus increase its total profit.

It turns out that the two-part tariff approach coordinates supply chains functioning in dynamic conditions as well. However, in contrast to the static models, the two-part tariff allows for different implementation strategies, which do not necessarily result in perfect coordination. That is, an optimal solution under vertical competition may not converge to the system-wide optimal solution. Specifically, if the supplier is the leader, he can set the wholesale price equal to his production cost, but charge the retailer with a fixed (possibly time-dependent) fee. With this dynamic version of the two-part tariff strategy, the supplier induces the retailer to order more products and regulates his share in the total supply chain profit without a special contract.

To show the effect of the dynamic two-part tariff on the supply chain, let the supplier be a leader who first sets the wholesale price $w(t) \equiv c(t)$, then the Hamiltonian of the retailer's problem takes the following form

$$H(t) = m(t)q(c(t)+m(t),t) - \psi_r(t)\gamma q(c(t)+m(t),t), \qquad (4.34)$$

where the co-state variable $\psi_r(t)$ is determined by

$$\dot{\psi}_r = (\psi_r \gamma - m)\frac{\partial q(c+m,t)}{\partial p}, \; \psi(T) = 0. \qquad (4.35)$$

Then the margin the retailer sets is found by differentiating the Hamiltonian with respect to m,

$$q(c+m,t) + (m - \psi_r \gamma)\frac{\partial q(c+m,t)}{\partial p} = 0. \qquad (4.36)$$

Comparing (4.12) and (4.36) we observe that the retailer orders a system-wide optimal quantity if in addition to $w(t) \equiv c(t)$, we have $\psi_r \equiv \psi$ (the retailer's shadow price is identical to the system-wide shadow price), which, with respect to (4.35) and (4.9), cannot hold. This is to say, the retailer accounts for a learning effect with shadow price $\psi_r < 0$ instead of $\psi < 0$. Accordingly, we have found that when setting $w(t) \equiv c(t)$ and charging fixed fees for orders, the supplier eliminates double marginalization and also induces the retailer to partially take into account the supplier's margin from cutting the production cost. However, the optimal retailer's response will never be equal to the system wide optimal solution. The explanation of the two-part tariff's partial efficiency is due to the cumulative memory of dynamics. Repeated setting of $w(t) = c(t)$ during a period of time, transforms the decision or control variable $w(t)$ into a state variable, identical to the state variable $c(t)$ whose dynamic properties are known and which can thus be accounted for by the retailer. This is in contrast to the memoryless static models which does not account for either previous settings or future effects.

We also note that, since $\psi_r(T)=\psi(T)=0$, ψ tends to ψ_r with time. This implies that time has a coordinating effect on the dynamic supply chain which becomes perfectly coordinated with the dynamic two-part tariff by the end of the production period. This passive way, however, is not the only way how the coordination can be improved with the two-part tariff.

An alternative way is to set the wholesale price equal to the system-wide, time-dependent, production cost, $w(t) \equiv c*(t)$. This time-variant two-part tariff strategy implies that the wholesale price is only a function of time rather than only of the learning experience. Consequently, $w(t)$ remains a decision variable and the supply chain can be perfectly coordinated. The disadvantage of this two-part tariff price however, is that since the wholesale price will follow exactly the evolution of the supplier production cost, the retailer may still interpret it as the dynamic two-part tariff. Consequently, the retailer may deviate from the system-wide optimal order quantity at some point of time. To prevent this (sort of) time-inconsistency, the supplier may choose another type of two-part tariff strategy. For example, the supplier, instead of choosing pure strategies with either $w(t) \equiv c(t)$ or $w(t) \equiv c*(t)$, may employ a mixed two-part tariff strategy. With such a strategy, the wholesale price could be selected randomly at constant levels around the production cost $c(t)$ over some fixed intervals of time. In such a case, $w(t)$ is announced as a deterministic function of time, $\hat{w}(t)$, rather than of the learning dynamics or of the optimal production cost and the retailer's optimality condition is reduced to:

$$q(\hat{w}+m,t)+(m-\psi_r\gamma)\frac{\partial q(\hat{w}+m,t)}{\partial p}=0.$$

As long as wholesale prices $\hat{w}(t)$ are not affected by the demand experience, the closer the price $\hat{w}(t)$ to $c(t)$, the more coordinated the supply chain will be. Further, the risk of viewing this strategy as a pure dynamic two-part tariff will also be reduced.

4.2.2 THE DIFFERENTIAL PRODUCTION GAME

In contrast to the previous section devoted to vertical competition, this section focuses on the effect of intertemporal horizontal competition on pricing and production decisions. We consider two manufacturers producing substitutable products over a period of time, T, and competing horizontally for the same customers through a single retailer that they supply. This is to say, the two manufacturers are the suppliers who have ample capacity and are able to deliver any quantity $q_1(t)$ and $q_2(t)$ per time unit to the retailer, where $q_1(t)$ and $q_2(t)$ are the quantities produced by supplier-one and supplier-two at time t respectively. The retailer, on the other hand, adopts the so-called vendor-managed inventory (VMI) policy, which implies that the suppliers decide on the quantities to deliver while the retailer simply charges the suppliers a fixed percentage from sales. In this way, the retailer does not take part in competition and consequently affects neither the system-wide optimal solution nor equilibrium order quantities. In contrast to the previous section, which assumes a Bertrand-Nash equilibrium and similar to the static production game of Chapter 2, Cournot behavior is assumed in the differential production game with the retail price being a function of customer demand. Specifically, the retailer faces the inverse demand function $p=p(Q)$ of total demand rate, $Q=q_1+q_2$. Note that since the products are fully substitutable, the inverse, downward slopping, demand function is symmetric in q_1 and q_2, i.e, $\dfrac{\partial p}{\partial q_1} = \dfrac{\partial p}{\partial q_2} < 0$ and $\dfrac{\partial^2 p}{\partial q_1^{\,2}} = \dfrac{\partial^2 p}{\partial q_2^{\,2}} = \dfrac{\partial^2 p}{\partial q_1 \partial q_2} \le 0$. Furthermore, the period during which the firms operate as a chain is sufficiently long so that the price p depends not only on the quantity sold per time unit Q but also on time t elapsed, $p=p(Q,t)$. In other words, $\dfrac{\partial p}{\partial t}$ is not necessarily equal to zero over the production period.

Similar to the previous section, we assume the suppliers' unit production costs $c_1(t)$ and $c_2(t)$ (state variables) decrease with experience, i.e., with cumulative quantity $\int\limits_0^t p(Q(s),s)ds$ produced. To simplify the discussion, we also assume that the initial production cost is the same for both suppliers, $c_1(0) = c_2(0)=C$, $p(0,t)>C$. This, however, does not ensure that the production costs remain identical over time as the ability to learn with experience can be different for the competing suppliers.

If the suppliers do not cooperate to maximize the overall supply chain profit over period T, their decisions $q_1(t)$ and $q_2(t)$ affect each other's revenues at every point of time resulting in a differential game described by the following problems.

The problem of supplier 1

$$\max_{q_1} J_1(q_1,q_2)=\max_{q_1} \ \beta \int\limits_0^T q_1(t)\big(p(q_1(t)+q_2(t),t)-c_1(t)\big)dt \qquad (4.37)$$

s.t.

$$\dot{c}_1(t) = -\gamma_1 q_1(t), \; c_1(0) = C, \tag{4.38}$$

$$q_1(t) \geq 0, \; p(q_1 + q_2, t) \geq c_1(t), \tag{4.39}$$

where γ_1 is the learning factor of the first supplier, i.e., the decrease in unit production cost per one more product produced, and $1-\beta$ is the percentage paid to the retailer by each manufacturer.

The problem of supplier 2

$$\max_{q_2} J_2(q_1, q_2) = \max_{q_2} \; \beta \int_0^T q_2(t) \big(p(q_1(t) + q_2(t), t) - c_2(t) \big) dt \tag{4.40}$$

s.t.

$$\dot{c}_2(t) = -\gamma_2 q_2(t), \; c_2(0) = C, \tag{4.41}$$

$$q_2(t) \geq 0, \; p(q_1 + q_2, t) \geq c_2(t), \tag{4.42}$$

where γ_2 is the learning factor of the second supplier.

If the supply chain is horizontally integrated, that is, if a single decision maker is in charge, then we have the following centralized problem as a benchmark.

The centralized problem

$$\max_{q_1, q_2} J(q_1, q_2) = \max_{q_1, q_2} [J_1(q_1, q_2) + J_2(q_1, q_2)] =$$

$$\max_{q_1, q_2} \; \beta \int_0^T [q_1(t) \big(p(q_1(t) + q_2(t), t) - c_1(t) \big) + q_2(t) \big(p(q_1(t) + q_2(t), t) - c_2(t) \big)] dt \tag{4.43}$$

s.t.

$$(4.38) \text{ - } (4.42).$$

System-wide optimal solution

To evaluate the best possible performance of the supply chain, we first study the centralized problem by employing the maximum principle. Specifically, the Hamiltonian for problem (4.43), (4.38) - (4.42) is

$$H(t) = \beta[q_1(t) \big(p(q_1(t) + q_2(t), t) - c_1(t) \big) + q_2(t) \big(p(q_1(t) + q_2(t), t) - c_2(t) \big)]$$

$$- \psi_1 \gamma_1 q_1(t) - \psi_2 \gamma_2 q_2(t), \tag{4.44}$$

where it is assumed that since the supply chain is profitable at each point of time and constraints $p(q_1 + q_2, t) \geq c_2(t)$, $p(q_1 + q_2, t) \geq c_1(t)$ are not binding, the co-state variables are determined by the co-state differential equations

$$\dot{\psi}_1(t) = \beta q_1(t), \; \psi_1(T) = 0; \; \dot{\psi}_2(t) = \beta q_2(t), \; \psi_2(T) = 0. \tag{4.45}$$

The Hamiltonian (4.44) can be interpreted as the instantaneous profit rate, which includes the values $\psi_1\dot{c}_1$ and $\psi_2\dot{c}_2$ of the negative increment in the unit production costs of the two suppliers due to the economy of scale. The co-state variable of a supplier is the shadow price, i.e., the net benefit from reducing production cost of the supplier by one more monetary unit at time t. The differential equations of (4.45) state that the marginal profit of a supplier from reducing its production cost at time t is equal to the portion of demand which is sold for the sole gain of the supplier at time t.

From (4.45) we have

$$\psi_1(t) = -\beta \int_t^T q_1(s)ds \text{ and } \psi_2(t) = -\beta \int_t^T q_2(s)ds . \qquad (4.46)$$

According to the maximum principle, the Hamiltonian is maximized by admissible controls $q_1(t)$ and $q_2(t)$ at each point of time. That is, by differentiating (4.44) with respect to $q_1(t)$ and $q_2(t)$, we obtain the following equations for an interior solution

$$\beta[p(q_1(t)+q_2(t),t)-c_1(t)]+\beta(q_1(t)+q_2(t))\frac{\partial p(q_1(t)+q_2(t),t)}{\partial q_1(t)}-\psi_1(t)\gamma_1 =0, (4.47)$$

$$\beta[p(q_1(t)+q_2(t),t)-c_2(t)]+\beta(q_1(t)+q_2(t))\frac{\partial p(q_1(t)+q_2(t),t)}{\partial q_2(t)}-\psi_2(t)\gamma_2 =0, (4.48)$$

where the production cost (state variable) for each supplier is found from (4.38) and (4.41)

$$c_1(t) = C-\gamma_1 \int_0^t q_1(s)ds \text{ ; } c_2(t) = C-\gamma_2 \int_0^t q_2(s)ds . \qquad (4.49)$$

Thus, if a solution, $q_1(t)$ and $q_2(t)$, of the system of two equations (4.47) and (4.48), is such that $q_1(t)\geq 0$, $p(q_1+q_2,t)\geq c_1(t)$, $q_2(t)\geq 0$, $p(q_1+q_2,t)\geq c_2(t)$ holds for $0\leq t\leq T$, then this solution determines the optimal production orders. Otherwise the supply chain may not always be profitable.

To gain further insights into the system-wide optimal solution, we henceforth assume that both suppliers have the same ability to learn, $\gamma_1=\gamma_2$. Then, similar to the static production game, the two problems become symmetric. Consequently, a symmetric solution to the system of equations (4.47)-(4.48), $q^*=q_1^*=q_2^*$, $\psi_1 =\psi_2$, $c_1=c_2=c$, satisfies the following equation

$$\beta[p(2q^*,t)-c]+\beta 2q^*\frac{\partial p(2q^*,t)}{\partial Q}-\psi\gamma=0 . \qquad (4.50)$$

Define the maximum order quantity, Q' (t), so that the supply chain is sustainable, $p(Q',t)=c(t)$. Then, differentiating the left-hand side of equation (4.50) with respect to q, we obtain

$$\frac{\partial^2 H}{\partial q^2} = 4\beta \frac{\partial p}{\partial Q} + 4\beta q \frac{\partial^2 p}{\partial Q^2} < 0. \tag{4.51}$$

This implies that the optimal solution $q*$, $q_1*+q_2*=2q* \in (0,Q')$ defined by (4.50) is unique and the supply chain is profitable if

$$\beta Q' \frac{\partial p(Q',t)}{\partial Q} - \psi\gamma < 0.$$

Assuming that the supply chain is profitable and comparing the system-wide dynamic optimality condition (4.50) with the optimality condition for the corresponding static formulation (2.20), we observe that the only difference is due to the product of the shadow price and learning factor γ present in the dynamic formulation. Consequently, referring to the static (short term) optimal solution at time point t as myopic, and taking into account that $\psi_1(t) = \psi_2(t) = \psi(t) \leq 0$ for $0 \leq t \leq T$ (see equations (4.46)), we find that the myopic approach leads to overpricing.

Proposition 4.6. *In intertemporal centralized production (4.38) - (4.43), if the suppliers make a profit, i.e., $Q'>q>0$, the myopic retail price will be greater and the myopic retailer's order less than the system-wide optimal (centralized) price and order quantity respectively.*
Proof: Comparing (2.20) and (4.50) and employing superscript M for myopic solution we observe that

$$\beta[p(2q*,t)-c]+\beta 2q* \frac{\partial p(2q*,t)}{\partial Q} - \psi\gamma = \beta[p(2q^M,t)-c]+\beta 2q^M \frac{\partial p(2q^M,t)}{\partial Q} = 0, \tag{4.52}$$

while taking into account that $\psi \leq 0$,

$$\beta[p(2q^M,t)-c]+\beta 2q^M \frac{\partial p(2q^M,t)}{\partial Q} < \beta[p(2q^M,t)-c]+\beta 2q^M \frac{\partial p(2q^M,t)}{\partial Q} - \psi\gamma. \tag{4.53}$$

Next, by denoting $f(q) = \beta[p(2q,t)-c]+\beta 2q \frac{\partial p(2q,t)}{\partial Q} - \psi\gamma$, and similar to

(4.51), we require that $\frac{\partial f(q)}{\partial q} < 0$.

Thus, from conditions (4.52) and (4.53) we have $f(q^M)>f(q*)$, which, with respect to the last inequality, requires that $q^M<q*$ and, hence, $p(q^M)>p(q*)$, as stated in this proposition.

According to Proposition 4.6, myopic pricing derived from static optimization is not optimal. This, however, does not mean that the long-term effect of economy of scale causes the optimal price to evolve with time. Specifically, if the demand does not explicitly depend on time, $q(p,t)=q(p)$, the optimal centralized pricing strategy is independent of time. Otherwise, an exogenous increase in prices, for example, naturally results in increased production, as stated in the following proposition under the assumption that if $\dfrac{\partial p(Q,t)}{\partial t} < 0$, then $\dfrac{\partial^2 p(Q,t)}{\partial Q \partial t} \leq 0$ and if $\dfrac{\partial p(Q,t)}{\partial t} > 0$, then

$\dfrac{\partial^2 p(Q,t)}{\partial Q \partial t} \geq 0$.

Proposition 4.7. *In intertemporal centralized production* (4.38) - (4.43), *if the supply chain is profitable, i.e.,* $Q'>q>0$, *and there is a demand time pattern,* $p(Q, t)$, *such that* $\dfrac{\partial p(Q,t)}{\partial t}$ *exists, then the system-wide optimal order quantity monotonically increases as long as* $\dfrac{\partial p(Q,t)}{\partial t} > 0$, *and decreases as long as* $\dfrac{\partial p(Q,t)}{\partial t} < 0$. *Otherwise, if* $\dfrac{\partial p(Q,t)}{\partial t} = 0$, *at an interval of time, then the system-wide optimal order quantity and price are constant at the interval.*

Proof: Differentiating (4.50), we have

$$\beta\frac{\partial p(2q,t)}{\partial t} + \beta\frac{\partial p(2q,t)}{\partial Q}2\dot{q} + \beta 2q\frac{\partial^2 p(2q,t)}{\partial Q^2}2\dot{q} + \beta 2q\frac{\partial^2 p(2q,t)}{\partial Q \partial t} + \beta 2\dot{q}\frac{\partial p(2q,t)}{\partial Q} = 0$$

and thus

$$\dot{q}*[4\beta\frac{\partial p}{\partial Q} + 4\beta q\frac{\partial^2 p}{\partial Q^2}] = -\beta\frac{\partial p}{\partial t} - \beta 2q\frac{\partial^2 p}{\partial Q \partial t}.$$

Recalling (4.51), we readily observe that $\dot{q}* > 0$ if $\dfrac{\partial p}{\partial t} > 0$, otherwise, $\dot{q}* \leq 0$.

Game analysis

Consider now a decentralized supply chain characterized by non-cooperative firms and assume that both suppliers decide on quantities to produce and supply to the retailer simultaneously at each time t. Using the maximum principle we construct the Hamiltonians for each supplier

$$H_1(t) = \beta[q_1(t)(p(q_1(t)+q_2(t),t) - c_1(t)) - \psi_1^1\gamma_1q_1(t) - \psi_2^1\gamma_2q_2(t),$$
$$H_2(t) = \beta[\,q_2(t)(p(q_1(t)+q_2(t),t) - c_2(t))] - \psi_1^2\gamma_1q_1(t) - \psi_2^2\gamma_2q_2(t).$$

Assuming that the supply chain is profitable at each point of time and thus constraints $p(q_1+q_2,t) \geq c_2(t)$, $p(q_1+q_2,t) \geq c_1(t)$ are not binding, the co-state variables are determined by the co-state differential equations

$$\dot{\psi}_1^{\,1}(t) = \beta q_1(t), \; \psi_1^{\,1}(T) = 0 \,; \; \dot{\psi}_2^{\,2}(t) = \beta q_2(t), \; \psi_2^{\,2}(T) = 0 \,, \; \psi_1^2 \equiv \psi_2^1 \equiv 0.$$

Note that in contrast to the centralized formulation, the Hamiltonian of a supplier includes either $\psi_1^1 \dot{c}_1$ or $\psi_2^2 \dot{c}_2$, which is the value of the negative increment in the unit production cost of the supplier due to the economy of scale. Similar to the centralized problem, the co-state variable of a supplier is the shadow price, with a differential equation that states that the marginal profit of the supplier from reducing his production cost at time t is equal to the portion of the demand at this point which is sold for the sole gain of the supplier.

Differentiating the two Hamiltonians with respect to $q_1(t)$ and $q_2(t)$, we find

$$\beta[p(q_1(t)+q_2(t),t) - c_1(t)] + \beta q_1(t) \frac{\partial p(q_1(t)+q_2(t),t)}{\partial q_1(t)} - \psi_1^{\,1}(t)\gamma_1 = 0, \quad (4.54)$$

$$\beta[p(q_1(t)+q_2(t),t) - c_2(t)] + \beta q_2(t) \frac{\partial p(q_1(t)+q_2(t),t)}{\partial q_2(t)} - \psi_2^{\,2}(t)\gamma_2 = 0, \quad (4.55)$$

Thus, if a solution of equations (4.54) and (4.55) is such that $q_1(t) \geq 0$, $p(q_1+q_2,t) \geq c_1(t)$, $q_2(t) \geq 0$, $p(q_1+q_2,t) \geq c_2(t)$, then this solution determines optimal production orders. Otherwise the supply chain may not be profitable at some time intervals and the optimal solution is not always interior with respect to the constraints.

Assuming again that both suppliers have the same ability to learn, $\gamma_1 = \gamma_2$, the two problems become symmetric, $\psi_1^{\,1} = \psi_2^{\,2} = \psi_s$, $c_1 = c_2 = c$. That is, the solution to the system of equations (4.54)-(4.55) is $q = q_1 = q_2$, and it satisfies the following equation

$$\beta[p(2q,t) - c] + \beta q \frac{\partial p(2q,t)}{\partial Q} - \psi_s \gamma = 0. \quad (4.56)$$

Using the arguments similar to those for the centralized optimality condition (4.50), we observe that if the derivative of the left-hand side of (4.56) is negative,

$$3\beta \frac{\partial p}{\partial Q} + 2\beta q \frac{\partial^2 p}{\partial Q^2} + \beta\gamma(T-t) < 0, \quad (4.57)$$

then the solution of (4.56) is unique. From (4.56) we observe by setting γ at zero that myopic suppliers produce less and the retail price is higher than those defined by the intertemporal production model. Assuming in addition to (4.57) that the supply chain is profitable and comparing (4.50) and (4.56), we conclude with the following result.

Proposition 4.8. *In horizontal competition of the differential production game with equal power players, if the suppliers profit at each t, the retail price will be lower and the quantities produced by the suppliers higher than the system-wide optimal price and production quantity respectively. Moreover, these gaps are even greater than those induced by the corresponding static production game.*

Proof: Comparing (4.50) and (4.56) we observe that if $q=q^*$, then $\psi=\psi_s$ and

$$\beta[p(2q,t)-c]+\beta q\frac{\partial p(2q,t)}{\partial Q}-\psi_s\gamma > \beta[p(2q^*,t)-c]+\beta 2q^*\frac{\partial p(2q^*,t)}{\partial Q}-\psi\gamma \quad (4.58)$$

while the derivative of the left-hand side of this inequality with respect to q is negative. Thus, $q>q^*$, which, in regard to the down-slopping price function $p(2q,t)$, means that $p(2q,t)<p(2q^*,t)$. Using the same arguments and accounting for the fact that the co-state variable is negative it is easy to observe that these gaps are even greater than those induced by the corresponding static production game.

Consequently, similar to the static production game, the overall efficiency of the supply chain deteriorates under intertemporal horizontal competition. Moreover, the profit margin from cutting the production cost (learning effect) may even worsen the situation by inducing further price reductions compared to the static (myopic) game. Therefore the suppliers' myopic attitude can become advantageous in terms of system performance.

Nash solution

Similar to Proposition 4.4, it is easy to verify that if (4.57) holds and

$$\beta[p(0,t)-c]\leq\psi_s\gamma,$$

then solution q^n of (4.56) in q is unique and thus the pair (q_1^n,q_2^n), which satisfies $q_1^n=q_2^n=q^n$, constitutes a unique open-loop Nash equilibrium of the differential pricing game with $0<q^n<Q'/2$.

By differentiating (4.56)

$$\beta[\frac{\partial p(2q,t)}{\partial t}+2\frac{\partial p(2q,t)}{\partial Q}\dot{q}+\gamma q]+\beta\dot{q}\frac{\partial p(2q,t)}{\partial Q}+\beta q[\frac{\partial^2 p(2q,t)}{\partial Q\partial t}+2\frac{\partial^2 p(2q,t)}{\partial Q^2}\dot{q}]-\beta q\gamma=0 \quad (4.59)$$

we next show that similar to the centralized supply chain, the production trajectory under intertemporal horizontal competition is monotonous where $\frac{\partial p(Q,t)}{\partial t}$ is monotonous. This is accomplished in the following proposition using the assumption of Proposition 4.7 that signs of $\dfrac{\partial p(2q,t)}{\partial t}$ and $\dfrac{\partial^2 p(2q,t)}{\partial Q \partial t}$ are the same.

Proposition 4.9. *In the differential production game, if the supply chain is profitable, and there is a demand time pattern $p(Q, t)$ such that $\dfrac{\partial p(Q,t)}{\partial t}$ exists, then the suppliers' order quantity monotonically increases as long as $\dfrac{\partial p(Q,t)}{\partial t} > 0$, and decreases as long as $\dfrac{\partial p(Q,t)}{\partial t} < 0$. Otherwise if $\dfrac{\partial p(Q,t)}{\partial t} = 0$ at an interval of time, then the Nash equilibrium does not depend on time at the interval.*

Proof: Rearranging terms in (4.59) we have

$$\dot{q}[3\beta\frac{\partial p(2q,t)}{\partial Q} + 2\beta q\frac{\partial^2 p(2q,t)}{\partial Q^2}] = -\beta\frac{\partial p(2q,t)}{\partial t} - \beta q\frac{\partial^2 p(2q,t)}{\partial Q \partial t}.$$

Recalling (4.57), we immediately observe from this expression that $\dot{q} > 0$ if $\dfrac{\partial p(q,t)}{\partial t} > 0$, otherwise, $\dot{q} \leq 0$.

We illustrate the results with an inverse demand function linear in production quantity.

Example 4.2.

Let price be linear in production quantity with customer demand potential a dependent on time, $p=a(t)-bQ$, $Q=q_1+q_2$, $p(0)=a>c$. Evidently, the requirements, $\dfrac{\partial p}{\partial q_1} = \dfrac{\partial p}{\partial q_2} = -b < 0$ and $\dfrac{\partial^2 p}{\partial q_1^{\,2}} = \dfrac{\partial^2 p}{\partial q_2^{\,2}} = \dfrac{\partial^2 p}{\partial q_1 \partial q_2} = 0$ are met for the selected function and the Nash condition (4.56) takes the following form

$$\beta[p(2q^n,t)-c]+\beta q^n\frac{\partial p(2q^n,t)}{\partial Q} - \psi_s\gamma = \beta(a-3bq^n-c)-\psi_s\gamma = 0.$$

Differentiating this equation, we find the specific form that equation (4.59) transforms into for the linear demand function

$$\beta(\dot{a}-3b\dot{q}^n-\dot{c})-\dot{\psi}_s\gamma = 0,$$

which, when taking into account that

$$\dot{c} = -\gamma q \text{ and } \dot{\psi} = \beta q$$

results in

$$\beta(\dot{a} - 3b\dot{q}'') = 0 .$$

Thus, $\dot{q}'' = \dfrac{\dot{a}}{3b}$ and $q''(t) = q''(T) - \dfrac{a(T)}{3b} + \dfrac{a(t)}{3b}$. Next employing (4.56) at $t=T$, i.e., $a(T) - 3bq''(T) - c(T) = 0$, we obtain

$$q''(t) = \frac{a(t) - c''(T)}{3b} \tag{4.60}$$

and

$$\int_0^t q''(\tau)d\tau = \frac{\displaystyle\int_0^t a(\tau)d\tau - c''(T)t}{3b} .$$

Denoting $A(t) = \displaystyle\int_0^t a(\tau)d\tau$, we find from the dynamic learning equation

$$c''(T) = C - \int_0^T \gamma q(t)dt = C - \frac{\gamma}{3b}(A(T) - c''(T))T .$$

Thus

$$c''(T) = \frac{C - \dfrac{\gamma}{3b}A(T)}{1 - \dfrac{\gamma T}{3b}}$$

and similar to the static production game, the portion of the profit, $1-\beta$, that the suppliers pay to the retailer does not affect the unique Nash solution, which is

$$q''(t) = \frac{a(t)}{3b} - \frac{1}{3b}\frac{C - \dfrac{\gamma}{3b}A(T)}{1 - \dfrac{\gamma}{3b}T} .$$

Assuming that $\gamma T < 3b$ and $\gamma A(T) < C$ to ensure non-negative terminal production cost $c(T)$, one can readily observe that the myopic ($\gamma=0$) Nash production quantity, $\dfrac{a-c}{3b}$, is lower and thus the retail price is higher compared to the corresponding Nash values of the differential game.

Subsequently, the centralized optimality condition (4.50) is

$$\beta[p(2q^*,t) - c] + \beta 2q^* \frac{\partial p(2q^*,t)}{\partial Q} - \psi\gamma = \beta[a - 4bq^* - c] - \psi\gamma = 0,$$

which, when differentiated, leads to

$$\dot{q}* = \frac{\dot{a}}{4b}.$$

Using (4.50) at $t=T$, i.e., $\beta[a(T) - 4bq*(T) - c(T)] = 0$, we find the optimal production quantity

$$q* = \frac{a - c(T)}{4b}. \qquad (4.61)$$

Comparing this system-wide optimal solution with the Nash quantity (4.60), we observe that $q* < q^n$ even if the terminal production cost is the same for both cases. However, if $q* < q^n$, then according to the dynamic learning equation, the Nash terminal production cost, $c^n(T)$, must be less than the system-wide production cost $c(T)$ which makes the inequality $q* < q^n$ even stronger (compare (4.60) and (4.61)). This sustains the fact that the supply chain performance deteriorates under horizontal competition.

Moreover, comparing $q*$ with the myopic Nash solution, $\dfrac{a - c}{3b}$, we observe that $q* = \dfrac{a - c(T)}{4b} \leq \dfrac{a - c}{3b}$ holds for any t if $\dfrac{a - C}{3} \geq \dfrac{a - c(T)}{4}$. This inequality, if for example $4C \leq a$, always holds and accordingly a myopic attitude of the suppliers is beneficial for the supply chain performance.

Consequently, substituting found $q*$ into the dynamic learning equation, we obtain

$$c(T) = C - \int_0^T \gamma q(t)dt = C - \frac{\gamma}{4b}(A(T) - c(T))T.$$

Thus

$$c(T) = \frac{C - \dfrac{\gamma}{4b}A(T)}{1 - \dfrac{\gamma T}{4b}} \quad \text{and} \quad q*(t) = \frac{a(t)}{4b} - \frac{1}{4b}\frac{C - \dfrac{\gamma}{4b}A(T)}{1 - \dfrac{\gamma}{4b}T}.$$

We next illustrate the Nash solution found in this example for the following system parameters: $a(t) = 15*e^{-0.1t}$, $T=8$, $C=9$, $b=0.2$, $\gamma=0.05$, $\beta=0.1$.

First, we define the demand potential $a(t)$ and its cumulative value over the production period T, $A(T)$

```
> a:=15*exp(-0.1*t);
```

$$a := 15\,e^{(-0.1\,t)}$$

```
> A:=int(a, t=0..T);
```

$$A := 150. - 150.\,e^{(-0.1000000000\,T)}$$

Consequently, the Nash solution is calculated in terms of production quantity q; retail price p; production cost c; and shadow price ψ. The chosen system parameters are then substituted and the results are plotted along with the system-wide optimal solution.

```
>q:=a/(3*b)-(1/(3*b))*(C-gamma*A/(3*b))/(1-gamma*   /
(3*b));
>qs:=a/(4*b)-(1/(4*b))*(C-gamma*A/(4*b))/(1-amma*T/
(4*b));
>p:=a-b*2*q;
>ps:=a-b*2*qs; (system-wide optimal price)
>qt:=subs(T=8, C=9, b=0.2, gamma=0.05, q);
>pt:=subs(T=8, C=9, b=0.2, gamma=0.05, p);
>c:=C-gamma*int(qt, t=0..t);
>ct:=subs(T=8, C=9, b=0.2, gamma=0.05, beta=0.1, c);
>plot([pt,ct,pst], t=0..8, legend=["Retail price,
p","Production cost, c1=c2=c", "System-wide price
p*"]);
```

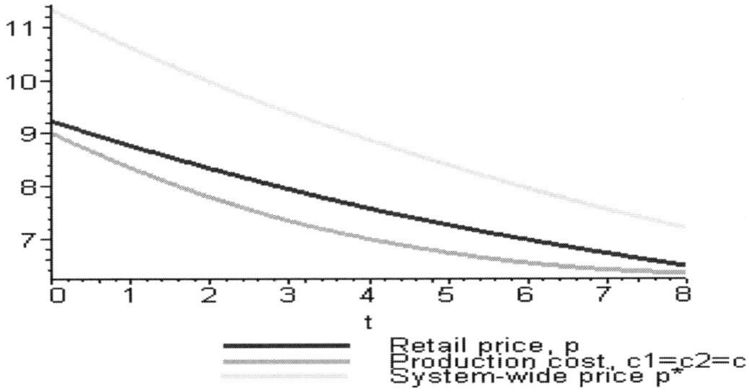

Figure 4.4. Evolution of the production cost, retail and system-wide optimal prices

```
>psi:=-beta*int(qt, t=t..T);
>psit:=subs(T=8,C=21,b=0.2,gamma=0.05,beta=0.1,psi);
>plot([psit, qt, qst],t=0..8, legend=["Co-state, psi",
"productionquantity,q","system-widequantity q*"]);
```

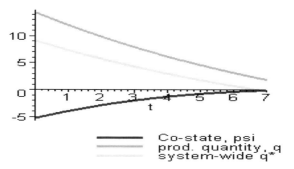

Figure 4.5. Evolution of the co-state, production and system-wide optimal order quantity

The effect of different learning abilities

Different learning abilities introduce an asymmetry into the choice of production quantities. If the supply chain is centralized, then it does not make sense to use an inefficient manufacturer. Therefore, the system-wide optimal solution will involve only the fastest learning supplier. Formally, this is found by adding the Lagrange multiplier $\mu(t) \geq 0$ to the optimality conditions. Then $Q=q_1>0$ and $q_2=0$ are defined by the optimality conditions (4.47)-(4.48), which take the following form

$$\beta[p(Q(t),t)-c_1(T)]+\beta Q(t)\frac{\partial p(Q(t),t)}{\partial Q}=0,$$

$$\beta[p(Q(t),t)-c_2(T)]+\beta Q(t)\frac{\partial p(Q(t),t)}{\partial Q}+\mu(t)=0.$$

From these equations we observe that if $\beta[p(Q(t),t)-c_1(T)]+$ $\beta Q(t)\frac{\partial p(Q(t),t)}{\partial Q}=0$ and $c_1(T)<c_2(T)$, then $\beta[p(Q(t),t)-c_2(T)]+$ $\beta Q(t)\frac{\partial p(Q(t),t)}{\partial Q}<0$ and thus $\mu>0$. That is, the suggested boundary solution is always feasible if $\gamma_1>\gamma_2\geq0$. On the other hand, if $c_1(T)>c_2(T)$, then $\mu<0$ and the solution is infeasible.

To illustrate the change in the interior Nash solution, we let $\gamma_1=\lambda\gamma_2$ and differentiate (4.54) and (4.55) to find that

$$\dot{q}_1[2\frac{\partial p}{\partial Q}+q_1\frac{\partial^2 p}{\partial Q^2}]=-\frac{\partial p}{\partial t}-q_1\frac{\partial^2 p}{\partial Q\partial t}$$

$$\dot{q}_2 [2\frac{\partial p}{\partial Q} + q_2 \frac{\partial^2 p}{\partial Q^2}] = -\frac{\partial p}{\partial t} - q_2 \frac{\partial^2 p}{\partial Q \partial t}.$$

This implies that although all properties of Proposition 4.9 hold, the dynamic changes in production quantities are not necessarily symmetric, $\dot{q}_1{}^n = \dot{q}_2{}^n$, unless either $\frac{\partial p}{\partial t} = 0$ (then simply $\dot{q}_1{}^n = \dot{q}_2{}^n = 0$), or $q_1{}^n = q_2{}^n$.

Furthermore, according to the optimality equations (4.54) and (4.55), the latter condition cannot hold for different learning abilities.

Assuming the former condition holds, $\frac{\partial p}{\partial t} = 0$, thereby $\dot{q}_1{}^n = \dot{q}_2{}^n = 0$, then $q_1{}^n(t) = q_1{}^n$, $q_2{}^n(t) = q_2{}^n$ and $Q(t)=Q$ for $0 \le t \le T$ and considering (4.54) and (4.55) at $t=T$ we have:

$$\beta[p(Q(T),T) - c_1(T)] + \beta q_1 \frac{\partial p(Q(T),T)}{\partial Q} = 0, \qquad (4.62)$$

$$\beta[p(Q(T),T) - c_2(T)] + \beta q_2 \frac{\partial p(Q(T),T)}{\partial Q} = 0. \qquad (4.63)$$

Then taking into account $Q(t)=Q$, $\frac{\partial p}{\partial t} = 0$, as well as the learning equations (4.38) and (4.41), we obtain

$$q_1{}^n = \frac{p(Q)}{\gamma_2 T \lambda - \frac{\partial p(Q)}{\partial Q}} \quad \text{and} \quad q_2{}^n = \frac{p(Q)}{\gamma_2 T - \frac{\partial p(Q)}{\partial Q}}.$$

Thus, in contrast to the system-wide optimal solution, if supplier one learns faster, $\lambda > 1$, then he produces less, $q_1{}^n < q_2{}^n$. From system (4.62)-(4.63) we also find

$$[q_2{}^n - q_1{}^n] \frac{\partial p(Q)}{\partial Q} = c_2(T) - c_1(T).$$

On the other hand, from learning equations (4.38) and (4.41), we obtain

$$c_2(T) - c_1(T) = \gamma_2 \int_0^T \lambda q_1(t)dt - \int_0^T \gamma_2 q_2(t)dt = \gamma_2 T(\lambda q_1{}^n - q_2{}^n).$$

Thus, $[q_2{}^n - q_1{}^n] \frac{\partial p(Q)}{\partial Q} = \gamma_2 T(\lambda q_1{}^n - q_2{}^n)$ and, since $q_1{}^n < q_2{}^n$, we conclude

that $q_2{}^n > \lambda q_1{}^n$. That is, the total production of supplier two will be more than λ times greater than that of supplier one and the greater the learning inequality of the suppliers the stronger the deterioration of the supply chain performance.

Coordination

Although consumers benefit from the non-cooperative behavior of the suppliers, which leads to a fall in retail prices and an increase in production as well as in consumption of the product, the horizontal competition has a negative effect on supply chain profits. As with the double marginalization effect, this happens because each manufacturer, when deciding on the quantity to produce, ignores the quantity which the other manufacturers produce, i.e., because of "double quantification". The major way to coordinate such a chain is to cooperate. The suppliers, by simply agreeing to set simultaneously their production quantities equal to the system-wide optimal quantity, rather than to the non-cooperative Nash quantities, will perfectly coordinate the supply chain as well as increase their profits with equal shares without any internal supply chain transfers. Transfers, however, will be needed if the cooperating suppliers differ in their ability to learn with production experience.

Another interesting point is the effect of myopia on horizontal competition in supply chains. The profit margin $\psi_s \gamma$ from cutting the production cost induces the suppliers to produce more, which decreases the prices compared to both the corresponding static (myopic) and system-wide optimal solutions. As a result, myopic suppliers, even if they are not able to cooperate, may make the supply chain more coordinated. This is in contrast to vertically competing firms where myopic behavior only worsens the performance of supply chains.

Finally, it is worth mentioning that even if no coordinating actions are exercised, the coordination improves with time as the shadow price of cutting the production cost tends to zero by the end of the production period T in both centralized and decentralized supply chains. By the same token, the less time that remains until the end of production period, the smaller the difference between an intertemporal model and the corresponding static model.

4.3 INTERTEMPORAL INVENTORY GAMES

The two previous sections were devoted to intertemporal production and pricing games with production learning dynamics. This section addresses the effect of inventory dynamics. We consider two types of games. One models pricing competition which accounts only for the retailer's dynamics. The other, which considers the supplier's and the retailer's inventory dynamics, focuses on vertical inventory competition.

4.3.1 THE DIFFERENTIAL INVENTORY GAME
WITH ENDOGENOUS DEMAND

In this section we consider a pricing game in a two-echelon supply chain consisting of a single supplier (manufacturer) selling a product type to a single retailer over a period of time, T. The supplier has ample capacity and can deliver any quantity at any time point. Similar to the pricing game of Section 4.2.1, the demand changes exogenously with time. In contrast to the pricing game considered in Section 4.2.1, the product type is characterized by a long lifecycle so that after a limited time period, the learning effect is negligible. Instead, the retailer has a finite processing capacity which may induce inventory accumulation to prepare for the time intervals where the demand exceeds processing capacity. The focus, then, in this section is on the effect of inventory dynamics rather than production experience on the supply chain.

Another important distinction from the differential pricing game is the demand pattern. In contrast to gradual evolution considered in the differential pricing game, we assume jumps in demand due to special business conditions such as national holidays. Furthermore, special business conditions frequently impact not only demand potential, but also customer sensitivity to prices. Empirical studies show that consumers are more price-sensitive during periods of high demand such as Christmas, Thanksgiving and weekends (see, for example, Chevalier et. al 2003; Bils, 1989; and Warner and Barsky 1995). The best response to such an instantaneous change often involves a limited-time promotion. In the UK, for example, Christmas sales of consumer electronics may reach up to 40% of the annual sales. A promotion during higher customer sensitivity can cause customers to buy more than they usually would, indeed, more than they would normally buy even during a regular promotion.

To illustrate this phenomenon and simplify further presentation, we will assume linear in price demand $d(p,t)$. One can view demand $d(p,t)$ for a product as a function of the current product price $p(t)$, the list price P and the customer price sensitivity $b(t)$, $d(p,t)=g(t)+b(t)(P-p(t))$, where $g(t)$ is the demand under anticipated list pricing, $p(t)=P$. Then, by denoting the demand potential, $a(t)=g(t)+b(t)P$, we observe that this function is equivalent to the standard linear demand function, $d(p,t)=a(t)-b(t)p(t)$ considered in most examples presented in this book. This is to say, if customer sensitivity $b(t)$ increases during a limited-time promotion, the demand potential $a(t)=g(t)+b(t)P$ may increase as well, even if $g(t)$ remains unchanged. This also implies that sales during a period of increased customer sensitivity and, as a result, increased demand elasticity, may become more efficient than those offered during regular times. For example, if customer price sensitivity $b(t)$ increases during a limited-time period by K units and $p(t) \leq P$, then the positive increment in demand, $b(t)(P\text{-}p(t))$, includes K additional product units for each dollar discounted in price $p(t)$ compared to sales offered at other times.

Although large manufacturers traditionally dominate trade deals, retailers armed with extensive data about profitability, product movement, and customer demand for a class of goods are developing sophisticated purchase and storage policies to take advantage of the trade promotions that manufacturers offer. A retailer, for instance, may engage in "forward buying", that is, purchasing more goods during a promotional period than he expects to sell (Zerrillo and Iacobucci 1995). In this section we derive such policies and provide formal rationales for complex purchase and inventory policies under increased customer sensitivity.

We assume there is a leader – a supplier or wholesaler – and a follower – the retailer. When the supplier sets a wholesale price, the retailer commits to purchase a certain quantity. Both desire to maximize their profits. The contract between these players is of the rolling-horizon type which implies that purchase orders can be periodically updated within certain limitations. If demand as well as the supply chain parameters is steady, then there is a static Stackelberg solution to this two-player game. However, if the demand changes, the Stackelberg strategy becomes dynamic.

Our notations include $X(t)$, which is the retailer's inventory level at time t, a state variable; $u(t)$ is the order quantity processed by the retailer at time t (processing rate), which can also be viewed as the quantity $q(t)=u(t)$ that the supplier delivers to the retailer at t, a decision (control) variable; and U is the retailer maximum processing rate. Other notations are similar to those employed in the previous sections: $p(t)$ is the retail price at time t, a decision (control) variable; $w(t)$ is the unit wholesale price charged by the supplier, a decision variable; h^+, h^- are the product unit holding and backlog costs respectively incurred per time unit by the retailer; and c_r, c_s are the product unit processing costs incurred by the retailer and supplier respectively.

A typical rolling horizon contract between a supplier and a retailer implies an infinite planning horizon and a minimum period, T, which characterizes the contract. During T, mutual supplier-retailer commitments cannot be revised. Specifically, the supplier sets a constant wholesale price for a period, T. In response, the retailer commits to order fixed quantities with minor variations to cope with demand fluctuations within the period. If the demand is steady, this type of supply chain results in a steady-state that the commitments determine, i.e., a constant wholesale price, as well as a constant retailer order quantity, inventory level and product price. This steady-state can be disrupted if a limited-time promotional sale is initiated.

With respect to this initiative, the supplier is expected to reduce the wholesale price from w_1 to w_2 for the promotional period of time $[t_s, t_f]$ to boost sales, i.e.,

$$w(t) = \begin{cases} w_1, & t < t_s \text{ and } t \geq t_f \\ w_2, & t_s \leq t < t_f \end{cases} ; \; w_1 \geq w_2. \tag{4.64}$$

This promotion is commonly coordinated with the retailer who also drops prices and increases order quantities. As a result, the chain is in a transient-state for a period of time comprising the interval, $[t_s, t_f]$. Furthermore, since the promotion dates are either advertised or coincide with especially sensitive seasons (e.g. holidays), the price sensitivity of the customers, $b(t)$, during these dates increases:

$$b(t) = \begin{cases} b_1, & t < t_s \text{ and } t \geq t_f \\ b_2, & t_s \leq t < t_f \end{cases}, \; b_2 \geq b_1.$$

As mentioned above, this increase in price sensitivity increases the demand potential $a(t)$ during the promotion as well,

$$a(t) = \begin{cases} a_1, & t < t_s \text{ and } t \geq t_f \\ a_2, & t_s \leq t < t_f \end{cases}.$$

That is, if $b_2 > b_1$, then from $a_1 = g + b_1 P$ and $a_2 = g + b_2 P$, we have $a_2 > a_1$. Since the effect of the customer sensitivity on demand potential is not necessarily linear, we relax the linearity and employ a more general assumption

$$\frac{a_1}{b_1} > \frac{a_2}{b_2},$$

which ensures that the demand elasticity, $-\dfrac{\partial d(t)}{\partial p(t)} \dfrac{p(t)}{d(t)} = \dfrac{p(t)}{a(t)\big/ b(t) - p(t)}$,

and thus the efficiency of price cuts increases. Note, that this assumption is always met for any linear function $a(t) = g + b(t)P$, if $b_2 > b_1$.

The effect of an increase in customer sensitivity occurs only if the promotional time interval, $[t_s, t_f]$, is much shorter than the regular contract period T, which is typically the case with limited-time promotions as well as national holidays. Therefore we consider a period of time $[0, T]$ such that the supply chain which was in the steady-state at the beginning of the period will have enough time after the promotional interval to return to this state by time T.

The supplier's problem

Since we assume that the supplier has ample capacity, his dynamics are straightforward: produce (supply) exactly according to retailer orders $u(t)$ to maximize expected profits by choosing regular, w_1, and promotional, w_2, wholesale prices:

$$\max_{w} J_s(w,u,p) = \max_{w} \int_0^T [w(t)u(t) - c_s u(t)]dt \qquad (4.65)$$

s.t.

$$w(t) \geq c_s, \qquad (4.66)$$

where the first term in the objective function (4.65) presents wholesale revenues over time and the other term presents supplier processing costs over time.

The retailer's problem

The retailer also wants to maximize profit by selecting proper order quantities and product prices $\{u(t), p(t), 0 \leq t \leq T\}$

$$\max_{u,p} J_r(w,u,p) = \max_{u,p} \int_0^T [p(t)(a(t)-b(t)p(t)) - c_r u(t) - w(t)u(t) - h(X(t))]dt \,(4.67)$$

s.t.

$$\dot{X}(t) = u(t) - (a(t) - b(t)p(t)); \qquad (4.68)$$
$$0 \leq u(t) \leq U; \qquad (4.69)$$
$$a(t)-b(t)p(t) \geq 0; \qquad (4.70)$$
$$p(t) \geq 0, \qquad (4.71)$$

where the first term in the objective function (4.67) presents revenues of the retailer from the sales $d(t)=a(t)-b(t)p(t)$; the second term reflects retailer processing costs; and the third is the cost of purchasing from the supplier at the wholesale price. The last term in (4.67) accounts for inventory costs $h(X(t))=h^+X^+(t)+h^-X^-(t)$, $X^+(t) = \max\{X(t),0\}$ and $X^-(t) = \max\{-X(t),0\}$, which are due to the bounded processing capacity (4.69) of the retailer. With respect to the inventory balance equation (4.68), if the cumulative processing rate at time t is greater than the cumulative demand at t, then the inventory holding cost is incurred at t, $h^+X(t)$, otherwise the backlog cost $h^-X(t)$ is incurred.

In this section we use the Stackelberg solution concept to solve the supplier and retailer problems with the supplier acting as the leader and the retailer acting as the follower. On the other hand, if the supply chain is vertically integrated or centralized so that a single decision maker is in charge, then we have the following single problem as a benchmark of the best supply chain performance.

The centralized problem

$$\max_{u,p} J_r(u,p) = \max_{u,p} \int_0^T [p(t)(a(t)-b(t)p(t))-c_r u(t)-c_s u(t)-h(X(t))]dt \quad (4.72)$$

s.t.

$$(4.68) - (4.71).$$

Objective function (4.72) is obtained by summing the retailer's and supplier's objective functions which eliminates the wholesale price since it represents transfer within the supply chain. In what follows we will distinguish between two different types of solutions: steady-state and transient-state. The former is related to the case when the dynamic production conditions of the supply chain transform into static conditions and, correspondingly, the differential pricing and inventory game into a static game. The latter is related to conditions when the supply chain behavior cannot be static.

System-wide optimal solution: steady-state conditions

We start off by considering static conditions and thus the centralized solution when the supply chain is in a steady-state characterized by constant wholesale prices, retailer orders and inventory levels which are naturally kept at zero level in such a case. This implies that we consider a sub-period during which customer sensitivity remains unchanged and no promotion initiative is expected.

We derive the optimal solution by maximizing the Hamiltonian of the centralized problem

$$H(t) = p(t)(a(t)-b(t)p(t))-(c_r+c_s)u(t)-h(X(t))+\psi(t)(u(t)-a(t)+b(t)p(t))$$

with respect to the retail price $p(t)$ and processing rate $u(t)$, where the co-state variable $\psi(t)$ is determined by the co-state differential equation

$$\dot{\psi}(t) = \begin{cases} h^+, & \text{if } X(t) > 0; \\ h^-, & \text{if } X(t) < 0; \\ h \in [-h^-, h^+], & \text{if } X(t) = 0. \end{cases} \quad (4.73)$$

Similar to the previous sections, the Hamiltonian can be interpreted as the instantaneous profit rate, which includes the profit $\psi\dot{X}$ from the increment in inventory level of the retailer created by processing u and pricing p.

The co-state variable is the shadow price, i.e., the net benefit from reducing inventory surplus/shortage by one more unit at time t. The differential equation (4.73) states that the marginal profit of the supply chain from reducing its inventory level at time t if there is a surplus at t, $X>0$ (or from reducing inventory shortage, if $X<0$) is equal to the product unit holding cost per time unit (or unit shortage cost , if $X<0$).

If the supply chain system is at the same steady-state at $t=0$ and $t=T$, i.e., it is characterized by steady demand potential $a(0)=a(T)$; customer sensitivity $b(0)=b(T)$; wholesale price $w(0)=w(T)$; and retailer's inventory state $X(0)=X(T)$. Then the co-state variable must also be the same at these points of time:

$$\psi(0) = \psi(T). \tag{4.74}$$

Maximizing the Hamiltonian with respect to $p(t)$, i.e., considering

$$H_p = p(t)(a(t) - b(t)p(t)) + \psi(t)b(t)p(t),$$

subject to (4.70) and (4.71) we readily find

$$p = \begin{cases} \dfrac{a}{b}, & \text{if } a + b\psi > 2a; \\[2mm] \dfrac{a + b\psi}{2b}, & \text{if } 0 \le a + b\psi \le 2a; \\[2mm] 0, & \text{if } a + b\psi < 0. \end{cases} \tag{4.75}$$

Similarly, by maximizing the $u(t)$-dependent part of the Hamiltonian,

$$H_u = (-c_r - c_s + \psi(t))u(t),$$

subject to (4.69), we find

$$u = \begin{cases} U, & \text{if } \psi > c_r + c_s; \\ 0, & \text{if } \psi < c_r + c_s; \\ a - bp, & \text{if } \psi = c_r + c_s. \end{cases} \tag{4.76}$$

Note, that the third condition in (4.76), which presents the case of an intermediate processing rate, is obtained by differentiating the singular condition,

$$\psi(t) = c_r + c_s,$$

along an interval of time where it holds. Then, by taking into account (4.73), we conclude that this condition holds only if $X=0$ along the interval, i.e., $u=d=a-bp$. Furthermore, this singular condition is feasible if in addition to the constraints (4.69)-(4.71), we have

$$d=a-bp \le U. \tag{4.77}$$

Consider a sub-period of time $\tau = [\hat{t}, \breve{t}] \subseteq [0,T]$ characterized by no-promotion, so that customer sensitivity $b(t)=b_1$ and potential $a(t)=a_1$ remain constant for a period of time, $t \in \tau$, rather than identical only at $t=0$ and $t=T$ as imposed by (4.74). As shown in the following proposition, if $X(0)=0$ this requirement implies that the dynamic system exhibits a static behavior characterized by constant retailer pricing and processing rates as well as zero inventory levels.

Proposition 4.10. *If $b(t)=b_1$, $a(t)=a_1$, , for $t \in \tau$, $\tau \subseteq [0,T]$, $X(\hat{t})=0$ and $0 \leq a_1 - b_1(c_r + c_s) \leq 2U$, then $X(t)=0$ for $t \in \tau$, and the system-wide optimal processing and pricing policies are:*

$$u*(t) = \frac{a_1 - b_1(c_r + c_s)}{2} \text{ and } p*(t) = \frac{a_1 + b_1(c_r + c_c)}{2b_1} \text{ for } t \in \tau,$$

respectively.

Proof: Consider the following solution for the state, co-state and decision variables:

$$X(t)=0, \ \psi(t) = c_r + c_s, \ p(t) = \frac{a_. + b_1(c_r + c_c)}{2b_1}, \ u(t) = \frac{a_1 - b_1(c_r + c_s)}{2} \text{ for } t \in \tau.$$

It is easy to observe that this solution satisfies the optimality conditions (4.74) - (4.76). Furthermore, this solution is always feasible if conditions (4.70) and (4.77) hold which is ensured by $0 \leq a_1 - b_1(c_r + c_s) \leq 2U$, as stated in the proposition. Finally, the centralized objective function involves only concave and piece-wise linear terms, which implies that the maximum-principle based optimality conditions are not only necessary, but also sufficient.

System-wide optimal solution: transient-state conditions

Transient-state conditions do not introduce much sophistication into the centralized supply chain. Indeed, it is easy to verify that if the change in demand parameters is such that $0 \leq a_2 - b_2(c_r + c_s) \leq 2U$ holds, then instantaneous change in customer sensitivity does not affect the form of the solution presented in Proposition (4.10). The price and the processing rate are simply adjusted to the changes as stated in the following proposition.

Proposition 4.11. *If $b(t)=b_1$, $a(t)=a_1$, for $t < t_s$, $t \geq t_f$, $X(\hat{t})=0$, $0 \leq a_1 - b_1(c_r + c_s) \leq 2U$, and $b(t)=b_2$, $a(t)=a_2$, for $t \geq t_s$, $t < t_f$, $0 \leq a_1 - b_1(c_r + c_s) \leq 2U$, then $X(t) \equiv 0$, and the system-wide optimal processing and pricing policies are:*

$$u*(t) = d*(t) = \frac{a_2 - b_2(c_r + c_s)}{2} \text{ and } p*(t) = \frac{a_2 + b_2(c_r + c_c)}{2b_2} \text{ for } t \geq t_s,$$

$t < t_f$ *respectively.*

Proof: The proof is very similar to that of Proposition 4.10.

Comparing statements of Propositions 4.11 and 4.10, we find that under our assumption, $\dfrac{a_1}{b_1} > \dfrac{a_2}{b_2}$, the optimal response of the centralized supply chain to increased customer price sensitivity for a period of time is a promotion during this interval. Denoting

$$u_1 = \frac{a_1 - b_1(c_r + c_s)}{2}, u_2 = \frac{a_2 - b_2(c_r + c_s)}{2},$$

$$p_1 = \frac{a_1 + b_1(c_r + c_c)}{2b_1}, p_2 = \frac{a_2 + b_2(c_r + c_c)}{2b_2},$$

one can straightforwardly verify the following statements.

Proposition 4.12. *If $b(t)=b_1$, $a(t)=a_1$, for $t < t_s$, $t \geq t_f$, $X(\widehat{t})=0$, $0 \leq a_1 - b_1(c_r + c_s) \leq 2U$, and $b(t)=b_2$, $a(t)=a_2$, for $t \geq t_s$, $t < t_f$, $0 \leq a_1 - b_1(c_r + c_s) \leq 2U$, $\dfrac{a_1}{b_1} > \dfrac{a_2}{b_2}$, then the system-wide optimal price decreases, while the demand and processing rate increase during transient period $t \geq t_s$, $t < t_f$, i.e., $p_1 > p_2$ and $u_1 < u_2$.*

To compare these results with the myopic attitude, we could set the shadow price at zero which is equivalent to disregarding dynamic differential equations. This approach provides standard static formulations in Sections 4.2.1 and 4.2.2 devoted to learning dynamics. However, this is not the case with the problem under consideration. Indeed, substituting ψ with zero in (4.75)-(4.76), we find that it is optimal not to process anything, $u=0$, and just to sell by backlogging and promising later deliveries (which will never come) at a lowered price, $p = \dfrac{a}{2b}$, compared to the system-wide optimal price. This policy, of course, has legal problems. On the other hand, if we assume that the retailer will process as many products as demanded by his customers, i.e., replace u with d, which is exactly what was assumed in all our deterministic static games. Then, when setting $\psi = 0$, we obtain a single optimality condition for the only variable, $p = \dfrac{a + b(c_r + c_s)}{2b}$. This expression, which was found for the static pricing game, does not come as much of surprise since, by setting $u=d$, we eliminate inventory dynamics and convert the dynamic game into the corresponding static pricing game. Consequently, similar to the previous sections, referring to the corresponding static model as myopic, we observe an interesting property:

The system-wide optimal solution is identical to the centralized myopic solution if the retailer processes as many products as demanded.

An immediate conclusion is that if the considered vertical supply chain with endogenous demand is centralized, then it exhibits static behavior so that it is not only performs best, but is also easily controlled with no dynamics or long-term effects that need to be accounted for.

In what follows we show that if the chain is not centralized and is in a transient-state, then its performance deteriorates and the control becomes sophisticated.

Game analysis: steady-state conditions

Given a wholesale price, $w(t)$, we first derive the retailer's optimal response for problem (4.67)-(4.71) by maximizing the Hamiltonian

$$H_r(t) = p(t)(a(t) - b(t)p(t)) - c_r u(t) - w(t)u(t) - h(X(t)) + \psi_r(t)(u(t) - a(t) + b(t)p(t))$$

with respect to the price $p(t)$ and processing rate $u(t)$, where the co-state variable $\psi_r(t)$ is determined by the co-state differential equation

$$\dot{\psi}_r(t) = \begin{cases} h^+, & \text{if } X(t) > 0; \\ h^-, & \text{if } X(t) < 0; \\ h \in [-h^-, h^+], & \text{if } X(t) = 0. \end{cases} \quad (4.78)$$

This equation, along with the co-state variable, has the same interpretation as in the centralized formulation. If the supply chain system is at the same steady-state at $t=0$ and $t=T$, i.e., it is characterized by the same demand potential $a(0)=a(T)$, customer sensitivity $b(0)=b(T)$, wholesale price $w(0)= w(T)$, and retailer inventory state $X(0)=X(T)$, then the co-state variable must be also the same at these points of time:

$$\psi_r(0) = \psi_r(T). \quad (4.79)$$

Maximizing the Hamiltonian with respect to $p(t)$ we readily find

$$p = \begin{cases} \dfrac{a}{b}, & \text{if } a + b\psi_r > 2a; \\ \dfrac{a + b\psi_r}{2b}, & \text{if } 0 \leq a + b\psi_r \leq 2a; \\ 0, & \text{if } a + b\psi_r < 0. \end{cases} \quad (4.80)$$

Note, that by using the same argument as in the analysis of the centralized system, we can say that if the retailer has a myopic attitude, then p is the only decision variable and $p = \dfrac{a + bc_r}{2b}$ is the optimal myopic price.

By maximizing the $u(t)$-dependent part of the Hamiltonian, we find

$$u = \begin{cases} U, & \text{if } \psi_r > c_r + w; \\ 0, & \text{if } \psi_r < c_r + w; \\ a - bp, & \text{if } \psi_r = c_r + w. \end{cases} \qquad (4.81)$$

Similar to the centralized approach, the third condition, which presents the case of an intermediate processing rate, is obtained by differentiating the singular condition, $\psi_r(t) = c_r + w_r(t)$, along an interval of time where the condition holds. Then, by taking into account (4.79), we conclude that this condition holds only if $X(t)=0$, i.e., $u(t)=d(t)=a(t)-b(t)p(t)$. Furthermore, this singular condition is feasible if, in addition to all constraints, (4.77) holds.

To derive the steady-state retailer's best response function, we assume steady sales at a sub-period of time $\tau = [\hat{t}, \check{t}] \subseteq [0,T]$ characterized by no-promotion, so that the customer sensitivity $b(t)=b_1$, potential $a(t)=a_1$ and wholesale price $w(t)=w_1$ remain constant for a period of time, $t \in \tau$. The following proposition states that this requirement implies static behavior characterized by constant pricing and processing rates as well as zero inventory levels.

Proposition 4.13. *If $b(t)=b_1$, $a(t)=a_1$, for $t \in \tau$, $\tau \subseteq [0,T]$, $X(\hat{t})=0$ and $0 \le a_1-b_1(c_r+ w) \le 2U$, then $X(t)=0$ for $t \in \tau$, and the best retailer's processing and pricing policies are:*

$$u(t) = \frac{a - b(c_r + w)}{2} \text{ and } p(t) = \frac{a + b(c_r + w)}{2b} \text{ for } t \in \tau \text{ respectively.}$$

Proof: The proof is very similar to that of Proposition 4.10.

Comparing statements of Proposition 4.10 and Proposition 4.13, we readily come up with the expected conclusion for static games:
if the supplier makes a profit, $w>c_s$, then in a steady-state vertical competition of the differential inventory game with endogenous demand, the retail price increases and the demand, along with the processing rate, decreases compared to the system-wide steady-state optimal solution.

Steady-state equilibrium

Proposition 4.13 determines the optimal retailer's strategy in a steady-state during a no-promotion period. To define the corresponding supplier's game in a steady-state over an interval of time, for example $[0,T]$, we substitute the best retailer's response for $\tau = [0,T]$ into the objective function (4.65):

$$\int_0^T [w(t)u(t) - c_s u(t)]dt = \frac{a_1 - b_1(c_r + w)}{2}(w - c_s)T . \qquad (4.82)$$

Note that the maximum of function (4.82) does not depend on the length of the considered interval T and can be determined by simply applying the first-order optimality conditions. Accordingly, we conclude with the following proposition for the supply chain which is in a steady-state along an interval, $[0,T]$.

Proposition 4.14. *If $b(t)=b_1$, $a(t)=a_1$ for $t \in [0,T]$, $X(0)=X$ and $0 \le a_1 - b_1(c_r + c_s) \le 4U$, then $X(t)=0$ for $t \in [0,T]$, the supplier's wholesale pricing policy* $w^s(t) = \dfrac{a_1 - b_1(c_r - c_s)}{2b_1}$, *and the retailer's processing*

$u^s(t) = \dfrac{a_1 - b_1(c_r + c_s)}{4}$ *and pricing* $p^s(t) = \dfrac{3a_1 + b_1(c_r + c_s)}{4b_1}$ *policies*

constitute the unique Stackelberg equilibrium for $t \in [0,T]$.

Proof: Since function (4.82) is concave in w, the first-order optimality condition applied to it results in a unique optimal solution $w^s(t) = \dfrac{a_1 - b_1(c_r - c_s)}{2b_1}$

which is feasible if $\dfrac{a}{b} \ge c_r + c_s$, as stated in this proposition. Substituting this result in the equations for $p(t)$ and $u(t)$ from Proposition 4.13 leads to the equilibrium equations stated in Proposition 4.14. Furthermore, $p^s(t)$ is feasible (meets (4.70)) due to the same condition, $\dfrac{a}{b} \ge c_r + c_s$. Finally, $u^*(t)$ is feasible if the condition, $0 \le a - b(c_r + w) \le 2U$, stated in Proposition 4.13 holds. Substitution of $w^s(t)$ into this condition as well completes the proof.

According to Propositions 4.13-4.14, the retailer's problem may have an optimal interior solution and the supply chain may be in a steady-state if the demand is non-negative in this state and the maximum processing rate is greater than the maximal demand

$$\frac{a}{b} \ge c_r + c_s \text{ and } a < U.$$

Game analysis: transient -state conditions

We assume first that since the promotion time is much shorter than the committed contract period T, the supplier chooses the wholesale price as determined in Proposition 4.14 to maintain a steady-state; a new wholesale price can only be selected at a predetermined date for a limited promotional period. In response, the retailer will change his policy accordingly. This changeover induces in the supply chain a transient-state in which both the supplier and retailer attempt to use increased customer sensitivity during the limited promotional period to increase sales.

We further assume that since T is longer than the promotion duration, the supply chain, which is in a steady-state (characterized by demand potential a_1 and sensitivity b_1) at time $t=0$, will return to this state by time $t=T$ after the promotion period, which starts at $t_s>0$ and ends at time $t_f<T$. This implies that the optimality conditions derived in the previous section remain the same, but that $w(t)$ is no longer constant and is defined by equation

(4.64), where $w_1 = w^s(t) = \dfrac{a_1 - b_1(c_r - c_s)}{2b_1}$, and w_2 is a decision variable.

To derive the retailer's best response function, we distinguish between two types of transient-states: brief and maximal changeover. The difference between the two is due to a temporal steady-state the supply chain may reach during the promotion. The presence of this temporal steady-state implies that the retailer has enough time to optimally reduce prices to a minimum level corresponding to the promotional wholesale price w_2. This phenomenon can be viewed as the maximum effect that a promotional initiative can cause, which is why we focus here on this type of transient-state, as discussed in the following theorem.

Theorem   4.1.   Let $a(t)-b(t)(c_r+w(t))\geq 0$, $w_1>w_2$ $d^*=\dfrac{a_1-b_1(c_r+w_1)}{2}$,

$d^{**}=\dfrac{a_2-b_2(c_r+w_2)}{2}$. *If* $t_1<t_s$, $t_2>t_s$, $t_3<t_f$, $t_4>t_f$, $t_2 \leq t_3$ *satisfy the following*

equations

$$U(t_2 - t_s) = \frac{1}{2}\left(a_1(t_s - t_1) + a_2(t_2 - t_s)\right) - \frac{1}{2}\left(b_1(t_s - t_1) + b_2(t_2 - t_s)\right)(c_r + w_1 + h^- t_1) +$$

$$+\frac{1}{4}h^-\left(b_1(t_s^2 - t_1^2) + b_2(t_2^2 - t_s^2)\right),\ h^-(t_2 - t_1) = w_1 - w_2,\qquad (4.83)$$

$$U(t_f - t_3) = \frac{1}{2}\left(a_1(t_4 - t_f) + a_2(t_f - t_3)\right) - \frac{1}{2}\left(b_1(t_4 - t_f) + b_2(t_f - t_3)\right)\left(c_r + w_2 - h^+ t_3\right) -$$

$$- \frac{1}{4}h^+\left(b_1(t_4^2 - t_f^2) + b_2(t_f^2 - t_3^2)\right), \; h^+(t_4 - t_3) = w_1 - w_2, \qquad (4.84)$$

then $X(t)=0$ for $0 \le t \le t_1$, $t_2 \le t \le t_3$, $t_4 \le t \le T$; $X(t)<0$ for $t_1 < t < t_2$, $X(t)>0$ for $t_3<t<t_4$; the optimal retailer's processing policy is
$u(t)=d^*$ for $0 \le t < t_1$ and $t_4 \le t \le T$, $u(t)=d^{**}$ for $t_2 \le t < t_3$,
$u(t)=U$ for $t_s \le t < t_2$ and $t_3 \le t < t_f$, $u(t)=0$ for $t_1 \le t < t_s$ and $t_f \le t < t_4$;
and the optimal retailer's pricing policy is

$$p(t) = \frac{a(t) + b(t)(c_r + w_1 - h^-(t - t_1))}{2b(t)} \; \text{for } t_1 \le t < t_2,$$

$$p(t) = \frac{a_2 + b_2(c_r + w_2)}{2b_2} \; \text{for } t_2 \le t < t_3,$$

$$p(t) = \frac{a_1 + b_1(c_r + w_1)}{2b_1} \; \text{for } 0 \le t < t_1, \; t_4 \le t \le T,$$

$$p(t) = \frac{a(t) + b(t)(c_r + w_2 + h^+(t - t_3))}{2b(t)} \; \text{for } t_3 \le t < t_4.$$

Proof: First note, that as mentioned before, the retailer's problem is a convex program, which implies that the necessary optimality conditions are sufficient.

Consider a solution which is characterized by four breaking points, t_1, t_2, t_3 and t_4 so that the retailer is in a steady-state between time points $t=0$ and $t=t_1$, between $t=t_2$ and $t=t_3$, and between $t=t_4$ and $t=T$, as described below:

$$X(t)=0 \text{ for } 0 \le t \le t_1, \; t_2 \le t \le t_3 \text{ and } t_4 \le t \le T; \qquad (4.85)$$

$$u(t)=d^* \text{ for } 0 \le t < t_1 \text{ and } t_4 \le t \le T, \; u(t)=d^{**} \text{ for } t_2 \le t < t_3, \quad (4.86)$$

$$u(t)=U \text{ for } t_s \le t < t_2, \; t_3 \le t < t_f, \; u(t)=0 \text{ for } t_1 \le t < t_s \text{ and } t_f \le t < t_4; (4.87)$$

$$\psi_r(t) = c_r + w_1 \text{ for } 0 \le t < t_1, \; t_4 \le t \le T, \; \psi_r(t) = c_r + w_2 \text{ for } t_2 \le t < t_3; (4.88)$$

$$\psi_r(t)=c_r + w_1 - h^-(t - t_1) \text{ for } t_1 \le t < t_2, \; \psi_r(t)=c_r + w_2 + h^+(t - t_3) \text{ for } t_3 \le t < t_4. (4.89)$$

It is easy to observe that the solution (4.85))-(4.89) meets optimality conditions ((4.76)) if $a(t)-b(t)(c_r+w(t)) \ge 0$, $a(t) \le U$ and there is sufficient time to reach a steady-state during the promotion period, i.e., $t_2 \le t_3$. Furthermore, the optimal pricing policy is immediately derived by substituting the co-state solution (4.88)-(4.89) into $p = \dfrac{a + b\psi_r}{2b}$ (see optimality conditions

(4.75)), as stated in the theorem. In turn, this solution is feasible if $p(t) \geq 0$ (which always holds) and $p(t) \leq \dfrac{a(t)}{b(t)}$ (see constraint (4.70)) or the same $d(t) \geq 0$. The latter holds because,

$$\frac{a_1}{b_1} > \frac{a_2}{b_2}, p(t) \leq \frac{a_1 + b_1(c_r + w_1)}{2b_1} \text{ and } w_1 = w^s = \frac{a_1 - b_1(c_r - c_s)}{2b_1}.$$

To complete the proof, we need to find the four breaking points and ensure that $t_2 \leq t_3$. Points t_1 and t_2, are found by solving a system of two equations (4.85) and (4.89). Specifically, from (4.85) and (4.68) we find that

$$X(t_2) = -\int_{t_1}^{t_2} (a(t) - b(t)p(t))dt + U(t_2 - t_s) = 0. \qquad (4.90)$$

By substituting found $p(t)$ into (4.90) we obtain

$$U(t_2 - t_s) = \frac{1}{2}(a_1(t_s - t_1) + a_2(t_2 - t_s)) - \frac{1}{2}(b_1(t_s - t_1) + b_2(t_2 - t_s))(c_r + w_1 + h^- t_1) +$$

$$+ \frac{1}{4}h^-(b_1(t_s^2 - t_1^2) + b_2(t_2^2 - t_s^2)),$$

which along with

$$c_r + w_2 = c_r + w_1 - h^-(t_2 - t_1)$$

from (4.88) and (4.89) results in the system of two equations (4.83) in unknowns t_1 and t_2 as stated in the theorem.

Similarly,

$$X(t_4) = U(t_f - t_3) - \int_{t_3}^{t_4} (a(t) - b(t)p(t))dt = 0,$$

which results in

$$U(t_f - t_3) = \frac{1}{2}(a_1(t_4 - t_f) + a_2(t_f - t_3)) - \frac{1}{2}(b_1(t_4 - t_f) + b_2(t_f - t_3))(c_r + w_2 - h^+ t_3) -$$

$$- \frac{1}{4}h^+(b_1(t_4^2 - t_f^2) + b_2(t_f^2 - t_3^2)). \qquad (4.91)$$

Considering (4.91) simultaneously with equation

$$c_r + w_2 + h^+(t_4 - t_3) = c_r + w_1$$

from (4.88) and (4.89) results in two equations (4.84) for t_3 and t_4 stated in the theorem. □

The solutions to equations (4.83)-(4.84) are unique and are as follows.

Solution of Equations (4.83)

$$t_1^* = t_s - A_1^* \text{ and } t_2^* = t_1^* + f_1,$$

where

$$f_1 = \frac{w_1 - w_2}{h^-}, f_2 = c_r + w_1,$$

$$A_1^* = \frac{U + \frac{1}{2}a_1 - \frac{1}{2}a_2 - \frac{1}{2}b_1 f_2 + \frac{1}{2}b_2 f + [D_1^*]^{\frac{1}{2}}}{\frac{1}{2}h^-[b_2 - b_1]}$$

$$D_1^* = U^2 + Ua_1 - Ua_2 - Ub_1 f_2 + Ub_2 f_2 + \frac{1}{4}a_1^2 + \frac{1}{4}a_2^2 - \frac{1}{2}a_1 a_2 - \frac{1}{2}a_1 b_1 f_2 + \frac{1}{2}a_1 b_2 f_2 -$$
$$\frac{1}{2}a_2 b_2 f_2 + \frac{1}{2}a_1 b_2 f_2 + \frac{1}{4}b_1^2 f_2^2 - \frac{1}{2}b_1 b_2 f_2^2 + \frac{1}{4}b_2^2 f_2^2 - (\frac{1}{2}h^- b_1 a_2 f_1 - \frac{1}{2}h^- b_1 b_2 f_1 f_2$$
$$+ \frac{1}{4}h^{-2} b_1 b_2 f_1^2 - h^- b_1 U f_1 - \frac{1}{2}h^- b_2 a_2 f_1 + \frac{1}{2}h^- b_2^2 f_1 f_2 - \frac{1}{4}h^{-2} b_2^2 f_1^2 + h^- b_2 U f_1)$$

Solution of Equations (4.84)

$$t_4^* = t_f + A_2^* \text{ and } t_3^* = t_4^* - f_1, \ f_1 = \frac{w_1 - w_2}{h^+},$$

where

$$f_2 = c_r + w_1,$$

$$A_2^* = \frac{U + \frac{1}{2}a_1 - \frac{1}{2}a_2 - \frac{1}{2}b_1 f_2 + \frac{1}{2}b_2 f_2 - \frac{1}{2}b_1 h^+ f_1 + \frac{1}{2}b_2 f_1 h^+ + [D_2^*]^{\frac{1}{2}}}{\frac{1}{2}h^+[b_2 - b_1]},$$

$$D_2^* = U^2 + Ua_1 - Ua_2 + \frac{1}{4}a_1^2 + \frac{1}{4}a_2^2 - \frac{1}{2}a_1 a_2 - b_1 f_2 U + b_2 f_2 U - \frac{1}{2}a_1 b_1 f_2 +$$
$$\frac{1}{2}a_1 b_2 f_2 + \frac{1}{2}a_2 b_1 f_2 - \frac{1}{2}a_2 b_2 f_2 - b_1 h^+ f_1 U + b_2 h^+ f_1 U + \frac{1}{4}b_1^2 f_2^2 + \frac{1}{4}b_2^2 f_2^2 -$$
$$- \frac{1}{2}a_1 b_1 f_1 h^+ - \frac{1}{2}a_2 b_2 f_1 h^+ + \frac{1}{2}a_1 b_2 f_1 h^+ + \frac{1}{2}a_2 b_1 f_1 h^+ + \frac{1}{2}b_2^2 f_1 f_2 h^+ -$$
$$- \frac{1}{2}b_1 b_2 f_1^2 h^{+2} - \frac{1}{2}b_1 b_2 f_2^2 + \frac{1}{2}b_1^2 f_1 f_2 h^+ - b_1 b_2 f_1 f_2 h^+ + \frac{1}{4}b_1^2 f_1^2 h^{+2} +$$
$$+ \frac{1}{4}b_2^2 f_1^2 h^{+2} - (-b_1 h^+ U f_1 + \frac{1}{2}b_1 h^+ a_2 f_1 - \frac{1}{2}b_1 h^+ b_2 f_1 f_2 - \frac{1}{4}b_1 h^{+2} b_2 f_1^2 +$$
$$+ b_2 h^+ U f_1 - \frac{1}{2}b_2 h^+ a_2 f_1 + \frac{1}{2}b_2^2 h^+ f_1 f_2 + \frac{1}{4}b_2^2 h^{+2} f_1^2).$$

The optimal solution derived in Theorem 4.1 is illustrated in Figure 4.6. According to this solution, it is beneficial for the retailer to change pricing and processing policies in response to a reduced wholesale price and increased customer price sensitivity during the promotion.

The change is characterized by instantaneous jumps upward in quantities ordered and downward in retailer prices at the point the promotion starts and vice versa at the point the promotion ends. Inventory surplus at the end of the promotion indicates that the retailer ordered more goods during the promotional period than he is able to sell (forward buying). Moreover, the retailer starts to lower prices even before the promotion starts. This strategy makes it possible to build greater demands by the beginning of the promotion period and to take advantage of the reduced wholesale price during the promotion. This is accomplished gradually so that a trade-off between the inventory backlog (surplus) cost and the wholesale price is sustained over time. Figure 4.6 shows that any reduction in wholesale price results first in backlogs and then surplus inventories. This is in contrast to a steady-state with no inventories being held.

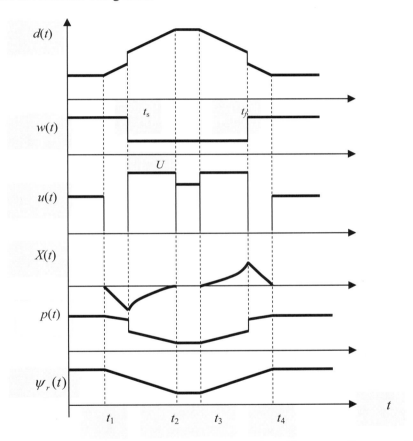

Figure 4.6. Optimal retailer policies under promotion (the case of symmetric costs, $h^+=h^-$).

There are two immediate conclusions emanating from Theorem 4.1. One is that the retailer's total order quantity increases with the decrease of the wholesale price as formulated in the following corollary.

Corollary 4.1. *If $a(t)-b(t)(c_r+w(t)) \geq 0$, the lower the promotional wholesale price, w_2, the greater the total order $\int_{t_1}^{t_4} u(t)dt = \int_{t_1}^{t_4} d(t)dt$ and the lower the overall product pricing $\int_{t_1}^{t_4} p(t)dt$.*

The other conclusion for transient conditions is drawn by comparing the maximum demand $d** = \dfrac{a_2 - b_2(c_r + w_2)}{2}$ and minimum price $p(t) = \dfrac{a_2 + b_2(c_r + w_2)}{2b_2}$ under non-cooperative solution with the corresponding demand $d_2 = u_2 = \dfrac{a_2 - b_2(c_r + c_s)}{2}$ and price $p_2 = \dfrac{a_2 + b_2(c_r + c_c)}{2b_2}$ under a centralized solution. The conclusion is straightforward and agrees with our previous results obtained for vertical competition in static conditions, as stated in the following proposition.

Corollary 4.2. *In the transient–state vertical competition of the differential inventory game with endogenous demand, if the supplier makes a profit, $w > cs$, then the retail price increases and the demand, along with the processing rate, decreases compared to the system-wide optimal price demand and processing rate respectively.*

As a result, neither the promotion prices nor the demand will be respectively that low or high as they should be in respect to the system-wide optimal setting. Furthermore, recalling that the myopic price at transient-state is $p(t) = \dfrac{a_2 + b_2(c_r + w_2)}{2b_2}$, we observe that this price is closer to the system-wide optimal price $p_2 = \dfrac{a_2 + b_2(c_r + c_c)}{2b_2}$ and even switches on and off at the same time. This implies that under some conditions the myopic attitude may coordinate the supply chain.

Another observation is that the myopic price is determined by the same equation $p(t) = \dfrac{a + b(c_r + w)}{2b}$ in both steady- and transient-state (only values of a and b change). Comparing this equation with the pricing policy determined by Theorem 4.1, we find the following property:

The myopic retail price does not exceed the corresponding price in the transient-state from $t=t_s$ to $t=t_f$ of vertical competition of the differential inventory game with endogenous demand.

This result is not typical since until now we have only observed over-pricing from a myopic approach. Overpricing does happen with a myopic approach but only at short time intervals $t_1<t<t_s$ and $t_f<t<t_4$. On the other hand, at intervals $t_s<t<t_2$ and $t_3<t<t_f$, the myopic price is strictly below the dynamic retail price.

Transient-state equilibrium

Theorem 4.1 identifies the best retailer's strategy in the presence of a transient-state during a promotion period. To define the corresponding supplier's strategy over interval $[0,T]$, we substitute the retailer's best response into objective function (4.65):

$$\int_0^T [w(t)u(t) - c_s u(t)]dt =$$

$$\int_0^{t_1}(w_1 - c_s)d * dt + \int_{t_s}^{t_2}(w_2 - c_s)Udt + \int_{t_2}^{t_3}(w_2 - c_s)d**dt +$$

$$\int_{t_3}^{t_f}(w_2 - c_s)Udt + \int_{t_4}^T(w_1 - c_s)d * dt .$$

That is,

$$\int_0^T [w(t)u(t) - c_s u(t)]dt =$$

$$(w_1 - c_s)d*(T - t_4 + t_1) + (w_2 - c_s)U(t_2 - t_s + t_f - t_3) + (w_2 - c_s)d**(t_3 - t_2) \quad (4.91)$$

Applying the first-order optimality conditions to this static function with respect to w_2 and denoting the result by $F(w_2)$, we obtain:

$$F(w_2) =$$

$$d*(w_1 - c_s)[t_1 - t_4]'_{w_2} + U(t_f - t_s) + U[(w_2 - c_s)(t_2 - t_3)]'_{w_2} + [d**(w_2 - c_s)(t_3 - t_2)]'_{w_2} = 0 \quad (4.92)$$

To show the uniqueness of the equilibrium for a transient-state, we need the property stated in the following proposition.

Proposition 4.15. Let $R_1 = -A_1^* - A_2^* + (w_1 - w_2)(\dfrac{1}{h^-} + \dfrac{1}{h^+})$, $a(t)-b(t)(c_r+w(t)) \geq 0$, $b_2 > b_1$,

$$R_2 = \frac{U(t_3-t_2)-d^*(w_1-c_s)\bigl[t_1-t_4\bigr]'_{w_2}-(U-\hat{d})(w_1-c_s)\left(\bigl[(t_1-t_4)\bigr]'_{w_2}-\left[\frac{1}{h^-}+\frac{1}{h^+}\right]\right)-(\hat{d}-\frac{b_2}{2}(w_1-c_s))(t_3-t_2)}{U}$$

and $\hat{d}=\dfrac{a_2-b_2(c_r+w_1)}{2}$, where A_1^* and A_2^* are determined by the solutions of (4.83)-(4.84). If $R_1 \le t_f - t_s \le R_2$, then equation (4.92) has only one root α, such that $c_s < \alpha < w_1 = \dfrac{a_1-b_1(c_r-c_s)}{2b_1}$.

Proof: First note that function (4.92) has a negative highest order (the third-order) term. Therefore, to prove that equation $F(w_2)=0$ has only one root $w_2=\alpha$ in the range of $c_s < \alpha < w_1$, it is sufficient to show that $F(c_s)>0$ and $F(w_1)<0$ (see Figure 4.7.)

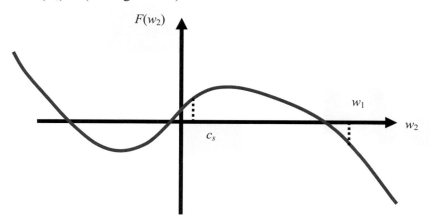

Figure 4.7. Analysis of the first-order optimality condition of the Stackelberg wholesale price

The fact that $F(c_s)>0$ is observed from (4.92) by substituting w_2 with c_s. This reduces (4.92) to

$$F(c_s)= d^*(w_1-c_s)\bigl[t_1-t_4\bigr]'_{w_2} +U(t_f-t_s)-U(t_3-t_2)+d^{**}(t_3-t_2),$$

which is positive if $\bigl[-t_4+t_1\bigr]'_{w_2}>0$.

Calculating the derivative of t_1 with respect to w_2, we obtain

$$\bigl[t_1\bigr]'_{w_2} =\Bigl[h^-(b_2-b_1)\sqrt{D_1^*}\Bigr]^{-1}\left(U-\frac{b_2(w_1-w_2)}{2}-(a_2-b_2(c_r+w_1))\right)(b_2-b_1),$$

which is always positive as $U \geq a_2$ and $\dfrac{b_2(w_1 - w_2)}{2} < b_2(c_r + w_1)$. Calcu-

lating the derivative of t_4 with respect to w_2 we find

$$\left[-t_4\right]_{w_2}' = \left[h^+(b_2 - b_1)\sqrt{D_2^*}\right]^{-1}\left(U - \frac{1}{2}a_2 + \frac{b_2(w_1 - w_2)}{2} + \frac{1}{2}b_2(c_r + w_1)\right)(b_2 - b_1),$$

which is always positive as well. Thus, we conclude $F(c_s) > 0$.

Similarly, from (4.92), we find

$$F(w_2) = d^*(w_1 - c_s)\left[t_1 - t_4\right]_{w_2}' + U(t_f - t_s) + U(w_2 - c_s)\left[(t_2 - t_3)\right]_{w_2}' - U(t_3 - t_2) +$$

$$+ d^{**}(w_2 - c_s)\left[(t_3 - t_2)\right]_{w_2}' - \frac{b_2}{2}(t_3 - t_2)(w_2 - c_s) + (t_3 - t_2)d^{**}.$$

Since $\left[t_2\right]_{w_2}' = \left[t_1\right]_{w_2}' - \dfrac{1}{h^-}$ and $\left[t_3\right]_{w_2}' = \left[t_4\right]_{w_2}' + \dfrac{1}{h^+}$, we have

$$F(w_2) = d^*(w_1 - c_s)\left[t_1 - t_4\right]_{w_2}' + U(t_f - t_s) + U(w_2 - c_s)\left(\left[(t_1 - t_4)\right]_{w_2}' - \left[\frac{1}{h^-} + \frac{1}{h^+}\right]\right) - U(t_3 - t_2) -$$

$$- d^{**}(w_2 - c_s)\left(\left[(t_1 - t_4)\right]_{w_2}' - \left[\frac{1}{h^-} + \frac{1}{h^+}\right]\right) - \frac{b_2}{2}(t_3 - t_2)(w_2 - c_s) + (t_3 - t_2)d^{**}.$$

Then substituting w_2 with w_1 and denoting

$$UR_2 = U(t_3 - t_2) - d^*(w_1 - c_s)\left[t_1 - t_4\right]_{w_2}' - (U - \hat{d})(w_1 - c_s)\left(\left[(t_1 - t_4)\right]_{w_2}' - \left[\frac{1}{h^-} + \frac{1}{h^+}\right]\right)$$

$$- (\hat{d} - \frac{b_2}{2}(w_1 - c_s))(t_3 - t_2),$$

where $\hat{d} = \dfrac{a_2 - b_2(c_r + w_1)}{2}$ and requiring $F(w_1) < 0$, we have $t_f - t_s \leq R_2$ as stated
in this proposition. Finally, recalling that according to Theorem 4.1, $t_3 \geq t_2$,
we find

$$t_3 - t_2 = t_f - t_s + A_1^* + A_2^* - (w_1 - w_2)(\frac{1}{h^-} + \frac{1}{h^+}) \geq 0.$$

As a result, denoting, $R_1 = -A_1^* - A_2^* + (w_1 - w_2)(\frac{1}{h^-} + \frac{1}{h^+})$, we require

$t_f - t_s \geq R_1$.

Thus, α is the wholesale equilibrium price in transient conditions. The
following proposition summarizes our results for both steady- and transient-
state conditions.

Proposition 4.16. *If $a_1 - b_1(c_r + c_s) \geq 0$, $a_2 - b_2(c_r + \alpha) \geq 0$, $R_1 \leq t_f - t_s \leq R_2$,*

then the supplier's wholesale pricing policy $w^s(t) = w_1^s = \dfrac{a_1 - b_1(c_r - c_s)}{2b_1}$

for $0 \leq t < t_s$, $t_f \leq t \leq T$ and $w^s(t) = w_2^s = \alpha$ for $t_s \leq t < t_f$, and the retailer's

processing $u(t)$ and pricing $p(t)$ policies, determined by Theorem 4.1, con-
stitute the unique Stackelberg equilibrium for $t \in [0, T]$.

Proof: The proof is immediate. According to Proposition 4.15, $F(w_2)=0$ has only one root in the feasible range of $c_s < \alpha < w_1$, therefore the optimal wholesale price it defines is unique. Furthermore, according to Theorem 4.1, $p(t)$ and $u(t)$ are unique and feasible if $t_3 \geq t_2$ and $a(t) - b(t)(c_r + w(t)) \geq 0$ hold. Substituting into the latter the corresponding values for $b(t)$ and $w(t)$, we obtain the conditions stated in this proposition.

The existence of equilibrium wholesale price $w^s(t)=w_2^s=\alpha$ stated in the previous proposition readily leads to the following corollary.

Corollary 4.3. *Let $a_1 - b_1(c_r + c_s) \geq 0$, $a_2 - b_2(c_r + \alpha) \geq 0$, and $R_1 \leq t_f - t_s \leq R_2$.*
If the customer sensitivity increases during the promotion period, $b_2 > b_1$,
then the wholesale price decreases $w_2 < w_1$.

From Corollaries 4.1 and 4.3, it immediately follows that during higher demand, the retail price falls (Corollary 4.1) when customer sensitivity increases (Corollary 4.3). Moreover, the retailer starts to lower prices even before the promotion starts (Theorem 4.1). This phenomenon has been widely observed in empirical studies of retail prices during and close to holidays (see, for example, Chevalier *et. al.* 2003; Bils, 1989 and Warner and Barsky, 1995).

Note, that one can view the optimal solution during the promotion conditions of Theorem 4.1 and Proposition 4.16 as a feedback policy. Indeed, the processing and pricing policies are such that inventory levels are kept at zero when the supply chain is in a new steady-state during the promotion, i.e., for $t_2 \leq t < t_3$. On the other hand, the remaining promotion time is characterized by a feedback, $\pi^0(X(t), t)$, where the upper index, 0, stands for the critical number $X=0$ (threshold) on which the feedback depends . This is summarized as:

$$u(t) = \pi_u^0(X(t), t) = \begin{cases} 0, \text{if } X(t) \leq 0 \text{ and } t_1 \leq t < t_s; \\ U, \text{if } X(t) < 0 \text{ and } t \geq t_s; \\ U, \text{if } X(t) \geq 0 \text{ and } t_3 \leq t < t_f; \\ 0, \text{if } X(t) > 0 \text{ and } t \geq t_f. \end{cases}$$

$$p(t) = \pi_p^0(X(t),t) = \begin{cases} \dfrac{a(t)+b(t)(c_r +w_1 -h^-(t-t_1))}{2b(t)}, & \text{if } X(t)<0; \\[4mm] \dfrac{a(t)+b(t)(c_r +w_2 +h^+(t-t_3))}{2b(t)}, & \text{if } X(t)>0. \end{cases}$$

As shown in Theorem 4.1, as well as in Corollaries 4.1-4.3, the optimal Stackelberg solution implies that if customer sensitivity increases during a promotional period, then both the retailer and the supplier increase their profits compared to a solution which disregards the change in customer sensitivity. This, however, does not necessarily mean that profits during the promotion will exceed those gained during regular operation at a steady-state. This is to say, on special occasions like Christmas, customer sensitivity may increase without any promotional initiative and the decentralized chain will have no other option than to respond. On the other hand, if a promotional initiative expected to impact customer sensitivity is assessed as not beneficial in regard to regular profits, then it can be abandoned in time. The necessary and sufficient condition with respect to the profitability of a limited-time, $R_1 \le t_f - t_s \le R_2$, promotion initiated by the leader is straightforwardly obtained from equation (4.91).

If $\theta_1(b_2) =$
$$(w_2 -c_s)U(t_2 -t_s +t_f -t_3)+(w_2 -c_s)d**(t_3 -t_2)-(w_1 -c_s)d*(t_4 -t_1)>0,$$

then the supplier (the leader) will gain an extra profit from the promotion compared to the regular (steady-state) profits under $d*$ for the same period of time. Similarly, from (4.67) one can define a gap function, $\theta_2(b_2)$, so that the retailer would have an extra profit if $\theta_2(b_2)>0$. Since these conditions involve extremely large expressions of the switching time points, we illustrate the evolution of profit gaps $\theta_1(b_2)$ and $\theta_2(b_2)$ quantitatively for different customer sensitivities and fixed promotion times in the following example. The interpretation is immediate – when both gaps are positive, the promotion is beneficial for both the leader and the follower.

Example 4.3.

We calculate wholesale equilibrium price as determined by Proposition 4.16 for $U=10000$, $a_1 = 2500$, $a_2 = 6000$ product units per time unit; $b_1 =10$ product units per dollar and time unit; $t_s =100$, $t_f = 300$ and $T=1000$ time units. The results are presented in Table 4.1.

From Table 4.1, we see that there is a bounded interval to the customer sensitivity values b_2 for which an equilibrium exists. The existence of the equilibrium starts from $b_2>24$ which ensures our general assumption of an increase in demand elasticity, $\dfrac{a_1}{b_1} > \dfrac{a_2}{b_2}$, and terminates at $b_2>52$ when the condition, $a_2-b_2(c_r+w_2)\geq 0$, of Theorem 4.1 no longer holds. More importantly, the range of values is such that the promotion gains extra profits for both the supplier and retailer (i.e., gaps $\theta_1(b_2)$ and $\theta_2(b_2)$ are both positive) from $b_2=28$ to $b_2=32$. This result is due to a non-linear relationship between the demand potential, a_2, which remains the same and sensitivity, b_2, which increases. The profitability range could be extended if, for example, a linear relationship, $a(t)=g+b(t)P$, were used in the example. Under such conditions, a_2 would always increase with b_2.

Coordination

So far, in our examples of supply chain games with endogenous demands, we assumed that only demand potential $a(t)$ may change with time. In this section we consider a differential inventory game where both customer demand potential $a(t)$ and customer sensitivity $b(t)$ change over time. As with other games that capture vertical competition in supply chains, we found that the prices increase and order quantities decrease compared to the corresponding system-wide optimal solutions. This deterioration in the performance is true regardless whether the supply chain is in a steady- or transient-state.

Customer-related dynamics, however, contribute some distinctive features to the supply chain performance. For example, although the equilibrium wholesale price changes instantaneously, the retail prices evolve in a more complex manner which includes both gradual and step-wise amendments which start even before the wholesale price drops and sometime after the wholesale promotion ends. Such a behavior is due to the fact that the retailer has additional instruments for a trade-off (compared to the corresponding static models) which are inventory-holding and backlogging over time. For example, by forward buying and storing some inventories during the wholesale promotion, the retailer may profit more compared to that under a system-wide solution. The system-wide optimal solution does not account for wholesale prices, viewing them as internal transfers thereby ignoring individual profits of each party. Due to inventory dynamics, the traditional two-part tariff is not as efficient as it is in static supply games. This occurs because the supplier when setting the wholesale price w_2, ignores not only the retailer's profit margin from sales, but also the profit margin from handling inventories, $\psi_r \dot{X}$.

Indeed, it is easy to observe from Theorem 4.1 (as well as Figure 4.6) that even if the supplier sets the wholesale price at the minimum level, $w_2=c_s$, (to earn profits during the promotion only from fixed contract costs), then the retail price and customer demand attain system-wide optimal levels only after an interval of time and will not remain at that level until the end of the promotion. Thus, though the two-part tariff during the promotion coordinates the supply chain, this policy is insufficient for perfect coordination.

As Theorem 4.1 demonstrates, the greater the shadow price rate of change, the faster the retail price (and therefore the demand) will attain the system-wide optimal level. This is not surprising since the rate of change of the co-state variable is the marginal profit from reducing inventories which the inventory holding/backlog costs determine. Consequently, the greater the holding and backlog costs, the less the retailer utilizes the inventory surplus/shortage and the more coordinated the supply chain becomes.

Table 4.1. Wholesale prices and profit gaps between transient and steady state (10^6\$)

	$h^+=1, h^-=2,$ $c_r=30, c_s=60$	$h^+=1, h^-=10,$ $c_r=30, c_s=60$	$h^+=1, h^-=2,$ $c_r=60, c_s=30$
$b_2=12$ to 24			
$\theta_1(b_2)\,(\theta_2(b_2))$	-	-	-
$w_2{}^*$	no equilibrium	no equilibrium	no equilibrium
$b_2=28$			
$\theta_1(b_2)\,(\theta_2(b_2))$	4.2342 (1.8058)	4.2540 (1.8669)	4.2342 (1.8058)
$w_2{}^*$	125.6560	125.2240	95.6560
$b_2=32$			
$\theta_1(b_2)\,(\theta_2(b_2))$	0.6568 (0.5914)	0.7292 (0.5289)	0.6568 (0.5914)
$w_2{}^*$	114.0560	113.2240	84.0560
$b_2=36$			
$\theta_1(b_2)\,(\theta_2(b_2))$	-2.0835 (0.037)	-1.9369 (-0.3428)	-2.0835 (0.0371)
$w_2{}^*$	104.5200	103.3680	74.5200
$b_2=40$			
$\theta_1(b_2)\,(\theta_2(b_2))$	-4.198 (-0.0935)	-3.9647 (-1.1398)	-4.198 (-0.0935)
$w_2{}^*$	96.5756	95.1680	66.5756
$b_2=44$			
$\theta_1(b_2)\,(\theta_2(b_2))$	-5.835 (-0.0398)	-5.5084 (-2.1196)	-5.835 (-0.0398)
$w_2{}^*$	89.8640	88.2800	58.8640

Interestingly, myopic centralized pricing is identical to the system-wide optimal solution during a steady-state. During the transient-time, despite vertical competition, myopic pricing is below the dynamic equilibrium pricing and above the system-wide optimal pricing. Moreover, the myopic price is even characterized by stepwise timing identical to the centralized solution. Thus, the myopic retailer's attitude may coordinate the supply chain. This, however, requires more precise analysis in each particular case to assess whether the overall profit of the supply chain improves or not.

Finally, a promising coordinating option for the supplier is to set a permanent wholesale price $w=c_s$, rather than a price for just a limited-time period when customer sensitivity changes. He then charges the retailer a fixed-cost per time unit. With such a two-part tariff, the retailer's problem becomes identical to the centralized problem and the supply chain is perfectly coordinated. However, with a rolling horizon contract, as assumed in this intertemporal inventory game, the supplier is giving up his profit from sales over an indefinite period of time and relying completely on fixed transfers of his share, which is equivalent to long-term cooperation between the supplier and retailer rather than competition.

4.3.2 THE DIFFERENTIAL INVENTORY GAME WITH EXOGENOUS DEMAND

Cycles and seasonal patterns in demand are frequently found in production and service operations. For example, housing starts and, thus, construction-related products tend to follow cycles. Automobile sales also tend to follow cycles (see, for example, Russell and Taylor 2000). In this section we study the effect of cyclic demands on supply chain operations.

Consider a production game in a two-echelon supply chain consisting of a single supplier (manufacturer) delivering a product type to a single retailer over a period of time, T. Similar to the game discussed in the previous section, the production horizon is infinite and there are periodic seasonal (instantaneous) changes in demand. Since the time between the seasons is sufficiently long, there is enough time for the supply chain to revert to the state it was in before the season began.

There are two major distinctive features of this supply chain game compared to that of the previous section. First, we consider exogenous customer demand that implies that the quantities produced and sold by this supply chain cannot affect the price level of the product. This simplifies the problem since price is no longer a decision variable. Moreover, we assume that the wholesale price is fixed and thus this decision variable is also excluded.

The second distinctive feature is linked to production capacity. In contrast to the inventory game with endogenous demand, the finite capacity of both the supplier and retailer implies that they produce, deliver and process at a rate not exceeding some predetermined maximum number of products per time unit. This complicates the problem by introducing multiple switching points which are induced by competing inventory decisions and capacity limitations. This is to say, as we look at differential inventory games with exogenous demand, we will be focusing on the sole effect of inventory dynamics on production decisions and associated costs.

We assume that both the supplier and the retailer have warehouses of infinite capacity for holding end-products. If, at a time point t, the cumulative number of products processed by the retailer exceeds the cumulative demand for the products, an inventory holding cost is incurred at t, h_r^+, per product and time unit. Otherwise, a backlog cost is incurred, h_r^-. The latter stipulation implies that all deficient products from the retailer's side will be backlogged and delivered to the customers when the retailer catches up with processing. This was also the case with the inventory game of the previous section. Similarly, if cumulative production by the supplier exceeds cumulative processing by the retailer, an inventory holding cost is incurred by the supplier, h_s^+. Otherwise there is a shortage cost paid, h_s^-. Any shortage of products at the supplier's side is immediately replenished by delivering products to the retailer from a safety stock. The safety stock will be restored as the supplier catches up with production, i.e., as soon as possible. We assume that the cost associated with the risk of depleting the safety stock is higher than that of holding the safety stock. Therefore, the adopted safety stock level, Q_s, is sufficiently high to cope with seasonal fluctuations in the retailer's orders.

The retailer's backlog cost is traditionally related to loss of customer goodwill. On the other hand, the supplier's shortage cost is related to the risk of depleting the safety stock. Indeed, if the cost, R, of risk associated with one product lacking in the safety stock for one time unit is greater than that of holding one unit in the safety stock for one time unit, h_S, then a shortage at time t, X_s^-, in the safety stock Q_s, $Q_s > X_s^-$, induces the following cost at t for one time unit

$$h_S(Q_s - X_s^-) + RX_s^- = h_S Q_s + (R - h_s)X_s^-.$$

Defining the difference between the risk and the holding costs, $R - h_S$, as the supplier's unit backlog or shortage cost $h_s^- = R - h_S$, we observe that due to the linearity of our model, the safety stock cost $h_S Q_s$ is a constant that does not affect the optimization.

Since the demand is periodic (seasonal), the objective of each party (the supplier and the retailer) is to find a cyclic production/processing rate, which minimizes all inventory-related costs over an infinite planning horizon.

The retailer's problem

$$\min_{u_r} J_r(u_r, u_s) = \min_{u_r} \int_{t_s}^{t_f} h_r(X_r(t)) dt \qquad (4.93)$$

s.t.

$$\dot{X}_r(t) = u_r(t) - d(t); \qquad (4.94)$$
$$0 \le u_r(t) \le U_r, \qquad (4.95)$$

where $X_r(t)$, $X_s(t)$ are the inventory levels of the retailer and supplier at time t respectively; $u_r(t)$, $u_s(t)$ are the retailer's and supplier's processing/production rates respectively; and U_r, U_s are the maximal production rates of the retailer and supplier respectively. The only term in (4.93) accounts for the retailer's inventory costs:

$$h_r(X_r(t)) = h_r^+ X_r^+(t) + h_r^- X_r^-(t), \ X^+(t) = \max\{X(t), 0\}, \ X^-(t) = \max\{-X(t), 0\}.$$

We assume that the customer demand rate for products, $d(t)$, is periodic and step-wise:

$$d(t) = d_2 > U_r, \ t_1^d + jT < t \le t_2^d + jT, \ j = 1, 2, \dots$$
$$d(t) = d_1 < U_s, \ t_2^d + (j-1)T < t \le t_1^d + jT, \ j = 1, 2, \dots.$$

Assume the system has reached the steady-state on an infinite planning horizon with its limit cycles T so that:

$$X_r(t_s) = X_r(t_f) = X_r \quad \text{and} \quad X_s(t_s) = X_s(t_f) = X_s, \qquad (4.96)$$

where t_s and t_f are the time points where a limit cycle starts and ends and $T = t_f - t_s$.

The supplier's problem

$$\min_{u_s} J_s(u_s, u_r) = \min_{u_s} \int_{t_s}^{t_f} h_s(X_s(t)) dt \qquad (4.97)$$

s.t.

$$\dot{X}_s(t) = u_s(t) - u_r(t); \qquad (4.98)$$
$$0 \le u_s(t) \le U_s, \qquad (4.99)$$

where

$$h_s(X_s(t)) = h_s^+ X_s^+(t) + h_s^- X_s^-(t)$$

is the supplier's inventory cost. It can be readily seen that both the supplier's and the retailer's problems are quite symmetric. The only difference seems to be between the dynamics of (4.98) and (4.94), where customer demand d in (4.94) is replaced with the retailer's processing rate u_r in (4.98). These dynamics, however, are symmetric as well because we assume that the processing rate of the retailer u_r is the retailer's demand (d_r) ordered from the supplier. Thus we could set demand for the supplier $d_r(t) = u_r(t)$ to make the dynamics symmetric.

The centralized problem

$$\min_{u_s, u_r}[J_s(u_s, u_r) + J_r(u_s, u_r)] = \min_{u_s}\{\int_{t_s}^{t_f}[h_s(X_s(t)) + h_r(X_r(t))]dt\}, \quad (4.100)$$

s.t.

$$(4.94)-(4.96), (4.98)-(4.99).$$

System-wide optimal solution

To study the centralized problem, we construct the Hamiltonian:

$$H(t) = -h_s(X_s(t)) - h_r(X_r(t)) + \psi_s(t)(u_s(t) - u_r(t)) + \psi_r(t)(u_r(t) - d(t)), \quad (4.101)$$

and the system of the co-state differential equations with co-state variables $\psi_s(t)$ and $\psi_r(t)$:

$$\dot{\psi}_s(t) = \frac{\partial\, h_s(X_s(t))}{\partial\, X_s(t)} \text{ and } \dot{\psi}_r(t) = \frac{\partial\, h_r(X_r(t))}{\partial\, X_r(t)} \qquad (4.102)$$

and the boundary constraints:

$$\psi_s(t_s) = \psi_s(t_f) = \psi_s \text{ and } \psi_r(t_s) = \psi_r(t_f) = \psi_r \quad . \qquad (4.103)$$

This leads to

$$\dot{\psi}_s(t) = \begin{cases} h_s^+, \text{ if } X_s(t) > 0 \\ -h_s^-, \text{ if } X_s(t) < 0 \\ h_s \in [-h_s^-, h_s^+], \text{ if } X_s(t) = 0 \end{cases}, \quad \dot{\psi}_r(t) = \begin{cases} h_r^+, \text{ if } X_r(t) > 0 \\ -h_r^-, \text{ if } X_r(t) < 0 \\ h_r \in [-h_r^-, h_r^+], \text{ if } X_r(t) = 0 \end{cases} \qquad (4.104)$$

Similar to the previous sections, the Hamiltonian is interpreted as the instantaneous profit rate. This includes the profit $\psi_s \dot{X}_s$ and $\psi_r \dot{X}_r$ from the increments in inventory level of the supplier and retailer respectively, which are created by processing u_r and producing u_s products. The co-state variables $\psi_r(t)$ and $\psi_s(t)$ are the shadow prices, i.e, the net benefits from reducing inventory surplus/shortage by one more unit on the part of the retailer and supplier respectively. Each differential equation of (4.104) states that the marginal profit of either the suppler or retailer (and thus the overall supply chain) from reducing his inventory level at time t, when there is a surplus (otherwise from reducing inventory shortage) is equal to the corresponding unit holding cost per time unit (or unit shortage cost).

Applying the maximum principle, we maximize the Hamiltonian at each time point with respect to the retailer's processing rate u_r and the supplier's production rate u_s. This results in the following optimality conditions.

$$u_s^*(t) = \begin{cases} U_s, \text{if } \psi_s(t) > 0, \text{ (working regime- PR);} \\ u \in [0, U_s], \text{if } \psi_s(t) = 0, \text{ (singular regime- SR);} \\ 0, \text{if } \psi_s(t) < 0, \text{ (idle regime- IR).} \end{cases} \quad (4.105)$$

$$u_r^*(t) = \begin{cases} U_r, \text{if } \psi_r(t) - \psi_s(t) > 0, \text{(working regime - PR);} \\ u \in [0, U_r], \text{if } \psi_r(t) - \psi_s(t) = 0, \text{ (singular regime - SR);} \\ 0, \text{if } \psi_r(t) - \psi_s(t) < 0 \text{(idle regime - IR).} \end{cases} \quad (4.106)$$

From (4.105)-(4.106) one can observe that in production/processing regimes PR as well as idling IR, the optimal production/processing rate is uniquely determined. The optimal control in the SR regime requires more analysis, as shown below. We will use the notations of the form $r \in SR$ and $s \in SR$ that say that the retailer (r) and the supplier (s) are in a SR regime at a specific time interval.

The optimal solution determined by conditions (4.105) and (4.106) depends on the relationship between the inventory costs. For different relationships there will be different optimal sequences of the regimes. We present here one possible solution by assuming that the unit inventory holding cost of the retailer is greater than the supplier's, while the backlog cost of the retailer is lower than the supplier's

$$h_r^- < h_s^-, \ h_r^+ > h_s^+, \ h_s^+ \neq h_r^-, \ h_s^- \neq h_r^+.$$

We will show that according to this assumption, the optimal solution ensures that there will be no backlog at the supplier's side (and thus no use for a safety stock), because the retailer takes into account the supplier's inventories in a centralized supply chain. In addition, we assume that the supplier's capacity is lower than the retailer's maximum production rate,

$$U_s < U_r.$$

Otherwise, the supplier simply follows the processing rate of the retailer and there are no inventory dynamics. In such a case, we only need to find the optimal solution for the retailer. Clearly, a cyclic solution to the problem exists if the supplier has enough capacity to satisfy the demand over each cycle of length T, as the following proposition states.

Proposition 4.17. *There always exists a cyclic solution if and only if*

$$T \geq \frac{d_2 - d_1}{U_s - d_1}(t_2^d - t_1^d). \quad (4.107)$$

Proof: If a cyclic solution exists, then the supplier, who has the smallest maximum production rate in the supply chain, should satisfy the demand. That is, his maximum production over the entire period, $U_s T$, should exceed the demand over the same period:

$$TU_s \geq d_1 \left(T - (t_2^d - t_1^d) \right) + d_2 (t_2^d - t_1^d). \tag{4.108}$$

By rearranging the terms in (4.108), inequality (4.107) is immediately obtained.

Next, assume that condition (4.107) is satisfied. We show that there is at least one cyclic solution. To simplify the discussion, we further assume that

$$TU_s = d_1 \left(T - (t_2^d - t_1^d) \right) + d_2 (t_2^d - t_1^d).$$

We then let $u_s(t) = u_r(t) = U_s$, $t_s \leq t \leq t_f$. So for the retailer we have

$$\int_{t_s}^{t_f} (u_r(t) - d(t)) dt = TU_s - \{d_2(t_2^d - t_1^d) + d_1(T - t_2^d + t_1^d)\} = 0.$$

Since the firms have the same production rate, they will have the same cumulative production and inventory will remain the same at the beginning of a cycle for both the supplier and retailer. Thus, we constructed a cyclic solution. The above argument would still be valid even if (4.108) were a strict inequality and we would be simply producing only for a part of the cycle.

We next study the singular regimes.

Proposition 4.18. If $r \in SR$ in a time interval $\tau \subset [0,T]$, then $X_r(t) = 0$ and/or $X_s(t) = 0$. If $s \in SR$ in a time interval $\tau \subset [0,T]$, then $X_s(t) = 0$, $t \in \tau$.

In case in a time interval $\tau \subset [0,T]$, $X_s(t) = 0$, then $u_s(t) = u_r(t)$, if $X_r(t) = 0$, then $u_r(t) = d(t)$ for $t \in \tau$.

Proof: By definition, in SR, $\psi_r(t) = \psi_s(t)$, $t \in \tau$ if $r \in SR$ and $\psi_s(t) = 0$ if $s \in SR$. Differentiating these equalities we have:

$$\dot{\psi}_r(t) = \dot{\psi}_s(t), \tag{4.109}$$

$$\dot{\psi}_s(t) = 0. \tag{4.110}$$

According to (4.104), the equalities (4.109) and (4.110) can be satisfied if and only if $X_r(t) = 0$ and/or $X_s(t) = 0$ for $r \in SR$ and if $X_s(t) = 0$ for $s \in SR$.

Finally, from the dynamic equations (4.98) and (4.94) we observe: in case $X_s(t) = 0$, then $u_s(t) = u_r(t)$, if $X_r(t) = 0$, then $u_r(t) = d(t)$.

To describe the results, we further partition the SR regime into SR1 if $u_r^* = d$, and SR2 otherwise.

We now use a constructive approach to solve the centralized problem. That is, we first propose a solution, and then we show this solution is indeed optimal. The optimal policy we are proposing is the following:

- Retailer: Use the SR1-PR-SR2-SR1 (producing/processing at the demand rate (SR1) first, then at the maximal rate (PR), then at the rate of the supplier (SR2), and finally again at the demand rate (SR1)) processing sequence with switching times t_1^r, t_2^r, t_2^s, $t_1^r \le t_2^r \le t_2^s$.

- Supplier: Use the SR1-PR-SR1 sequence with switching times t_1^s, t_2^s, $t_1^r \ge t_1^s$.

This policy, illustrated in Figure 4.8, is more rigorously defined in the following proposition.

Proposition 4.19. *The control policy:*

(i) $u_r(t) = u_s(t) = d_1$, $X_r(t) = X_s(t) = 0$, $\psi_r(t) = \psi_s(t) = 0$, $t \in [t_s, t_1^s]$ *and* $t \in [t_2^s, t_f]$; $X_r(t) = 0$, $\psi_r(t) = 0$, $t \in [t_s, t_f]$.

(ii) $u_r(t) = d_1$, $t \in [t_1^s, t_1^r]$; $u_r(t) = U_r$, $t \in [t_1^r, t_2^r]$; $u_r(t) = U_s$, $t \in [t_2^r, t_2^s]$. $X_r(t) = 0$, $t \in [t_1^s, t_1^r]$, $X_r(t_3^r) = 0$, $\psi_r(t) = h_s^+$, $t \in [t_1^s, t_1^r]$, $\psi_r(t) = h_r^+$, $t \in [t_1^r, t_3^r]$, $\psi_r(t) = -h_r^-$, $t \in [t_3^r, t_2^s]$.

(iii) $u_s(t) = U_s$, $t \in [t_1^s, t_2^s]$; $X_s(t) = 0$, $t \in [t_2^r, t_2^s]$; $\psi_r(t) = \psi_s(t) = h_s^+$, $t \in [t_1^s, t_2^r]$, $\dot\psi_r(t) = \dot\psi_s(t) = -h_r^-$, $t \in [t_2^r, t_2^s]$.

provides the system-wide optimal solution.

Proof: Consider $t \in [t_s, t_1^s]$. According to (i) of Proposition 4.19, $u_r(t) = u_s(t) = d_1$, $X_r(t) = X_s(t) = 0$, $\psi_r(t) = \psi_s(t) = 0$. Therefore (4.94)-(4.95), (4.98)-(4.99), and (4.104)-(4.106) are satisfied and (4.100) is maximized.

Now consider $t \in [t_1^s, t_1^r]$. In this interval $u_r(t) = d_1$, $u_s(t) = U_s$, $X_s(t) > 0$, $\psi_r(t) = \psi_s(t) = h_s^+$, $X_r(t) = 0$. Again it is easy to check that co-state equations (104) are satisfied and that (4.100) is maximized. If $t \in [t_1^r, t_3^r]$, then, recalling our assumptions on the relationships between inventory costs, we find,

$$\psi_r(t) = \psi(t_1^r) + h_r^+ t > \psi_s(t) = \psi_s(t_1^r) + h_s^+ t, \text{ for } t_1^r < t \le t_3^r,$$

that is, the optimality conditions (4.106) are satisfied.

The proof of the proposition for the remaining time intervals is similar and therefore omitted.

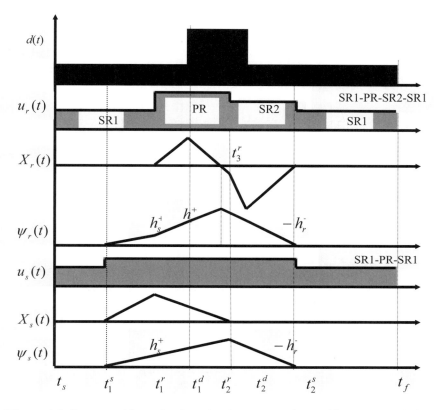

Figure 4.8. System-wide optimal production, processing and inventory policies

Drawing upon Proposition 4.19 we can detail the system-wide optimal solution by finding the switching time points. To proceed, we first integrate the retailer's differential equation with boundary conditions (4.96) to obtain:

$$(U_r - d_1)(t_1^d - t_1^r) + (U_r - d_2)(t_2^r - t_1^d) + (U_s - d_2)(t_2^d - t_2^r) + (U_s - d_1)(t_2^s - t_2^d) = 0$$
(4.111)

Similarly for the supplier we find:

$$(U_s - d_1)(t_1^r - t_1^s) = -(U_s - U_r)(t_2^r - t_1^r).$$
(4.112)

The following equation for t_3^r, at which the inventory of the retailer reaches zero level, is derived from equation (4.111):

$$(U_r - d_1)(t_1^d - t_1^r) + (U_r - d_2)(t_3^r - t_1^d) = 0 . \tag{4.113}$$

By integrating the co-state differential equations, we find for the retailer

$$h_s^+(t_1^r - t_1^s) + h_r^+(t_3^r - t_1^r) - h_r^-(t_2^s - t_3^r) = 0 , \tag{4.114}$$

and for the supplier:

$$h_s^+(t_2^r - t_1^s) - h_r^-(t_2^s - t_2^r) = 0 . \tag{4.115}$$

Solving system (4.111)-(4.115) we find:

$$t_1^r = \frac{\dfrac{d_2 - d_1}{U_r - d_2}(h_r^+ + h_r^-)t_1^d + \dfrac{d_2 - d_1}{U_s - d_1}h_r^-(t_2^d - t_1^d) + t_1^s(h_s^+ + h_r^-)}{h_s^+ - h_r^+ + \dfrac{U_r - d_1}{U_r - d_2}(h_r^+ + h_r^-)} ; \tag{4.116}$$

$$t_2^r = \frac{d_2 - d_1}{U_s - d_1}\frac{c_r^-}{c_r^+ + c_r^-}(t_2^d - t_1^d) + t_1^s ; \tag{4.117}$$

$$t_2^s = \frac{d_2 - d_1}{U_s - d_1}(t_2^d - t_1^d) + t_1^s ; \tag{4.118}$$

$$t_1^s = \frac{\dfrac{\dfrac{d_2 - d_1}{U_r - d_2}(h_r^+ + h_r^-)t_1^d + \dfrac{d_2 - d_1}{U_s - d_1}h_r^-(t_2^d - t_1^d)}{h_s^+ - h_r^+ + \dfrac{U_r - d_1}{U_r - d_2}(h_r^+ + h_r^-)}(U_r - d_1) - \dfrac{d_2 - d_1}{U_s - d_1}\cdot\dfrac{c_r^-}{c_s^+ + c_r^-}(t_2^d - t_1^d)(U_r - U_s)}{U_r - d_1 - \dfrac{h_s^+ + h_r^-}{h_s^+ - h_r^+ + \dfrac{U_r - d_1}{U_r - d_2}(h_r^+ + h_r^-)}} .$$

$$\tag{4.119}$$

An interesting observation follows from these results and Proposition 4.19. In the centralized supply chain the retailer processes products at maximum rate when he has an inventory surplus; when there is a shortage, the retailer reduces the processing rate to that equal to the maximum production rate of the supplier (see regime SR2 in Figure 4.8). This allows the supplier to catch up and prevents backlogs at the supplier's facilities. The explanation is straightforward; the retailer's backlog cost is lower than the supplier's. The cooperative retailer evidently takes this into account by reducing his orders. On the other hand, since the inventory holding cost of the retailer is higher than that of the supplier, in cases of surplus, the retailer does not reduce the processing rate. This induces the supplier (who has a smaller capacity) to increase his inventories in advance. As we show below in the game analysis, the supply chain performance deteriorates because the non-cooperative retailer ignores the supplier's backlogs.

Similar to the previous section, to obtain a static analogue of the centralized inventory problem, we eliminate inventory dynamics. This is accomplished by replacing $X_r(t)$ with $u_r(t)-d(t)$ and $X_s(t)$ with $u_s(t)- u_r(t)$ in the objective function. The myopic solution for the low demand d_1 is trivial – produce and process products at the rate of d_1. We thus have a simple problem with piece-wise linear objective function for high demand d_2 and three possible solutions to check out,

$u_r= U_r$ and $u_s= U_s$, then $J(u_s, u_r)=- h_r^- (U_r - d_2 - X_r^-)- h_s^- (U_s - U_r - X_s^-)$;

$u_r= U_s$ and $u_s= U_s$, then $J(u_s, u_r)=- h_r^- (U_s - d_2 - X_r^-)$;

$u_r= 0$ and $u_s= 0$, then $J(u_s, u_r)= h_r^- d_2 - h_r^- X_r^-$,

while the other possible combinations are obviously not optimal. Since $d_2>U_s$ and $h_r^- < h_s^-$, we readily find that the minimum cost is $J(u_s, u_r)= - h_r^- (U_s - d_2 - X_r^-)$, that is,

when demand is high, the myopic retailer's order rate is lower than the system-wide optimal retailer's processing rate. Both the retailer and the supplier have no inventory surplus, i.e., inventories decrease and thus the retailer's backlog increases. The supplier has no backlog.

Accordingly, the myopic centralized supply chain does not employ possibility of inventory accumulation, which naturally increases backlogs when demand is high. Since the supplier's backlog cost is higher, the retailer takes all backlogs on himself by reducing his processing rate.

The multi-echelon effect

Based on the optimal policies proven in Proposition 4.19 for a two-echelon centralized supply chain, we can outline the solution for a multi-echelon supply chain.

To facilitate the presentation, we now reformulate the centralized problem and present the notion of the restricting firm or agent. Let a multi-echelon supply chain contain I firms. Then the inventory dynamics of all firms can be described by the following differential equations:

$$\dot{X}_i(t) = u_i(t) - u_{i+1}(t), \quad i=1,2,..,I-1;$$
$$\dot{X}_i(t) = u_i(t) - d(t), \quad i = I. \tag{4.120}$$

where $X_i(t)$ is the inventory level accumulated by firm i at time t and $u_i(t)$, $i=1,2,...,I$ is the production rate for firm i,

$$0 \le u_i(t) \le U_i, \tag{4.121}$$

with U_i being the maximal production (processing, distribution and so on) rate of firm i. We assume that firm $i=I$ is the retailer, while firms $i=1,..,I-1$ are consecutive suppliers and/or distributors.

The steady-state conditions on limit cycles T are:

$$X_i(t_s) = X_i(t_f) = X, \quad i = 1,\ldots,I, \tag{4.122}$$

The objective is to minimize the total inventory-related cost:

$$\int_{t_s}^{t_f} \sum_{i=1}^{I} h(X_i(t)) dt \to \min \tag{4.123}$$

Then the Hamiltonian for this problem is:

$$H = -C\,(X_i(t)) + \sum_{i<I} \psi_i\,(t)(u_i(t) - u_{i+1}(t)) + \psi_I(t)(u_I(t) - d(t)), \tag{4.124}$$

and the system of the co-state differential equations with co-state variables $\psi_i(t)$:

$$\dot{\psi}_i(t) = \begin{cases} h_i^+, & \text{if } X_i(t) > 0 \\ -h_i^-, & \text{if } X_i(t) < 0 \\ h_i \in [-h_i^-, h_i^+], & \text{if } X_i(t) = 0 \end{cases}, \quad i = 1,..,I \tag{4.125}$$

with the boundary constraints:

$$\psi_i(t_s) = \psi_i(t_f) = \psi_i. \tag{4.126}$$

We employ the same assumptions

$$h_{i+1}^- < h_i^-, \ h_{i+1}^+ > h_i^+, \ h_i^+ \neq h_{i+1}^-, \ h_i^- \neq h_{i+1}^+ \ \ i = 1,2,..,I-1$$

and the following definition.

Definition 4.1

Firm i' is restricting if either $i' = I$, or $U_{i'} < U_i$ for all $i > i'$, $i' \neq I$.

With respect to this definition, Figure 4.9 shows the system-wide optimal solution for a four-echelon supply chain which contains only two restricting firms. Note, that since non-restricting firms simply follow the production plan of the adjacent downstream restricting firms, we only need to find the optimal solution for the restricting firms.

Based on the optimal policies of Proposition 4.19 for two restricting firms, it is now easy to conjecture the optimal behavior for the system with an unlimited number of restricting firms. Figure 4.10 illustrates a case involving three restricting firms, which satisfies all conditions of the maximum principle. There are two changes compared to the case with two restricting firms. The first is that there is a time lag when switching from the SR1 to the PR (early switching) for every restricting firm i' relative to its adjacent downstream restricting firm $i > i'$ or to the last firm (the retailer I).

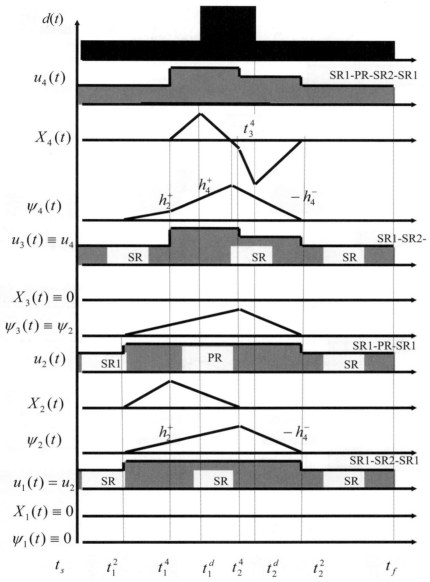

Figure 4.9. System-wide optimal solution of four-echelon supply chain with two restricting firms, $h_1^+ < h_2^+ < h_3^+ < h_4^+$ and $h_1^- > h_2^- > h_3^- > h_4^-$

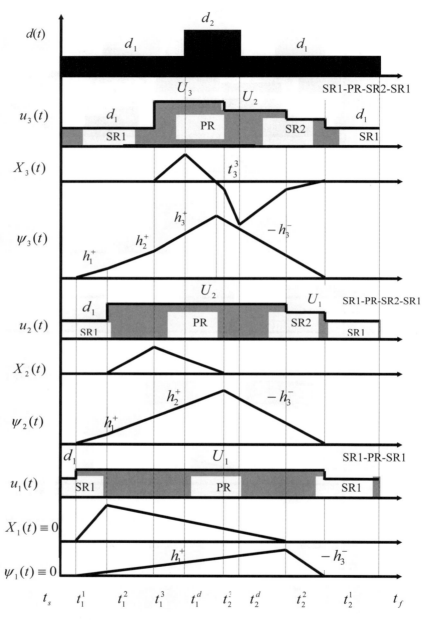

Figure 4.10. System-wide optimal solution of three-echelon supply chain with three restricting firms, $h_1^+ < h_2^+ < h_3^+$ and $h_1^- > h_2^- > h_3^-$

The other change is that the SR2 for every restricting firm, i', is split into a number of steps. The control value of every step is equal to the maximal production rate of the corresponding upstream restricting firm which prevents these firms from experiencing higher backlog costs.

Game analysis

Since we have assumed from the very beginning that the supplier always provides timely deliveries by means of available inventories and, if necessary, from the safety stock, the retailer's decisions are independent of those made by the supplier. Thus the parties have asymmetric power and the game between them is as follows.

The retailer first analyzes the demand and determines his processing rate or equivalently the quantities per time unit to order. Next, the supplier analyzes the retailer's orders during the production cycle and determines his production plan to ensure timely deliveries. The supplier then produces and delivers the products which the retailer processes and sells. In this sequence of decisions, the retailer is the Stackelberg leader and the supplier is the follower. We address first the supplier's best response.

To optimize the supplier's production plan, we construct the Hamiltonian:

$$H_s(t) = -h_s(X_s(t)) + \psi_s^s(t)(u_s(t) - u_r(t)) + \psi_r^s(t)(u_r(t) - d(t)), \quad (4.127)$$

and the system of the co-state differential equations

$$\dot{\psi}_s^{\,s}(t) = \begin{cases} h_s^+, & \text{if } X_s(t) > 0 \\ -h_s^-, & \text{if } X_s(t) < 0 \\ h_s \in [-h_s^-, h_s^+], & \text{if } X_s(t) = 0 \end{cases} \quad \text{and } \dot{\psi}_r^{\,s}(t) = 0 \qquad (4.128)$$

with the boundary constraints:

$$\psi_s^{\,s}(t_s) = \psi_s^{\,s}(t_f) = \psi_s^{\,s}, \ \psi_r^{\,s}(t_s) = \psi_r^{\,s}(t_f) = \psi_r^{\,s}. \qquad (4.129)$$

Note that since the constant ψ_r^s does not affect the optimization it can be set at zero. This implies that the Hamiltonian's instantaneous profit rate includes only the profit $\psi_s^s \dot{X}_s$ from the increment in the supplier's inventory level.

Maximizing the Hamiltonian at each time point with respect to the supplier's production rate u_s we obtain the condition similar to (4.105):

$$u_s(t) = \begin{cases} U_s, \text{if } \psi_s^{\,s}(t) > 0, & \text{(working regime - PR)}; \\ u \in [0,U_s], \text{if } \psi_s^{\,s}(t) = 0, & \text{(singular regime - SR)}; \\ 0, \text{if } \psi_s^{\,s}(t) < 0, & \text{(idle regime - IR)}. \end{cases} \quad (4.130)$$

Analogous to Proposition 4.18, the singular regime condition is resolved by differentiating $\psi_s^{\,s}(t) = 0$, which readily results in $X_s(t) = 0$, and therefore, $u_s(t) = u_r(t)$.

Assuming that the retailer's order $u_r(t)$ is characterized by a constant maximum rate at an interval of time $[t_{s1},t_{s2})$, we now construct an optimal production policy as the SR1-PR-SR1 sequence with switching points t_1, t_2 and show that it is optimal.

Proposition 4.20. *Let $u_r(t)=d_1$ for $t_s \le t < t_{s1}$ and $t_{s2} \le t \le t_f$, $u_r(t)=U_r$ for $t_{s1} \le t < t_{s2}$. The supplier's production policy: $u_s(t) = d_1$ for $t_s \le t < t_1$ and $t_2 \le t \le t_f$, $u_s(t)=U_s$ for $t_1 \le t < t_2$, where*

$$t_2 = t_1 + \frac{U_r - d_1}{U_s - d_1}(t_{s2} - t_{s1}), t_1 = \frac{U_r - d_1}{U_s - d_1}t_{s1} + \frac{U_s - U_r}{U_s - d_1}t',$$

$$t' = t_{s1} + \frac{h_s^-}{h_s^+ + h_s^-}(t_{s2} - t_{s1})$$

is optimal

Proof: Consider the following solution for the state and co-state variable:

(i) $u_s(t) = d_1$, $X_s(t) = 0$, $\psi_s^{\,s}(t) = 0$, for $t_s \le t < t_1$ and $t_2 \le t \le t_f$;

(ii) $u_s(t) = U_s$, for $t_1 \le t < t_2$, $X_s > 0$ and $\dot{\psi}_s^{\,s}(t) = h_s^+$ for $t_1 < t < t'$, $X_s < 0$ and $\dot{\psi}_s^{\,s}(t) = -h_r^-$ for $t' < t < t_2$.

This solution evidently meets the optimality conditions (4.130). To find the switching points, we first integrate the co-state equation from (ii) with initial and terminal conditions from (i). This results in

$$h_s^+(t_1 - t') - h_s^-(t' - t_2) = 0. \quad (4.131)$$

Similarly, by integrating state equation (4.98) with controls defined by (ii) and initial as well terminal conditions from (i) we have

$$(U_s - d_1)(t_{s1} - t_1 + t_2 - t_{s2}) + (U_s - U_r)(t_{s2} - t_{s1}) = 0. \quad (4.132)$$

$$(U_s - d_1)(t_{s1} - t_1) + (U_s - U_r)(t' - t_{s1}) = 0. \quad (4.133)$$

Equations (4.131)-(4.133) constitute a system of three equations in three unknowns t_1, t' and t_2, whose solution is feasible, $t_2 > t' > t_1$ and is as stated in this proposition.

Next, to optimize the retailer's processing plan, we construct the Hamiltonian:

$$H_r(t) = -h_r(X_r(t)) + \psi_s^r(t)(u_s(t) - u_r(t)) + \psi_r^{\ r}(t)(u_r(t) - d(t)) \quad (4.134)$$

and the system of the co-state differential equations

$$\dot{\psi}_r^{\ r}(t) = \begin{cases} h_r^+, & \text{if } X_r(t) > 0 \\ -h_r^-, & \text{if } X_r(t) < 0 \\ h_r \in [-h_s^-, h_s^+], & \text{if } X_r(t) = 0 \end{cases} \quad \text{and } \dot{\psi}_s^{\ r}(t) = 0 \qquad (4.135)$$

with the boundary constraints:

$$\psi_r^{\ r}(t_s) = \psi_s^{\ r}(t_f) = \psi_s^{\ r} \text{ and } \psi_r^{\ r}(t_s) = \psi_r^{\ r}(t_f) = \psi_r^{\ r}. \qquad (4.136)$$

Note that since the constant $\psi_s^{\ r}$ can be set at zero, the optimization is not affected. This implies that, the Hamiltonian's instantaneous profit rate includes only the profit $\psi_r^{\ r} \dot{X}_r$ from the increment in the retailer's inventory level.

Applying the maximum principle, we obtain the familiar condition:

$$u_r(t) = \begin{cases} U_r, & \text{if } \psi_r^{\ r}(t) > 0, \text{ (working regime-PR)}; \\ u \in [0, U_r], & \text{if } \psi_r^{\ r}(t) = 0, \text{ (singular regime-SR)}; \\ 0, & \text{if } \psi_r^{\ r}(t) < 0, \text{ (idle regime-IR)}, \end{cases} \quad (4.136)$$

where the singular regime condition is resolved by differentiating $\psi_r^{\ r}(t) = 0$, which readily results in $X_r(t) = 0$, and therefore, $u_r(t) = d(t)$.

The retailer's best solution depends only on the demand function. It can easily be seen that the following proposition is very similar to Proposition 4.20 and can be derived by simply replacing the corresponding variables.

Proposition 4.21. *The retailer's processing policy:* $u_r(t) = d_1$ *for* $t_s \leq t < t_1''$ *and* $t_2'' \leq t \leq t_f$, $u_r(t) = U_r$ *for* $t_1'' \leq t < t_2''$, *where*

$$t_2'' = t_1'' + \frac{d_2 - d_1}{U_r - d_1}(t_2^{\ d} - t_1^{\ d}), t_1'' = \frac{d_2 - d_1}{U_r - d_1}t_1^{\ d} + \frac{U_r - d_2}{U_r - d_1}t'',$$

$$t'' = t_1^{\ d} + \frac{h_r^-}{h_r^+ + h_r^-}(t_2^d - t_1^d)$$

is optimal.

Equilibrium

The next result immediately follows from Propositions 4.17, 4.20-4.21 by simply setting $t_{s2}=t_2$" and $t_{s1}=t_1$".

Proposition 4.22. *The supplier's production rate* $u_s^s(t) = d_1$ *for* $t_s \leq t < t_1$ *and* $t_2 \leq t \leq t_f$, $u_s^s(t) = U_s$ *for* $t_1 \leq t < t_2$, *and the retailer's processing rate* $u_r^s(t) = d_1$ *for* $t_s \leq t < t_1$" *and* t_2"$\leq t \leq t_f$, $u_r^s(t) = U_r$ *for* t_1"$\leq t < t_2$",*where all switching points are determined by Propositions 4.20 and 4.21 with* $t_{s1}=t_1$" *and* $t_{s2}=t_2$", *constitute the unique Stackelberg equilibrium for the differential inventory game.*

Comparing the equilibrium with the system-wide optimal solution of Proposition 4.19 and referring to the time intervals $[t_1, t_2)$ and $[t_1$", t_2"), as the periods of response to higher demand (d_2) by the supplier and retailer respectively, we conclude with the following observation.

Proposition 4.23. *In vertical inventory competition, the retailer reduces the response period to higher demand and increases the processing rate to the maximum compared to the system-wide optimal solution; the supplier's response period does not change and, in contrast to the system-wide optimal solution, the supplier incurs a backlog.*

Proof: The proof is immediately apparent since the retailer's equilibrium solution no longer involves the SR2 compared to the centralized solution. That is, the retailer no longer reduces his processing rate during the response period to higher demand and does not allow the supplier to catch up without backlogging. The supplier's response period remains unchanged since neither the total retailer's order nor the supplier's production policy (production at maximum rate) changes.

Figure 4.11 presents the equilibrium solution. Proposition 4.23 implies that the supply chain performance deteriorates under the vertical intertemporal competition since the retailer, when ordering products, ignores the supplier's inventory level and associated costs.

Using the same argument as for the system-wide myopic solution, the performance of the supply chain will further deteriorate if a myopic approach is employed since neither inventory accumulation nor storage is involved. As a result, inventory backlogs increase on both sides.

Example 4.4.

Consider a three-echelon supply chain with system parameters presented in Table 4.2. The demand rates are $d_1 = 1.5$ and $d_2 = 4.5$, $t_s = 0$, $t_1^d = 5.0$,

$t_2^d = 8.0$ and $T = 12$. Firms indexed by $i=1$ and $i=2$ are the suppliers and $i=3$ is the retailer.

Table 4.2. System parameters

Firm Index	Unit inventory costs		Maximal production rate
I	h_i^+	h_i^-	U_i
1	0.4	1.1	3.5
2	0.5	1.0	2.5
3	0.7	0.8	3.0

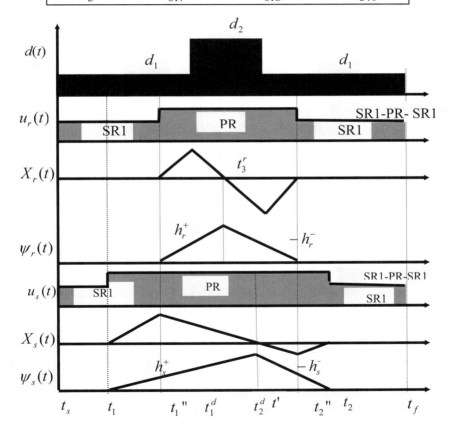

Figure 4.11. Stackelberg equilibrium production, processing and inventory policies

We first identify the restricting firms. According to Definition 4.1, these are firms $i=3$ and $i=2$ (the retailer, $i=3$, is referred as r and the nearest upstream supplier, $i=2$ is referred as s). In addition we verify the necessary and sufficient condition of the limit cycle (4.107):

$$\frac{d_2 - d_1}{U_s - d_1}(t_2^d - t_1^d) = \frac{4.5 - 1.5}{2.5 - 1.5}(8 - 5) = 9.$$

Since $T=12>9$, we conclude that the limit cycle exists (see Proposition 4.17).

Next, we employ Proposition 4.21 to find the retailer's switching points

$$t'' = t_1^d + \frac{h_r^-}{h_r^+ + h_r^-}(t_2^d - t_1^d) = 5.0 + \frac{0.8}{0.7 + 0.8}(8.0 - 5.0) = 6.6,$$

$$t_1'' = \frac{d_2 - d_1}{U_r - d_1}t_1^d + \frac{U_r - d_2}{U_r - d_1}t'' = \frac{4.5 - 1.5}{3 - 1.5}5 + \frac{3 - 4.5}{3 - 1.5}6.6 = 3.4,$$

$$t_2'' = t_1'' + \frac{d_2 - d_1}{U_r - d_1}(t_2^d - t_1^d) = 3.4 + \frac{4.5 - 1.5}{3 - 1.5}(8 - 5) = 9.4,$$

and the retailer's Stackelberg processing policy

$$u_r^s(t) = 1.5 \text{ for } 0 \le t < 3.4 \text{ and } 9.4 \le t \le 12, \ u_r^s(t) = 3 \text{ for } 3.4 \le t < 9.4.$$

Similarly, by setting $t_{s1}=t_1''$ and $t_{s2}=t_2''$ with respect to Proposition 4.22 and employing Proposition 4.20, we find the supplier's switching points

$$t' = t_{s1} + \frac{h_s^-}{h_s^+ + h_s^-}(t_{s2} - t_{s1}) = 3.4 + \frac{1}{0.5 + 1}(9.4 - 3.4) = 7.4,$$

$$t_1 = \frac{U_r - d_1}{U_s - d_1}t_{s1} + \frac{U_s - U_r}{U_s - d_1}t' = \frac{3 - 1.5}{2.5 - 1.5}3.4 + \frac{2.5 - 3}{2.5 - 1.5}7.4 = 1.4,$$

$$t_2 = t_1 + \frac{U_r - d_1}{U_s - d_1}(t_{s2} - t_{s1}) = 1.4 + \frac{3 - 1.5}{2.5 - 1.5}(9.4 - 3.4) = 10.4,$$

and the supplier's equilibrium production policy

$$u_s^s(t) = 1.5 \text{ for } 0 \le t < 1.4 \text{ and } 10.4 \le t \le 12, \ u_s^s(t) = U_s \text{ for } 1.4 \le t < 10.4.$$

Next, to determine the system-wide optimal solution, we return to indexes $i=1,2,3$. From equation (4.119) we find $t_1^2 = 0.97$. Then by substituting this in equations (4.116)-(4.118), we obtain $t_2^2 = 9.97$, $t_1^3 = 3.85$, and $t_2^3 = 5.77$.

Consequently, according to Proposition 4.19, the system-wide optimal production for the retailer ($i=3$) is the following:

$$u_3{}^*(t) = 1.5 \text{ for } 0 \le t < 3.85, \ u_3{}^*(t) = 3 \text{ for } 3.85 \le t < 5.77,$$
$$u_3{}^*(t) = 2.5 \text{ for } 5.77 \le t < 9.97, \ u_3{}^*(t) = 1.5 \text{ for } 9.97 \le t < 12.$$

The system-wide optimal production for supplier $i=2$ is defined as:

$$u_2*(t) = 1.5 \text{ for } 0 \leq t < 0.97, \quad u_2*(t) = 2.5 \text{ for } 0.97 \leq t < 9.97,$$

$$u_2*(t) = 1.5 \text{ for } 9.97 \leq t < 12.$$

Finally, the system-wide optimal control for the non-restricting firm, $i=1$, replicates that of supplier $i=2$:

$$u_1*(t) = 1.5 \text{ for } 0 \leq t < 0.97, \quad u_1*(t) = 2.5 \text{ for } 0.97 \leq t < 9.97,$$

$$u_1*(t) = 1.5 \text{ for } 9.97 \leq t < 12.$$

Comparing the corresponding switching points of the Stackelberg equilibrium and the system-wide optimal solution, we observe the reduction of the equilibrium response period of the retailer with respect to the centralized solution from 9.97-3.85=6.12 to 9.4-3.4=6. The supplier's response period remains the same as shown in Proposition 4.23.

Coordination

First of all, note that if we assume that the retailer pays for the supplier's inventory-related costs, then the Hamiltonian (4.134) becomes analogous to the centralized Hamiltonian (4.101). As a result, the optimality condition (4.136) transforms into the condition symmetric to (4.106). This implies that when setting $\psi_s^r(t) = \psi_s^s(t)$, the equilibrium solution satisfies all conditions of the maximum principle and is identical to the centralized solution.

Consequently, if the retailer accounts for the supplier's inventories, the Stackelberg equilibrium turns into the system-wide optimal solution, i.e., the supply chain becomes perfectly coordinated. That is to say, we have a similar effect to that found in the previous inventory game with endogenous demand. The effect in that game arose from ignoring the profit margin from handling inventories and was analogous to ignoring the other party's profit margin from sales (double marginalization) or from production cost reduction (learning effect). In contrast to the previous inventory game in Section 4.3.1, a sort of two-part tariff will perfectly coordinate the supply chain. This can be accomplished if the supplier: (i) sets a fixed premium (rather than cost) which the retailer can get by the end of each production cycle and (ii) requests (variable) payment from the retailer at each time point for the inventory costs the supplier incurs. Thus, the retailer will gain the fixed premium with the supplier's inventory costs deducted "just in time". Since a constant premium does not affect the optimization while the variable inventory cost does affect it, the retailer will no longer ignore the supplier's inventories and the supply chain performance will not deteriorate. Moreover, if the supplier sets the premium equal to his system-wide optimal inventory costs over the production cycle, then the retailer covers all system costs and still continues to order a system-wide optimal quantity. Thus, as with the two- part tariff, the profit (cost) shares of the firms are balanced by the proper choice of the fixed premium.

4.4 INTERTEMPORAL SUBCONTRACTING COMPETITION

In this section we consider two intertemporal games: production balancing, which involves subcontracting, and outsourcing. These games are natural extensions of the static stocking and outsourcing games discussed in Chapter 2. The stocking and outsourcing games are based on the classic newsvendor model; similarly, the basis for the extensions is the dynamic newsvendor model. However, in contrast to these static games where the retailer's (manufacturer's) inventories are accounted for only at the beginning of period or selling season, inventory dynamics or expected inventories are accounted for at each time point along the season.

4.4.1 THE PRODUCTION BALANCING GAME

In this section we consider a supply chain consisting of a single manufacturer (retailer) and single supplier (subcontractor); a case of multiple suppliers in an outsourcing context is considered in the next section. The supplier is characterized by ample capacity and thus the inventory dynamics are trivial. On the other hand, since the manufacturer, has a limited capacity, his decisions are dependent on available inventories. Moreover, the demand for products is random and its realization is known only by the end of a selling season. The short selling season and significant leadtime make it impossible or difficult to subcontract during this time. Therefore, the manufacturer can only place an advance order to the supplier for an initial inventory of end-products which is then used to balance production along with in-house capacity.

There are two new features involved in this production balancing problem. The first is that the precise inventory level is observed only once per period. In the differential inventory games discussed so far, we assumed that inventories are continuously reviewed during the production period. This, however, is not always possible and many large retailers, especially supermarkets, know exact inventory levels only once per certain period at time points of full inventory review. Such a policy is referred to in the literature as a periodic inventory review. This however does not imply that inventory costs are not incurred during this period. In practice, although inventory levels and losses associated with inventories may not be known precisely for a period of time, inventory costs (as well as costs of book-keeping, material tracking, transportation, space, material transformation, labor, depreciation, etc.) are incurred continuously. At the end of the period, it is possible to determine exactly how much was spent, when, and why. This explains the use of the expected inventory cost as the objective function in inventory control models.

The other feature is related to the available initial inventory, which is now a decision variable. Until now, we assumed that the initial inventory level was known and that it was possible to have instant deliveries along the production or selling period. In the present formulation (as mentioned above), we assume that the selling season is short, as is frequently the case in the fashion industry, and that it is too late to issue an order during the selling season. In-house capacity, however, is sufficient for a limited adjustment to demand fluctuations. Consequently, the initial inventory level is an important decision variable which is a part of the production balancing game, where the supplier is the Stackelberg leader who sets a wholesale price. The manufacturer is the follower who, in response to the wholesale price, selects an order quantity to be shipped by the beginning of the selling season as well as intertemporal production policy along the selling season.

The described conditions involve two well-known problems: production control and newsboy. The inventory game considered in the previous section relies on production control and the fact that both demand and inventories are known along the production horizon. If, however, the demand is unknown until the end of a selling season and production smoothing during the selling season has negligible efficiency, then an optimal choice of advance inventory orders is referred to as a newsboy or newsvendor model, as discussed in Chapter 2. In production control models, on the other hand, the main focus is on efficient production adjustment during the season.

The model we employ in this section combines both features: it allows for advance ordering or subcontracting of products and continuous-time in-house production adjustment during the selling season. As in production control problems (see, for a review, Maimon et al. 1998), the production rate is controllable along the selling season. As in the classical single-period newsboy problem, the probability distribution of the demand is known while exact realization of the determined is revealed only by the end of the selling season.

The supplier's problem

The supplier or subcontractor maximizes profits from the advance order quantity $X(0)$:

$$\max_{w} J_s(X(0), u, w) = \max_{w}(w - c_s)X(0), \qquad (4.136)$$

s.t.

$$w \geq c_s, \qquad (4.137)$$

where c_s is the supplier's unit production cost and w is the wholesale price.

The manufacturer's problem

The manufacturer produces products of the same type and has a facility for storing finished products during the production (selling season) T. The production process is described by the following balance equation

$$\dot{X}(t) = u(t) - a(t)d, \qquad (4.138)$$

where $X(t)$ is the surplus level in the storage by time t; $a(t)d$ is the demand rate; and $u(t)$ is the production rate at time t. The dynamic process (4.138) is determined by two decision variables, production rate, $u(t)$, which is bounded by the maximum capacity of the manufacturer U

$$0 \leq u(t) \leq U, \qquad (4.139)$$

and the initial inventory level $X(0)$ being stored. The initial level $X(0)$ is due to advance orders contracted out and delivered at unit cost w by the beginning of the selling season

$$X(0) \geq 0. \qquad (4.140)$$

The demand, $a(t)d$, is a time-dependent parameter representing at time t the amount of the product-type required per time unit, where $a(t)$, $a(t) > 0$ for $0 \leq t < T$, is a known demand shape and d is a random demand amplitude. For a selling season T, there will be a single realization of d, D, which is known only by time T. Exact inventories are observed only once per period as well. Therefore, a decision has to be made, under these uncertain conditions before production starts, based on probability density $\varphi(D)$ and cumulative distribution $\Phi(D)$ functions. We assume $\varphi(D)$ is differentiable.

Similar to the inventory dynamics considered so far, the difference between the cumulative production and the cumulative demand described by differential equation (4.138) is the surplus level. If the cumulative demand exceeds the cumulative production (shortage), $X(t) < 0$, i.e., the surplus is negative, a penalty, h^-, will have to be paid for each backlogged unit. Otherwise, an inventory holding cost is incurred for each product unit, h^+. Furthermore, production costs per product unit, c, are incurred at points t when the manufacturer is not idle. Since the current trend in industry is to find the cheapest manufacturer for contracting out advance orders (e. g., in the Far East), we assume here, that the in-house production cost is greater than the unit cost of advance orders, $c_s < c$.

The objective is to find such a production rate $u(t)$ and advance order $X(0)$ that satisfy constraints (4.138)- (4.140) while minimizing the following expected cost over the selling season T:

$$\min_{X(0),u} J_m(X(0),u,w) = \min_{X(0),u} E\left[\int_0^T \left(cu(t) + h^+ X^+(t) + h^- X^-(t)\right)dt + wX(0)\right], (4.141)$$

where

$$X^+(t) = \max\{0, X(t)\}, \quad X^-(t) = \max\{0, -X(t)\}.$$

To facilitate the analysis, let us substitute (4.138) into the objective ((4.141). Given probability density $\varphi(D)$ of the demand, denoting

$$Y(t) = X(0) + \int_0^t u(\tau)d\tau, \tag{4.142}$$

and using conditional expectation, we find:

$$J_m = cY(T) + wY(0) +$$

$$+ \int_0^T \left[\int_{-\infty}^{\frac{Y(t)}{A(t)}} h^+ (Y(t) - DA(t))\varphi(D)dD - \int_{\frac{Y(t)}{A(t)}}^{\infty} h^- (Y(t) - DA(t))\varphi(D)dD\right]dt, \tag{4.143}$$

where

$$A(t) = \int_0^t a(\tau)d\tau. \tag{4.144}$$

The new objective (4.143) is subject to constraints (4.139)-(4.140) and (4.142), which together constitute a deterministic problem equivalent to the stochastic problem (4.138)- (4.141).

The centralized problem

The centralized formulation is independent of the wholesale price which represents an internal transfer in the supply chain.

$$\min_{u,X(0)} J(X(0),u) = \min_{u,X(0)} \{J_m(X(0),u,w) - J_s(X(0),u,w)\} =$$

$$\min_{X(0),u} E\left[\int_0^T \left(cu(t) + h^+ X^+(t) + h^- X^-(t)\right)dt + c_s X(0)\right] \tag{4.145}$$

s.t.(4.138)- (4.140).

Similar to the manufacturer's problem, the deterministic equivalent of problem (4.138) - (4.140) and (4.145) includes the following objective function

$$J_m = cY(T) + c_s Y(0) +$$

$$+ \int_0^T \left[\int_{-\infty}^{\frac{Y(t)}{A(t)}} h^+ (Y(t) - DA(t))\varphi(D)dD - \int_{\frac{Y(t)}{A(t)}}^{\infty} h^- (Y(t) - DA(t))\varphi(D)dD \right] dt . \quad (4.146)$$

and constraints (4.139)- (4.140) and (4.142).

System-wide optimal solution

To study the equivalent deterministic centralized problem, we construct the Hamiltonian

$$H(t) = -cu(t) + \psi(t)u(t) -$$

$$\int_{-\infty}^{\frac{Y(t)}{A(t)}} h^+ (Y(t) - DA(t))\varphi(D)dD + \int_{\frac{Y(t)}{A(t)}}^{\infty} h^- (Y(t) - DA(t))\varphi(D)dD,$$

where the co-state variable $\psi(t)$ represents the margin gained by producing one more product unit at time t. The co-state variable satisfies the following co-state equation

$$\dot{\psi}(t) = -\frac{\partial H(t)}{\partial Y(t)},$$

that is,

$$\dot{\psi}(t) = (h^+ + h^-)\Phi(\frac{Y(t)}{A(t)}) - h^- \qquad (4.147)$$

with transversality (boundary) constraint

$$\psi(T) = 0, \qquad (4.148)$$

and complementary slackness

$$\psi(0) = c_s, \text{ if } Y(0) > 0 \text{ and } \psi(0) \le c_s, \text{ if } Y(0) = 0 . \qquad (4.149)$$

By rearranging only $u(t)$-dependent terms of the Hamiltonian, we obtain:

$$H_u(t) = u(t)(\psi(t) - c) . \qquad (4.150)$$

Since this term is linear in $u(t)$, the optimal production rate that maximizes the Hamiltonian is

$$u(t) = \begin{cases} U, & \text{if } \psi(t) > c; \\ a \in [0,U], & \text{if } \psi(t) = c; \\ 0, & \text{if } \psi(t) < c. \end{cases} \tag{4.151}$$

Thus, under the optimal solution, the manufacturer may be: idle (when $\psi(t) < c$); working at his maximum production rate ($\psi(t) > c$); or entering the singular regime ($\psi(t) = c$) which is characterized by an intermediate production rate between 0 and U. The ambiguity of the last condition in terms of the production rate is resolved in the following proposition.

Proposition 4.24. *Let there exist* b, $0 \le ba(t) \le U$, $0 \le t \le T$, *such that*

$$\Phi(b) = \frac{h^-}{h^+ + h^-}.$$ *If* $\psi(t) = c$ *over an interval of time, then* $u(t) = ba(t)$ *and*

$$\frac{Y(t)}{A(t)} = b$$ *along this interval.*

Proof: To find the production rate which meets the singular regime condition $\psi(t) = c$ along an interval of time, we differentiate it, $\dot{\psi}(t) = 0$, and substitute (4.147), which results in

$$\Phi\left(\frac{Y(t)}{A(t)}\right) = \frac{h^-}{h^+ + h^-}. \tag{4.151}$$

Next, by choosing b so that

$$\Phi(b) = \frac{h^-}{h^+ + h^-}, \tag{4.152}$$

we have along this interval

$$\frac{Y(t)}{A(t)} = b. \tag{4.153}$$

Differentiating this condition along the same time interval, we find $\dot{Y}(t) = ba(t)$ and with respect to (4.142), $u(t) = ba(t)$, as stated in the proposition.

Note that a relationship between production control and the newsboy problem discussed in the beginning of this section is sustained with equation (4.152), which is the classical newsboy solution discussed in Chapter 2. Thus, if inventory (surplus/shortage) costs are viewed as momentary overage and underage costs, then an optimal production regime may satisfy the classical equation over a part of the selling season.

If production can be profitable at all, then $b > 0$, as is assumed henceforth.

Given b, $0 < ba(t) \le U$, which meets $\Phi(b) = \frac{h^-}{h^+ + h^-}$, then, according to

Proposition 4.24, the optimality conditions (4.151) take the following form:

$$u(t)=\begin{cases} U, & \text{if } \psi(t)>c; \\ ba(t), & \text{if } \psi(t)=c; \\ 0, & \text{if } \psi(t)<c. \end{cases} \qquad (4.154)$$

However, if $ba(t)>U$, then the singular regime, $\psi(t)=c$, cannot hold at an interval of time and the optimality conditions (4.151) take the following form

$$u(t)=\begin{cases} U, & \text{if } \psi(t)\geq c; \\ 0, & \text{if } \psi(t)<c. \end{cases} \qquad (4.155)$$

Accordingly, the system-wide optimal solution depends on the relationship between system parameters. Therefore, in what follows, we determine the solution separately for different cases. First, we consider the case when production activity is not cost efficient, that is, both advance orders and production are not justified.

Proposition 4.25. *If $h^- \leq \dfrac{c_s}{T}$, then the system-wide optimal solution is given by $X^*(0)=0$ and $u^*(t)=0$ for $0 \leq t \leq T$.*

Proof: Consider the following solution

$$Y(t)=0 \text{ and } u(t)=0 \text{ for } 0 \leq t \leq T. \qquad (4.156)$$

If $Y(0)=0$ and thus according to (4.142), $X(0)=0$, then with respect to (4.149), $\psi(0) \leq c_s < c$. Since

$$\frac{Y(t)}{A(t)}=0<b, \ \dot{\psi}(t)=(h^+ +h^-)\Phi(\frac{Y(t)}{A(t)})-h^- <0,$$

i.e., the optimality condition $\psi(t)<c$ from (4.154) holds for $0 \leq t \leq T$. As a result, if solution (4.156) is feasible with respect to (4.148), $\psi(T)=0$, then it is optimal. Taking into account (4.147), equation (4.148) transforms into

$$\psi(0) + \int_0^T [(h^+ +h^-)\Phi(0)-h^-]dt=0 .$$

This, with respect to $\psi(0) \leq c_s$ and $\Phi(0)=0$, results in the condition stated in the proposition.

The next proposition treats the case when subcontracting (an advance order) is more advantageous than utilizing the manufacturer's production capacity.

Proposition 4.26. *Let* t_1 *and* e *satisfy the following system of equations*

$$A(t_1) = \frac{e}{b} \text{ and } c_s + \int_0^T [(h^+ + h^-)\Phi(\frac{e}{A(t)}) - h^-]dt = 0.$$

If $\frac{c_s}{T} < h^- \le \dfrac{\displaystyle\int_{t_1}^{T}(h^+ + h^-)\Phi(\frac{e}{A(t)})dt + c}{T - t_1}$, *then the system-wide optimal advance*

order is $X^*(0)=e$, *and the system-wide optimal production rate is* $u^*(t)=0$
for $0 \le t \le T$.

Proof: Consider the following solution

$$X(0)=bA(t_1), \ \psi(t_1)=0, \ Y(t)=X(0) \text{ and } u(t)=0 \text{ for } 0 \le t \le T. \quad (4.157)$$

If $X(0){>}0$, then with respect to (4.149), $\psi(0) = c_s < c$. Since $A(t)$ is an increas-

ing function of time and $\dfrac{X(0)}{A(0)} > b$, there can be a point t_1 such that

$$\frac{Y(t_1)}{A(t_1)} = \frac{X(0)}{A(t_1)} = b, \ \dot\psi(t) = (h^+ + h^-)\Phi(\frac{X(0)}{A(t)}) - h^- > 0 \text{ for } 0 \le t < t_1 \text{ and}$$

$$\dot\psi(t) = (h^+ + h^-)\Phi(\frac{X(0)}{A(t)}) - h^- < 0 \text{ for } t_1 < t \le T.$$

Thus, if $t_1{>}0$ and $\psi(t_1) \le c$, then solution (4.157) is feasible and meets the
optimality condition, $\psi(t) < c$, from (4.154) for $0 \le t \le T$. Using conditions $\psi(T) = 0$ and $\psi(0) = c_s$, we find an equation for $X(0)$:

$$c_s + \int_0^T [(h^+ + h^-)\Phi(\frac{X(0)}{A(t)}) - h^-]dt = 0 \quad (4.158)$$

and for $\psi(t_1)$

$$\psi(t_1) = - \int_{t_1}^T [(h^+ + h^-)\Phi(\frac{X(0)}{A(t)}) - h^-]dt \quad (4.159)$$

If $X(0){>}0$, then $A(t_1) = \dfrac{X(0)}{b} > 0$ which with respect to (4.158) results in

$h^- > \dfrac{c_s}{T}$, as stated in the proposition. The other condition stated in the

proposition is immediately obtained by substituting (4.159) into $\psi(t_1) \le c$
and setting $X(0)=e$.

The following two propositions treat two cases when both subcontracting
and producing with the manufacturer's own production capacity is beneficial.
In the first case, a singular regime holds over an interval of time, that is,

$ba(t) \leq U$ (See Proposition 4.24) and the optimal solution is determined by conditions (4.154). The other case arises when a singular regime cannot occur, $ba(t) > U$, and therefore the optimal solution is determined by (4.155).

Proposition 4.27. *Let* $0 < ba(t) \leq U$ *for* $t_1 \leq t < t_2$, t_1 *and* t_2 *satisfy the following equations*

$$c_s + \int_0^{t_1} [(h^+ + h^-)\Phi(\frac{bA(t_1)}{A(t)}) - h^-]dt = c \; ; \; - \int_{t_2}^{T} [(h^+ + h^-)\Phi(\frac{bA(t_2)}{A(t)}) - h^-]dt = c$$

respectively and e be determined by

$$c_s + \int_0^{T} [(h^+ + h^-)\Phi(\frac{e}{A(t)}) - h^-]dt = 0 \; .$$

If $h^- > \dfrac{\int_0^{T}(h^+ + h^-)\Phi(\frac{e}{A(t)})dt + c}{T - t_1}$ *, then the system-wide optimal advance order is*

$X^*(0) = bA(t_1)$ *and the optimal production rate is* $u^*(t) = 0$ *for* $0 \leq t < t_1$; $u^*(t) = ba(t)$ *for* $t_1 \leq t < t_2$ *and* $u^*(t) = 0$ *for* $t_2 \leq t \leq T$.

Proof: If condition $h^- \leq \dfrac{\int_{t_1}^{T}(h^+ + h^-)\Phi(\frac{e}{A(t)})dt + c}{T - t_1}$ of Proposition 4.26 is not

met, then there must be two switching points, t_1 and t_2, $0 < t_1 < t_2 < T$, such that
$$\dot{\psi}(t) > 0 \; , \; Y(t) = X(0) \text{ and } u(t) = 0 \text{ for } 0 \leq t < t_1 \; ;$$
$$\psi(t) = c \; , \; Y(t) = X(0) + b(A(t) - A(t_1)) \text{ and } u(t) = ba(t) \text{ for } t_1 \leq t < t_2 \; ;(4.160)$$
$$\dot{\psi}(t) < 0 \; , \; Y(t) = X(0) + b(A(t_2) - A(t_1)) \text{ and } u(t) = 0 \text{ for } t_2 \leq t \leq T \; .$$
It is easy to observe that solution (4.160) meets optimality conditions (4.154). As with Proposition 4.26, we find from (4.160), (4.147) and (4.148) the equations for the two switching points:

$$\psi(t_1) = c_s + \int_0^{t_1} [(h^+ + h^-)\Phi(\frac{X(0)}{A(t)}) - h^-]dt = c \; ; \quad (4.161)$$

$$\psi(t_2) = - \int_{t_2}^{T} [(h^+ + h^-)\Phi(\frac{X(0) + b(A(t_2) - A(t_1))}{A(t)}) - h^-]dt = c. \quad (4.162)$$

Finally, taking into account (4.153),
$$\frac{Y(t_1)}{A(t_1)} = \frac{X(0)}{A(t_1)} = b$$

and substituting it into (4.161) and (4.161), we obtain the equations stated in this proposition.

Proposition 4.28. *Let* $ba(t) > U$, $\bar{t}_1 \leq t < \bar{t}_2$, $\bar{t}_1$ *and* $\bar{t}_2$ *be determined by*

$$c_s + \int_0^{t_1}[(h^+ + h^-)\Phi(\frac{bA(\bar{t}_1)}{A(t)}) - h^-]dt = c \;,\; -\int_{t_2}^T[(h^+ + h^-)\Phi(\frac{bA(\bar{t}_2)}{A(t)}) - h^-]dt = c;$$

$X(0)$, t_1, t_2 *and* t_3 *satisfy the following equations*

$$c_s + \int_0^{t_1}[(h^+ + h^-)\Phi(\frac{X(0)}{A(t)}) - h^-]dt = c \;;\; -\int_{t_2}^T[(h^+ + h^-)\Phi(\frac{X(0) + U(t_2 - t_1)}{A(t)}) - h^-]dt = c;$$

$$\int_{t_1}^{t_3}[(h^+ + h^-)\Phi(\frac{X(0 + U(t - t_1)}{A(t)}) - h^-]dt = -\int_{t_3}^{t_2}[(h^+ + h^-)\Phi(\frac{X(0) + U(t - t_3)}{A(t)}) - h^-]dt \;;$$

$$\frac{X(0) + U(t_3 - t_1)}{A(t_3)} = b$$

and e be determined by

$$c_s + \int_0^T[(h^+ + h^-)\Phi(\frac{e}{A(t)}) - h^-]dt = 0 \;.$$

If $h^- > \dfrac{\int_{t_1}^T (h^+ + h^-)\Phi(\dfrac{e}{A(t)})dt + c}{T - t_1}$ *, then the system-wide optimal advance order is*

$X^*(0) = bA(t_3) - U(t_3 - t_1)$ *and the system-wide optimal production rate is* $u^*(t) = 0$ *for* $0 \leq t < t_1$*;* $u^*(t) = U$ *for* $t_1 \leq t < t_2$ *and* $u^*(t) = 0$ *for* $t_2 \leq t \leq T$ *.*

Proof: If condition $h^- \leq \dfrac{\int_{t_1}^T (h^+ + h^-)\Phi(\dfrac{e}{A(t)})dt + c}{T - t_1}$ of Proposition 4.26 is not

met and $ba(t) > U$, $\bar{t}_1 \leq t < \bar{t}_2$, then there must be a break point, t_3, such that $\dfrac{Y(t_3)}{A(t_3)} = b$, $\dot{\psi}(t) > 0$ for $0 \leq t < t_3$, $\dot{\psi}(t_3) = 0$ and $\dot{\psi}(t) < 0$ for $t_3 < t \leq T$. That

is, there are two switching points, t_1 and t_2, $0 < t_1 < t_2 < T$, such that

$$\psi(t) < c \;,\; Y(t) = X(0) \text{ and } u(t) = 0 \text{ for } 0 \leq t < t_1\;;$$
$$\psi(t) \geq c \;,\; Y(t) = X(0) + U(t - t_1) \text{ and } u(t) = U \text{ for } t_1 \leq t < t_2;\quad (4.163)$$
$$\psi(t) < c \;,\; Y(t) = X(0) + U(t_2 - t_1) \text{ and } u(t) = 0 \text{ for } t_2 \leq t \leq T \;.$$

We immediately see that solution (4.163) meets optimality conditions (4.155). As with Proposition 4.27, we find from (4.163), (4.147) and (4.148) the equations for the two switching points, co-state break point and advance order as stated in the proposition

$$\psi(t_1) = c_s + \int_0^{t_1} [(h^+ + h^-)\Phi(\frac{X(0)}{A(t)}) - h^-]dt = c \ ;$$

$$\psi(t_2) = - \int_{t_2}^{T} [(h^+ + h^-)\Phi(\frac{X(0) + U(t_2 - t_1)}{A(t)}) - h^-]dt = c;$$

$$\psi(t_3) - c = \int_{t_1}^{t_3} [(h^+ + h^-)\Phi(\frac{X(0 + U(t - t_1)}{A(t)}) - h^-]dt = - \int_{t_3}^{t_2} [(h^+ + h^-)\Phi(\frac{X(0) + U(t - t_3)}{A(t)}) - h^-]dt \ ;$$

$$\frac{X(0) + U(t_3 - t_1)}{A(t_3)} = b \ ,$$

as stated in the proposition.

Proposition 4.25-4.28 presents closed form solutions for various production conditions. These solutions are globally optimal as shown in the following theorem.

Theorem 4.2. *A solution determined by Propositions 4.25-4.28 is the globally optimal solution of problem (4.139)-(4.140), (4.142) and (4.143).*

Proof: The necessary optimality conditions (4.147)-(4.151) utilized by Propositions 4.25-4.28 are sufficient if the problem (4.139)-(4.140), (4.142)-(4.143) is convex. To verify the convexity, first note that constraints (4.139)-(4.140) and (4.142) are linear. The objective function (4.143) consists of three terms. The first two terms are linear as well. The third term is non-linear. To analyze the third term,

$$L = \int_0^T \left[\int_{-\infty}^{\frac{Y(t)}{A(t)}} h^+ (Y(t) - DA(t))\varphi(D)dD - \int_{\frac{Y(t)}{A(t)}}^{\infty} h^- (Y(t) - DA(t))\varphi(D)dD \right] dt \ ,$$

we consider the second derivative with respect to $Y(t)$:

$$\frac{\partial^2 L}{\partial Y(t)^2} = (h^+ + h^-)\frac{\partial \Phi(z)}{\partial z}\frac{1}{A(t)} \ ,$$

where $z = \dfrac{Y(t)}{A(t)}$.

It is easy to observe that it is non-negative, that is, L and thus the objective function (4.143) is convex.

Sensitivity Considerations

In this section we assume that the demand amplitude is random, but that the demand shape is known, as is often the case with fashion goods. In light of these assumptions, an important question arises as to what happens if the demand shape changes in a limited way? It turns out that the solutions determined in Propositions 4.25-4.28 do not always depend on the demand shape. Specifically, the solution defined by Propositions 4.25, which depends on the system costs, is completely insensitive to the demand shape. On the other hand, the solutions of Propositions 4.26-4.28 do not depend on the demand shape at each point of time, but do depend on a cumulative value,

$\int_{t_a}^{t_b} \Phi(\frac{X(0)}{A(t)}) dt$ over some points t_a and t_b. For example, if $\Phi(e) = \frac{e}{R}$ (the

uniform distribution of the demand amplitude), then according to Propositions 4.25, $t_a=0$, $t_b=T$, and the optimal advance order is determined by

$$c_s + \int_0^T [(h^+ + h^-)\Phi(\frac{X(0)}{A(t)}) - h^-] dt = c_s + (h^+ + h^-) \frac{X(0)}{R} \int_0^T \frac{dt}{A(t)} - h^- = 0 .$$

This implies that the demand shape, $a(t)$, or $A(t) = \int_0^t a(\tau) d\tau$, may change at

each point of time. This change does not affect the optimal solution as long

as the cumulative value, $\int_0^T \frac{dt}{A(t)}$, does not change.

Finally, the most demand-sensitive solution is that defined by Proposition 4.27. Indeed, according to Proposition 4.27, even if the corresponding cumulative values of the demand shape do not change, the optimal production rate, $u(t)=ba(t)$, changes for $t_1 \le t < t_2$, when $a(t)$ changes. Therefore, it is especially important to review sales along this particular interval of time. The more accurate the information on the demand shape, the closer the production rate can be to the optimal solution for $t_1 \le t < t_2$.

Note that if the manufacturer has a myopic attitude and does not take into account inventory dynamics, then he will contract out all expected demand since in-house production is more expensive than subcontracting. Consequently, the myopic order quantity, $X(0)$ is determined by the newsboy formula discussed in Chapter 2:

$$\Phi(b^M) = \frac{h^- - c_s}{h^- + h^+}, \quad b^M = \frac{X(0)}{a(0)} . \tag{4.164}$$

Noting that $b > b^M$, and assuming that the manufacturer has sufficient capacity, $ba(t) \le U$, and comparing (4.164) with the corresponding expressions from Propositions 4.25-4.27, we conclude that *if the unit in-house production cost is greater than the unit supplier's cost, then the myopic manufacturer does not produce and his advance order is always less than the system-wide optimal order.*

A low supply cost (or wholesale price) is the major reason for the full-production outsourcing that is frequently observed nowadays. Manufacturers also tend to order less, frequently ignoring the fact that the inventories deplete as the selling season progresses and thus the true inventory-related costs are lower over the season than those accounted for when ordering in the beginning of the season. Finally, production smoothing during the selling season allows for a trade-off between initial order costs and dynamic inventory costs. If the overall production is outsourced, then the trade-off benefits disappear.

Game analysis

Consider first the equivalent deterministic manufacturer's problem (4.139)-(4.143). Applying the maximum principle, we find the same Hamiltonian as for the centralized problem (with c_s replaced with w) and co-state differential equation (4.147) with boundary condition (4.149), where we simply replace $\psi(t)$ with $\psi_m(t)$. The complementary slackness condition is similar as well with c_s replaced with w:

$$\psi_m(0) = w, \text{ if } Y(0) > 0 \text{ and } \psi_m(0) \le w, \text{ if } Y(0) = 0 \ . \qquad (4.165)$$

Thus, the optimality conditions as well as all propositions derived for the centralized formulation hold for non-cooperative parties and the only change we have is c_s replaced with w and $\psi(t)$ with $\psi_m(t)$. For example, Proposition 4.27 can be restated as follows.

Proposition 4.29. *Let $0 < ba(t) \le U$ for $t_1 \le t < t_2$, t_1 and t_2 satisfy the following equations*

$$w + \int_0^{t_1} [(h^+ + h^-)\Phi(\frac{bA(t_1)}{A(t)}) - h^-]dt = c \ ; \ - \int_{t_2}^{T} [(h^+ + h^-)\Phi(\frac{bA(t_2)}{A(t)}) - h^-]dt = c$$

respectively and e be determined by

$$w + \int_0^{T} [(h^+ + h^-)\Phi(\frac{e}{A(t)}) - h^-]dt = 0 \ .$$

If $h^- > \dfrac{\int_{t_1}^{T} (h^+ + h^-)\Phi(\frac{e}{A(t)})dt + c}{T - t_1}$, then the manufacturer's best response with respect to the advance order is $X(0) = bA(t_1)$ and with respect to the production *rate is $u(t)=0$ for $0=t<t_1$; $u(t)=ba(t)$ for $t_1=t<t_2$ and $u(t)=0$ for $t_2=t=T$.*

We now analyze the most likely production conditions (Proposition 4.29) characterized by sufficient manufacturing capacity $ba(t) \leq U$ and where both subcontracting and producing are carried out at the manufacturer's own production facility. Comparing the equations for the first switching point of the manufacturer's optimal response (see Proposition 4.29) and those of the system-wide optimal solution (see Proposition 4.27) we conclude with the following proposition.

Proposition 4.30. *In regard to the subcontracting competition of the production balancing game, if the supplier makes profit, i.e., $w>c_s$, then the manufacturer produces more in-house and subcontracts less than the system-wide optimal solution.*

Proof: The proof is straightforward. Comparing the equations for the first switching point of the manufacturer's optimal response (see Proposition 4.29) with the corresponding point of the system-wide optimal solution (see Proposition 4.27)

$$w+\int_0^{t_1}[(h^+ +h^-)\Phi(\frac{bA(t_1)}{A(t)})-h^-]dt=c \tag{4.166}$$

$$c_s+\int_0^{t_1}[(h^+ +h^-)\Phi(\frac{bA(t_1)}{A(t)})-h^-]dt=c, \tag{4.167}$$

we observe that since $w>c_s$ and $\Phi(.)$ is a cumulative function, t_1 found from (4.166) is smaller than that from (4.167). Thus, the manufacturer starts production earlier under competition. On the other hand, the equations for the second switching point t_2 and the production rates are the same with respect to Propositions 4.29 and 4.27. Thus the manufacturer produces more if the supply chain is not centralized. In addition, since $X(0)=bA(t_1)$ and $A(.)$ is a cumulative function, the manufacturer subcontracts less.

The result, of course, does not come as a surprise since subcontracting competition of the production balancing game is a vertical competition where double marginalization results in a decreased quantity which the manufacturer orders from the supplier, thus enhancing the incentive for in-house production. Note that a myopic manufacturer also orders less than the system-wide optimal quantity but assumes no in-house production if the subcontractor's wholesale price is below the in-house production cost, $w<c$. Consequently, a myopic attitude can hardly improve supply chain performance.

Equilibrium

Consider now the supplier's problem. Applying the first-order optimality condition to the supplier's objective function (4.136), we find that the optimal wholesale price w is defined by the equation:

$$(w-c_s)\frac{dX(0)}{dw}+X(0)=0. \tag{4.168}$$

Then, with respect to Proposition 4.27 and $A(t_1)=\int_0^{t_1}a(t)dt$, this implies

that $X(0)=bA(t_1)$ and equation (4.168) transforms into

$$(w-c_s)\frac{bdA(t_1)}{dw}+bA(t_1)=(w-c_s)\frac{bdt_1}{dw}a(t_1)+bA(t_1)=0,$$

where t_1 is determined by

$$c_s+\int_0^{t_1}[(h^+ +h^-)\Phi(\frac{bA(t_1)}{A(t)})-h^-]dt=c. \tag{4.169}$$

Using implicit differentiation of (4.169) and the fact that $\Phi(b)=\dfrac{h^-}{h^+ +h^-}$,

we find that

$$\frac{dt_1}{dw}=-\frac{1}{\int_0^{t_1}[(h^+ +h^-)\frac{b}{A(t)}a(t_1)\varphi(\frac{bA(t_1)}{A(t)})]dt}, \tag{4.170}$$

which implies that *the greater the wholesale price, the earlier the manufacturer will start using his in-house capacity.* Moreover, this also means that a solution, w, which satisfies the optimality condition, $(w-c_s)\frac{bdt_1}{ds}a(t_1)+bA(t_1)=0$, is greater than c_s.

Let $t_1=A^{-1}\left(\dfrac{X(0)}{b}\right)$, then equation (4.170) takes the following form

$$\frac{dt_1}{dw}=-\frac{1}{\int_0^{A^{-1}\left(\frac{X(0)}{b}\right)}[(h^+ +h^-)\frac{b}{A(t)}a(A^{-1}\left(\frac{X(0)}{b}\right))\varphi(\frac{X(0)}{A(t)})]dt}, \tag{4.171}$$

which by substituting into the first-order optimality condition results in

$$-\frac{(w-c_s)ba\left(A^{-1}\left(\dfrac{X(0)}{b}\right)\right)}{A^{-1}\left(\frac{X(0)}{b}\right)} + X(0) = 0.$$
$$\int_0^{A^{-1}\left(\frac{X(0)}{b}\right)} [(h^+ + h^-)\frac{b}{A(t)}a(A^{-1}\left(\frac{X(0)}{b}\right))\varphi(\frac{X(0)}{A(t)})]dt$$

We thus conclude with the following proposition.

Proposition 4.31. *Let all conditions of Proposition 4.27 be met,* $\dfrac{dt_1}{ds}$ *be defined by (4.171),* $\dfrac{\partial\Phi(z)}{\partial z} > 0$ *and,* δ *and* λ *satisfy the following equations*

$$\lambda + \int_0^{A^{-1}\left(\frac{\delta}{b}\right)}[(h^+ + h^-)\Phi(\frac{\delta}{A(t)}) - h^-]dt = c , \qquad (4.172)$$

$$-\frac{(\lambda-c_s)ba\left(A^{-1}\left(\dfrac{\delta}{b}\right)\right)}{\int_0^{A^{-1}\left(\frac{\delta}{b}\right)}[(h^+ + h^-)\frac{b}{A(t)}a(A^{-1}\left(\frac{\delta}{b}\right))\varphi(\frac{\delta}{A(t)})]dt} + \delta = 0. \qquad (4.173)$$

If

$$(\lambda - c_s)\left(\frac{bd^2t_1}{dw^2}a(t_1) + \frac{bdt_1}{dw}a'(t_1)\right) + \frac{bdt_1}{dw}a(t_1) + ba(t_1) < 0, \qquad (4.174)$$

then the wholesale price $w^s = \lambda < c$, *the manufacture's advance order* $X^s(0) =$
δ *and production policy* $u^s(t)=0$ *for* $0 \le t < t_1$; *$u^s(t)=ba(t)$ *for* $t_1 \le t < t_2$;*
$u^s(t)=0$ *for* $t_2 \le t \le T$ *constitute the unique Stackelberg equilibrium in the*
differential production balancing game.

Proof: To prove the proposition, it is sufficient to verify the secondorder optimality condition, which immediately results in condition (4.174) stated in the proposition.

The following example illustrates the results for demands following a uniform distribution.

Example 4.5.

The following example is based on a problem faced by a large supplier of fashion goods, where demand is quite steady, i.e., $a(t)=1$, but the amplitude d is a random parameter characterized by the uniform distribution,

$$\varphi(D) = \begin{cases} \dfrac{1}{R}, & \text{for } 0 \leq D \leq R; \\ 0, & \text{otherwise.} \end{cases}$$

Then $\Phi(e) = \dfrac{e}{R}$ for $e \leq R$ and $\Phi(e) = 1$ for $e > R$ and with respect to (4.152),

$$b = R \dfrac{h^-}{h^+ + h^-} .$$

Given $b < U$, the input data for the supply chain are presented in Table 4.3.

Table 4.3. System parameters.

c_S	h^+	h^-	C	R	T	b
3.0	1.0	4.0	6.0	20.0	100	16.0

We start by calculating the system-wide optimal solution. Since $A(t) = t$,

$$h^- > \dfrac{\int_{t_1}^{T}(h^+ + h^-)\Phi(\frac{e}{t})dt + c}{T - t_1} \quad \text{and } b \leq U, \text{ we apply Proposition 4.27 to find all swit-}$$

ching points, $0 < t_1 < t_2 < T$ as follows:

(i) $X(0) = bt_1$;

(ii) $c_s + \int_0^{t_1}[(h^+ + h^-)\Phi(\frac{bt_1}{t}) - h^-]dt = c$, i.e.,

$$c_s - c + \int_0^{\frac{bt_1}{R}}[(h^+ + h^-)\cdot 1 - h^-]dt + \int_{\frac{bt_1}{R}}^{t_1}[(h^+ + h^-)\frac{bt_1}{Rt} - h^-]dt = 0,$$

and thus

$$c_s - c + (h^+ + h^-)\frac{bt_1}{R}(1 + \ln t_1 - \ln\frac{bt_1}{R}) - h^- t_1 = 0,$$

which, by taking into account that $b = R\dfrac{h^-}{h^+ + h^-}$, results in

$$t_1 = \dfrac{c - c_s}{h^- \ln\dfrac{h^+ + h^-}{h^-}} ; \tag{4.175}$$

(iii) $\int_{t_2}^{T}[(h^+ + h^-)\Phi(\frac{bt_2}{t}) - h^-]dt - c = 0$, i.e.,

$$\int_{t_2}^{T} [(h^+ + h^-)(\frac{bt_2}{Rt}) - h^-]dt - c = 0 \text{ , and thus}$$

$$-(h^+ + h^-)\frac{bt_2}{R}(\ln T - \ln t_2) + h^-(T - t_2) - c = 0 \text{ .} \tag{4.176}$$

Using data from Table 4.3, equations (4.175) and (4.174) result in

$$t_1 = \frac{6-3}{4\ln\dfrac{1+4}{4}} = 3.3610$$

$$-4t_2(\ln 100 - \ln t_2) + 4(100 - t_2) - 6 = 0 \text{ ,} \tag{4.177}$$

respectively. Solving equation (4.177) in t_2, we find that $t_2 = 83.1862$. Thus, the system-wide optimal advance order quantity is

$$X^*(0) = bt_1 = 53.7770$$

and the system-wide optimal production rate (see Figure 4.12.) is

$$u^*(t) = 0 \text{ for } 0 \leq t < 3.361; \; u^*(t) = 16 \text{ for } 3.361 \leq t < 83.1862$$
$$u^*(t) = 0 \text{ for } 83.1862 \leq t \leq 100 \text{ .}$$

Next, to find the Stackelberg equilibrium, we employ Proposition 4.31. Specifically, we first solve equation (4.172), which results in

$$X^s(0) = \frac{b(c - w^s)}{h^- \ln\dfrac{h^+ + h^-}{h^-}} \text{ ,} \tag{4.178}$$

where $b = R\dfrac{h^-}{h^+ + h^-}$. Next solving (4.173) we have

$$-\frac{(w^s - c_s)}{\int_0^{\frac{X^s(0)}{b}} [(h^+ + h^-)\frac{1}{tR}]dt} + X^s(0) = -\frac{(w^s - c_s)R}{(h^+ + h^-)\ln\dfrac{X^s(0)}{b}} + X^s(0) = 0 \text{ .}$$

Substituting (4.178) into the last expression, we obtain the equation for the equilibrium wholesale price w^s:

$$\frac{w^s - c_s}{\ln\left(\dfrac{c - w^s}{h^- \ln R/b}\right)} = \frac{c - w^s}{\ln(R/b)} \text{ .} \tag{4.179}$$

Consequently, plugging the data from Table 4.3 into (4.179) results in the Stackelberg wholesale price w^s=**4.769642**<c=6. Substituting this value into (4.178) provides equilibrium advance order $X^s(0)$= 22.055. Then t_1=$X(0)/b$=**1.378438**, while t_2 = 83.1862 remains unchanged. Thus the equilibrium production rate is $u^s(t)$=0 for $0 \leq t < 1.3784$; $u^s(t)$=16 for $1.378438 \leq t < 83.1862$ and $u^s(t)$=0 for $83.1862 \leq t \leq 100$.

Finally, the uniqueness of the found wholesale price (condition (4.174)) can be straightforwardly verified by differentiating expression $\dfrac{c-w^s}{\ln(R/b)} - \dfrac{w^s - c_s}{\ln\left(\dfrac{c-w^s}{h^- \ln R/b}\right)}$ from (4.179), which results in a negative expression. Comparing the system-wide optimal solution with the equilibrium solution, we observe that $X^s(0) < X^*(0)$ and $t_1{}^s < t_1{}^*$, as stated in Proposition 4.30.

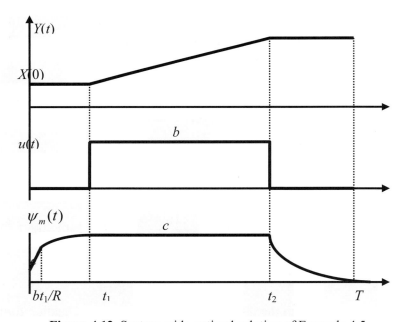

Figure 4.12. System-wide optimal solution of Example 4.5

Coordination

According to Proposition 4.30, system performance deteriorates if the firms are non-cooperative. This result is to be expected since balancing production with advance orders presents a case of vertical competition.

Comparing Propositions 4.27 and 4.31, it is readily seen that if the supplier sets the wholesale price equal to his marginal cost, then the advance order quantity becomes equal to the system-wide optimal quantity and the manufacturer's production rate converges to the system-wide optimal policy. Thus, the production balancing game is an example of when double marginalization not only decreases the order quantity but also affects production and inventory dynamics. Therefore, perfect coordination, is straightforwardly obtained by the two-part tariff. The supplier sets the wholesale price equal to his marginal cost and makes a profit by choosing the appropriate fixed transformation cost.

4.4.2 THE DIFFERENTIAL OUTSOURCING GAME

In this section we consider a supply chain consisting of one producer (manufacturer) and multiple suppliers of limited capacity. The suppliers (or service providers) are the leaders and the individual producer is a follower. To compensate for the suppliers' power asymmetry and capacity restrictions, the producer may use a number of potential suppliers. Accordingly, in contrast to the balancing production game described in the previous section, this outsourcing game involves the decision to select a number of external suppliers (or service providers) for contingent future demands. Furthermore, the orders can be issued at any point of time rather than only once before the selling season starts. Similar to the production balancing problem and in contrast to the static outsourcing game considered in Chapter 2, there is no fixed or setup cost incurred when in-house production is launched. Nor are in-house production costs necessarily lower than the suppliers' wholesale prices.

We assume that the producer maintains the contingent and bounded in-house capacity to produce, at a known cost, quantities over time, $u(t)$. The demand consists of two components. One component reflects the regular, relatively low demand – of a known, steady level – which is traditionally met by in-house production and thus does not affect the optimization. As a result, this component (which was modeled in the production balancing problem) is not introduced explicitly in our model. It is accounted for by reduced (with respect to the regular demand consumption) in-house maximum capacity U. The other demand component represents peak demands (a new feature compared to the production balancing problem) and is introduced explicitly as a random variable. For example, oil and gas contracts are often negotiated and in some cases implemented well before energy demands are revealed. In such cases, home-heating firms may tend to build-up supplies for the winter, preventing problems associated with high demands (whether expected or not) Similarly, universities build up Internet server capacities by entering early into contractual agreements with Internet suppliers, building thereby an optional capacity to meet potential and future demand for services. In some cases, the firms, in addition to their own limited capacity, use external suppliers, relying thereby on outside capacities to meet future demands for products and service. Some extreme cases involve, of course, an outsourcing problem which consists in transferring activities that were previously in-house to a third party (Gattorna 1988; La Londe and Cooper 1989; Razzaque and Sheng 1998).

Since supply chain management frequently relies on sequential transmissions of information (Malone 2002), the problem of production-supply outsourcing is set as a hierarchical game where suppliers are sequential leaders while the producer is a follower. In this framework, the producer uses a demand estimate for some future date T (for example, the demand for oil at the beginning of the winter), selects time-sensitive production and a supply policy which is time-consistent with the firm's cost-minimizing objectives. The supply policy implies that given N potential suppliers, a subset of them is selected by the producer. Based on the producer's rational outsourcing decisions, each supplier selects a wholesale price to offer while the producer orders a certain product quantity $v_n(t)$ from the nth, $n=1,2,\ldots,N$ supplier at the stated price.

The producer's problem

Assume a firm producing a single product-type (commodity or service) to satisfy an exogenous demand, d, for the product-type at the end of a planning horizon, T. Inventories (or service capacity) are stored until the selling season starts, i.e., until $t=T$:

$$\dot{X}(t) = u(t) + \sum_n v_n(t), \quad X(0) = X^0. \tag{4.180}$$

where $X(t)$ is the surplus level of inventories by time t; $u(t)$ is the producer in-house production rate at time t; $v_n(t)$ is the supply rate of ordered and received from supplier n products; and X^0 is a constant. Both self-production and supplier capacity are bounded:

$$0 \leq u(t) \leq U, \tag{4.181}$$

$$0 \leq v_n(t) \leq V_n, \, n=1,..,N \tag{4.182}$$

where U is the producer's capacity and V_n is the capacity of supplier n.

The demand d at the end of planning period T is a random variable given by probability density and cumulative distribution $\varphi(D)$ and

$\Phi(a) = \int\limits_0^a \varphi(D)dD$ functions respectively. For each planning horizon T, there will be a realization D of d, which is known only at time T. Equation (4.180) presents the flow of products determined by production and supply rates from all engaged suppliers. The difference between the cumulative supply and production of the product and its demand, $X(T)$-D, is a surplus. If the demand exceeds the cumulative production and supplies, a penalty is paid. On the other hand, if $X(T)$-D>0, an overproduction cost is incurred at the end of the planning horizon. Furthermore, production costs are incurred at time t when the producer is not idle; holding costs are incurred when inventory levels are positive, $X(t) > 0$. Note that (4.180) implies that $X(t) \geq 0$ always holds.

The producer's objective is to find such a production program, $u(t)$, and supply schedule rates, $v_n(t)$ (outsourcing program), that satisfy constraints (4.180)- (4.182) while minimizing the following expected cost over the planning horizon T:

$$\min_{u, v_1, \dots, v_N} J_m(v_1, \dots, v_N, u, w_1, \dots, w_N) =$$

$$\min_{u, v_1, \dots, v_N} E\left[\int\limits_0^T \left(\sum_n w_n(t)v_n(t) + cu(t) + hX(t)\right)dt + P(X(T) - D)\right], \quad (4.183)$$

where $w_n(t)$ is the supplier n unit wholesale price at time t; h is the inventory holding cost of one product per time unit; and a piece-wise linear cost function is used for the surplus/shortage costs,

$$P(Z) = p^+ Z^+ + p^- Z^-, \quad (4.184)$$

where $Z^+ = \max\{0, Z\}$, $Z^- = \max\{0, -Z\}$, p^+ and p^- are the costs of one product surplus and shortage respectively. Substituting (4.184) into the objective (4.183), we have:

$$J_m = \int\limits_0^T \left(\sum_n w_n(t)v_n(t) + cu(t) + hX(t)\right)dt +$$

$$+ \int\limits_0^{X(T)} p^+ (X(T) - D)\varphi(D)dD + \int\limits_{X(T)}^\infty p^- (D - X(T))\varphi(D)dD \to \min. \quad (4.185)$$

Objective (4.185) is subject to constraints (4.180)- (4.182), which together constitute a deterministic problem equivalent to the stochastic problem (4.180)-(4.183).

The supplier's problem

Let the suppliers be ranked and then numbered with respect to their marginal costs, s_n, so that $s_{n-1} < s_n$ for $n=2,..,N$. The information on their wholesale prices, w_n, is obtained sequentially, starting from the highest rank supplier $n=N$ (see Kubler and Muller, 2004 for known examples and experimental evidence of sequential price setting). We assume that wholesale prices depend on the marginal costs and that the rank of a supplier, n ($1 < n < N$), is not reconsidered if $w_{n-1} \le w_n \le w_{n+1}$ holds.

Each supplier operates without inventories, supplying just-in-time at maximum rate V_n. Therefore the supplier's inventory dynamics is trivial. The nth supplier objective is:

$$\min_{w_n} J_s^n(v_1,..,v_N,u,w_1,..w_N) = \min_{w_n} \int_0^T (s_n v_n(t) - w_n(t)v_n(t))dt,\ n=1,2,..,N, \quad (4.186)$$

where $s_n v_n(t)$ – the supplier n expenditure rate and $w_n(t)v_n(t)$ – the supplier n revenue rate from wholesales at time t. Naturally, for the supplier to be sustainable and maintain his ranking, we require that

$$w_n(t) \ge s_n,\ n=1,2,..,N \text{ and } w_n(t) \le w_{n+1}(t),\ n=1,2,..,N\text{-}1. \quad (4.187)$$

The centralized problem

The centralized formulation excludes vertical competition by replacing the wholesale (transfer) prices with the corresponding marginal costs:

$$\min_{u,v_1,..,v_N} J(v_1,..,v_N,u,w_1,..,w_N) =$$

$$\min_{u,v_1,..,v_N} E\left[\int_0^T \left(\sum_n s_n(t)v_n(t) + cu(t) + hX(t) \right)dt + P(X(T)-D) \right] \quad (4.188)$$

subject to constraints (4.180)- (4.182).

Similar to the producer's problem, substituting (4.184) into the objective (4.188), we have:

$$J = \int_0^T \left(\sum_n s_n(t)v_n(t) + cu(t) + hX(t) \right)dt +$$

$$+ \int_0^{X(T)} p^+(X(T)-D)\varphi(D)dD + \int_{X(T)}^\infty p^-(D-X(T))\varphi(D)dD \to \min. \quad (4.189)$$

Objective (4.189) is subject to constraints (4.180)- (4.182), which together constitute a deterministic problem equivalent to the stochastic centralized problem (4.180) - (4.182) and (4.188).

System-wide optimal solution

The Hamiltonian for problem (4.189), (4.180)- (4.182) is as follows

$$H(t) = -\sum_n s_n v_n(t) - cu(t) - hX(t) + \psi(t)(u(t) + \sum_n v_n(t)) \, . \, (4.190)$$

The co-state variable $\psi(t)$ is the shadow price or margin gained by producing/outsourcing one more product unit at time t. According to the maximum principle, $\psi(t)$ satisfies the following co-state equation:

$$\dot{\psi}(t) = -\frac{\partial H(t)}{\partial X(t)} ,$$

with transversality (boundary) condition:

$$\psi(T) = -\frac{\partial \left[\int_0^{X(T)} p^+ (X(T) - D)\varphi(D)dD + \int_{X(T)}^{\infty} p^- (D - X(T))\varphi(D)dD \right]}{\partial X(T)}$$

$$= -\int_0^{X(T)} p^+ \varphi(D)dD + \int_{X(T)}^{\infty} p^- \varphi(D)dD \, .$$

That is,

$$\dot{\psi}(t) = h , \qquad\qquad (4.191)$$

$$\psi(T) = -p^+ \Phi(X(T)) + p^- (1 - \Phi(X(T))) \, . \qquad (4.192)$$

Rearranging only the decision variable-dependent terms of the Hamiltonian, we obtain:

$$H_u(t) = (\psi(t) - c)u(t) , \qquad\qquad (4.192)$$

$$H_v(t) = \sum_n (\psi(t) - s_n)v_n(t) \, . \qquad\qquad (4.192)$$

Thus, the optimal production and supply rates that maximize the Hamiltonian are:

$$u(t) = \begin{cases} U, & \text{if } \psi(t) > c; \\ a \in [0,U], & \text{if } \psi(t) = c; \\ 0, & \text{if } \psi(t) < c. \end{cases} \qquad (4.193)$$

$$v_n(t) = \begin{cases} V_n, & \text{if } \psi(t) > s_n; \\ b \in [0,V], & \text{if } \psi(t) = s_n; \\ 0, & \text{if } \psi(t) < s_n(t). \end{cases} \qquad (4.194)$$

An immediate insight from equations (4.193) and (4.194) is: (i) it is optimal to either not produce or produce only at maximum rate U, (ii) if it is optimal to use a supplier n for outsourcing, then it must be accomplished at a maximum rate, V_n, as shown in the following proposition.

Proposition 4.32. *If* $h \neq 0$, *then* $u(t) \in \{0,U\}$ *and* $v_n(t) \in \{0, V_n\}$, $n=1,..,N$ *and* $0 \leq t \leq T$.

Proof: The proof is by contradiction. Assume that production at an intermediate rate can be optimal. According to the optimality condition (4.193), the singular regime, $\psi(t) = c$, is the only regime along which intermediate values of the production rate are possible at a measurable time interval, τ. Therefore, assuming the singular regime condition holds over τ and differentiating this condition, we find:

$$\dot{\psi}(t) = 0,$$

which contradicts the co-state equation (4.191), $\dot{\psi}(t) = h \neq 0$. Thus no intermediate production rate is optimal, i.e., $u(t) \in \{0,U\}$. Similarly, one can verify that an intermediate outsourcing rate is not feasible.

An additional observation follows from optimality conditions (4.193) and (4.194) as well as from the linearity of the co-state variable (4.191). Specifically, if the producer's own unit production cost is lower than that of all suppliers, then the producer will first use his capacity to produce, starting from a time point, say t_0, and then seek supplies at a maximum rate beginning with the least costly supplier, say $n=1$, starting from a point in time, t_1. Next, he will seek supplies from the second less costly supplier and so on. This type of supply is advantageous when the producer's own capacity is relatively low while the expected demands are high and thus can be dealt with by just-in-time supply deliveries. On the other hand, if supply marginal costs are lower than the producer unit cost, then consecutive supplies will be sought first; self-production will be the last refuge, if at all.

To consider the most general conditions, we assume that there are M, $M<N$, suppliers for which marginal costs are below the producer's own cost, c, i.e., $s_n(t) < c$ for $n=0,1,..,M$ and $N-M$ suppliers with $s_n(t) > c$ for $n=M,M+1,..,N$. We next distinguish between various types of optimal solutions. First we delineate the conditions when the expected demand is low relative to system costs and initial inventories.

Proposition 4.33. *If*

$$\Phi(X^0) \geq \frac{p^- - s_1}{p^- + p^+}, \tag{4.195}$$

then it is not optimal to produce or to seek supplies, i.e., $u(t)=0$ *and* $v_n(t)=0$ *for* $0 \leq t \leq T$, $n=1,...,N$.

Proof: Consider the following solution for the co-state variable:

$$\psi(T) = -p^{+}\Phi(X^{0}) + p^{-}(1 - \Phi(X^{0})), \ \psi(t) = \psi(T) - h(T - t).$$

This solution implies $u(t)=0$ and $v_n(t) = 0$ for $0 \le t \le T$, $n=1,\ldots,N$ and thus $X(T)=X^0$, if the optimality conditions (4.193)- (4.194) are met, i.e., $\psi(t) < s_1$ for $0 \le t \le T$, as stated in the proposition.

If condition (4.195) is not met, then the supply rate from a supplier, n, can be optimal starting at time, t_n, while the manufacturer's in-house production starts from time t_0 (see Figure 4.13), as shown in the following proposition.

Proposition 4.34. *Let* $h \ne 0$, $\Phi(X^0) < \dfrac{p^{-} - s_1}{p^{-} + p^{+}}$ *and the breaking time*

point t_1 *satisfies the following equation for integer* K, $M < K \le N$

$$\Phi[X^0 + U(T - \left(\frac{c - s_1}{h} + t_1\right)) + \sum_{n=1}^{K} V_n(T - \left(\frac{s_n - s_1}{h} + t_1\right))] = \frac{p^{-} - s_1 - h(T - t_1)}{p^{-} + p^{+}}. \quad (4.196)$$

If $0 \le t_1$, $t_K < T$, *then* $t_n = \dfrac{1}{h}(s_n - s_1) + t_1$ *for* $n=2,\ldots,K$, *and*

$t_0 = \dfrac{1}{h}(c - s_1) + t_1$, *the system-wide optimal solution is unique and is given*

by: $v_n(t) = 0$ *for* $0 \le t < t_n$; $v_n(t) = V$ *for* $t_n \le t \le T$, $n=1,\ldots,K$ *and u(t)=0 for* $0 \le t < t_0$; *u(t)=U for* $t_0 \le t \le T$.

Proof: First note that since problem (4.180)- (4.182), (4.185) is convex, the maximum principle-based necessary optimality conditions are sufficient. Moreover, according to Proposition 4.32, when $h \ne 0$, the solution which meets the optimality conditions (4.193) and (4.194) is unique. For the state (4.180) and co-state (4.191) equations, consider the following solution which is determined by $K+1$ breaking points and satisfies the optimality conditions (4.193) and (4.194):

$$v_n(t) = 0 \ \text{for} \ 0 \le t < t_n; \ v_n(t) = V \ \text{for} \ t_n \le t \le T, n=1,\ldots,K; u(t)=0 \ \text{for}$$
$$0 \le t < t_0;$$

$$u(t)=U \ \text{for} \ t_0 \le t \le T; \ X(T) = X^0 + U(T - t_0) + \sum_{n=1}^{K} V_n(T - t_n); \ \psi(t_0) = c;$$

$\psi(t) = \psi(t_1) + h(t - t_1)$, $t \ge t_1$, $\psi(T) = s_1 + h(T - t_1)$; $\psi(t_n) = s_n$, $n=1,\ldots,K$.

If this solution is feasible, then it is also an optimal solution. To verify feasibility, we first determine the breaking points, $t_n = \dfrac{1}{h}(s_n - s_1) + t_1$, $n>1$. By substituting in the terminal inventory expression, $X(T)$, we obtain:

$$X(T) = X^0 + U(T - \left(\frac{c - s_1}{h} + t_1\right)) + \sum_{n=1}^{K} V_n(T - \left(\frac{s_n - s_1}{h} + t_1\right)).$$

Next, by taking into account the transversality condition (4.192) and $\psi(T) = s_1 + h(T - t_1)$, we find

$$- p^+ \Phi(X(T)) + p^- (1 - \Phi(X(T))) = s_1 + h(T - t_1).$$

Finally, by substituting $X(T)$ into the last expression, we determine equation (4.196) in unknown t_1. The feasibility of this solution is ensured by $0 \le t_1$, $t_K < T$, as stated in the proposition.

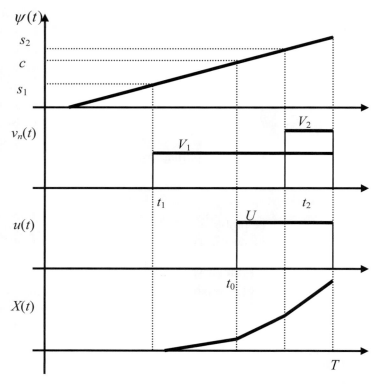

Figure 4.13. The system-wide optimal solution: the case of two suppliers and in-house production

There are some important observations from Proposition 4.34. First, even though the optimal number of suppliers, K, is not known in advance, one can easily find it by solving equation (4.196) repeatedly for $K=1,2,$ and so on until a feasible (and thus optimal) solution is found, i.e, $0 \le t_1$,

$t_K < T$. Furthermore, if K does not exist such that $t_1 \geq 0$, then the expected demand is too high or production/outsourcing capacity is too low and production/supply must be started from the very beginning of the planning horizon. Henceforth, we assume that this is not the case and focus on the broadest production conditions.

Game analysis

We next derive the best producer's response to wholesale prices $w_n(t)$ for $n=1,2,...N$ set by the suppliers sequentially starting from the highest rank $n=N$. To apply the maximum principle to the deterministic equivalent of the producer's problem we construct the Hamiltonian

$$H_m(t) = -\sum_n w_n(t)v_n(t) - cu(t) - hX(t) + \psi_m(t)(u(t) + \sum_n v_n(t)), (4.197)$$

where the co-state variable $\psi_m(t)$, the margin which the producer gains from producing/outsourcing one more product unit, satisfies the following co-state equation:

$$\dot{\psi}_m(t) = h , (4.198)$$

with transversality (boundary) condition:

$$\psi_m(T) = -p^+ \Phi(X(T)) + p^-(1 - \Phi(X(T))). (4.199)$$

Consequently, the optimal production and supply rates that maximize the Hamiltonian are:

$$u(t) = \begin{cases} U, & \text{if } \psi_m(t) > c; \\ a \in [0,U], & \text{if } \psi_m(t) = c; \\ 0, & \text{if } \psi_m(t) < c. \end{cases} (4.200)$$

$$v_n(t) = \begin{cases} V_n, & \text{if } \psi_m(t) > w_n(t); \\ b \in [0,V], & \text{if } \psi_m(t) = w_n(t); \\ 0, & \text{if } \psi_m(t) < w_n(t). \end{cases} (4.201)$$

Comparing conditions (4.198)-(4.201) with the corresponding conditions found for the centralized problem, we observe that they are symmetric and are obtained by replacing ψ with ψ_m and s_n with w_n. Thus, with the replacement provided, all results derived for the centralized problem can be restated for the producer's problem. Specifically, it is optimal to either not produce or produce only at maximum rate U. And if it is optimal to use suppliers, then it must be accomplished at a maximum rate, V_n, unless the rate of increase of the unit wholesale price is equal to the unit holding cost rate, h, as shown in the following proposition. The new requirement to the rate of change of the wholesale prices is due to the fact that while the marginal costs s_n are constant, the wholesale $w_n(t)$ price may change with time.

Proposition 4.35. *If* $\dot{w}_n(t) \neq h$ *and* $h \neq 0$, *then* $u(t) \in \{0, U\}$ *and* $v_n(t) \in \{0, V_n\}$, $n=1,..,N$ *and* $0 \leq t \leq T$.

Although we assume that the wholesale price is a differentiable function of time, all subsequent results can easily be presented for arbitrary $w_n(t)$.

We first assume that the suppliers' pricing policy, w_n, does not affect their rating and then verify this. Again, to consider the most general conditions, we assume that there are M, $M<N$, suppliers for which wholesale prices are below the producer's own cost, c, i.e., $w_n(t)<c$ for $n=0,1,..,M$ and $N-M$ suppliers with $w_n(t)>c$ for $n=M,M+1,..,N$. Then replacing s_1 with w_1 we restate the no production/outsourcing conditions derived in Proposition 4.33.

Proposition 4.36. *If*

$$\Phi(X^0) \geq \frac{p^- - w_1(t) - h(T-t)}{p^- + p^+}, \text{ for } 0 \leq t \leq T, \qquad (4.202)$$

then it is not optimal to produce or to seek supplies, i.e., $u(t)=0$ *and* $v_n(t) = 0$ *for* $0 \leq t \leq T$, $n=1,..,N$.

If condition (4.202) is not met, then the supply rate from a supplier, n, can be optimal starting at time, t_n, while the producer's production may start optimally from time t_0. The following proposition presents the producer's optimal response (see Figure 4.14) when the contract between the producer and a supplier has no flexibility. That is, supplier n commits to a steady wholesale price $w_n(t)=w_n$, while the producer orders a constant quantity $V_n(T-t_n)$. In such a case, the result of Proposition 4.37 is obtained by replacing s_1 with w_1 in Proposition 4.34.

Proposition 4.37. *Let* $\dot{w}_n(t) \neq h$, $h \neq 0$, $w_n(t)=w_n$, $0 \leq t \leq T$, *there exist time point t, such that* $\Phi(X^0) < \dfrac{p^- - w_1(t) - h(T-t)}{p^- + p^+}$ *and the breaking time point* t_1 *satisfies the following equation for integer* K, $M < K \leq N$

$$\Phi\left[X^0 + U\left(T - \left(\frac{c - w_1}{h} + t_1\right)\right) + \sum_{n=1}^{K} V_n\left(T - \left(\frac{w_n - w_1}{h} + t_1\right)\right)\right] = \frac{p^- - w_1 - h(T - t_1)}{p^- + p^+} \qquad (4.203)$$

If $0 \le t_1$, $t_K < T$, *then* $t_n = \dfrac{1}{h}(w_n - w_1) + t_1$ *for* $n=2,...,K$, *and*

$t_0 = \dfrac{1}{h}(c - w_1) + t_1$, *the optimal producer's response is unique and is given*

by: $v_n(t) = 0$ *for* $0 \le t < t_n$; $v_n(t) = V$ *for* $t_n \le t \le T$, $n=1,...,K$ *and u(t)=0*
for $0 \le t < t_0$; *u(t)=U for* $t_0 \le t \le T$.

Denoting the total order

$$X = U\left(T - \left(\frac{c - w_1}{h} + t_1\right)\right) + \sum_{n=1}^{K} V_n\left(T - \left(\frac{w_n - w_1}{h} + t_1\right)\right)$$

and taking into account $t_n = \dfrac{1}{h}(w_n - w_1) + t_1$, equation (4.203) results in

$$\Phi[X^0 + X] = \frac{p^- - w_K - h(T - t_K)}{p^- + p^+}. \tag{4.204}$$

If we set unit inventory holding cost at zero, thereby disregarding inventory dynamics, then equation (4.204) takes the familiar newsboy form,

$$\Phi[X^0 + X] = \frac{p^- - w_K}{p^- + p^+}.$$

We thus conclude, *given wholesale prices, a myopic producer will order more than he needs to account for inventory dynamics.*

This is to say, that a primary difference emanates from the dynamics of inventory accumulation. The time-dependent solution of Proposition 4.37 implies that the producer may not order until a certain breaking point in time which depends on the problem's parameters and the inventory holding cost.

As we shall show in Proposition 4.39 below, the greater the wholesale price of a supplier, the longer the producer waits before he orders. This observation is justified by the differential game since the producer has an advantage over a supplier up to and until the breaking point (for outsourcing to this suppler) is reached. This implies as well that a supplier may not be able to wait for an order at the stated price and thus may reduce his wholesale price—inducing the producer to order earlier. After a breaking point has been reached and an order placed, the situation changes and the supplier has an advantage over the producer as shown in Proposition 4.38 below. This result confirms the observed behavior of suppliers who tend to increase contracted wholesale prices over time when supply contracts are flexible and market conditions change.

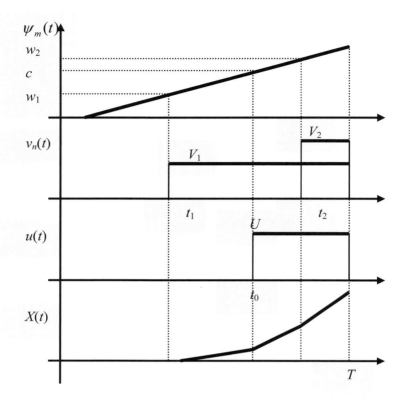

Figure 4.14. The producer's optimal response: the case of two suppliers and in-house production

Unlike the steady-price assumption of Proposition 4.37, we shall distinguish next between two cases: when the wholesale price $w_n(t)$ of supplier n changes before supplies start, i.e., before t_n, and after this breaking point. It turns out that in the latter case there are certain bounds such that changes in $w_n(t)$ for $t_n < t \le T$ do not affect the optimal response of the producer as shown in the following proposition.

Proposition 4.38. *Let* $w_n(t_n) = w_n$, *and maintain conditions of Proposition 4.37. The solution determined by Proposition 4.37 remains optimal if* $w_n(t) < w_n(t_n) + h(t - t_n)$ *for* $t_n < t \le T$.

Proof: According to the co-state solution of Proposition 4.37, we have
$$\psi_m(t) = \psi_m(t_n) + h(t - t_n) = w(t_n) + h(t - t_n) .$$

Substituting this in the optimality condition, $v_n(t)=V_n$ if $\psi_m(t)>w_n(t)$, from (4.201) we find, $w_n(t)<w_n(t_n)+h(t-t_n)$, $t_n<t\leq T$.

Note, that according to the co-state solution of Proposition 4.37 (see the proof of Proposition 4.34), we have

$$\psi_m(t)=\psi_m(t_1)+h(t-t_1)=w(t_1)+h(t-t_1).$$

Therefore, if $w_n(t)$ is differentiable, then taking into account $w_n(t_n)=w_n=w_1(t_1)+h(t-t_1)$, Proposition 4.38 is transformed into a simple-to-use sufficient condition: if $\dot{w}_n(t)<h$ for $t_n\leq t\leq T$, then the optimal solution does not change.

Proposition 4.38 indicates that a supplier's wholesale price can drop arbitrarily after the supplier begins delivery. This will not affect the producer's ordering policy with respect to the supplier since the supplier already delivers products at a maximum rate. Further, a wholesale price increase is bounded by the unit inventory holding cost per time unit. Indeed, if this condition does not hold, the producer may be better off keeping lower inventory levels rather than accumulating inventories resulting from a supplier's fulfilling order.

Finally, if a wholesale price changes before supplies are dispatched, the corresponding breaking point changes. Below, we show that an increase in the wholesale price, $w_n(t)=w_n$, of a supplier before the breaking point results in delay in supply orders, i.e., the optimal solution changes no matter how small the price change.

Proposition 4.39. *If all conditions of Proposition 4.37 hold, then* $\dfrac{dt_n}{dw_n}>0$,

where

$$\frac{dt_n}{dw_n}=\frac{\dfrac{U}{h}\varphi(Y)+\dfrac{1}{p^-+p^+}}{(U+V_n)\varphi(Y)+\dfrac{h}{p^-+p^+}}, \tag{4.205}$$

$$Y=X^0+U(T-\left(\frac{c-w_1}{h}+t_1\right))+\sum_{n=1}^{K}V_n(T-\left(\frac{w_n-w_1}{h}+t_1\right)). \tag{4.206}$$

Proof: Taking into account that $t_n=\dfrac{1}{h}(w_n-w_1)+t_1$ (see Proposition 4.37),

equation (4.204) transforms to

$$F = \Phi[X^0 + U(T - \left(\frac{c - w_1}{h} + t_n - \frac{w_n - w_1}{h}\right)) + \sum_{n=1}^{K} V_n(T - t_n)] -$$

$$\frac{p^- - w_1 - h(T - t_n + \frac{w_n - w_1}{h})}{p^- + p^+} = 0.$$

To show $\frac{dt_n}{dw_n} > 0$, we employ implicit differentiation of the last

expression, $\frac{dt_n}{dw_n} = -\frac{\partial F / \partial w_n}{\partial F / \partial t_n}$, which, using notation (4.206), results in:

$$\frac{dt_n}{dw_n} = -\frac{\frac{U}{h}\varphi(Y) + \frac{1}{p^- + p^+}}{-(U + V_n)\varphi(Y) - \frac{h}{p^- + p^+}} > 0.$$

An immediate corollary from this lemma is that the greater the wholesale prices, the later the producer places outsourcing orders thereby outsourcing smaller quantities. With respect to the centralized supply chain this implies: *If the suppliers profit by setting $w_n(t) > s_n$, for $0 \le t \le T$, then the outsourcing order quantity decreases compared to the system-wide optimal quality; the greater the wholesale prices, the larger the difference between the competitive solution of the differential outsourcing game and the system-wide optimal solution.*

Equilibrium

The Stackelberg strategy is frequently associated with time-inconsistency, namely, even though such an equilibrium can be formulated at a given time, it might not be sustainable over time (Jorgenson and Zaccour 2004). The intuition for the time-inconsistency is the following. In a Stackelberg game, the leader seeks to influence the follower's choice of strategy for making the most profits. For this purpose, the leader sets at time zero the strategy for the entire horizon. However, if at a time point t the leader finds a more profitable strategy over the remaining part of the horizon, the leader has no reason to follow the initially announced strategy. In the context of this outsourcing game, the supplier's decision to eventually deviate from the initially announced wholesale price depends on the type of contract between the producer and the supplier. This implies that the time-inconsistency may affect the equilibrium if a contract with flexibility is preferred to the total commitment type of contract (e.g., minimum total quantity commitment and periodical commitment; see, for example, Anupinidi and Bassok 1999). The leaders then can use the flexibility to change the prices. In what follows we consider the game that results in such an equilibrium.

To determine the Stackelberg (or hierarchical) strategy, we substitute the producer's optimal response into the supplier n objective function which is minimized. According to Propositions 4.37 and 4.38, $v_n(t) = 0$ for $0 \le t < t_n$; $v_n(t) = V$ for $t_n \le t \le T$, and t_n is a function of $w_n(t)$, i.e., $t_n = t_n(w_n(t))$. Therefore using the first-order optimality condition, we obtain:

$$\frac{dJ_s^n}{dw_n} = \frac{d\left[\int_{t_n}^{T}(s_n v_n(t) - w_n(t)v_n(t))dt\right]}{dw_n} = \frac{dV_n\left(s(T - t_n) - \int_{t_n}^{T}w_n(t)v_n(t)dt\right)}{dw_n} = 0,$$

$$n = 1, 2, .., K,$$

and thus

$$\frac{dt_n}{dw_n(t)}(w_n(t_n) - s_n) - (T - t_n) = 0. \tag{4.206}$$

Recalling Proposition 4.38 and 4.39, we substitute (4.204) into (4.206) to find that if $\dot{w}_n(t) < h$, $t_n \le t \le T$ (while other conditions of Proposition 4.37 hold), then an optimal wholesale price satisfies the following equation:

$$\frac{\frac{U}{h}\varphi(Y) + \frac{1}{\frac{p^- + p^+}{h}}}{(U + V_n)\varphi(Y) + \frac{h}{p^- + p^+}}(w_n(t_n) - s_n) - (T - t_n) = 0. \tag{4.207}$$

Using breaking points $t_n = \frac{1}{h}(w_n(t_n) - w_1(t_1)) + t_1$ (see Proposition 4.37),

equation (4.205) is reduced to:

$$Y = X^0 + U\left(T - \left(t_n - \frac{w_n(t_n) - c}{h}\right)\right) + \sum_{n=1}^{K}V_n(T - t_n). \tag{4.208}$$

Let the solution to equation (4.207)-(4.208) in $w_n(t_n)$ and Y be β_n and γ respectively. If the contract is not flexible, the wholesale price cannot be changed, and therefore $w_n(t) = w_n(t_n) = \beta_n$ for $t_n \le t \le T$, $n = 1, 2, ..., K$, as shown in the following proposition.

Proposition 4.40. Let all conditions of Proposition 4.37 be met, $\beta_n \le \beta_{n+1}$, $n = 1, .., K-1$ and $\beta_K \le s_{K+1}$. If

$$\left(\frac{U+V_n}{\left((U+V_n)\varphi(\gamma)+\dfrac{h}{p^-+p^+}\right)}-\frac{U}{U\varphi(\gamma)+\dfrac{h}{p^-+p^+}}\right)\left((U+V_n)\frac{dt_n}{dw_n}-\frac{U}{h}\right)\varphi_Y'\,(\beta_n-s_n)+2>0,$$

$$(4.209)$$

then the supplier's wholesale pricing policy is $w_n^s(t)=\beta_n$, and the producer's production $u^s(t)$ and outsourcing $v_n^s(t)$ policies, $n=1,..,K$ are determined by Proposition 4.37. These policies constitute the unique Stackelberg equilibrium for $t \in [0,T]$ in the differential outsourcing game.

Proof: To prove the proposition, it is sufficient to show that the second-order optimality condition holds, that is, the derivative of the left-hand side of (4.207) is positive; and that $\beta_n \geq s_n$, i.e., constraint (4.187) holds. The fact that the latter is true is immediately observed from (4.207) which can be met only if $w_n(t_n)=\beta_n>s_n$. To show the former, we differentiate (4.207) with respect to $w_n(t_n)$:

$$\frac{d^2 J_s^n}{dw_n^2}=$$

$$-\frac{\dfrac{U}{h}\varphi(Y)+\dfrac{1}{p^-+p^+}}{\left((U+V_n)\varphi(Y)+\dfrac{h}{p^-+p^+}\right)^2}\left(\frac{U}{h}-(U+V_n)\frac{dt_n}{dw_n}\right)(U+V_n)\varphi_Y'\,(w_n-s_n)+$$

$$\frac{\dfrac{U}{h}\varphi'(Y)(\dfrac{U}{h}-(U+V_n)\dfrac{dt_n}{dw_n})}{(U+V_n)\varphi(Y)+\dfrac{h}{p^-+p^+}}(w_n-s_n)+\frac{\dfrac{U}{h}\varphi(Y)+\dfrac{1}{p^-+p^+}}{(U+V_n)\varphi(Y)+\dfrac{h}{p^-+p^+}}+\frac{dt_n}{dw_n}),$$

Taking into account (4.205) and requiring that $\dfrac{d^2 J_s^n}{dw_n^2}>0$ we obtain condition (4.209) as stated in this proposition.

Note that condition (4.209) does not necessarily hold for every probability distribution. If this is the case, the equilibrium may not be unique.

Example 4.6.

Consider a supply chain system characterized by two suppliers $N=2$ each of which supplies the manufacturer, i.e., $K=2$ and by the uniform demand distribution

$$\varphi(D) = \begin{cases} \dfrac{1}{A}, & \text{for } 0 \le D \le A; \\ 0, & \text{otherwise.} \end{cases} \quad \text{and} \quad \Phi(a) = \frac{a}{A}.$$

First we observe that $\varphi'(D) = 0$ and thus condition (4.209) of Proposition 4.40 is met. Next, if supplier $n=1$ has a wholesale price lower than the producer's production cost and supplier $n=2$ is costlier than the producer, then from (4.204) we have:

$$\frac{1}{A}[X^0 + U(T - (\frac{c - w_1}{h} + t_1)) + \sum_{n=1}^{2} V_n(T - (\frac{w_n - w_1}{h} + t_1))] = \frac{p^- - w_1 - h(T - t_1)}{p^- + p^+}.$$

That is,

$$t_1 = \frac{X^0 + U(T - \dfrac{c - w_1}{h}) + \displaystyle\sum_{n=1}^{2} V_n(T - \dfrac{w_n - w_1}{h}) - A\dfrac{p^- - w_1 - hT}{p^- + p^+}}{\displaystyle\sum_{n} V_n + U + \dfrac{Ah}{p^- + p^+}},$$

and

$$t_0 = \frac{1}{h}(c - w_1) + t_1 \; ; \; t_2 = \frac{1}{h}(w_2 - w_1) + t_1 \, .$$

Accordingly, if $t_1 > 0$ and $t_2 < T$, then according to Proposition 4.37 $v_1(t) = 0$ for $0 \le t < t_1$; $v_1(t) = V_1$ for $t_1 \le t \le T$; $v_2(t) = 0$ for $0 \le t < t_2$; $v_2(t) = V_2$ for $t_2 \le t \le T$ and $u(t)=0$ for $0 \le t < t_0$; $u(t)=U$ for $t_0 \le t \le T$.

Finally, substituting the breaking points t_1, t_2 and t_0 into (4.207) we obtain two linear equations with two unknowns w_1, w_2:

$$\frac{\dfrac{U}{Ah} + \dfrac{1}{p^- + p^+}}{\dfrac{(U + V_n)}{A} + \dfrac{h}{p^- + p^+}}(w_n - s_n) - (T - t_n) = 0 \, , \text{ for } n=1, 2.$$

Let $p^+ = p^- = \$5$/product unit; $c = \$0.7$ per product unit; $h = \$0.1$ per product unit and time unit; $U=10$, $V_1 = 5$, $V_2 = 15$ product units per time unit; $s_1 = 0$, $s_2 = \$0.8$, $T=10$ time units; and $A=200$ product units. This implies that the sequence of suppliers is $n=1$ ($s_1 = 0$), in-house production ($c=0.7$) and $n=2$ ($s_2 = 0.8$). Inserting these in the last equations we have:

$$7.06w_1 - (10-t_1) = 0, \; 0.48(w_2 - 0.8) - (10-t_1) + 10(w_2 - w_1) = 0;$$

$t_1 = 6.65 + 5.63w_1 - 4.68w_2, \; t_0 = 10(0.7 - w_1) + t_1$, and $t_2 = 10(w_2 - w_1) + t_1$.

Solving these equations, we find the breaking points t_1, t_0 and t_2 at which the producer begins to outsource to the first supplier at the rate V_1 of 5 per time unit; self-production at $U = 10$ per time unit; and outsourcing to the second supplier at $V_2 = 15$ per time unit respectively:

$$t_1 = 5.14, \; t_0 = 5.34, \text{ and } t_2 = 9.74.$$

In addition, the Stackelberg wholesale prices of the two suppliers are:

$$w_1^s = 0.68 \text{ and } w_2^s = 1.14.$$

Recalling that $c = 0.7$, we verify that $w_2^s > c > w_1^s$.

Similarly, from Proposition 4.34 we find the system-wide optimal solution by solving the following equations

$$t_1 = \frac{X^0 + U(T - \dfrac{c - s_1}{h}) + \sum_{n=1}^{2} V_n (T - \dfrac{s_n - s_1}{h}) - A \dfrac{p^- - s_1 - hT}{p^- + p^+}}{\sum_n V_n + U + \dfrac{Ah}{p^- + p^+}} = 0.94,$$

$$t_0 = \frac{1}{h}(c - s_1) + t_1 = 7.94; \quad t_2 = \frac{1}{h}(s_2 - s_1) + t_1 = 8.94.$$

Comparing this result with the Stackelberg policy, we observe the expected effect of vertical competition. The total production order $X^*(T) = U(T - t_0) + V_1(T - t_1) + V_2(T - t_2)$ of the centralized supply chain $X^*(T) = 81.8$ is greater than that under equilibrium competition $X^s(T) = 74.8$.

The effect of time-inconsistency

The equilibrium wholesale price determined by Proposition 4.40 is embodied within the total commitment type of contract between all parties. However, in real life it is often observed that once a contract with a degree of flexibility has been signed, and deliveries initiated (i.e., after breaking point t_n), suppliers may use numerous excuses (e.g., service extensions, increased labor costs and raw material prices and so on) to raise their prices. This is particularly the case for Internet and telecom providers, who add various pay services and limitations to increase in the course of time, explicitly as well as implicitly, their initial wholesale price. The equilibrium as a result may in practice be problematic.

To understand this, we employ the sensitivity analysis conducted with respect to Proposition 4.38. Indeed, according to Proposition 4.38, a producer's optimal response does not change if the initial wholesale price $w_n(t_n)$ increases after t_n at a rate slower than the inventory holding cost h. Accordingly, if supplier n is cunning enough to properly increase the initial price, the optimal solution in terms of the producer will remain the same. This implies that since the producer will not change the order quantity while paying a higher price, the supplier will collect a greater profit. This phenomenon is referred to as the time-inconsistency of the equilibrium.

A Stackelberg strategy is frequently associated with time-inconsistency which usually causes the equilibrium to fall apart. In our case, however, this does not happen! All players show a steady behavior and supply contracts are not abandoned. Moreover, if suppliers increase wholesale prices at a constant rate, $\dot{w}_n(t)$, which tends to h (but never equal to it), then they will gain a maximum profit under the same producer's response. This is to say, the supply chain will attain a new equilibrium which is time-consistent as the following proposition states.

Proposition 4.41. *Let all conditions of Proposition 4.37 be met except for $w_n(t)=w_{n\cdot}$.*

If $\beta_n \leq \beta_{n+1}$, $n=1,..,K-1$, $\beta_K \leq s_{K+1}$ and

$$\frac{1}{\left((U+V_n)\varphi(\gamma) + \dfrac{h}{p^- + p^+} \right)} \left((U+V_n)\frac{dt_n}{dw_n} - \frac{U}{h} \right)\varphi_Y' \, (\beta_n - s_n) + 2 > 0,$$

then the supplier's wholesale pricing policy is $w_n^s(t)=\beta_n+\lambda(t-t_n)$ for $t_n \leq t \leq T$, $\lambda \to h-$, and the producer's production $u^s(t)$ and outsourcing $v_n^s(t)$ policies, $n=1,..,K$ are determined by Proposition 4.37. These policies constitute a Stackelberg equilibrium in the differential outsourcing game with unique, initial wholesale prices $w_n^s(t_n)$ and unique production and outsourcing policies for $t \in [0,T]$.

Proof: The proof immediately follows from Propositions 4.38 and 4.40.

We next illustrate the new, time-consistent equilibrium with the same example.

Example 4.6. (continued)

Returning to our example, we note that with respect to Proposition 4.41 the time-consistent equilibrium wholesale prices are:

$$w_1^s(t)=0.68+\lambda(t-5.14) \text{ for } 5.14 \leq t \leq 10 \text{ and } w_2^s(t)=1.14 + \lambda(t-9.74)$$
$$\text{for } 9.74 \leq t \leq 10,$$

where $\lambda \to h$ -0 and $h=0.1$.

This, for example, implies that the equilibrium wholesale price of the first supplier can reach the initial wholesale price, 1.14, of the second supplier by the end of the planning horizon. The essential implication of the equilibrium is: if λ attains $h=0.1$, then, according to Proposition 4.35, the producer has multiple optimal responses and can select an order quantity $v_n(t)$ less than the maximum one, implying that the maximum profit is no longer assured for supplier n. Moreover, if $\lambda>0.1$, then optimality conditions (4.201) induce the producer to completely stop ordering, causing suppliers to lose profits by an increase in wholesale prices. Thus, suppliers may increase wholesale prices at a rate very close, but never equal to 0.1.

Coordination

Building-up a supply capacity to meet future and uncertain demands for products and services is a costly strategic issue which involves decisions being made in the course of time with the sole purpose of meeting a demand in real-time that may outstrip an available capacity. Of course, firms may build-up their self-capacity and thereby meet demands when they occur, but such an approach is often deemed far too costly. Therefore firms use multiple suppliers who can provide an added supply capacity as well as goods that may be stored to meet prospective demands.

We assume that the producer lacks the capacity to meet peak demands at known specific times and therefore depends on suppliers. This results in vertical outsourcing competition with a Stackelberg equilibrium solution different from that for the corresponding centralized supply chain. The deterioration of the supply chain performance is due to double marginalization as is often the case with vertical competition. Similar to the supply chain games with underlying vertical competition discussed in this book, the two-part tariff efficiently coordinates the system. Indeed, by comparing Propositions 4.34 and 4.37, we observe that if all suppliers set the wholesale prices equal to their marginal costs, the solution becomes system-wide optimal and the suppliers can get their share of the profits by setting fixed costs of supplies. The difference between this approach and the other two-part tariff applications discussed so far is related to time-inconsistency. As shown in Proposition 4.41, if the contract between the producer and suppliers allows for some level of flexibility, the suppliers will be tempted in time to gradually increase wholesale prices above the marginal costs at a rate close but less than the unit holding cost of the producer. As long as this condition holds, the producer's best response does not change (see Proposition 4.38). This implies that the overall supply chain profit does not change as well. Therefore the performance of the supply chain does not deteriorate and the amendments in the wholesale prices are just internal transfers of the chain. Thus, the two-part tariff in this intertemporal system has a "third part" which is dynamic. Specifically, the suppliers first set the wholesale price equal to the marginal cost at their breaking point (supply time) $w_n(t_n)=s_n$ and fixed transaction cost. Then they gradually increase the price, $w_n^s(t)=s_n+\lambda(t-t_n)$ for $t_n \leq t \leq T$, $\lambda \to h$ -0. The result is that the supply chain is perfectly coordinated and the dynamic increase in the wholesale prices provides the suppliers (the Stackelberg leaders) with additional profits that constitute a bargaining tool that the suppliers may use to reduce the fixed supply cost.

4.5 INTERTEMPORAL CO-INVESTMENT IN SUPPLY CHAINS

This section considers investment in a supply chain infrastructure using an inter-temporal model. We assume that firms' capital is essentially the supply chain's infrastructure. As a result, firms' policies consist in selecting an optimal level of employment as well as the level of co-investment in the supply chain infrastructure. So far we have mainly discussed open-loop equilibrium solutions of competing firms. This section presents both open-loop and feedback solutions for non-cooperating firms, as well as, long- and short-run investment cooperation.

4.5.1 THE DIFFERENTIAL INVESTMENT AND LABOR GAME

Consider N-firms operating in a supply chain, each characterized by its output price $p_j(t)$ at time t, labor force $L_j(t)$, investment policy $I_j(t)$ and an aggregate production function $Q=f(K,L_j)$, $\dfrac{\partial f}{\partial K} \geq 0$, $\dfrac{\partial f}{\partial L_j} > 0$ for $L \neq 0$,

$$\frac{\partial f(K,0)}{\partial L_j} = 0, \quad \frac{\partial^2 f}{\partial L_j^2} \leq 0, \quad \frac{\partial^2 f}{\partial K^2} \leq 0.$$

Dynamic Model for Co-Investment in Infrastructure

We let $K(t)$ be the level of current supply chain infrastructure capital, deteriorating at the rate δ. The process of capital accumulation is then given by:

$$\frac{dK(t)}{dt} = -\delta K(t) + \sum_{j=1}^{N} I_j(t), \ K(0){=}K_0, \ I_j(t) \ge 0, j = 1,;;;,N \ . \ (4.210)$$

The firms' objective consists in maximizing the discounted profit by selecting an optimal employment policy on the one hand and a co-investment in supply chain infrastructure (contributing thereby to all firms potential revenues) on the other. The objective is specified by:

$$\underset{L_j,I_j}{Max} \int_0^{\infty} e^{-r_j t} [p_j(t) f(K(t),L_j(t)) - c_j(t) L_j(t) - C_{Ij}((1-\theta) I_j(t))] dt, j{=}1,..,N, \ (4.211)$$

where $c_j(t)$ is the labor cost and $C_{Ij}(.)$ is a continuous, twice differ-

entiable and increasing investment cost function, $\dfrac{\partial C_{Ij}}{\partial I_j} > 0$, $\dfrac{\partial^2 C_{Ij}}{\partial I_j^{\ 2}} \ge 0$,

mitigated by a proportion which is subsidized and given by θ. To study the problem, we construct the Hamiltonians:

$$H_j(t){=}e^{-r_j t}[p_j(t) f(K(t),L_j(t)) - c_j(t) L_j(t) - C_{Ij}((1{-}\theta) I_j(t))] {+} \psi_j(t)(\sum_{j=1}^{N} I_j(t) {-} \delta K(t)), \ (4.212)$$

where the co-state variables are determined by

$$\dot{\psi}_j(t) = -e^{-r_j t} p_j(t) \frac{\partial f(K(t),L_j(t))}{\partial K(t)} + \delta \psi_j(t), \ \lim_{t \to \infty} \psi_j(t) = 0 . \ (4.213)$$

The Hamiltonian (4.212) can be interpreted as the instantaneous profit rate of firm j, which includes the firm j value $\psi_j(t)\dot{K}(t)$ of increment $\dot{K}(t)$ in the infrastructure capital. The co-state variable $\psi_j(t)$ is the shadow price, i.e., the net benefit of firm j from investing one more monetary unit at time t. The differential equation (4.213) states that the marginal opportunity cost $\delta \psi_j(t)$ of investment of firm j in infrastructure should equal the

(discounted) marginal profit $e^{-r_j t} p_j(t) \dfrac{\partial f(K(t),L_j(t))}{\partial K(t)}$ from increased pro-

ductivity and from the capital gain $\dot{\psi}_j(t)$.

Optimal policies are found by maximizing the Hamiltonians with respect to investments, $I_j(t)$, and labor, $L_j(t)$, which yields:

$$L_j(t) = \begin{cases} l_j(t), \text{if } \dfrac{\partial f(K(t),L_j(0))}{\partial L_j} \le \dfrac{c_j(t)}{p_j(t)}, \\ 0, \text{otherwise.} \end{cases} \ (4.214)$$

where $l_j(t)$ is determined by

$$\frac{\partial f(K(t),l_j(t))}{\partial L_j} - \frac{c_j(t)}{p_j(t)} = 0; \tag{4.215}$$

and

$$I_j(t) = \begin{cases} i_j(t), \text{if } \psi_j(t) \geq \dfrac{\partial C_{Ij}(0)}{\partial I_j} e^{-r_j t} \\ 0, \text{otherwise} \end{cases}, \tag{4.216}$$

with $i_j(t)$ determined by

$$\psi_j(t) = \frac{\partial C_{Ij}((1-\theta)i_j(t))}{\partial I_j} e^{-r_j t}. \tag{4.217}$$

Since the objective function (4.211) is concave and constraints (4.210) are linear, conditions (4.214)-(4.217) are necessary and sufficient for optimality and will be considered next in detail, providing specific insights regarding the investment process in supply chain infrastructure.

The N-Firms Open-Loop Nash Strategies

The Nash equilibrium for each firm is obtained by optimizing simultaneously all N Hamiltonians (4.213). This straightforwardly results in the following proposition.

Proposition 4.42. *If a(t), b$_j$(t) and i$_j$(t), j=1,..,N satisfy the following system of equations*

$$\frac{da(t)}{dt} = -\delta a(t) + \sum_{j=1}^{N} i_j(t), \ a(0) = K_0, \ , b_j(t) = \frac{\partial C_{Ij}((1-\theta)i_j(t))}{\partial I_j} e^{-r_j t},$$

$$\frac{\partial f(a(t),l_j(t))}{\partial L_j} = \frac{c_j(t)}{p_j(t)},$$

$$\dot{b}_j(t) = -e^{-r_j t} p_j(t)\frac{\partial f(a(t),l_j(t))}{\partial a(t)} + \delta b_j(t), \ \lim_{t\to\infty} b_j(t) = 0, \ j=1,..,N,$$

then, the pair of dynamic strategy sets {Ij^n(t)=i$_j$(t), j=1,..,N} and {Lj^n(t)= l$_j$(t), j=1,..,N}, t ≥ 0 is a Nash equilibrium in the supply chain co-investment and labor force differential game.

The implications of this proposition are best examined through an example which assumes a Cobb-Douglas Production function and a quadratic investment cost.

Example 4.7.

Let the aggregate production function be a Cobb-Douglas function, $f(K, L_j) = aK^{\alpha}L_j^{\beta}$, with $\alpha + \beta = 1$, $C_{Ij}(I) = c_{Ij}(1-\theta)I_j^2$ and, let the labor cost increase slower than the price index raised to power β so that:

$$\omega_j(t) = \left[\frac{p_j(t)}{c_j^{\beta}(t)}\right]^{\frac{1}{1-\beta}} = e^{\varepsilon t}, \ \varepsilon < \min\{r_j, j=1,..,N\}.$$

Using (4.215), we have $aK^{\alpha}(t)\beta l_j^{\beta-1}(t) - \dfrac{c_j(t)}{p_j(t)} = 0$, and thus,

$$L_j(t) = l_j(t) = \left[\frac{1}{a\beta}\frac{c_j(t)}{p_j(t)K^{\alpha}(t)}\right]^{\frac{1}{\beta-1}}.$$

Note, that $\dfrac{\partial L_j(t)}{\partial K(t)} > 0$, if $\beta < 1$ and $\dfrac{\partial L_j(t)}{\partial K(t)} < 0$, if $\beta > 1$. Next, from Proposition 4.42, we have

$$\dot{b}_j(t) = -e^{-r_j t} p_j(t)a\alpha a^{\alpha-1}(t)l_j^{\beta}(t) + \delta b_j(t) =$$

$$= -e^{-r_j t}\left[\frac{p_j(t)}{c_j^{\beta}(t)}\right]^{\frac{1}{1-\beta}}\frac{\alpha}{\beta}[a\beta]^{\frac{1}{1-\beta}}a^{\frac{\alpha}{1-\beta}-1}(t) + \delta b_j(t),$$

which with respect to $\alpha + \beta = 1$ and $\omega_j(t) = \left[\dfrac{p_j(t)}{c_j^{\beta}(t)}\right]^{\frac{1}{1-\beta}} = e^{\varepsilon t}$, results in

$$\dot{b}_j(t) = -e^{-(r_j-\varepsilon)t}\frac{\alpha}{\beta}[a\beta]^{\frac{1}{1-\beta}} + \delta b_j(t).$$

Noting that $\lim_{t\to\infty} b_j(t) = 0$, we find:

$$b_j(t) = \frac{e^{-(r_j-\varepsilon)t}}{r_j - \varepsilon + \delta}\frac{\alpha}{\beta}[a\beta]^{\frac{1}{1-\beta}}.$$

Solving $b_j(t) = \dfrac{\partial C_I((1-\theta)i_j(t))}{\partial I_j}e^{-r_j t}$ in $i_j(t)$ we find optimal investment strategies for each of the firms, j:

$$I_j^n(t) = i_j(t) = \frac{e^{\varepsilon t}}{r_j - \varepsilon + \delta}\frac{\alpha}{2(1-\theta)c_{Ij}\beta}[a\beta]^{\frac{1}{1-\beta}}, j=1,...,N.$$

The total supply chain capital is then obtained from (4.210), by:

$$\frac{da(t)}{dt} = -\delta a(t) + \sum_{j=1}^{N} i_j(t), \text{ or } \dot{a}(t) = -\delta a(t) + \sum_{j=1}^{N} \frac{e^{a}}{r_j - \varepsilon + \delta} \frac{\alpha}{2(1-\theta)c_{1j}\beta}[\alpha\beta]^{\frac{1}{1-\beta}}.$$

The solution of this differential equation yields the supply chain capital explicitly given by:

$$K^n(t) = a(t) = \frac{1}{(\varepsilon + \delta)} \frac{\alpha}{2(1-\theta)\beta}[\alpha\beta]^{\frac{1}{1-\beta}} \sum_{j=1}^{N} \frac{1}{2(r_j - \varepsilon + \delta)c_{1j}} e^{a} + A e^{-\delta t},$$

where A is determined by the boundary condition $a(0) = K_0$,

$$A = K_0 - \frac{1}{(\varepsilon + \delta)} \frac{\alpha}{2(1-\theta)\beta}[\alpha\beta]^{\frac{1}{1-\beta}} \sum_{j=1}^{N} \frac{1}{2(r_j - \varepsilon + \delta)c_{1j}}.$$

This solution implies that the growth of equilibrium investments over time is inversely proportional to the firms' discount rates and investment costs. This strategy compensates the effect of price index increases over weighted labor costs as shown in Figure 4.15. Further, we have $\frac{\partial I_j}{\partial \theta} > 0, \frac{\partial^2 I_j}{\partial \theta^2} > 0$, meaning that the larger the subsidies the larger the co-investments, growing then at an increased rate. For this reason, supply chain support for individual member firm co-investment is indeed important and may justify in some cases a "centralized control" which dictates to member firms the intensity of their investment. The level of capital will thus increase as well as a function of the support parameter.

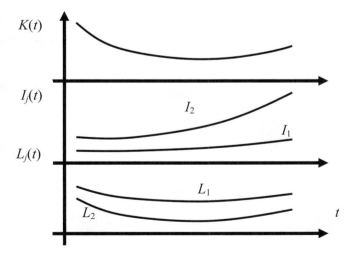

Figure 4.15. The Nash equilibrium over time, the case of $N=2$

Figure 4.15 above points out to the optimal equilibrium over time when

$\omega(t)=e^{st}$, $r_1>r_2$, $c_{l1}>c_{l2}$, and $\left[\dfrac{p_1(t)}{c_1^{\beta}(t)}\right]<\left[\dfrac{p_2(t)}{c_2^{\beta}(t)}\right]$.

Thus, if firms comprising the supply chain differ in their basic parameters and at least some of the parameters (price index, labor cost, investment costs and so on) change in time, then firms' investments not only differ and change over time, but their co-investment shares in the overall infrastructure capital may diverge in time as well. Such a change in investment strategies thus makes it difficult to plan future capital development of the supply chain. For this reason, a strategy that imposes steady co-investment shares by firms might lead to results that are viable. We shall turn our attention next to this special case.

Proposition 4.43. *Let* $p_j(t)\dfrac{\partial f(\overline{K},l_j(t))}{\partial K}\Big/\dfrac{\partial C_{Ij}((1-\theta)\overline{i}_j)}{\partial I_j}=\vartheta_j$ *and* $\overline{K},\overline{i}_j,l_j(t)$,

j=1,..,N, satisfy the following equations for $t\geq 0$.

$$p_j(t)\frac{\partial f(\overline{K},l_j(t))}{\partial K}-\frac{\partial C_{Ij}((1-\theta)\overline{i}_j)}{\partial I_j}(r_j+\delta)=0,\ \sum_{j=1}^{N}\overline{i}_j=\delta\overline{K},$$

$$\frac{\partial f(\overline{K},l_j(t))}{\partial L_j}=\frac{c_j(t)}{p_j(t)},\ j=1,..,N$$

If $K_0=\overline{K}$, *then there exists the pair of strategy sets: static investment* $\{I_j^n(t)=\overline{i}_j,\ j=1,..,N\}$ *and employment* $\{L_j^n(t)=l_j(t),\ j=1,..,N\}$, $t\geq 0$ *which is a Nash equilibrium in the supply chain investment and labor force differential game.*
Proof: Consider the solution for the state and co-state variables, which is characterized by a constant level of capital $K(t)=\overline{K}$:

$$\dot{\psi}_j(t)=-e^{-r_jt}p_j(t)\frac{\partial f(K(t),l_j(t))}{\partial K(t)}+\delta\psi_j(t),\ \lim_{t\to\infty}\psi_j(t)=0,$$

$$\frac{dK(t)}{dt}=-\delta K(t)+\sum_{j=1}^{N}I_j(t)=0.$$

This solution is optimal if $l_j(t)$ and $i_j(t)$ from (4.214) and (4.216),

$$\frac{\partial f(K(t),l_j(t))}{\partial L_j}-\frac{c_j(t)}{p_j(t)}=0;\ \psi_j(t)=\frac{\partial C_{Ij}((1-\theta)i_j(t))}{\partial I_j}e^{-r_jt},$$

are non-negative which is evidently true as $\dfrac{\partial f}{\partial L_j} \geq 0$ for $L \neq 0$,

$$\frac{\partial f(K,0)}{\partial L_j} = 0, \quad \frac{\partial C_{Ij}}{\partial I_j} \geq 0 \text{ and } \psi_j(t) = \frac{\partial C_{Ij}((1-\theta)i_j(t))}{\partial I_j}e^{-r_jt} \geq \frac{\partial C_{Ij}(0)}{\partial I_j}e^{-r_jt}.$$

Differentiating $\psi_j(t) = \dfrac{\partial C_{Ij}((1-\theta)i_j(t))}{\partial I_j}e^{-r_jt}$ and substituting the co-state

equation we find that

$$\dot{\psi}_j(t) = -\frac{\partial C_{Ij}((1-\theta)\bar{i}_j)}{\partial I_j}r_je^{-r_jt} = -e^{-r_jt}p_j(t)\frac{\partial f(\overline{K},l_j(t))}{\partial K} + \delta\frac{\partial C_{Ij}((1-\theta)\bar{i}_j)}{\partial I_j}r_je^{-r_jt},$$

If $p_j(t)\dfrac{\partial f(\overline{K},l_j(t))}{\partial K} \Big/ \dfrac{\partial C_{Ij}((1-\theta)\bar{i}_j)}{\partial I_j} = \vartheta_j$ is constant, then capital $\overline{K}$ and

investment $\sum\limits_{j=1}^{N}\bar{i}_j = \delta\overline{K}$ policies are constant as well, as stated in the propo-

sition. Finally, we straightforwardly verify the boundary condition,

$$\lim_{t\to\infty}\psi_j(t) = \lim_{t\to\infty}\frac{\partial C_{Ij}((1-\theta)\bar{i}_j)}{\partial I}e^{-r_jt} = 0.$$

Below, we consider an example to highlight the effects of a constant co-investment strategy.

Example 4.8.

Let the aggregate production function be again the Cobb-Douglas function, $f(K,L_j) = aK^{\alpha}L_j^{\beta}$, with $\alpha+\beta=1$, $C_{Ij}(I)=c_{Ij}(1-\theta)I_j^2$, $r_j=r$ for $j=1,..,N$,

$K_0=\overline{K}$, and $\omega_j = \left[\dfrac{p_j(t)}{c_j^{\beta}(t)}\right]^{\frac{1}{1-\beta}}$. Using $\dfrac{\partial f(\overline{K},l_j(t))}{\partial L_j} = \dfrac{c_j(t)}{p_j(t)}$ of Proposition

4.43, we have

$$L_j(t)=\bar{l}_j(t) = \left[\frac{1}{a\beta}\frac{c_j(t)}{p_j(t)\overline{K}^{\alpha}}\right]^{\frac{1}{\beta-1}} = \overline{K}\left[\frac{1}{a\beta}\frac{c_j(t)}{p_j(t)}\right]^{\frac{1}{\beta-1}}.$$

Taking into account

$$p_j(t)\frac{\partial f(\overline{K},l_j(t))}{\partial K} + \frac{\partial C_{Ij}((1-\theta)\bar{i}_j)}{\partial I_j}(r+\delta) = 0 \text{ and } \sum_{j=1}^{N}\bar{i}_j = \delta\overline{K}$$

of Proposition 4.43, we obtain a system of $N+1$ equations in $N+1$ unknowns, $\bar{i}_j$, $j=1,..,N$, $\overline{K}$:

$$w_j \frac{\alpha}{\beta}[\alpha\beta]^{\frac{1}{1-\beta}}\overline{K}^{\frac{\alpha}{1-\beta}-1} - 2c_{Ij}(1-\theta)(r+\delta)\bar{i}_j = 0, \ \sum_{j=1}^{N}\bar{i}_j = \delta\overline{K}.$$

Summing all equations we have

$$\sum_{j=1}^{N}\left[\frac{w_j}{c_{Ij}}\right]\frac{\alpha}{\beta}[\alpha\beta]^{\frac{1}{1-\beta}}\overline{K}^{\frac{\alpha}{1-\beta}-1} - 2(1-\theta)(r+\delta)\delta\overline{K} = 0,$$

which together with $\alpha+\beta=1$ results a constant co-investment strategy when $K_0=\overline{K}$:

$$\overline{K} = \frac{\sum_{j=1}^{N}\left[\frac{w_j}{c_{Ij}}\right]\frac{\alpha}{\beta}[\alpha\beta]^{\frac{1}{1-\beta}}}{2(1-\theta)(r+\delta)\delta} \text{ and } \bar{i}_j = \frac{w_j\frac{\alpha}{\beta}[\alpha\beta]^{\frac{1}{1-\beta}}}{2c_{Ij}(1-\theta)(r+\delta)}, j=1,..,N.$$

Interestingly, note that $\frac{\partial\overline{K}}{\partial\theta} > 0$ and $\frac{\partial\bar{i}}{\partial\theta} > 0$, which points out to a growth of capital and co-investment when investment subsidies increase.

What if a sustainable $\overline{K}$ exists and is attainable, but $K_0 \neq \overline{K}$? There are two possible approaches to dealing with this problem.

The first approach is non-cooperative and readily emanates from Propositions 4.42 and 4.43. Indeed, one can assume that there is a point in time, say t^*, at which the supply chain attains a stationary investment equilibrium (described in Proposition 4.43) as t^* tends to infinity. To determine whether such an equilibrium exists, the system of equations stated in Proposition 4.42 can be resolved for the terminal condition, $K(t^*)=\overline{K}$. The solution that Proposition 4.42 thus provides is a dynamic, open-loop Nash equilibrium. While such a solution allows to gain some insights, it is difficult to implement. If the firms do not collaborate, then a closed-loop solution may be more viable. We develop such a solution next.

The other approach consists in using open-loop policies for cooperating, which can be short- and long-run as discussed in Section 4.5.2.

The N-Firms Feedback Nash Strategy

In this section, we show how to obtain a closed-loop equilibrium in the conditions of Proposition 4.43, i.e., when a stationary investment equilibrium is attainable. The derivation is accomplished by employing an equivalent formulation of the maximum principle. Specifically, let $\Psi_j(t) = \psi_j(t)e^{r_j t}$. Then $\psi_j(t) = \Psi_j(t)e^{-r_j t}$ and $\dot{\psi}_j(t) = e^{-r_j t}(\dot{\Psi}_j(t) - r_j\Psi_j(t))$. Using these notations in conditions of Proposition 4.43, the co-state equation (4.213) and the optimality condition (4.217) take the following form respectively:

$$\dot{\Psi}_j(t) - r_j \Psi_j(t) = -p_j(t)\frac{\partial f(K(t), L_j(t))}{\partial K(t)} + \delta \Psi_j(t), \ \lim_{t \to \infty} \Psi(t)e^{-r_j t} = 0; \quad (4.218)$$

$$\Psi_j(t) = \frac{\partial C_{Ij}((1-\theta)i_j(t))}{\partial I_j}. \quad (4.219)$$

Denote the solution of equation (4.219) as $i_j(t) = F_j(\Psi_j(t))$. To simplify the presentation we next suppress index t wherever the dependence on time is obvious. Consequently, the stationary investment conditions are $\dot{K} = 0$ and $\dot{\Psi} = 0$, and from (4.218) the static co-state value $\overline{\Psi}_j$ of the co-state variable is fined by,

$$\overline{\Psi}_j = \frac{p_j}{\delta + r_j}\frac{\partial f(\overline{K}(, \overline{l}_j)}{\partial K}, \quad (4.220)$$

as well as the steady-state capital is equal to $\overline{K}$ (see Proposition 4.43). Let us introduce a new function, $\Phi_j(.)$,

$$\Psi_j(t) = \Phi_j(K(t)). \quad (4.221)$$

Denote the solution of equation (4.215) as $l_j = F_{Lj}(K)$. Differentiating (4.221) we have $\dot{\Psi}_j = \Phi'_j(K)\dot{K}$, which when substituting the state (4.210) and co-state (4.218) equations leads to

$$-p_j\frac{\partial f(K, F_{Lj}(K))}{\partial K} + (\delta + r_j)\Phi_j(K) = \Phi'_j(K)[\sum_{j=1}^N F_j(\Phi_j(K)) - \delta K], \ \Phi_j(\overline{K}) = \overline{\Psi}_j.$$

Thus, we have proved the following theorem.

Theorem 4.3. *If* $p_j\dfrac{\partial f}{\partial K}\Big/\dfrac{\partial C_{Ij}}{\partial I_j}$ *does not explicitly depend on time for* $j=1,..,N$, *then investment* $\{I^n_j = F_j(\Phi_j(K)), \ j=1,...,N\}$ *and employment* $\{L_j^n(t) = F_{Lj}(K), j=1,...,N\}$, $t \geq 0$ *constitute a feedback Nash equilibrium in the supply chain investment and labor force differential game, where* $\Phi_j(K), j=1,..,N$ *satisfy the following differential equations,*

$$\Phi_j(K)[\sum_{j=1}^N F_j(\Phi_j(K)) - \delta K] + p_j\frac{\partial f(K, F_{Lj}(K))}{\partial K} - (\delta + r_j)\Phi_j(K) = 0, \Phi_j(\overline{K}) = \overline{\Psi}_j, j=1,..,N. (4.222)$$

The following example illustrates the results of Theorem 4.3.

Example 4.9.

Assume the conditions of Example 4.8, except $K_0 < \overline{K}$ and $\alpha + \beta < 1$. Then we have

$$I_j^n = \frac{\Psi_j}{2c_{Ij}(1-\theta)} = \frac{\Phi_j(K)}{2c_{Ij}(1-\theta)}, \quad \overline{K} = \left[\frac{\left[\sum\limits_{j=1}^{N} \left[\dfrac{p_j}{c_{Ij}} \right] \left[a\beta \dfrac{p_j}{c_j} \right]^{\frac{\beta}{1-\beta}} \right]^{\frac{1-\beta}{2-2\beta-\alpha}}}{2\delta(1-\theta)(r+\delta)} \right]$$

and $L_j^n = K^{\frac{\alpha}{1-\beta}} \left[\dfrac{1}{a\beta} \dfrac{c_j}{p_j} \right]^{\frac{1}{\beta-1}}$.

As a result, the system of backward differential equations (4.222) takes the following form:

$$\Phi_j'(K)\left[\sum\limits_{j=1}^{N} \frac{\Phi_j(K)}{2(1-\theta)c_{Ij}} - \delta K \right] + \xi_j K^{\frac{\alpha}{1-\beta}-1} - (\delta+r)\Phi_j(K) = 0. \quad (4.223)$$

$$\Phi_j(\overline{K}) = \frac{\xi_j}{\delta+r} \overline{K}^{\frac{\alpha}{1-\beta}-1}, \quad j=1,..,N,$$

where $\xi_j = p_j a\alpha \left[\dfrac{1}{a\beta} \dfrac{c_j}{p_j} \right]^{\frac{\beta}{\beta-1}}$.

We solve this system of equations with Maple for two firms, $N=2$, $a=1$, $\alpha=0.1$, $\beta=0.1$, $\theta=0.4$, $\delta=0.04$, $r_1=r_2=0.002$, $c_1=0.4$, $c_2=0.5$, $c_{I1}=0.2$, $c_{I2}=0.3$, $p_1=p_2=7$.

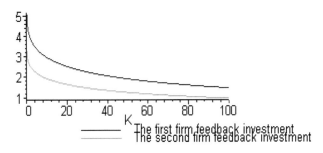

The first firm feedback investment
The second firm feedback investment

Figure 4.16. The feedback equilibrium investments as a function of capital, $I_1^n = F_1(\Phi_1(K))$ and $I_2^n = F_2(\Phi_2(K))$

The resultant feedback policies of the two firms, $I_1^n = \dfrac{\Phi_1(K)}{2c_{I_1}(1-\theta)}$ and $I_2^n = \dfrac{\Phi_2(K)}{2c_{I_2}(1-\theta)}$, is illustrated graphically in Figure 4.16. The corresponding evolution in time of the capital and investments for the case of $K(0)=0.2<\overline{K}=69.91217939$ are depicted in Figures 3 and 4, respectively.

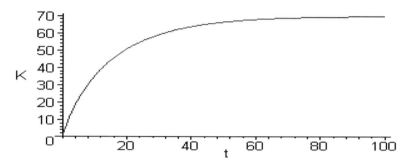

Figure 4.17. Evolution of the capital over time, $K^n(t)$

From Figures 2 - 4 we observe that the greater the capital, the lower the investments. When the infrastructure capital is greater (smaller) than the steady-state level $\overline{K}$, it is optimal to invest in total by all firms less (more) than $\delta \overline{K} = 2.796487176$, so that the overall accumulated capital decreases (increases) towards the stationary investment equilibrium. Furthermore, the investments decrease much faster when the capital exceeds the static level compared to the rate of their decrease when the capital is lower than the static level (see Figure 4.16).

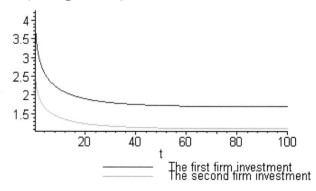

————— The first firm investment
················· The second firm investment

Figure 4.18. Evolution of the investment over time, $I_1^n(t)$ and $I_2^n(t)$

Naturally, the first firm, which has lower investment cost, $c_{I1}=0.2<c_{I2}=0.3$ invests more than the second firm (see Figure 4.18).

Since $\dfrac{c_j}{p_j}$ is constant in the example, the static capital, $\overline{K}$, induces the equilibrium employment to attain a static level as well, $\overline{L}_j = \overline{K}^{\frac{\alpha}{1-\beta}} \left[\dfrac{1}{\alpha\beta} \dfrac{c_j}{p_j} \right]^{\frac{1}{\beta-1}}$. The evolution in time of the equilibrium employment for the two firms is shown in Figures 5.

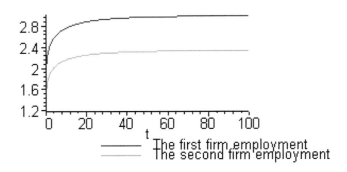

Figure 4.19. Evolution of the employment over time, $L_1{}^n(t)$ and $L_2{}^n(t)$

From Figure 4.19 we observe that the employment increases with the capital and it tends to the static level for the firms, $\overline{L}_1 = 2.98533$ and $\overline{L}_2 = 2.329779$, which is higher for the first firm as its wages are lower, $c_1 = 0.4<c_2=0.5$. Since employment is proportional to the infrastructure capital, the rate of employment changes much faster when the infrastructure capital is low.

4.5.2 SHORT-RUN AND LONG-RUN COOPERATION

If all parties are interested in a stationary co-investment strategy for the supply chain (and therefore are seeking a stationary equilibrium, see Proposition 4.43), it can be implemented by determining jointly time t^*, and collaborative investment policies, $I_j(t)$ for $0 \le t \le t^*$, $j=1,..,N$. This is accomplished by requiring that the joint-capital, $K(t)$, will reach the desired optimal level $\overline{K}$ by $t*$, i.e., $K_0 + \sum_{j=1}^{N} \int_{0}^{t^*} I_j(t)dt = \overline{K}$ (as shown in Figure 4.20).

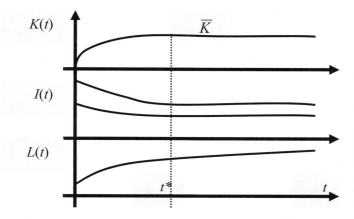

Figure 4.20. The equilibrium over time, the case of $K_0 < \overline{K}$ and $N=2$

Short-Run Cooperation

An ultimate way of short-run cooperation is a one-time partnership. Assume that firms agree to cooperate until a common point $t*$ to reach the stationary equilibrium in minimum time. Optimal control theory shows that in such a case one time, high level investments, $\hat{I}_j$, $j=1,..,N$ will be optimal, so that

$$\sum_{j=1}^{N} \hat{I}_j = \overline{K} - K_0.$$

To model an instantaneous investment, $\hat{I}_j$, we employ the Dirac delta function $\Delta(t)$, $I_j(t) = \hat{I}_j \Delta(t)$. Specifically, the optimal policy for reaching the desired equilibrium in minimum time must be a solution of the following optimization problem,

$$\min_{I_j} t * \qquad (4.224)$$

s.t.

$$\frac{dK(t)}{dt} = -\delta K(t) + \sum_{j=1}^{N} I_j(t), \ K(0)=K_0, \ K(t^*)=\overline{K} \ ,$$

$$I_j(t) \geq 0, j = 1,..,N \ , \ 0 \leq t \leq t * .$$

Using the maximum principle we construct the Hamiltonian

$$H(t) = \lambda(t)\left(-\delta K(t) + \sum_{j=1}^{N} I_j(t)\right), \tag{4.225}$$

where the co-state variable is determined by

$$\dot{\lambda}(t) = \delta\lambda(t). \tag{4.226}$$

Furthermore, since t^* is unknown, an additional necessary transversality condition is that, $H(t)=1$, i.e.,

$$\lambda(t)\left(-\delta K(t) + \sum_{j=1}^{N} I_j(t)\right) = 1. \tag{4.227}$$

Maximizing the Hamiltonian (4.225) we readily observe that if $\lambda(t)<0$, then no investment is optimal, if $\lambda(t)=0$, then the investments are arbitrary non-negative values. Otherwise, if $\lambda(t)>0$, then $I_j(t) \to +\infty$, $j=1,..,N$. Based on these properties, we have the following result.

Proposition 4.44. *The optimal solution of problem (4.224) is $t^*=0$ and $I_j(t^*) = \hat{I}_j \Delta(t^*)$, $j=1,..,N$, where $\Delta(t^*)$ is Dirac delta function at t^* and*

$$\sum_{j=1}^{N} \hat{I}_j = \overline{K} - K_0.$$

Proof: The proposition is proved by contradiction. Let us assume that the optimal solution be obtained for $t^*>0$. Then $I_j(t)$, $i=1,..,N$ are finite (otherwise any capital can be reached in no time which contradicts $t^*>0$) and thus $\lambda(t) \le 0$. On the other hand, according to the maximum principle, if $\lambda(t)<0$, $0 \le t \le t^*$, then no capital is invested and thereby $K(t)$ will never reach $\overline{K}$. Accordingly, the only case left is $\lambda(t)=0$ for a measurable interval of time. However, condition (4.226) never holds during this interval, if $\lambda(t)=0$. Thus we get a contradiction and $t^*=0$.

To meet the boundary condition $K(t^*)=\overline{K}$ we need

$$K_0 + \int_0^{0+} [-\delta K(t) + \sum_{j=1}^{N} I_j(t)]dt = \overline{K}.$$

Thus, $I_j(t^*) = \hat{I}_j \Delta(t^*)$ and substituting this into the last expression we find an equation for unknowns $\hat{I}_j$, as stated in the proposition.

Note that a stationary equilibrium can be viewed as both open- and closed-loop equilibrium. From Proposition 4.44 it follows that, if *competing* firms are able to *cooperate* in setting their one-time investments, then the firms can reach the stationary Nash equilibrium in *no-time* and stay there infinitely long as summarized in the following theorem.

Theorem 4.4. *Assume that at some point of time t=t* there exists a sustainable level of the capital K(t*). Let* $p_j(t)\dfrac{\partial f(\overline{K},l_j(t))}{\partial K}\Big/\dfrac{\partial C_{1j}((1-\theta)\overline{i}_j)}{\partial I_j}=\vartheta_j$

and $\overline{K},\overline{i}_j,l_j(t),j=1,..,N$, *satisfy the following equations for* $t \geq t*$:

$$p_j(t)\frac{\partial f(\overline{K},l_j(t))}{\partial K}-\frac{\partial C_{1j}((1-\theta)\overline{i}_j)}{\partial I_j}(r_j+\delta)=0,\ \sum_{j=1}^{N}\overline{i}_j=\delta\overline{K},$$

$$\frac{\partial f(\overline{K},l_j(t))}{\partial L_j}=\frac{c_j(t)}{p_j(t)},j=1,..,N.$$

If K(t*)= $\overline{K}$ *, then there exists the pair of strategy sets: static investment* $\{Ij^n(t)=\overline{i}_j, j=1,..,N\}$ *and employment* $\{Lj^n(t)=l_j(t), j=1,..,N\}$, $t \geq 0$ *which is a Nash equilibrium in the supply chain investment and labor force differential game for* $t \geq t*$.

Long-Run Cooperation and the Centralized Solution

In this section we shall consider a centralized, supply chain co-investment strategy, which turns out, expectedly, to be different than the Nash strategy obtained earlier. In this organizational mode the supply chain "controller" will dictate to member firms how much to invest in infrastructure to maximize centralized profits. Subsequently, we shall discuss inducement for firms to cooperate and thereby reach a sustainable centralized investment strategy in the long run. In particular an example will be treated in detail. The centralized supply chain investment problem is formulated as a control problem:

$$\underset{L_j,I_j}{Max}\int_0^\infty\sum_{j=1}^n e^{-r_jt}[p_j(t)f(K(t),L_j(t))-c_j(t)L_j(t)-C_{1j}((1-\theta)I_j(t))]dt \quad (4.228)$$

s.t. (4.210),
whose Hamiltonian is:

$$H(t)=\sum_{j=1}^N e^{-r_jt}[p_j(t)f(K(t),L_j(t))-c_j(t)L_j(t)-C_{1j}((1-\theta)I_j(t))]$$

$$+\psi(t)(\sum_{j=1}^N I_j(t)-\delta K(t)), \quad (4.229)$$

where the co-state variable $\psi(t)$ is determined by

$$\dot{\psi}(t)=-\sum_{j=1}^N e^{-r_jt}p_j(t)\frac{\partial f(K(t),L_j(t))}{\partial K(t)}+\delta\psi(t),\ \lim_{t\to\infty}\psi(t)=0. \quad (4.230)$$

Maximizing the Hamiltonian with respect to investments, $I_j(t)$, and labor, $L_j(t)$, we note that the optimal employment remains the same as that of the non-cooperative game, (4.214) and (4.215), while the investment strategy now depends on a single co-state variable:

$$I_j(t) = \begin{cases} i_j(t), \text{if } \psi(t) \geq \dfrac{\partial C_{1j}(0)}{\partial I_j} e^{-r_j t} \\ 0, \text{otherwise} \end{cases} \tag{4.231}$$

where $i_j(t)$ is determined by

$$\psi(t) = \frac{\partial C_{1j}((1-\theta)i_j(t))}{\partial I_j} e^{-r_j t}. \tag{4.232}$$

This change implies that the optimal co-investment in the centralized chain is different from that for the corresponding decentralized chain. As with Proposition 4.42 we outline first the general case, summarized by Proposition 4.45 below.

Proposition 4.45. *If $a(t)$, $b_j(t)$ and $i_j(t)$, $j=1,..,N$ satisfy the following system of equations for $t \geq 0$,*

$$\frac{da(t)}{dt} = -\delta a(t) + \sum_{j=1}^{N} i_j(t), \; a(0)=K_0, \; b(t) = \frac{\partial C_{1j}((1-\theta)i_j(t))}{\partial I_j} e^{-r_j t},$$

$$\frac{\partial f(a(t), l_j(t))}{\partial L_j} = \frac{c_j(t)}{p_j(t)}, \; j=1,..,N$$

$$\dot{b}(t) = -\sum_{j=1}^{N} e^{-r_j t} p_j(t) \frac{\partial f(a(t), l_j(t))}{\partial a(t)} + \delta b(t), \; \lim_{t \to \infty} b(t) = 0,$$

then, the strategy pair $\{Ij^*(t)=i_j(t), \; j=1,..,N\}$ and $\{Lj^*(t)=l_j(t), \; j=1,..,N\}$, $t \geq 0$ is optimal for the centralized supply chain problem (4.228) and (4.210).

Coordination

The centralized optimal solution is, of course, more profitable. As a result, a centralized investment strategy may be desirable but it may also be difficult to implement. A sustainable cooperative solution where the profits of centralization "are appropriately" distributed among firms would provide a self enforceable procedure that allows the implementation of such a solution. To attain such a self enforced cooperation we consider a special static-investments case. Our results are summarized by Proposition 4.46 below and by some examples.

Proposition 4.46. Let $\sum_{i=1}^{N} e^{-r_i + r_j} p_j(t) \dfrac{\partial f(\hat{K}, \hat{l}_j(t))}{\partial K} \Big/ \dfrac{\partial C_{Ij}((1-\theta)\hat{i}_j)}{\partial I_j} = \vartheta_j$

and $\hat{K}, \hat{i}_j, \hat{l}_j(t), j=1,..,N, \ t \geq 0$, be such that:

$$\frac{\partial C_{Ij}((1-\theta)\hat{i}_j)}{\partial I_j}(r_j + \delta)e^{-r_j t} = \sum_{i=1}^{N} e^{-r_i t} p_i(t) \frac{\partial f(\hat{K}, \hat{l}_i(t))}{\partial K} ,$$

$$\sum_{j=1}^{N} \hat{i}_j = \delta \hat{K} , \quad \frac{\partial f(\hat{K}, \hat{l}_j(t))}{\partial L_j} = \frac{c_j(t)}{p_j(t)}, j=1,..,N .$$

If $K_0 = \hat{K}$, then there exists the pair of strategy sets: static investment $\{Ij*(t) = \hat{i}_j, j=1,..,N\}$ and employment $\{Lj*(t) = \hat{l}_j(t), j=1,..,N\}$, $t \geq 0$ which is optimal for the centralized supply chain problem (4.210) and (4.228).

Proof: Consider the solution for the state and co-state variables, which is characterized by a constant level of capital $K(t) = \hat{K}$:

$$\dot{\psi}(t) = -\sum_{j=1}^{N} e^{-r_j t} p_j(t) \frac{\partial f(K(t), l_j(t))}{\partial K(t)} + \delta \psi(t), \ \lim_{t \to \infty} \psi(t) = 0 ,$$

$$\frac{dK(t)}{dt} = -\delta K(t) + \sum_{j=1}^{N} I_j(t) = 0.$$

This solution is optimal if $l_j(t)$ and $i_j(t)$ satisfy (4.214) and (4.216),

$$\frac{\partial f(K(t), l_j(t))}{\partial L_j} - \frac{c_j(t)}{p_j(t)} = 0; \ \psi(t) = \frac{\partial C_{Ij}((1-\theta)i_j(t))}{\partial I_j} e^{-r_j t} .$$

Differentiating $\psi(t) = \dfrac{\partial C_{Ij}((1-\theta)i_j(t))}{\partial I_j} e^{-r_j t}$ we find that

$$\dot{\psi}(t) = -\frac{\partial C_{Ij}((1-\theta)\hat{i}_j)}{\partial I_j} r_j e^{-r_j t} = -\sum_{i=1}^{N} e^{-r_i t} p_i(t) \frac{\partial f(\hat{K}, \hat{l}_i(t))}{\partial K} + \delta \psi(t), j=1,..,N.$$

That is,

$$\frac{\partial C_{Ij}((1-\theta)\hat{i}_j)}{\partial I_j}(r_j + \delta)e^{-r_j t} = \sum_{i=1}^{N} e^{-r_i t} p_i(t) \frac{\partial f(\hat{K}, \hat{l}_i(t))}{\partial K}, j=1,..,N.$$

If $\sum_{i=1}^{N} e^{-r_i + r_j} p_j(t) \dfrac{\partial f(\hat{K}, \hat{l}_j(t))}{\partial K} \Big/ \dfrac{\partial C_{Ij}((1-\theta)\hat{i}_j)}{\partial I_j} = \vartheta_j$ is constant then the capi-

tal and the investment policies are constant as well and are given by $\hat{K}$

and $\sum_{j=1}^{N} \hat{i}_j = \delta \hat{K}$, as stated in this proposition. Finally, we straightforwardly

verify the boundary condition, $\lim_{t \to \infty} \psi(t) = \lim_{t \to \infty} \sum_{j=1}^{N} \dfrac{\partial C_{lj}((1-\theta)\hat{i}_j)}{\partial I} e^{-r_j t} = 0$.

Setting $r_j = r$ for $j=1,..,N$ for comparing Proposition 4.43 and Proposition 4.46, we observe that the only difference is that instead of the equation,

$$\frac{\partial C_{lj}((1-\theta)\bar{i}_j)}{\partial I_j}(r+\delta) = p_j(t)\frac{\partial f(\bar{K}, l_j(t))}{\partial K} \qquad (4.233)$$

determined by Proposition 4.43, the following equation results from Proposition 4.46,

$$\frac{\partial C_{lj}((1-\theta)\hat{i}_j)}{\partial I_j}(r+\delta) = \sum_{i=1}^{N} p_i(t)\frac{\partial f(\hat{K}, \hat{l}_i(t))}{\partial K}. \qquad (4.234)$$

Consequently, we can conclude that the difference between a centralized and a decentralized supply chain is that *in a centralized supply chain, investments by each firm are proportional to the total supply chain production rate per capital unit. On the other hand, in a decentralized supply chain, investments by firms are only proportional to firms' production rate per capital unit. Thus, the more firms cooperate and invest proportionally to the overall supply chain production rate, the closer the decentralized investment strategy is to the centralized one.* The incentive for such cooperation is evident: firms should share in the total supply chain profits such that their profit will increase comparatively to the non-cooperative (decentralized) solution. An example to this effect is considered next.

Example 4.10.

Consider again Example 4.8 with $f(K, L_j) = aK^{\alpha}L_j^{\beta}$, with $\alpha+\beta=1$, $C_{lj}(I)=$

$c_{lj}(1-\theta)I_j^{2}$, $r_j = r$ for $j=1,..,N$, $K_0 = \bar{K}$, and $\omega_j = \left[\dfrac{p_j(t)}{c_j^{\beta}(t)}\right]^{\frac{1}{1-\beta}}$.

Using $\dfrac{\partial f(\hat{K}, \hat{l}_j(t))}{\partial L_j} = \dfrac{c_j(t)}{p_j(t)}$ of Proposition 4.46, we have

$L_j(t) = \hat{l}_j(t) = \left[\dfrac{1}{a\beta}\dfrac{c_j(t)}{p_j(t)\hat{K}^{\alpha}}\right]^{\frac{1}{\beta-1}}$ which is identical to the labor condition found in Example 4.8. Next, taking into account (4.234) and $\sum_{j=1}^{N} \hat{i}_j = \delta \hat{K}$ of Proposition 4.46, we obtain the algebraic system of $N+1$ equations in $N+1$ unknowns, $\hat{i}_j$, $j=1,..,N$ and $\hat{K}$:

$$\sum_{i=1}^{N} [w_i] \frac{\alpha}{\beta} [\alpha\beta]^{\frac{1}{1-\beta}} \hat{K}^{\frac{\alpha}{1-\beta}-1} - 2c_{lj}(1-\theta)(r+\delta)\hat{i}_j = 0, \quad \sum_{j=1}^{N} \hat{i}_j = \delta \hat{K}.$$

Summing all equations we have

$$\sum_{j=1}^{N} \frac{1}{c_{lj}} \sum_{j=1}^{N} [w_j] \frac{\alpha}{\beta} [\alpha\beta]^{\frac{1}{1-\beta}} \hat{K}^{\frac{\alpha}{1-\beta}-1} - 2(1-\theta)(r+\delta)\delta\hat{K} = 0,$$

which with respect to $\alpha+\beta=1$ results in

$$\hat{K} = \frac{\sum_{j=1}^{N} \frac{1}{c_{lj}} \sum_{j=1}^{N} [w_j] \frac{\alpha}{\beta} [\alpha\beta]^{\frac{1}{1-\beta}}}{2(1-\theta)(r+\delta)\delta} \quad \text{and} \quad \hat{i}_j = \frac{\frac{\alpha}{\beta} [\alpha\beta]^{\frac{1}{1-\beta}} \sum_{i=1}^{N} [w_i]}{2c_{lj}(1-\theta)(r+\delta)}, j=1,..,N.$$

Note in this case that subsidizing investments in a centralized supply chain can provide the same results (or better) as those obtained for a decentralized supply chain, i.e., $\frac{\partial \hat{i}_j}{\partial \theta} > 0, \frac{\partial^2 \hat{i}_j}{\partial \theta^2} > 0$ and $\frac{\partial \hat{K}}{\partial \theta} > 0$.

Comparing the result of Example 4.10 with that of Example 4.8, we observe that if $c_{lj}=c_l$ for $j=1,..,N$, the profit of the centralized supply chain is due to the fact that the optimal centralized chain capital $\hat{K}$ increases N times and employment increases $N^{\frac{1}{\beta-1}+(\beta-1)}$ times compared to the decentralized solution. This increase is provided by higher investment by each firm, j, in the centralized chain which is now proportional to the total weighted ratio of the price index and the labor cost, $\sum_{i=1}^{N} [w_i] = \sum_{j=1}^{N} \left[\frac{p_j(t)}{c_j^{\beta}(t)} \right]^{\frac{1}{1-\beta}}$ over all firms, rather than to the individual ratio for each firm, j, $\omega_j = \left[\frac{p_j(t)}{c_j^{\beta}(t)} \right]^{\frac{1}{1-\beta}}$.

This however does not guarantee that if the firms decide to cooperate by investing as required by the centralized solution, then all firms will benefit individually without a reallocation of the overall supply chain profits.

REFERENCES

Anupindi R, Bassok Y (1998), Supply contracts with quantity commitments and stochastic demand, in *Quantitative Models for Supply Chain Management* edited by S. Tayur, R. Ganeshan and M. J. Magazine, Kluwer.

Anupinidi R, Bassok Y (1999). *Supply Contracts*, in Quantitative Models for Supply Chain Management, Tayur, S., R. Ganeshan, and M. Magazine (eds.), International Series in Operations Research and Management Science, Boston, Kluwer.

Basar T, Olsder GL (1982). *Dynamic Noncooperative Game Theory*. Academic Press, London.

Gattorna J (Ed.), (1998). *Strategic Supply Chain Alignment*, Chap. 27, Gower, Aldershot.

Bils M (1989), Pricing in a Customer Market, *Quarterly Journal of Economics*, 104(4): 699-717.

Cachon G (2003) Supply chain coordination with contracts. *Handbooks in Operations Research and Management Science: Supply Chain Management*. edited by Steve Graves and T. de Kok. North Holland.

Cachon G, Netessine S (2004) Game theory in Supply Chain Analysis in *Handbook of Quantitative Supply Chain Analysis: Modeling in the eBusiness Era*. edited by D. Simchi-Levi, S. D. Wu and Z.-J. Shen, Kluwer.

Chan ZJ, Shen M, Simchi-Levi D, Swann J (2003) Coordination of Pricing and Inventory Decisions: A survey and Classification. Working paper, Georgia Institute of Technology, Atlanta, GA.

Chevalier JA, Kashyap AK, Rossi PE (2003) Why Don't Prices Rise During Periods of Peak Demand? Evidence from Scanner Data, *American Economic Review* 93(1): 15-37.

Desai VS (1992) Marketing-production decisions under independent and integrated channel structures. *Annals of Operations Research* 34: 276-306.

Desai VS (1996) Interactions between members of a marketing-production channel under seasonal demand. *European Journal of Operational Research* 90: 115-141.

Eliashberg J, Steinberg R (1987) Marketing-production decisions in an industrial channel of distribution. *Management Science*, 33: 981-1000.

Elmaghraby W, Keskinocak P (2003) Dynamic Pricing in the Presence of Inventory Considerations: Research Overview, Current Practice, and Future Directions. *Management Science* 49(10): 1287-1309.

Feichtinger G, Jorgenson S (1983) Differential game models in Management Science. *European Journal of Operational Research* 14: 137- 155.

Jorgenson S (1986) Optimal production, purchasing and pricing: A differential game approach. *European Journal of Operational Research* 23: 64-76.

Jorgensen S, Zaccour G (2004) *Differential Games in Marketing*, Boston, Kluwer.

Kubler D, Muller W (2002) Simulteneous and sequential competition in heterogenous duopoly markets: Experimental evidence. International *Journal of Industrial Organization* 20(10): 1437-1460.

La Londe B, Cooper M (1989) *Partnership in providing customer service:a third-party perspective*, Council of Logistics Management, Oak Brook, IL.

Maimon O, Khmelnitsky E, Kogan K (1998) *Optimal Flow Control in Manufacturing Systems: Production Planning and Scheduling*, Kluwer Academic Publisher, Boston.

Malone R (2002) The Search for e-Business Enlightenment, *Supply Chain Technology*, Inbound Logistics.

Razzaque MA, Sheng CC (1998) Outsourcing of logistics functions: a literature survey. *International Journal of Physical Distribution & Logistics Management* 28: 89-107.

Ritchken P, Tapiero CS (1986) Contingent Claim Contracts and Inventory Control. *Operations Research* 34: 864-870.

Russell RS, Taylor BW (2000) *Operations Management*, Third Edition, Prentice Hall, Upper Saddle River, New Jersey.

Stackelberg, HV (1952) *The Theory of the Market Economy*, Translated by Peacock AT, William Hodge and CO., London.

Warner EJ, Barsky RB (1995) The Timing and Magnitude of Retail Store Markdowns: Evidence from Weekends and Holidays. *Quarterly Journal of Economics* 110(2): 321-352.

Yano CA, Gilbert SM (2002) Coordinating Pricing and Production/ Procurement decisions: A review. Chakravarty A, Eliashberg J eds. *Managing Business Interfaces: Marketing, Engineering and Manufacturing Perspectives*, Kluwer.

Zerrillo P, Iacobucci D (1995) Trade promotions: a call for a more rational approach, *Business Horizons* 38(4): 69-76.

5 SUPPLY CHAIN GAMES: MODELING IN AN INTERTEMPORAL FRAMEWORK WITH PERIODIC REVIEW

In this chapter we consider firms which employ a periodic review policy and handle products characterized by a relatively short life-cycle, so that a single inventory update at a predetermined point on the production horizon (i.e., two-period approach) may suffice to identify demand over the remaining part of the production horizon. Fisher et al. (2000) reported an example of this approach in the apparel industry where highly accurate demand forecasts were made after observing only 11% of demand. Fisher and Raman (1996) reported further examples in which very accurate forecasts were obtained after observing 20% of demand.

The point of update is assumed to be chosen from previous experience and may generally involve different considerations, such as the change from high- to off-season, expected customer fatigue towards the product, or the impact of competitors catching up with production of a similar product. As a result, demand realization determined for the second part of the production horizon may be different from that observed for the first part. In contrast to the previous chapter, where customer demand along with inventories was observed either continuously or only by the end of production horizon, we assume here that an update is possible before the end of the horizon and that the probability distribution of demand for the second part of the production horizon depends on demand realization over the first part of the production horizon (see also Kogan et al. 2004, 2007; Kogan and Herbon 2007).

5.1 TWO-PERIOD INVENTORY OUTSOURCING

Large manufacturers are continuously striving to reduce inventory costs related to both raw materials and finished goods. In this chapter we address inventory outsourcing to a selected distributor or large retailer as one way of reducing inventories. Specifically, we consider a supply chain which involves a single producer or manufacturer and a distributor. The producer has a small warehouse characterized by a fixed cost which affects neither optimal production nor inventory policy. Instead, the distributor handles just-in-time all inventory-related operations, including transportation from the producer to the distributor's warehouses and to customers. The distributor sets a price, $h^+(t)$, for handling product units, while the producer decides on the number of products, $u(t)$ to produce and store $X^+(t)$ at the distributor's warehouse during the production horizon, T. Since both parties are operating constantly (producing and handling inventories), they are continuously incurring related costs even though the exact demands are unknown until a certain point of time where the inventories are reviewed and the demand is updated.

Producer's Problem

Exogenous demands and the following inventory dynamics characterize the periodic review, production control problem:

$$\dot{X}(t) = u(t) - d_i, X(0) \text{ is given}, 0 \le t \le T,$$

$$i=1 \text{ for } 0 \le t < \tau \text{ and } i=2 \text{ for } \tau \le t < T, \qquad (5.1)$$

where demand rate d_i is a random variable which has a constant realization D_1 before a point of review, $t=\tau$, and a constant realization after τ, D_2. Both realizations are known only by time $t=\tau$, $0<\tau<T$ and are not necessarily equal. Exact inventory level is observed from time $t=\tau$. We assume that the demand at the second part of the production horizon depends on that realized at the first part and denote the probability density functions as $f_i(.)$ and the corresponding cumulative functions as $F_i(.)$, $i=1,2$, where

$$f_1(D_1) \text{ and } F_1(a) = \int_0^a f_1(D_1)dD_1$$

are univariate distribution functions;

$$f_2(D_2|D_1) \text{ and } F_2(a|D_1) = \int_0^a f_2(D_2|D_1)dD_2$$

are conditional distribution functions.

The production control is bounded

$$0 \le u(t) \le U. \qquad (5.2)$$

The producer's goal is to minimize expected inventory-related costs

$$\min_u J_p(u,h^+) = \min_u E\left[\int_0^T C(X(t))dt \right], \qquad (5.3)$$

where

$$C(X(t)) = h^+(t)\max\{X(t),0\} + h^-\max\{-X(t),0\}, \qquad (5.4)$$

$h^+(t)$- is the unit inventory holding cost and h^- is the unit backlog cost. Henceforth we assume that the inventory unit cost may change or be reconsidered only once. This can occur only at a point of inventory review, $t=\tau$, i.e.,

$$h^+(t) = \begin{cases} h_1^+, & \text{if } 0 \le t < \tau; \\ h_2^+, & \text{if } \tau \le t \le T. \end{cases}$$

In what follows, we assume that the probability distributions $f_i(.)$ and $F_i(.)$ are continuously differentiable functions and that demand does not exceed the production capacity, $0 \le d_i \le U$. The latter assumption does not affect the approach and is made to reduce the number of awkward mathematical expressions.

Deterministic Component of the Problem

Consider the second time interval $[\tau, T]$. At this interval, problem (5.1)-(5.4) takes the following deterministic form:

$$J_{det} = \int_\tau^T C(X(t))dt \to \min \tag{5.5}$$

s.t.

$$\dot{X}(t) = u(t) - D_2, \ \tau \le t \le T, X(\tau) \text{ is given,}$$
$$0 \le u(t) \le U.$$

An optimal solution for problem (5.5) and thus the best producer's response to an inventory holding price $h^+(t)$ can be straightforwardly found without applying any specific optimization technique. It is formalized in the following two propositions. Proposition 5.1 treats the case that occurs when there is an inventory surplus at the beginning of the second period.

Proposition 5.1. *Let $X(\tau) \ge 0$, then the optimal solution for problem (5.5) and the optimal value of the objective function are*

If $D_2 < \dfrac{X(\tau)}{T-\tau}$, then $u(t)=0$ for $\tau \le t \le T$ and

$$J_{det} = h_2^+ \left(X(\tau)(T-\tau) - \frac{D_2}{2}(T-\tau)^2 \right),$$

otherwise, if $\dfrac{X(\tau)}{T-\tau} \le D_2$, then $u(t)=0$, $\tau \le t < \tau + \dfrac{X(\tau)}{D_2}$ and $u(t) = D_2$ for

$\tau + \dfrac{X(\tau)}{D_2} \le t \le T$, *and*

$$J_{det} = h_2^+ \frac{X(\tau)^2}{2D_2}.$$

Proof: The minimum value of the objective function of problem (5.5) is obviously zero which is attainable only when $X(t) \equiv 0$. Thus, if $\frac{X(\tau)}{T-\tau} \leq D_2$, there does not exist a better control than $u(t)=0$ for $\tau \leq t < \tau + \frac{X(\tau)}{D_2}$ and $u(t)=D_2$ for $\tau + \frac{X(\tau)}{D_2} \leq t \leq T$, so that $X(t) = 0$ for $\tau + \frac{X(\tau)}{D_2} \leq t \leq T$. Since this provides zero cost over interval $\tau + \frac{X(\tau)}{D_2} \leq t \leq T$ and no production, $u(t)=0$, for $\tau \leq t < \tau + \frac{X(\tau)}{D_2}$, inventory $X(t)$ can be deleted to zero as fast as possible. Any increase of such a control at a time point t, $\tau \leq t < \tau + \frac{X(\tau)}{D_2}$ would result in increased inventory holding cost of J_{det} and reduce the period during which $X(t) \equiv 0$. Similarly, since no production $u(t)=0$ for $\tau \leq t \leq T$ is optimal when the demand is too low relative to the available inventory, $D_2 < \frac{X(\tau)}{T-\tau}$, no shortage can occur. Production at a time point can only increase the holding cost.

With the found optimal solution, we next calculate the optimal value for the objective function. That is, for $D_2 < \frac{X(\tau)}{T-\tau}$, we have

$$J_{det} = \int_\tau^T h_2^+ (X(\tau) - D_2(t-\tau))dt = h_2^+ (X(\tau)(T-\tau) - \frac{D_2}{2}(T-\tau)^2).$$

Otherwise,

$$J_{det} = \int_\tau^{\tau + \frac{X(\tau)}{D_2}} h_2^+ (X(\tau) - D_2(t-\tau))dt =$$

$$h_2^+ (\frac{X(\tau)^2}{D_2} - \frac{D_2}{2}(\frac{X(\tau)}{D_2})^2) = h_2^+ \frac{X(\tau)^2}{2D_2}.$$

The following proposition deals with the case of inventory shortage when there is one period to go.

Proposition 5.2. *Let* $X(\tau) < 0$, *then the optimal solution for problem (5.5) and the optimal value of the objective function are*

If $D_2 < U + \dfrac{X(\tau)}{T - \tau}$, *then* $u(t) = U$ *for* $\tau \le t < \tau - \dfrac{X(\tau)}{U - D_2}$ *and* $u(t) = D_2$ *for*

$\tau - \dfrac{X(\tau)}{U - D_2} \le t \le T$, *and*

$$J_{det} = h^- \frac{X(\tau)^2}{2(U - D_2)},$$

otherwise if $D_2 \ge U + \dfrac{X(\tau)}{T - \tau}$, *then* $u(t) = U$ *for* $\tau \le t \le T$ *and*

$$J_{det} = -h^- \left(X(\tau)(T - \tau) + \frac{U - D_2}{2}(T - \tau)^2 \right).$$

Proof: The proof is similar to that of Proposition 5.1. This time, we initially have a shortage. The fastest way to get rid of it is to produce as much as possible, $u(t) = U$, until either the end of the production horizon or $X(t)$ attains zero, which is the most desirable inventory level.

The optimal value then for the objective function is $D_2 < U + \dfrac{X(\tau)}{T - \tau}$.

$$J_{det} = -\int_{\tau}^{\tau - \frac{X(\tau)}{U - D_2}} h^-(X(\tau) + (U - D_2)(t - \tau))dt = -h^-\left(\frac{X(\tau)^2}{U - D_2} + \frac{U - D_2}{2}(\frac{X(\tau)}{U - D_2})^2\right) = h^- \frac{X(\tau)^2}{2(U - D_2)}.$$

Otherwise,

$$J_{det} = -\int_{\tau}^{T} h^-(X(\tau) + (U - D_2)(t - \tau))dt = -h^-(X(\tau)(T - \tau) + \frac{U - D_2}{2}(T - \tau)^2).$$

Stochastic Component of the Problem

Given an optimal solution over the second interval $[\tau, T]$, we are now interested in an optimal solution and accordingly, the best producer's response over the first time interval $[0, \tau]$. First we split the objective function (5.3) into two parts with respect to the two periods:

$$J_p(u, h^*) = E\left[\int_0^T C(X(t))dt\right] = E\left[\int_0^{\tau} C(X(t))dt\right] + E\left[\int_{\tau}^T C(X(t))dt\right]. \quad (5.6)$$

Applying conditional expectation and accounting for equations (5.1) and (5.4) we have

$$J_p(u,h^+) = \int_0^\tau \{ \int_0^{X(0)+\int_0^t u(s)ds} h^+\left(X(0)+\int_0^t u(s)ds - D_1 t\right) f_1(D_1)dD_1 -$$

$$\int_{X(0)+\int_0^t u(s)ds}^{U} h^-\left(X(0)+\int_0^t u(s)ds - D_1 t\right) f_1(D_1)dD_1 \} dt +$$

$$+ E\left[\int_\tau^T h(X(t))dt\right] \tag{5.7}$$

Using Propositions 5.1-5.2, the last term in (5.7) can be readily determined as a sum of J_{det} obtained for each particular case multiplied by the corresponding probability. Specifically, from Proposition 5.1 for $X(\tau) \geq 0$, we have

$$J_{det} = h_2^+\left(X(\tau)(T-\tau) - \frac{D_2}{2}(T-\tau)^2\right), \text{ when } D_2 < \frac{X(\tau)}{T-\tau} \text{ and } J_{det} = h_2^+ \frac{X(\tau)^2}{2D_2}$$

when $\dfrac{X(\tau)}{T-\tau} \leq D_2$. Taking into account equation (5.1), $X(\tau) = X(0) + \int_0^\tau u(s)ds - d_1\tau \geq 0$, which by definition of the demand distribution $f(.)$,

occurs with probability $\displaystyle\int_0^{\frac{X(0)+\int_0^\tau u(s)ds}{\tau}} f_1(D_1)dD_1$, we conclude

$$E_{X(\tau)\geq 0}\left[\int_\tau^T C(X(t))dt\right] = \int_0^{\frac{X(0)+\int_0^\tau u(s)ds}{\tau}} f_1(D_1)[\int_0^{\frac{X(\tau)}{T-\tau}} h_2^+(X(\tau)(T-\tau) - \frac{D_2}{2}(T-\tau)^2)f_2(D_2|D_1)dD_2 +$$

$$\int_{\frac{X(\tau)}{T-\tau}}^{U} h_2^+ \frac{X(\tau)^2}{2D_2} f(D_2|D_1)dD_2]dD_1 .$$

Similarly, from Proposition 5.2 we have for $X(\tau)<0$:

$$E_{X(\tau)<0}\left[\int_\tau^T C(X(t))dt\right] = \int_\tau^{\frac{X(0)+\int_0^\tau u(s)ds}{\tau}} f_1(D_1)[\int_0^{U+\frac{X(\tau)}{T-\tau}} h^- \frac{X(\tau)^2}{2(U-D_2)} f_2(D_2|D_1)dD_2 -$$

$$- \int_{U+\frac{X(\tau)}{T-\tau}}^{U} h^-(X(\tau)(T-\tau)+\frac{U-D_2}{2}(T-\tau)^2)f_2(D_2|D_1)dD_2]dD_1.$$

Consequently,

$$E\left[\int_{\tau}^{T}C(X(t))dt\right] = \underset{X(\tau)\geq 0}{E}\left[\int_{\tau}^{T}C(X(t))dt\right] + \underset{X(\tau)<0}{E}\left[\int_{\tau}^{T}C(X(t))dt\right]. \qquad (5.8)$$

Let us introduce a new variable, $Y(t)$:

$$\dot{Y}(t) = u(t), \; Y(0)=X(0), \; 0 \leq t \leq \tau. \qquad (5.9)$$

Then, by substituting $X(\tau) = X(0) + \int_0^\tau u(s)ds - D_1\tau = Y(\tau)-D_1\tau$ into (5.8) and taking into account (5.9), the objective function (5.7) takes the following form

$$J_p(u,h^+) = \int_0^\tau \{ \int_0^{\frac{Y(t)}{t}} h_1^+(Y(t)-D_1t)f_1(D_1)dD_1 - \int_{\frac{Y(t)}{t}}^{U} h^-(Y(t)-D_1t)f_1(D_1)dD_1\}dt + \varphi(Y(\tau)), (5.10)$$

where

$$\varphi(Y(\tau)) = \int_0^{\frac{Y(\tau)}{\tau}} f_1(D_1)[\int_0^{\frac{Y(\tau)-D_1\tau}{T-\tau}} h_2^+((Y(\tau)-D_1\tau)(T-\tau)-\frac{D_2}{2}(T-\tau)^2)f_2(D_2|D_1)dD_2$$

$$+ \int_{\frac{Y(\tau)-D_1\tau}{T-\tau}}^{U} h_2^+ \frac{(Y(\tau)-D_1\tau)^2}{2D_2} f_2(D_2|D_1)dD_2]dD_1 +$$

$$\int_{\frac{Y(\tau)}{\tau}}^{U} f(D_1)[\int_0^{U+\frac{Y(\tau)-D_1\tau}{T-\tau}} h^- \frac{(Y(\tau)-D_1\tau)^2}{2(U-D_2)} f_2(D_2|D_1)dD_2$$

$$- \int_{U+\frac{Y(\tau)-D_1\tau}{T-\tau}}^{U} h^-((Y(\tau)-D_1\tau)(T-\tau)+\frac{U-D_2}{2}(T-\tau)^2)f_2(D_2|D_1)dD_2]dD_1. \qquad (5.11)$$

We thus proved the following theorem.

Theorem 5.1. *Control u(t), which is optimal for deterministic problem (5.5) when $\tau \leq t \leq T$ and for deterministic problem (5.2),(5.9)-(5.11) when $0 \leq t < \tau$ is optimal for stochastic problem (1)-(4) for $0 \leq t \leq T$.*

Problem (5.9)-(5.11) and (5.2) is a canonical, deterministic, optimal control problem which can be studied with the aid of the maximum principle. Since all constraints are linear, the maximum principle-based optimality conditions are not only necessary but also sufficient if the objective function

(5.10) is convex. Moreover, this problem has a unique solution if the objective function is strictly convex, which evidently holds if

$$\frac{\partial F_1(D)}{\partial D} > 0 \text{ and } \frac{\partial^2 \varphi(Y(\tau))}{\partial Y^2} > 0 . \tag{5.12}$$

Accordingly, we next use the maximum principle by first constructing the Hamiltonian

$$H = -\int_0^{\frac{Y(t)}{t}} h_1^+ \left(Y(t) - D_1 t\right) f(D_1) dD_1 + \int_{\frac{Y(t)}{t}}^U h^- \left(Y(t) - D_1 t\right) f(D_1) dD_1 + \psi(t) u(t), \tag{5.13}$$

where the co-state variable $\psi(t)$ is determined by the co-state differential equation

$$\dot{\psi}(t) = -\frac{\partial H}{\partial Y(t)} = (h_1^+ + h^-) F_1 \left(\frac{Y(t)}{t}\right) - h^- \tag{5.14}$$

with boundary (transversality) condition

$$\psi(T) = -\frac{\partial \varphi(Y(\tau))}{\partial Y(\tau)} . \tag{5.15}$$

According to the maximum principle, the optimal control maximizes the Hamiltonian, that is,

$$u(t) = \begin{cases} U, \text{if } \psi(t) > 0; \\ 0, \text{if } \psi(t) < 0; \\ b \in [0, U], \text{if } \psi(t) = 0. \end{cases} \tag{5.16}$$

We resolve the ambiguity of the third condition from (5.16) in the following proposition.

Proposition 5.3. Let $F_1(\beta) = \dfrac{h^-}{h_1^+ + h^-}$ and $\psi(t) = 0$ at a measurable interval, τ. If $\beta \le U$, then $Y(t) = \beta t$ and $u(t) = \beta$ for $t \in \tau$.

Proof: Differentiating the condition $\psi(t) = 0$ over τ and taking into account (5.14), we find

$$F_1 \left(\frac{Y(t)}{t}\right) = \frac{h^-}{h_1^+ + h^-} . \tag{5.17}$$

Thus $\beta = \dfrac{Y(t)}{t}$ and, therefore, $Y(t) = \beta t$ and $\dot{Y}(t) = u(t) = \beta$ for $t \in \tau$.

We next introduce two switching points t_a and t_b which satisfy

$$\int_{t_a}^{\tau} [(h_1^+ + h^-) F_1 \left(\frac{\beta t_a + U(t - t_a)}{t}\right) - h^-] dt = -\frac{\partial \varphi(Y(\tau))}{\partial Y(\tau)} \Big|_{Y(\tau) = \beta t_a + U(\tau - t_a)}, \tag{5.18}$$

$$\int_{t_b}^{\tau}[(h_1^+ + h^-)F_1\left(\frac{\beta t_b}{t}\right) - h^-]dt = -\frac{\partial\varphi(Y(\tau))}{\partial Y(\tau)}\bigg|_{Y(\tau)=\beta t_b} \qquad (5.19)$$

respectively and assume that the production system has sufficient capacity, i.e., $\beta \le U$.

Similar to the optimal solution for the second period, an optimal solution for the first period can be structured into a number of cases depending on the parameters of the production system. We study first two general cases of a non-negative initial inventory level which are described in the following two propositions (see Figures 5.1 - 5.4).

Proposition 5.4. Let $X(0) \ge 0$ and $X(0) < \beta\tau$. If $t_a > \frac{X(0)}{\beta}$ and

$$\frac{\partial\varphi(Y(\tau))}{\partial Y(\tau)}\bigg|_{Y(\tau)=\beta t_a + U(\tau - t_a)} < 0, \text{ then the optimal production control is}$$

$$u(t)=0 \text{ for } 0 \le t < \frac{X(0)}{\beta}, u(t)=\beta \text{ for } \frac{X(0)}{\beta} \le t < t_a, \text{ and}$$

$$u(t) = U \text{ for } t_a \le t \le \tau.$$

Proof: Consider the following solution for the state variables

$$u(t)=0, Y(t)=X(0) \text{ for } 0 \le t < \frac{X(0)}{\beta}; u(t)=\beta,$$

$$Y(t)=\beta t \text{ for } \frac{X(0)}{\beta} \le t < t_a; u(t) = U, Y(t)=\beta t_a + U(t-t_a) \text{ for } t_a \le t \le \tau$$

and co-state variables

$$\psi(t) = -\int_{t}^{\frac{X(0)}{\beta}}[(h_1^+ + h^-)F_1\left(\frac{X(0)}{t}\right) - h^-]dt \text{ for } 0 \le t < \frac{X(0)}{\beta};$$

$$\psi(t) = 0 \text{ for } \frac{X(0)}{\beta} \le t < t_a;$$

$$\psi(t) = \int_{t_a}^{t}[(h_1^+ + h^-)F_1\left(\frac{\beta t_a + U(t-t_a)}{t}\right) - h^-]dt \text{ for } t_a \le t \le \tau.$$

If this solution is feasible and satisfies optimality conditions (5.16), then it is optimal.

The feasibility, $t_a > \frac{X(0)}{\beta}$ and $\frac{\partial\varphi(Y(\tau))}{\partial Y(\tau)}\bigg|_{Y(\tau)=\beta t_a + U(\tau - t_a)} < 0$, is imposed by

the statement of this proposition. The optimality conditions are verified straightforwardly. Specifically, it is easy to observe that from $X(0) < \beta\tau$ and

$F_1(\beta) = \dfrac{h^-}{h_1^+ + h^-}$, we have $(h_1^+ + h^-)F_1\left(\dfrac{X(0)}{t}\right) - h^- < 0$ for $0 \le t < \dfrac{X(0)}{\beta}$ and

thus $\psi(t) = -\int\limits_{t}^{\frac{X(0)}{\beta}} [(h_1^+ + h^-)F_1\left(\dfrac{X(0)}{t}\right) - h^-]dt > 0$, i.e., the second optimality condition,

$u(t) = 0$, from (5.16) holds. Similarly, $\psi(t) = \int\limits_{t_a}^{t} [(h_1^+ + h^-)F_1\left(\dfrac{\beta t_a + U(t - t_a)}{t}\right) -$

$h^-]dt > 0$ for $t_a < t \le \tau$ and thus the first optimality condition, $u(t) = U$, from (5.16) holds. The third condition from (5.16) is explicit, $\psi(t) = 0$ for $\dfrac{X(0)}{\beta} \le t < t_a$ and therefore $u(t) = \beta$.

Note that the optimal control described in Proposition 5.4 consists of three different trajectories. If the feasibility requirement $\dfrac{\partial \varphi(Y(\tau))}{\partial Y(\tau)}\bigg|_{Y(\tau) = \beta t_a + U(\tau - t_a)} < 0$ is not met, then the third trajectory implies no-production instead of production at maximum rate as stated in the following proposition.

Proposition 5.5. *Let* $X(0) \ge 0$ *and* $X(0) < \beta\tau$. *If* $t_b > \dfrac{X(0)}{\beta}$ *and*

$\dfrac{\partial \varphi(Y(\tau))}{\partial Y(\tau)}\bigg|_{Y(\tau) = \beta t_b} > 0$, *then the optimal production control is*

$u(t) = 0$ *for* $0 \le t < \dfrac{X(0)}{\beta}$, $u(t) = \beta$ *for* $\dfrac{X(0)}{\beta} \le t < t_b$, *and*

$u(t) = 0$ *for* $t_b \le t \le \tau$.

Similarly we can state the two general cases when the initial inventory level is negative as shown in the following two propositions (see Figures 5.3-5.4). Proofs for Proposition 5.5 as well as for the next two propositions are similar and therefore omitted. Of course, special cases readily emanate from Propositions 5.4 and 5.5 when one of the switching points or both vanish.

Proposition 5.6. *Let* $X(0) < 0$ *and* $X(0) > -\tau(U - \beta)$. *If* $t_a > \dfrac{-X(0)}{U - \beta}$ *and*

$\dfrac{\partial \varphi(Y(\tau))}{\partial Y(\tau)}\bigg|_{Y(\tau) = \beta t_a + U(\tau - t_a)} < 0$, *then the optimal production control is*

$$u(t)=U \text{ for } 0 \le t < \frac{-X(0)}{U-\beta}, \; u(t)=\beta \text{ for } \frac{-X(0)}{U-\beta} \le t < t_a, \text{ and}$$

$$u(t) = U \text{ for } t_a \le t \le \tau.$$

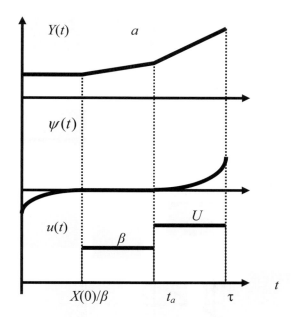

Figure 5.1. Optimal control over the first period for $X(0)>0$ when $\dfrac{\partial\varphi(Y(\tau))}{\partial Y(\tau)} < 0$

Proposition 5.7. *Let $X(0)<0$ and $X(0) > -\tau(U-\beta)$. If $t_b > \dfrac{-X(0)}{U-\beta}$ and*

$\dfrac{\partial\varphi(Y(\tau))}{\partial Y(\tau)}\Big|_{Y(\tau)=\beta t_b} > 0$, *then the optimal production control is $u(t)=U$ for*

$0 \le t < \dfrac{-X(0)}{U-\beta}$, *$u(t)=\beta$ for $\dfrac{-X(0)}{U-\beta} \le t < t_b$, and $u(t)=0$ for $t_b \le t \le \tau$.*

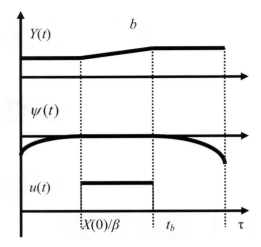

Figure 5.2. Optimal control over the first period for $X(0)>0$ when $\dfrac{\partial \varphi(Y(\tau))}{\partial Y(\tau)} > 0$

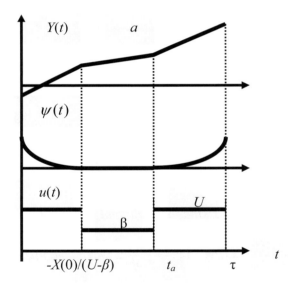

Figure 5.3. Optimal control over the first period for $X(0)<0$ when $\dfrac{\partial \varphi(Y(\tau))}{\partial Y(\tau)} < 0$

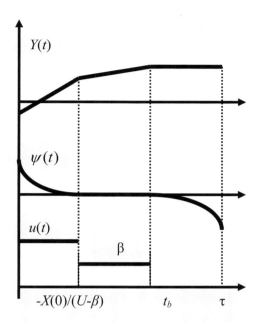

Figure 5.4. Optimal control over the first period for $X(0)<0$ when $\dfrac{\partial \varphi(Y(\tau))}{\partial Y(\tau)} > 0$

Example 5.1

In this example we derive explicit equations for $\psi(T) = -\dfrac{\partial \varphi(Y(\tau))}{\partial Y(\tau)}$ and $\varphi(Y(\tau))$. The example is motivated by the goods which have a short life-cycle during which it is likely that the demand has a single realization, i.e., $D_1 = D_2$, which is estimated by time τ of inventory review. Therefore, after deriving a general optimal solution, we focus on the example which is based on the conditional density function

$$f_2(D_2|D_1) = \delta(D_2 - D_1),$$

where $\delta(D_2 - D_1)$ is a Dirac function. When substituting this conditional distribution function into equation (5.11), $\varphi(Y(\tau))$ simplifies to:

$$\varphi(Y(\tau)) = \int_0^{\frac{Y(\tau)}{T}} f_1(D_1) h_2^+ ((Y(\tau) - D_1\tau)(T - \tau) - \frac{D_1}{2}(T - \tau)^2) dD_1 +$$

$$\int_{\frac{Y(\tau)}{T}}^{\frac{Y(\tau)}{T}} f_1(D_1) h_2^+ \frac{(Y(\tau) - D_1\tau)^2}{2D_1} dD_1 + \int_{\frac{Y(\tau)}{\tau}}^{\frac{Y(\tau)}{T} + U\frac{T-\tau}{T}} f_1(D_1) h^- \frac{(Y(\tau) - D_1\tau)^2}{2(U - D_1)} dD_1$$

$$- \int_{\frac{Y(\tau)}{T} + U\frac{T-\tau}{T}}^{U} f_1(D_1) h^- ((Y(\tau) - D_1\tau)(T - \tau) + \frac{U - D_1}{2}(T - \tau)^2) dD_1 . \tag{5.20}$$

This affects only boundary condition (5.15) as follows

$$\psi(T) = -\frac{\partial\varphi(Y(\tau))}{\partial Y(\tau)} = (h^+ f_1(\frac{Y(\tau)}{T}) \frac{Y(\tau)}{2T^2}(T - \tau)^2 + \int_0^{\frac{Y(\tau)}{T}} f_1(D_1) h^+ (T - \tau) dD_1 -$$

$$- h^+ f_1(\frac{Y(\tau)}{T}) \frac{Y(\tau)}{2T^2}(T - \tau)^2 + \int_{\frac{Y(\tau)}{T}}^{\frac{Y(\tau)}{\tau}} f_1(D_1) h^+ \frac{(Y(\tau) - D_1\tau)}{D_1} dD_1 +$$

$$+ h^- f_1(\frac{Y(\tau)}{T} + U\frac{T-\tau}{T}) \frac{(T - \tau)^2}{2T^2}(U\tau - Y(\tau)) + \int_{\frac{Y(\tau)}{\tau}}^{\frac{Y(\tau)}{T} + U\frac{T-\tau}{T}} f_1(D_1) h^- \frac{Y(\tau) - D_1\tau}{U - D_1} dD_1$$

$$- h^- f_1(\frac{Y(\tau)}{T} + U\frac{T-\tau}{T}) \frac{(T - \tau)^2}{2T^2}(U\tau - Y(\tau)) - \int_{\frac{Y(\tau)}{T} + U\frac{T-\tau}{T}}^{U} f_1(D_1) h^- (T - \tau) dD_1).$$

That is,

$$\psi(T) - h_2^+ (T - \tau) F_1(\frac{Y(\tau)}{T}) - h_2^+ \int_{\frac{Y(\tau)}{T}}^{\frac{Y(\tau)}{\tau}} f_1(D_1) \frac{Y(\tau) - D_1\tau}{D_1} dD_1 -$$

$$- h^- \int_{\frac{Y(\tau)}{\tau}}^{\frac{Y(\tau)}{T} + U\frac{T-\tau}{T}} f_1(D_1) \frac{Y(\tau) - D_1\tau}{U - D_1} dD_1 + h^- (T - \tau)(1 - F_1(\frac{Y(\tau)}{T} + U\frac{T-\tau}{T})). \tag{5.21}$$

Distributor's Problem

The distributor handles at each time point t all inventory-related operations. For each time unit of this service, the distributor charges the producer per item holding cost, $h^+(t)$. The distributor's goal is to minimize his expected cost (or, which is the same in this case, to maximize his profit):

$$\min_{h^+} J_d(u,h^+) = \min_{h^+} E\left[\int_0^T (m - h^+(t))X^+(t)dt\right] \qquad (5.22)$$

s.t.

$$(5.1)\text{-}(5.2)$$
$$m \le h^+(t) \le M, \qquad (5.23)$$

where m is the distributor's marginal cost and M is the maximum inventory holding cost so that the producer will not explore the market for another competing distributor.

Using the same approach as in the previous section, we first analyze the deterministic part of the problem, i.e., when there is only one period to go.

Deterministic Component of the Problem

Consider time interval $[\tau, T]$. At this interval, problem (5.1)-(5.2), (5.22) and (5.23) takes the following deterministic form:

$$I_{det} = \int_\tau^T (m - h_2^+)X^+(t)dt \to \min \qquad (5.24)$$

s.t.

$$(5.1),(5.2) \text{ and } (5.23).$$

An optimal solution for problem (5.24) and thus the best distributor's response to any producer's production and inventory policy is straightforward. Indeed, the dynamic equation (5.1) depends neither on $h^+(t)$ explicitly, nor the switching points in Propositions 5.1 and 5.2. As a result, the optimal solution for this problem is trivial, which is to charge the producer for handling the inventories as much as possible, $h_2^+ = M$. Similar to Propositions 5.1 and 5.2, the objective function value is then

$$I_{det} = 0 \text{ if } X(\tau) \le 0; I_{det} = (m - h_2^+)(X(\tau)(T - \tau) - \frac{D_2}{2}(T - \tau)^2),$$

$$\text{if } X(\tau) > D_2(T - \tau),$$

$$I_{det} = (m - h_2^+)\frac{X(\tau)^2}{2D_2} \text{ if } 0 < X(\tau) \le D_2(T - \tau). \qquad (5.25)$$

In what follows, we summarize these observations with the aid of the fact that the optimal production policy, $u(t)$, from Propositions 5.1 and 5.2 can be presented in a time-dependant (integral) feedback form, $\pi^*(X(\tau),t)$.

5.2 GAME ANALYSIS

The game between the producer and the distributor is as follows. At each period, the distributor selects a price, h^+, to charge the producer for handling inventories during one period while the producer decides how many products $u(t)$ to produce and outsource (store) $X^+(t)$ at each time point t of period. The manufacturer then produces the products and the distributor handles the surplus.

We consider different relationships between the players – the supply chain parties. The equal strength of the parties characterizes one case in which the decisions are made simultaneously (Nash strategy). The other cases are due to the presence of a leader (Stackelberg strategy). If the leader is a distributor who sets a charge for carrying inventories, then the manufacturer is a follower who responds to the charge with a production policy and, consequently, an inventory outsourcing policy as well. A reverse situation is also possible, when the manufacturer is the leader. Since an optimal solution of the producer does not depend on the distributor's charge at the last period (see Propositions 5.1 and 5.2), the equilibrium at this period is straightforward, regardless of leadership in the supply chain, as stated in the following proposition.

Proposition 5.8. *Given inventory level $X(\tau)$, equilibrium in a two-period inventory outsourcing differential game for $\tau \leq t \leq T$ does not depend on whether there is a leader in the supply chain or not. Specifically, a Nash equilibrium as well as a Stackelberg equilibrium when the distributor is the leader and when the producer is the leader are described by the same unique production feedback policy $u(t)=\pi^*(X(\tau),t)$,*

$$
\pi*(X(\tau),t) = \begin{cases} 0, & \text{if } X(\tau) > D_2(T-\tau), \ \tau \leq t \leq T; \\[2mm] 0, & \text{if } 0 \leq X(\tau) \leq D_2(T-\tau), \ \tau \leq t < \tau + \dfrac{X(\tau)}{D_2}; \\[2mm] D_2, & \text{if } 0 \leq X(\tau) \leq D_2(T-\tau), \ \tau + \dfrac{X(\tau)}{D_2} \leq t \leq T; \\[2mm] D_2, & \text{if } (D_2-U)(T-\tau) < X(\tau) \leq 0, \ \tau - \dfrac{X(\tau)}{U-D_2} \leq t \leq T; \\[2mm] U, & \text{if } (D_2-U)(T-\tau) < X(\tau) \leq 0, \ \tau \leq t < \tau - \dfrac{X(\tau)}{U-D_2}; \\[2mm] U, & \text{if } X(\tau) \leq (D_2-U)(T-\tau), \ \tau \leq t \leq T \end{cases}
$$

and inventory holding price $h_2^+ = M$.

Note, that if full commitment and no flexibility characterize the contract between the manufacturer and the distributor, then the inventory holding price cannot be reconsidered at the inventory review point. Thus the stochastic component of the problem determines the equilibrium price, which will be $h_2^+ = h_1^+$ instead of $h_2^+ = M$.

Stochastic Component of the Problem

Given an optimal solution over the second interval $[\tau, T]$, we are now interested in an optimal solution and the best distributor's response over time for the first interval $[0, \tau]$ of problem (1),(2), (22) and (23). Applying conditional expectation to the objective function (5.22) we obtain:

$$J_d(u,h^+) = \int_0^\tau \int_0^t (m-h_1^+)\left(X(0)+\int_0^t u(s)ds - D_1 t\right) f_1(D_1)dD_1 dt + E\left[\int_\tau^T (m-h_2^+)X^*(t)dt\right], (5.26)$$

where

$$E\left[\int_\tau^T (m-h_2^+)X^*(t)dt\right] = \int_0^{\frac{X(0)+\int_0^\tau u(s)ds}{\tau}} f_1(D_1)(m-h_2^+)\left[\int_0^{\frac{X(\tau)}{T-\tau}} (X(\tau)(T-\tau)-\frac{D_2}{2}(T-\tau)^2)f_2(D_2|D_1)dD_2 + \int_{\frac{X(\tau)}{T-\tau}}^U \frac{X(\tau)^2}{2D_2}f(D_2|D_1)dD_2\right]dD_1 . \tag{5.27}$$

Taking into account equation (5.9) and

$$X(\tau) = X(0) + \int_0^\tau u(s)ds - D_1\tau = Y(\tau) - D_1\tau,$$

objective function (5.26) transforms into

$$J_d(u,h^+) = \int_0^\tau \int_0^t (m-h_1^+)(Y(t) - D_1 t)f_1(D_1)dD_1 dt + \eta(Y(\tau)) \to \min_{h_1^+} \tag{5.28}$$

where

$$\eta(Y(\tau)) = \int_0^{\frac{Y(\tau)}{\tau}} f_1(D_1)(m-h_2^+)\left[\int_0^{\frac{Y(\tau)-D_1\tau}{T-\tau}} ((Y(\tau)-D_1\tau)(T-\tau)-\frac{D_2}{2}(T-\tau)^2)f_2(D_2|D_1)dD_2 \right.$$

$$\left. + \int_{\frac{Y(\tau)-D_1\tau}{T-\tau}}^U \frac{(Y(\tau)-D_1\tau)^2}{2D_2}f_2(D_2|D_1)dD_2\right]dD_1 . \tag{5.29}$$

As with Theorem 5.1, we conclude with the following theorem.

Theorem 5.2. *Control $h^+(t)$, which is optimal for deterministic problem (5.24) when $\tau \leq t \leq T$ and for deterministic problem (5.2), (5.9), (5.23), (5.28) and (5.29) when $0 \leq t < \tau$ is optimal for stochastic problem (5.1)-(5.2), (5.22) and (5.23) for $0 \leq t \leq T$.*

Similar to the deterministic case, when $\tau \leq t \leq T$, we readily observe from (5.1),(5.28)-(5.29) that the inventory dynamics do not depend on price h_1^+ unless the distributor is the Stackelberg leader and is therefore able to take into account the best producer's response to a price he offers. Since function (5.28) is linear in h_1^+, in cases where there are no leaders or the producer is the leader, the optimal distributor's response is again to charge the producer as much as possible for handling his inventories, $h_1^+ = M$.

Proposition 5.9. *Given inventory level X(0), if $\dfrac{\partial^2 \varphi(Y(\tau))}{\partial Y^2} > 0$, $\dfrac{\partial F_1(D)}{\partial D} > 0$, and that the supply chain has no leader or the producer is the leader, then the production policy determined by Propositions 5.4-5.7 and inventory holding price $h_1^+ = M$ constitute a unique Nash equilibrium as well as a Stackelberg equilibrium under the producer's leadership in the two-period differential inventory outsourcing game for $0 \leq t \leq \tau$.*

The situation changes if a large distributor is the leader and the manufacturer is the follower. Then the optimal manufacturer's production policy defined in Propositions 5.4-5.7 is substituted into the distributor's problem (5.2), (5.9),(5.23), (5.28) and (5.29). This is to say that, in such a case, the distributor assumes production $u(t)$ to be endogenous. Propositions 5.4-5.7 identify four general types of optimal solutions and a number of sub-cases. These cases depend on the system parameters and whether the manufacturer has an initial shortage of products or a surplus. Each of these cases thus induces a corresponding equilibrium. To avoid awkward expressions, we here focus only on two cases, both of which are due to the standard assumption that the initial inventory level is zero, $X(0)=0$. Then the first switching point in Propositions 5.4-5.7 vanishes and the best producer's response takes the following form.

Proposition 5.10. *The optimal production policy is:*
Low demand expectation

$$if\ h^- \tau \leq \frac{\partial \varphi(Y(\tau))}{\partial Y(\tau)}\bigg|_{Y(\tau)=0},\ then\ u(t)=0\ \ for\ 0 \leq t \leq \tau,$$

$$if \left.\frac{\partial \varphi(Y(\tau))}{\partial Y(\tau)}\right|_{Y(\tau)=0} < h^- \tau \; but \; \left.\frac{\partial \varphi(Y(\tau))}{\partial Y(\tau)}\right|_{Y(\tau)=\beta\tau} > 0, \; then \; u(t)=\beta \; for \; 0 \le t < t_b$$

$$and \; u(t)=0 \; for \; t_b \le t \le \tau,$$

Average demand expectation

$$if \left.\frac{\partial \varphi(Y(\tau))}{\partial Y(\tau)}\right|_{Y(\tau)=\beta\tau} = 0, \; then \; u(t)=\beta \; for \; 0 \le t \le \tau;$$

High demand expectation

$$if \; ((h_1^+ + h^-)F(U) - h^-)\tau > -\left.\frac{\partial \varphi(Y(\tau))}{\partial Y(\tau)}\right|_{Y(\tau)=U\tau} \; but \; \left.\frac{\partial \varphi(Y(\tau))}{\partial Y(\tau)}\right|_{Y(\tau)=\beta\tau} < 0,$$

$$then \; u(t)=\beta \; for \; 0 \le t < t_a \; and \; u(t)=U \; for \; t_a \le t \le \tau, \; otherwise$$

$$if \; ((h_1^+ + h^-)F(U) - h^-)\tau \le -\left.\frac{\partial \varphi(Y(\tau))}{\partial Y(\tau)}\right|_{Y(\tau)=U\tau}, \; then \; u(t)=U \; for \; 0 \le t \le \tau.$$

To find the Stackelberg equilibrium, we substitute (5.9) along with the first production policy from Proposition 5.10 (induced by a low demand expectation) into (5.28) and (5.29). This converts the dynamic problem (5.2), (5.9), (5.23), (5.28) and (5.29) into a static one, to which we can apply the first-order optimality condition,

$$\frac{\partial J_d}{\partial h_1^+} = 0. \tag{5.30}$$

Let us denote a solution of equation (5.30) in h_1^+ as γ. We thus conclude with the following proposition.

Proposition 5.11. *Given inventory level $X(0)=0$, $\dfrac{\partial F_1(D)}{\partial D} > 0$, $\dfrac{\partial^2 \varphi(\beta t_b)}{\partial Y^2} > 0$ and the distributor is the leader in the supply chain. If $\dfrac{\partial^2 J_d}{(\partial h_1^+)^2} > 0$, $\dfrac{\partial \varphi(0)}{\partial Y} < h^- \tau$, $\dfrac{\partial \varphi(\beta\tau)}{\partial Y} > 0$ and $m \le \gamma \le M$, then the production policy $u(t)=\beta$ for $0 \le t < t_b$, $u(t)=0$ for $t_b \le t \le \tau$, and inventory holding price $h_1^+ = \gamma$ constitute a unique Stackelberg equilibrium in the two-period differential inventory outsourcing game for $0 \le t \le \tau$.*

Similarly, we can determine an equilibrium for the case of high demand expectation by substituting the corresponding production policy from Proposition 5.10 into equations (5.28) and (5.29). Denoting the solution to equation (5.30) for such a case as ρ, we conclude with the following proposition:

Proposition 5.12. *Given inventory level $X(0)=0$, $\dfrac{\partial F_1(D)}{\partial D}>0$,*

$\dfrac{\partial^2 \varphi(\beta t_a + U(\tau - t_a))}{\partial Y^2} > 0$ *and the distributor is the leader in the supply*

chain. If $\dfrac{\partial^2 J_d}{(\partial h_1^+)^2}>0$, $\dfrac{\partial \varphi(\beta \tau)}{\partial Y}<0$, $\dfrac{\partial \varphi(U\tau)}{\partial Y} > (h^- - (h_1^+ + h^-)F(U))\tau$,

and $m \le \rho \le M$, *then the production policy $u(t)=\beta$ for $0 \le t < t_a$, $u(t)=U$*
for $t_a \le t \le \tau$, and inventory holding price $h_1^+ = \rho$ constitute a unique
Stackelberg equilibrium in the two-period differential inventory outsourc-
ing game for $0 \le t \le \tau$.

The following example illustrates Proposition 5.11 for specific probability
distribution functions.

Example 5.2.

Consider a uniform distribution, $f_1(D)=1/A$, $A<U$ and $f_2(D_2|D_1)=\delta(D_2-D_1)$.
Then the objective function (5.28) takes the following form:

$$J_d(u,h^+) = \int_0^\tau \int_0^{\frac{Y(t)}{t}} (m-h_1^+)(Y(t)-D_1 t)f_1(D_1)dD_1 dt + \eta(Y(\tau)) =$$

$$\int_0^\tau \int_0^{\frac{Y(t)}{t}} (m-h_1^+)(Y(t)-D_1 t)\frac{1}{A}dD_1 dt + \eta(Y(\tau)) = \int_0^\tau (m-h_1^+)\frac{Y^2(t)}{2tA}dt + \eta(Y(\tau)),$$

where

$$\eta(Y(\tau)) = \int_0^{\frac{Y(\tau)}{T}} f_1(D_1)(m-h_2^+)((Y(\tau)-D_1\tau)(T-\tau)-\frac{D_1}{2}(T-\tau)^2)dD_1 +$$

$$+ \int_{\frac{Y(\tau)}{T}}^{\frac{Y(\tau)}{\tau}} f_1(D_1)(m-h_2^+)\frac{(Y(\tau)-D_1\tau)^2}{2D_1}dD_1 =$$

$$\int_0^{\frac{Y(\tau)}{T}} \frac{1}{A}(m-h_2^+)((Y(\tau)(T-\tau)-D_1\tau(T-\tau)-\frac{D_1}{2}(T-\tau)^2)dD_1 +$$

$$\int_{\frac{Y(\tau)}{T}}^{\frac{Y(\tau)}{\tau}} \frac{1}{A}(m-h_2^+)[\frac{Y^2(\tau)}{2D_1} - \frac{Y(\tau)D_1\tau}{D_1} + \frac{(D_1\tau)^2}{2D_1}]dD_1 =$$

$$\frac{1}{A}(m-h_2^+)(T-\tau)\frac{Y^2(\tau)}{T}[1-\frac{\tau}{2T}-\frac{(T-\tau)}{4T}]+$$

$$\frac{1}{A}(m-h_2^+)[\frac{Y^2(\tau)}{2}(\ln\frac{Y(\tau)}{\tau}-\ln\frac{Y(\tau)}{T})-Y(\tau)\tau(\frac{Y(\tau)}{\tau}-\frac{Y(\tau)}{T})+\frac{\tau^2}{4}(\frac{Y^2(\tau)}{\tau^2}-\frac{Y^2(\tau)}{T^2})]=$$

$$=\frac{1}{A}(m-h_2^+)\frac{Y^2(\tau)}{4}[\frac{(3T-\tau)(T-\tau)}{T^2}+2\ln\frac{T}{\tau}-4(1-\frac{\tau}{T})+(1-\frac{\tau^2}{T^2})].$$

Next, if $\left.\frac{\partial\varphi(Y(\tau))}{\partial Y(\tau)}\right|_{Y(\tau)=0}<h^-\tau$ and $\left.\frac{\partial\varphi(Y(\tau))}{\partial Y(\tau)}\right|_{Y(\tau)=\beta\tau}>0$, then $u(t)=\beta$ for

$0\le t<t_b$ and $u(t)=0$ for $t_b\le t\le\tau$, (see Proposition 5.10). Therefore with the aid of the last expression we have

$$J_d=\int_0^\tau(m-h_1^+)\frac{Y^2(t)}{2tA}dt+\eta(Y(\tau))=\int_0^{t_b}(m-h_1^+)\frac{\beta^2 t}{2A}dt+\int_{t_b}^\tau(m-h_1^+)\frac{\beta^2 t_b^2}{2tA}dt+\eta(Y(\tau))=$$

$$=(m-h_1^+)\frac{\beta^2 t_b^2}{2A}(\frac{1}{2}+\ln\tau-\ln t_b)+$$

$$\frac{1}{A}(m-h_2^+)\frac{\beta^2 t_b^2}{4}[\frac{(3T-\tau)(T-\tau)}{T^2}+2\ln\frac{T}{\tau}-4(1-\frac{\tau}{T})+(1-\frac{\tau^2}{T^2})],\quad(5.31)$$

where the switching point t_b is a function of h^+ as defined by equation (5.19). Using the same uniform distribution equation (5.19) takes the following form

$$\int_{t_b}^\tau[(h_1^++h^-)\frac{\beta t_b}{At}-h^-]dt=-\left.\frac{\partial\varphi(Y(\tau))}{\partial Y(\tau)}\right|_{Y(\tau)=\beta t_b},\quad(5.32)$$

where $\varphi(Y(\tau))$ is determined by (5.20) (or, equivalently, $-\frac{\partial\varphi(Y(\tau))}{\partial Y(\tau)}$ is found in (5.21)). Substituting the uniform distribution into (5.21) we find:

$$-\left.\frac{\partial\varphi(Y(\tau))}{\partial Y(\tau)}\right|_{Y(\tau)=\beta t_b}=\psi(T)=-h_2^+(T-\tau)\frac{\beta t_b}{AT}-h_2^+\int_{\frac{\beta t_b}{T}}^\tau\frac{\beta t_b-D_1\tau}{AD_1}dD_1-$$

$$h^-\int_{\frac{\beta\beta}{\tau}}^{\frac{\beta\beta}{T}+U\frac{T-\tau}{T}}\frac{1}{A}\frac{\beta t_\beta-D_1\tau}{U-D_1}dD_1+h^-(T-\tau)(1-(\frac{\beta t_b}{AT}+U\frac{T-\tau}{AT})).$$

Thus equation (5.32) takes the following form

$$\int_{t_b}^{\tau}[(h_1^+ +h^-)\frac{\beta t_b}{At}-h^-]dt=-h_2^+(T-\tau)\frac{\beta t_b}{AT}-h_2^+\int_{\frac{\beta t_b}{T}}^{\frac{\beta t_b}{T}}\frac{\beta t_b-D_1\tau}{AD_1}dD_1-$$

$$h^-\int_{\frac{\beta t_\beta}{\tau}}^{\frac{\beta t_\beta}{T}+U\frac{T-\tau}{T}}\frac{1}{A}\frac{\beta t_\beta-D_1\tau}{U-D_1}dD_1+h^-(T-\tau)(1-(\frac{\beta t_b}{AT}+U\frac{T-\tau}{AT})),$$

After integration we have

$$(h_1^+ +h^-)\frac{\beta t_b}{A}(\ln\tau-\ln t_b)-h^-(\tau-t_b)=$$

$$h_2^+(T-\tau)\frac{\beta t_b}{AT}-h_2^+\frac{\beta t_b}{A}(\ln\frac{\beta t_b}{\tau}-\ln\frac{\beta t_b}{T})+h_2^+\frac{\tau}{A}\beta t_b(\frac{1}{\tau}-\frac{1}{T})+$$

$$\frac{h^-}{A}\beta t_\beta[\ln(U-\frac{\beta t_\beta}{T}-U\frac{T-\tau}{T})-\ln(U-\frac{\beta t_\beta}{\tau})]$$

$$-\frac{h^-\tau}{A}\frac{\beta t_\beta}{T}+U\frac{T-\tau}{T}-\frac{\beta t_\beta}{\tau}+\ln(U-\frac{\beta t_\beta}{T}-U\frac{T-\tau}{T})-\ln(U-\frac{\beta t_\beta}{\tau})]+$$

$$h^-(T-\tau)(1-(\frac{\beta t_b}{AT}+U\frac{T-\tau}{AT})), \tag{5.33}$$

Although equation (5.33) is transcendental and cannot be resolved explicitly in t_b, one can easily derive a closed-form expression for $\frac{dt_b}{dh_1^+}$ by implicit differentiation of (5.33) with respect to h_1^+ and assuming that t_b depends on h_1^+. Then, given $\frac{dt_b}{dh_1^+}$, one can straightforwardly find the optimal price $h_1^+ = \gamma$ by solving the equation $\frac{\partial J_d}{\partial h_1^+} = 0$ in h_1^+. Specifically, differentiating the objective function J_d (defined by (5.31)), we obtain

$$\frac{\partial J_d}{\partial h_1^+}=-\frac{\beta^2 t_b^2}{2A}(\frac{1}{2}+\ln\tau-\ln t_b)+(m-h_1^+)\frac{\beta^2 t_b}{A}(\frac{1}{2}+\ln\tau-\ln t_b)\frac{dt_b}{dh_1^+}$$

$$-(m-h_1^+)\frac{\beta^2 t_b}{2A}+\frac{1}{A}(m-h_2^+)\frac{\beta^2 2t_b}{4}\frac{(3T-\tau)(T-\tau)}{T^2}\frac{dt_b}{dh_1^+}$$

$$-\frac{\beta t_b^2}{A}\frac{1}{2}+\ln\tau-\ln t_b)\frac{d\beta}{dh_1^+}+(m-h_1^+)\frac{2\beta t_b}{A}(\frac{1}{2}+\ln\tau-\ln t_b)\frac{dt_b}{dh_1^+}\frac{d\beta}{dh_1^+}$$

$$-(m-h_1^+)\frac{\beta t_b}{A}\frac{d\beta}{dh_1^+}+\frac{1}{A}(m-h_2^+)\beta t_b\frac{(3T-\tau)(T-\tau)}{T^2}\frac{dt_b}{dh_1^+}\frac{d\beta}{dh_1^+}=0,\ (5.34)$$

where $\dfrac{d\beta}{dh_1^+}=-h^-\left(\dfrac{1}{h_1^++h^-}\right)^2\Bigg/ f_1(\beta)=-\dfrac{h^-}{\left(h_1^++h^-\right)^2}A$.

If the contract between the parties does not allow for any change in inventory holding price within the production horizon, then h_1^+ is set equal to h_2^+ when solving (5.34). Otherwise, if a change is possible at the end of the first period (a contract with flexibility), then h_2^+ should be set at maximum, i.e., $h_2^+=M$ (see Proposition 5.8) when solving (5.34). This implies that the equilibrium price depends not only on leadership in the supply chain, but also on the type of contract.

Furthermore, we note that equation (5.34) is a closed-form expression for Stackelberg price h_1^+ , where t_b is determined by (5.33). Thus, equations (5.33) and (5.34) form a system of two algebraic equations in two unknowns h_1^+ and t_b. Once h_1^+ has been found, it is then verified that $m\le h_1^+\le M$ and that $\dfrac{\partial\varphi(0)}{\partial Y}<h^-\tau$ and $\dfrac{\partial\varphi(\beta\tau)}{\partial Y}>0$, as required by Proposition 5.11.

Finally, the cases defined in Proposition 5.10, which are described by boundary controls and no switching points, are immediate: the equilibrium charge should be at the maximum value.

We next compare the competitive solutions found for the dynamic differential game with a centralized approach.

System-wide Optimal Solution

If the supply chain is vertically integrated, then an inventory holding price is a transfer cost which does not affect a centralized solution. This implies that the centralized problem is identical to the producer's problem, the only difference being that the inventory holding price is replaced with the distributor's marginal cost m. With respect to Proposition 5.8, which states that the optimal production policy at the second period is independent of the inventory holding price, we readily conclude with the following result.

Proposition 5.13. *In competition of the two-period differential inventory outsourcing game, the producer's production quantity u(t) is identical to the system-wide optimal solution over the second period, i.e., $\tau\le t\le T$.*

This result implies that the distributor, by choosing the maximum holding price to charge the producer for each product unit, affects only his profit share for $\tau\le t\le T$, but the overall profit of the supply chain remains the same as in the corresponding centralized system. The situation is, however, different at the first period, i.e., when $0 \le t \le \tau$. If the distributor makes profit, $m > h^+$, then the production rate $u(t) = \beta$, $F_1(\beta) = \dfrac{h^-}{h_1^+ + h^-}$ decreases, compared to the system-wide optimal rate determined by

$$F_1(\beta^*) = \frac{h^-}{m + h^-}.$$

Coordination

According to Proposition 5.14, the supply chain is perfectly coordinated at the second period $\tau \le t \le T$, regardless of power asymmetry (leadership). During the first period, $0 \le t \le \tau$, however, double marginalization of vertically competing firms impacts the system in a similar way to what was observed in single-period problems (Chapters 2 and 4). Evidently, the distributor, by setting the holding price equal to his marginal cost, $h^+ = m$, and charging, instead, a fixed transaction cost, converts the producer's problem into the corresponding centralized problem, thereby ensuring the producer will follow the system-wide optimal production policy. Thus, as in many other cases of vertical competition, with the two-part tariff, the supply chain becomes perfectly coordinated during the first period as well.

REFERENCES

Fisher M, Raman A, McClelland A (2000) Rocket science retailing is almost here-Are you ready? *Harvard Business Review* 78: 115-124.

Fisher M, Raman A (1996) Reducing the cost of demand uncertainty through accurate response to early sales. *Operations Research* 44: 87-99.

Kogan K, Herbon A (2007) Optimal Inventory Control under Periodic Demand Update. *EJOR*, to appear.

Kogan K, Lou S, Tapiero C (2007) Supply Chain With Inventory Review and Dependent Demand Distributions: Dynamic Inventory Outsourcing, *Working paper*.

Kogan K, Shu C, Perkins J (2004) Optimal Finite-Horizon Production Control in a Defect-Prone environment. *IEEE Transactions on Automatic Control* 49(10): 1795-1800.

6 SUSTAINABLE COLLABORATION IN SUPPLY CHAINS

The intertemporal supply chain models presented in Chapters 4 and 5 focus on inventory, production and pricing relationships between a supplier and a retailer according to different types of demands. However, in reality, there are many other factors that affect members of a supply chain. For example, uncertainty can be associated not only with demands but also with production yields. The firms may utilize common resources such as energy, raw materials, budget and logistics infrastructure, which can be limited or delivered by a supplier of bounded capacity. Furthermore, the firms may choose to expand their outsourcing activities to include repair and maintenance operations rather than just production or inventory.

In this chapter we extend our attention to broader issues and consider supply chains in which the parties collaborate to gain centralized control over decision-making. We are thus interested in reexamining system-wide optimal production and inventory policies to account for additional constraints and conditions imposed on supply chains. Special attention is paid here to production control of multiple manufacturers sharing limited supply chain resources.

6.1 MULTI-ECHELON SUPPLY CHAINS WITH UNCERTAINTY

This section addresses a multi-echelon, continuous-time extension to the classical single-period newsboy problem (for a review on the classical formulation and its extensions see, for example, Khouja 1999; Silver et. al. 1998). Products flow from one echelon to the next. We assume that we don't know the demand during the planning horizon, but we do know the *cumulative* demand at the end of the planning horizon. This is the same assumption made in classical newsboy problems discussed in Chapter 2. Forecast updates are not available along the horizon. The objective is to adjust the production rates during the planning horizon in order to minimize total expected costs. The total costs include shortage or surplus costs

occurring at the end of the planning horizon for the last downstream echelon (as considered in the classical newsboy problem), as well as the surplus
costs in the other echelons during the planning horizon.

Problem formulation

Consider a supply chain containing I vertically connected producers or
manufacturers. The chain produces a single (aggregate) product-type to
satisfy a cumulative demand D for the product-type by the end of a planning horizon, T. The product undergoes consecutive production and supply
stages and thus transforms from a raw material to an end-product.

The following differential equations describe this system:

$$\dot{X}_i(t) = u_i(t) - u_{i+1}(t), \quad X_i(0) = 0, \quad i = 1,2,..,I-1;$$

$$\dot{X}_i(t) = u_i(t), \quad X_i(0) = 0, \quad i = I, \tag{6.1}$$

where $X_i(t)$ is the surplus level at the warehouse located after the i-th
manufacturer (denoted m_i); $u_i(t)$ is the production rate of m_i; $u_i(t)$ is the
control variable whose value can be instantly set within certain bounds:

$$0 \le u_i(t) \le U_i, \quad i = 1,2,...,I, \tag{6.2}$$

with U_i being the maximal production rate of manufacturer i. The product
demand D is a random variable representing yield amount of the product-
type and characterized by probability density $\varphi(D)$ and cumulative distribution $\Phi(a) = \int_0^a \varphi_d(D)dD$ functions respectively. For each planning horizon T, there will be a single realization of D which is known only by time
T. Therefore, the decision has to be made under these uncertain conditions
before production starts.

Equations (6.1) present the flow of products through a warehouse placed
between two consecutive manufacturers. If the warehouse is intermediate,
this flow is determined at each point of time by the difference between the
current production rates of the two consecutive manufaturers. If the warehouse is located after the last downstream manufacturer and is intended for
the finished products, then the flow is determined by the production rate of
the last manufacturer. The products are accumulated in this warehouse in
order to be delivered to the customers at the end of the production horizon.
The difference between the cumulative production and the cumulative
demand, $X_I(T) - D$, is the surplus level of the last manufacturer m_I.
If the cumulative demand exceeds the cumulative production of m_I, i.e.,
if the surplus is negative, a penalty will have to be paid for the lost sales.

On the other hand, if $X_I(T) - D > 0$ an overproduction cost is incurred at the end of the planning horizon. Furthermore, inventory costs are incurred when warehouse levels of the manufacturers are positive, $X_i(t) > 0$, $i=1,2,..I$. Negative warehouse levels are prohibited:

$$X_i(t) \geq 0 \quad i=1,2,\ldots,I\text{-}1. \tag{6.3}$$

Note, that (6.1) implies that $X_I(t) \geq 0$ always holds.

The objective is to find such controls $u_i(t)$ that satisfy constraints (6.1)-(6.3) while minimizing the following expected cost over the planning horizon T:

$$J = E\left[\int_0^T \sum_i C_i(X_i(t))dt + P(X_I(T) - D)\right] \to \min. \tag{6.4}$$

Similar to the previous chapters, linear and piece-wise linear cost functions are used for the inventory and surplus/backlog costs respectively,

$$C_i(X_i(t)) = c_i X_i(t), \tag{6.5}$$

$$P(Z) = p^+ Z^+ + p^- Z^-, \tag{6.6}$$

where $Z^+ = \max\{0, Z\}$ and $Z^- = \max\{0, -Z\}$.

Analysis of the problem

Let us substitute (6.5) and (6.6) into the objective (6.4). Then, given probability density $\varphi(D)$ of the demand, we find:

$$J = \int_0^T \sum_i c_i X_i(t)dt + \int_0^\infty p^+ \max\{0, X_I(T) - D\}\varphi(D)dD +$$

$$+ \int_0^\infty p^- \max\{0, D - X_I(T)\}\varphi(D)dD =$$

$$\int_0^T \sum_i c_i X_i(t)dt + \int_0^{X_I(T)} p^+(X_I(T) - D)\varphi(D)dD + \int_{X_I(T)}^\infty p^-(D - X_I(T))\varphi(D)dD . \tag{6.7}$$

The new objective (6.7) is subject to constraints (6.1) - (6.3) which together constitute a deterministic problem equivalent to the stochastic problem (6.1)-(6.6).

Proposition 6.1. *Problem (6.1)-(6.3), (6.7) is a convex program.*
Proof: Since constraints (6.1)-(6.3) are linear, cost functions $C_i(X_i(t))$ are linear and the sum of convex functions is a convex function, the proof is straightforwardly obtained by verifying whether the second term of objective function (6.7)

$$R = \int_0^{X_I(T)} p^+(X_I(T)-D)\varphi(D)dD + \int_{X_I(T)}^{\infty} p^-(D-X_I(T))\varphi(D)dD$$

is convex with respect to $X_I(T)$:

$$\frac{\partial^2 R}{\partial X(T)^2} = (p^+ + p^-)\frac{\partial \Phi(X_I(T))}{\partial X_I(T)} \geq 0.$$

To study the equivalent deterministic problem, we formulate a dual problem with co-state variables $\psi_i(t)$ satisfying the following co-state equations

$$d\psi_i(t) = c_i dt - d\mu_i(t), \quad i = 1,2,...,I-1 \text{ and } \dot{\psi}_I(t) = c_I \qquad (6.8)$$

with transversality (boundary) constraints:

$$\psi_i(T+0) = 0, \quad i = 1,2,...,I-1;$$

$$\psi_I(T) = -\frac{\partial\left[\int_0^{X_I(T)} p^+(X_I(T)-D)\varphi(D)dD + \int_{X_I(T)}^{\infty} p^-(D-X_I(T))\varphi(D)dD\right]}{\partial X_I(T)} =$$

$$= \int_0^{X_I(T)} p^+\varphi(D)dD + \int_{X_I(T)}^{\infty} p^-\varphi(D)dD.$$

That is,

$$\psi_I(T) = -p^+\Phi(X_I(T)) + p^-(1-\Phi(X_I(T))). \qquad (6.9)$$

Left-continuous functions of bounded variation, $\mu_i(t)$, are due to the state constraint (6.3) and present possible jumps of the corresponding co-state variables when $X_i(t) = 0$. These jumps satisfy the non-negativity

$$d\mu_i(t) \geq 0 \qquad (6.10)$$

and complementary slackness condition

$$\int_0^T X_i(t)d\mu_i(t) = 0. \qquad (6.11)$$

The Hamiltonian is the objective for the dual problem, which is maximized by the optimal controls according to the maximum principle:

$$H = -\sum_i c_i X_i(t) + \sum_{i\neq I}\psi_i(t)(u_i(t)-u_{i+1}(t)) + \psi_I(t)u_I(t) \rightarrow \max. \qquad (6.12)$$

By rearranging only control dependent terms of the Hamiltonian we obtain:

$$H(t) = \sum_{i>1} u_i(t)(\psi_i(t)-\psi_{i-1}(t)) + u_1(t)\psi_1(t).$$

Since this term is linear in $u_i(t)$, it can be easily verified that the optimal production rate that maximizes $H(t)$ is

$$u_i(t) = \begin{cases} U_i, & \text{if } \psi_i(t) - \psi_{i-1}(t) > 0, \forall i > 1 \text{ and} \psi_i(t) > 0, \ i = 1 \text{ (production regime-PR)}; \\ w \in [0, U_i], & \text{if } \psi_i(t) - \psi_{i-1}(t) = 0, \forall i > 1 \text{ and} \psi_i(t) = 0, \ i = 1 \text{ (singular regime-SR)} \\ 0, & \text{if } \psi_i(t) - \psi_{i-1}(t) < 0, \forall i > 1 \text{ and} \psi_i(t) < 0, \ i = 1 \ \text{ (idle regime-IR)}. \end{cases}$$

(6.13)

Accordingly, under the optimal control, the ith manufacturer m_i can either be idle (denoted $m_i \in IR$), working with its maximal production rate ($m_i \in PR$), or enter the singular regime ($m_i \in SR$). Since the primal problem is convex (see Proposition 6.1), the maximum principle provides not only the necessarily, but also the sufficient conditions of optimality. Therefore, all triplets $(u_i(t), X_i(t), \psi_i(t))$ that satisfy the primal (6.1)-(6.3), the dual (6.8)-(6.12), and (6.13) will minimize the objective function (6.7).

We next study the singular regime as its underlying controls are not uniquely determined in optimality conditions (6.13). To ensure the uniqueness of the solutions over this regime, we need the following assumption:: $c_i \neq c_{i+1}$ for all $1 \leq i < I$.

Proposition 6.2. (i) If $m_i \in SR$ in a time interval τ then $X_{i-1}(t) = 0$ and/or $X_i(t) = 0$. (ii) If in a time interval τ $X_i(t) = 0$, $1 \leq i < I-1$ then $u_i(t) = u_{i+1}(t)$, and if $i = I$, $u_i(t) = 0$.

Proof: By definition, in SR $\psi_i(t) = \psi_{i-1}(t)$, $t \in \tau$. Differentiating this equality we have:

$$d\psi_i(t) = d\psi_{i-1}(t), i > 1, \tag{6.14}$$

$$d\psi_i(t) = 0, \quad i = 1. \tag{6.15}$$

By substituting the corresponding co-state equations, we then find:

$$c_i dt - d\mu_i(t) = c_{i-1} dt - d\mu_{i-1}(t), \ i > I;$$

$$c_i dt - d\mu_i(t) = 0, \ i = 1.$$

By taking into account (6.10) and (6.11), we conclude that the last equalities can be satisfied if and only if $X_{i-1}(t) = 0$ and or $X_i(t) = 0$ for $i \neq I$.

The second statement of this proposition is immediately observed from the system equation (6.1).

System-wide optimal solution

To study optimal behavior of the supply chain system, we need to distinguish between two types of manufacturers and warehouses. The distinction is due to the capacity limitations of the manufacturers. Similar to Chapter 4 we define a restricting manufacturer.

Definition 6.1. *Manufacturer* i', *is a restricting manufacturer if either* $i' = I$ *or* $U_{i'} < U_i$, *for all* $I \geq i > i'$, $i' \neq I$.

In addition, we identify a restricting warehouse.

Definition 6.2. *Warehouse placed after manufacturer* i', *is a restricting warehouse if either* $i' = I$ *or* $c_{i'} < c_i$, *for all* $I \geq i > i'$, $i' \neq I$.

Similar to the deterministic multi-echelon supply chain discussed in Section 4.3.2, the most important question is to derive the optimal behavior of the restricting manufacturers. The behavior of the non-restricting manufacturers followed by non-restrictive warehouses, is completely determined by the restricting manufacturers. According to Definitions 6.1 and 6.2, inventory costs and manufacturer maximum production rates determine whether the corresponding manufacturers and/or warehouses are restricting. These notions provide an important insight on ranking manufacturers in terms of their optimal control. Based on this insight, we assume in what follows that the supply chain consists only of restricting manufacturers. Once an optimal solution for such a system is found, we then generalize it for all types of manufacturers. Moreover, in order for the problem to be tractable, we assume: if manufacturer i', is a non-restricting manufacturer, then its warehouse is also non-restricting.

We now use a constructive approach to solve the problem. That is, we first propose a solution, and then we show that this solution is indeed optimal. To formalize the solution we denote by J the total number of restricting warehouses and by $R(j), j=1,2,...,J$ their indexes.

The optimal control policy we are proposing for each subset $S_j = \left\{ m_{R(j)}, m_{R(j)-1},..., m_{R(j-1)+1} \right\}$, $R(0)=0$, $j=1,2,...,J$ of the restricting manufacturers is the following.

Use the IR-PR (no production and then production at the maximum rate of $m_{R(j-1)+1}$) production sequence for $m_{R(j-1)+1}$ and IR-SR (no production and then singular production at the maximum rate of $m_{R(j-1)+1}$ with the same switching time t_j for each m_i, $i=R(j),R(j)-1,...,R(j-1)+2$.

This policy is more rigorously defined in the following.

Policy A: Consider a system with I restricting manufacturers, J restricting warehouses and J switching points, $0 \le t_1 \le t_2 \le ... \le t_J \le T$. The behavior we are proposing for each subset of manufacturers $S_j = \{m_{R(j)}, m_{R(j)-1}, ..., m_{R(j-1)+1}\}$, $R(0)=0, j=1,2,...,J$ is:

(i) $u_i(t) = 0$ for $0 \le t < t_j$, $u_i(t) = U_{R(j-1)+1}$ for $t_j \le t \le T$, $i=R(j), R(j)-1, ..., R(j-1)+1$.

(ii) $X_i(t) = 0$ for $0 \le t \le T$, $i=R(j)-1, ..., R(j-1)+1$; $X_{R(j)}(t) = 0$ for $0 \le t \le t_j$, $X_{R(j)}(t) > 0$ for $t_j < t < T$, $X_{R(j)}(T) = 0$ for $R(j) \ne I$.

(iii) $\psi_i(t) = \psi_{R(j-1)}(t)$, for $0 \le t \le t_j$, $\dot{\psi}_i(t) = c_{R(j)}$, for $t_j \le t \le T$, $\psi_0(t) = \psi_1(t)$, $i=R(j), R(j)-1, ..., R(j-1)+1$, $j > 1$; $\psi_i(t) = \psi_{R(1)}(t)$, $\psi_i(t_1) = 0$, $\dot{\psi}_i(t) = c_{R(1)}$ for $0 \le t \le T$, $\psi_i(t_1) = 0$, $i=R(1), R(1)-1, ..., 1$, $j = 1$.

We now show the proposed behavior for restricting manufacturers satisfies the co-state equations (6.8)-(6.11) and the maximum principle based optimality conditions (6.13).

Proposition 6.3. *If all manufacturers are restricting, then Policy A provides the optimal solution.*

Proof: First note that according to Policy A, $X_i(T) = 0$ for $i=1,2,...,I-1$, which with respect to (6.10) and (6.11) implies that the transversality constraints $\psi_i(T+0) = 0$ are satisfied with instant jumps $d\mu_i(T)$.

Consider the first subset of manufacturers, $S_1 = \{m_{R(1)}, m_{R(1)-1}, ..., m_1\}$. According to (i) of Policy A, $u_i(t) = 0$ for $0 \le t < t_1$, $u_i(t) = U_1$ for $t_1 \le t \le T$, $i=R(1), R(1)-1, ..., 1$. This control is feasible since the production system consists of only restricting manufacturers, that is $U_1 < U_i$.

Next, according to (ii) of Policy A, $X_i(t) = 0$ for $0 \le t \le T$, $i=R(1)-1, ..., 1$; $X_{R(1)}(t) = 0$ for $0 \le t \le t_1$, $X_{R(1)}(t) > 0$ for $t_1 < t < T$, $X_{R(1)}(T) = 0$, which evidently satisfies the state equations (6.1) if $u_{R(1)+1}(t) = 0$ for $0 \le t < t_2$, $u_{R(1)+1}(t) = U_{R(1)+1}$ for $t_2 \le t \le T$, as stated in Policy A(ii) and

$$t_1 U_1 = t_2 U_{R(1)+1} . \qquad (6.16)$$

Consider now a solution for the co-state variables. According to Policy A(iii), $\psi_i(t) = \psi_{R(1)}(t)$, $\psi_i(t_1) = 0$ and $\dot{\psi}_i(t) = c_{R(1)}$ for $0 \le t \le T$, $i=R(1)$,

$R(1)$-1,…,1, $j = 1$, which meets co-state equations (6.8) for the determined behavior of the state variables. This also implies that $\psi_i(t) < 0$ (idle regime, $u_i(t) = 0$ according to (6.13)) for $0 \le t < t_1$, and $\psi_i(t) > 0$ (full production $u_i(t) = U_1$ according to (6.13)) for $t_1 \le t \le T$, i=1. Furthermore, condition $\psi_i(t) = \psi_{R(1)}(t)$ satisfies the singular regime from (6.13), Proposition 6.2 conditions for i=$R(1)$,$R(1)$-1,…,1. Thus (6.1), (6.8)-(6.11) are satisfied and (6.12) is maximized.

Similarly, by considering subsequent subsets of manufacturers, $S_j = \{m_{R(j)}, m_{R(j)-1},…, m_{R(j-1)+1}\}$, j=2,…,J, one can verify that the proposed solution satisfies the state and co-state equations and the optimality conditions (6.13) if the switching times are set as:

$$t_j U_{R(j-1)+1} = t_{j+1} U_{R(j)+1}, \quad j < J.$$ (6.17)

The only difference is that the last manufacturer I =$R(J)$ does not have a predecessor and his switching point, $t_J = \dfrac{X_I(T)}{U_{R(J-1)+1}}$ is determined so that $X_I(T)$ satisfies the corresponding transversality condition (6.9).

Based on Policy A, we can solve the two-point boundary value problem (6.1),(6.2),(6.8)-(6.11) to find the switching time points.

We first note that Policy A(ii) condition $X_i(T) = 0$ for i=1,2,..,I-1 along with (6.17) implies:

$$t_j = T - \frac{X_I(T)}{U_{R(j-1)+1}} \quad \text{for } j\text{=1,2,…,}J.$$ (6.18)

Then, by integrating the co-state equations (6.8) with boundary conditions from Policy A(iii), we find:

$$\psi_I(T) = \sum_{1<j\le J} c_{R(j-1)}(t_j - t_{j-1}) + c_I(T - t_J).$$ (6.19)

Finally, by substituting (6.18) into (6.19) and taking into account the corresponding transversality condition from (6.9), we obtain the following equation in unknown $X_I(T) = 0$:

$$\sum_{1<j\le J} c_{R(j-1)}\left(\frac{1}{U_{R(j-2)+1}} - \frac{1}{U_{R(j-1)+1}}\right) + c_I \frac{1}{U_{R(J-1)+1}} = \frac{-p^+ \Phi(X_I(T)) + p^-(1-\Phi(X_I(T)))}{X_I(T)}.$$ (6.20)

Equation (6.20) allows determining optimal production or order quantity $X_I(T)$ for the multi-echelon supply chain. Note, that by setting all inventory costs at zero in equation (6.20), one can now obtain the classical, single-stage newsboy problem solution:

$$\text{If } c_i = 0 \text{ for } i=1,\ldots,I, \text{ then } \Phi(X_I(T)) = \frac{p^-}{p^+ + p^-}.$$

We now study the effect of non-restricting manufacturers accompanied by non-restricting warehouses on the optimal behavior of the supply chain. The approach is similar to that for the restricting manufacturers. We denote by K the total number of the restricting manufacturers and by $Q(k)$, $k=1,2,\ldots,K$ their indexes. Next, we propose an optimal control policy for each subset $S_k = \{m_{Q(k)-1}, m_{Q(k)-2}, \ldots, m_{Q(k-1)+1}\}$, $Q(0)=0$, $k=1,2,\ldots,K$ of the non-restricting manufacturers as follows.

If $k>1$, use SR (the singular production at the rate of adjacent upstream restricting manufacturer $m_{Q(k-1)}$) for each m_i, $i=Q(k)-1,Q(k)-2,\ldots,Q(k-1)+1$.

If $k=1$, use SR (the singular production at the rate of adjacent downstream restricting manufacturer $m_{Q(k)}$) for each m_i, $i=Q(k)-1,Q(k)-2,\ldots,Q(k-1)+1$.

This policy is more rigorously defined as follows.

Policy B: Consider a system with I restricting manufacturers, J restricting warehouses and J switching points, $0 \le t_1 \le t_2 \le \ldots \le t_J \le T$. The behavior we are proposing for each subset of non-restricting manufacturers $S_k = \{m_{Q(k)-1}, m_{Q(k)-2}, \ldots, m_{Q(k-1)+1}\}$, $Q(0)=0$, $k=1,2,\ldots,K$ is:

(i) if $k>1$, $u_i(t) = u_{Q(k-1)}(t)$, otherwise $u_i(t) = u_{Q(k)}(t)$ for $0 \le t \le T$, $i=Q(k)-1,Q(k)-2,\ldots,Q(k-1)+1$;

(ii) $X_i(t) = 0$ for $0 \le t \le T$, $i=Q(k)-1,Q(k)-2,\ldots,Q(k-1)+1$;

(iii) if $k>1$, $\psi_i(t) = \psi_{Q(k-1)}(t)$, otherwise $\psi_i(t) = 0$ for $0 \le t \le T$, $i=Q(k)-1,Q(k)-2,\ldots,Q(k-1)+1$.

We now show that the proposed behavior for non-restricting manufacturers satisfies the co-state equations (6.8)-(6.11), the maximum principle-based optimality conditions (6.13), and does not effect the optimal behavior of the restricting manufacturers determined by Policy A.

Proposition 6.4. *If all restricting manufacturers satisfy Policy A and all non-restricting manufacturers satisfy Policy B, then these policies provide the optimal solution.*

Proof: The proof is straightforward. The solution described in Policy B satisfies Proposition 6.2, that is, it satisfies the system (6.8) – (6.11) and conditions (6.13) for the non-restricting manufacturers. Moreover, one can readily observe, that the co-state variables of non-restricting manufacturers for $k>1$ are simply identical to those for the adjacent upstream restricting manufacturers while the optimality conditions (6.13) for the restricting manufacturers do not change. Finally, copying for $k>1$, $\psi_i(t) = \psi_{Q(k-1)}(t)$ is feasible because of two facts: the assumption, which implies $c_i \geq c_{q(k-1)}$ and $X_i(t) = 0$ for $0 \leq t \leq T$, which implies any jumps $d\mu_i(t) \geq 0$ of the co-state variables are allowed.

Algorithm

We summarize our findings with an algorithm. The algorithm is straight-forward and immediately follows from Policies A and B as described below.

INPUT: I; c_i, U_i, $i=1,...,I$; p^+, p^-.

Step 1. Use Definitions 6.1 and 6.2 to determine and number restricting manufacturers $Q(k)$, $k=1,...,K$ and restricting warehouses $R(j)$, $j=1,...,J$.

Step 2. Consider only restricting manufacturers. Use (6.20) to calculate $X_I(T)$. Form subsets $S_j = \{m_{R(j)}, m_{R(j)-1},..., m_{R(j-1)+1}\}$, $R(0)=0$, $j=1,2,...,J$.

Step 3. Use equation (6.18) to calculate J switching points. Use Policy A to set the optimal solution for all subsets of the restricting manufacturers.

Step 4. Consider all manufacturers. Form subsets of the non-restricting manufacturers

$$S_k = \{m_{Q(k)-1}, m_{Q(k)-2},..., m_{Q(k-1)+1}\}, \quad Q(0)=0, \quad k=1,2,...,K.$$

Step 5. Use Policy B to set the optimal solution for all subsets of the non-restricting manufacturers.

OUTPUT: Optimal controls $u_i(t)$ for $0 \leq t \leq T$, $i=1,2,...,I$.

From this algorithm we readily conclude with the following proposition.

Proposition 6.5. *Given $X_I(T)$ which satisfies equation (6.20), if*
$$\frac{X_I(T)}{U_{Q(1)}} \leq T, \text{ then problem (6.1)-(6.6) is solvable in O(I) time.}$$

Proof: According to Propositions 6.2-6.4, the solution presented by Policies A and B is optimal if the switching points determined by (6.18) are feasible, i.e., there is enough capacity to produce $X_I(T)$ by the end of the

planning horizon. The feasibility is straightforwardly provided by requiring $t_1 \geq 0$. This inequality, by taking into account equation (6.18) and the
fact that the first restricting manufacturer $R(0)+1$ is $Q(1)$, results in
$\dfrac{X_I(T)}{U_{Q(1)}} \leq T$, as stated in the proposition. Moreover, due to Proposition 6.1,
this solution is globally optimal. Finally, provided that a solution can be
found for the optimal order quantity equation (6.20), each step of the algorithm evidently requires only $O(I)$ operations.

Remark. Proposition 6.5 estimates the complexity of solving problem
(6.1)-(6.6) provided that an optimal order equation (6.20) can be resolved
analytically. However, an analytical solution is not always available. One
can readily observe that equation (6.20) is monotone in the unknown
$X_I(T)$, which implies that it can be easily solved numerically to any desired precision.

Example 6.1.

Consider the uniform distribution, $\varphi(D) = \begin{cases} \dfrac{1}{d}, & \text{for } 0 \leq D \leq d; \\ 0, & \text{otherwise.} \end{cases}$. Then

$\Phi(a) = \dfrac{a}{d}$ and equation (6.20) takes the following form:

$$X_I(T) = \frac{p^-}{\dfrac{\left(p^+ + p^-\right)}{d} + \displaystyle\sum_{1 < j \leq J} c_{R(j-1)}\left(\dfrac{1}{U_{R(j-2)+1}} - \dfrac{1}{U_{R(j-1)+1}}\right) + c_I \dfrac{1}{U_{R(J-1)+1}}}. \quad (6.21)$$

To illustrate each step of the algorithm, we consider a small, five-echelon
supply chain system. Table 6.1 presents the input data for such a system.
In addition, $T=5$ time units, $d=24$ product units, $p^+ = 1\,\$$ per product unit
and $p^- = 2\,\$$ per product unit.

Table 6.1. Parameters of the five-echelon supply chain system

Parame-ters	Manufac-turer 1	Manufac-turer 2	Manufac-turer 3	Manufac-turer 4	Manufac-turer 5
U_i	3	2	4	7	6
c_i	2.5	0.5	3	2	1

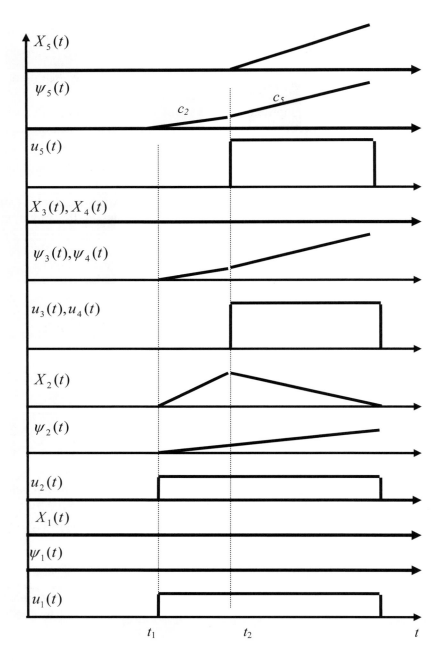

Figure 6.1. Optimal control of the five-echelon supply chain

Step 1 identifies sets of three restricting manufacturers ($K=3$) as $Q(1)=2$, $Q(2)=3$, $Q(3)=5$ and two restricting warehouses ($J=2$) as $R(1)=2$, $R(2)=5$, respectively. Considering only restricting manufacturers 2,3 and 5 at Step 2 results in

$$X_5(T) = \frac{p^-}{\dfrac{(p^+ + p^-)}{d} + c_2 \left(\dfrac{1}{U_2} - \dfrac{1}{U_3} \right) + c_5 \dfrac{1}{U_5}} = 4.8 \text{ product units,}$$

$$S_1 = \{m_2\} \text{ and } S_2 = \{m_5, \ m_3\}.$$

Two switching points are calculated at Step 3, as $t_1 = T - \dfrac{X_5(T)}{U_2} = 2.6$

time units and $t_2 = T - \dfrac{X_5(T)}{U_3} = 3.8$ time units. Then, based on Policy A, the optimal solution is set for restricting manufacturers (see Figure 6.1):

$$u_2(t) = 0 \text{ for } 0 \le t < 2.6 \text{ and } u_2(t) = 2 \text{ for } 2.6 \le t \le 5$$

$$u_3(t) = u_5(t) = 0 \text{ for } 0 \le t < 3.8 \text{ and } u_3(t) = u_5(t) = 4 \text{ for } 3.8 \le t \le 5.$$

Next, Step 5 forms subsets of non-restricting manufacturers as $S_1 = \{m_1\}$, $S_2 = \{\varnothing\}$ and $S_3 = \{m_4\}$. Finally, according to Policy B, the optimal solution for the non-restricting manufacturers is set at Step 5:

$$u_1(t) = u_2(t) \text{ and } u_4(t) = u_3(t) \text{ for } 0 \le t \le 5.$$

Finally, we emphasize that with the aid of the maximum principle, the problem of centralized control of a multi-echelon supply chain is reduced to determining optimal production order quantity; ranking manufacturers and warehouses; calculating a limited number of switching time points; and assigning production rates over the switching points with respect to the manufacturer and warehouse ranks.

6.2 SUPPLY CHAINS WITH LIMITED RESOURCES

Sharing resources is common in industrial applications and can involve energy and natural resources, production equipment, logistics infrastructure and information systems. Advances in information technology have challenged Internet and database suppliers with the problem of providing a high level of service in the face of permanently growing demands. Specifically, the explosive growth of the Internet and the World Wide Web has brought a dramatic increase in the number of customers that compete for the shared resources of distributed system environments (Liu at al. 2000).

In this section we address two problems of centralized control over parallel production under limited resources. Multiple manufacturers produce different products for the same retailer. Since the retailer gains a fixed percentage from sales, control over the supply chain is unaffected. The manufacturers do not compete in terms of products as the products are not substitutable. However, the production utilizes the same resources, which implies that an increase in production rate of one of the manufacturers may induce the others to reduce their production rate (Kogan et al. 2002).

The difference between the two intertemporal problems considered in this section is due to the production conditions. One case involves uncertain demands for products that the manufacturers face. The other problem is concerned with preventive maintenance service of a resource the supply chain depends upon, such as logistics infrastructure, transportation and production equipment. This service is outsourced to an independent firm of bounded capacity. The firm receives a fixed transfer cost for periodic maintenance of the resource and as a result the objective of the overall supply chain is unaffected. However, during the maintenance service of the resource, the production of the manufacturers is interrupted which means that the system dynamics are affected.

6.2.1 PRODUCTION CONTROL OF PARALLEL PRODUCERS WITH RANDOM DEMANDS FOR PRODUCTS

In this section we consider a horizontal supply chain operating under uncertain demands subject to a renewable resource. The demand for items, D_n, $n=1,2,...,N$ is not known until the end of the horizon. The goal is to minimize item holding costs while meeting the demand as close as possible in terms of shortage and surplus costs of the items incurred by the end of the planning horizon by each manufacturer. Since demands do not arrive until the end of the production horizon and the production is controllable and takes time (i.e., it is a continuous-time, dynamic process), a decision has to be made before the production starts. This implies that this problem also can be viewed as a dynamic extension of the newsvendor problem, specifically, of the multi-item, single-period newsvendor problem. Therefore, similar to the newsboy problem, the model can be applied to evaluating and allocating advanced orders in manufacturing and retailing as well as in service industries dealing with items that become obsolete quickly, spoil quickly, or have a future that is uncertain beyond a single period. Contrary to the classical newsboy problem, the resources are constrained and the dynamic extension enables us to determine not only optimal order quantities, but also optimal sequencing and timing of them.

Problem formulation

Consider a manufacturing supply chain system consisting of N producers which utilizes *a single* resource to produce N corresponding (not substitutable) product types and a warehouse for each product type to collect finished products. The system produces to satisfy a cumulative demand, D_n, for product-type n by the end of a planning horizon, T. The production process can be described by the following differential equations:

$$\dot{X}_n(t) = k_n r_n(t), \quad X_n(0) = 0, \quad n = 1,2,...,N, \tag{6.22}$$

where $X_n(t)$ is the surplus level of product-type n produced by manufacturer n and stored in its warehouse by time t; and $k_n r_n(t)$ is the production rate of product n at time t which is linearly proportional with coefficient k_n to the resource usage $r_n(t)$ at time t. Naturally, the resource usage is bounded from below

$$r_n(t) \geq 0, \ n=1,2,...,N \tag{6.23}$$

and from above by the maximal level of the resource usage

$$\sum_n w_n r_n(t) \leq M, \tag{6.24}$$

where weights w_n present the resource consumption per unit of product n.

Furthermore, the product demands D_n, $n=1,2,...,N$ are random variables characterized by probability density functions $\varphi(D_n)$ and cumulative distribution functions $\Phi(a_n) = \int_0^{a_n} \varphi(D_n) dD_n$. For each planning horizon T, there will be a single realization of D_n which is known only by time T. Therefore, the decision has to be made under these uncertain conditions before the production starts.

Similar to the problem considered in Section 6.1, the difference between the cumulative production and the demand, $X_n(T) - D_n$, is the surplus level. If the cumulative demand exceeds the cumulative production, i.e., if the surplus is negative (shortage of needed items), a penalty will have to be paid for backlogs. On the other hand, if $X_n(T) - D_n > 0$ (extra items which were never used) over-production cost is incurred at the end of the planning horizon. Furthermore, inventory holding costs are incurred at points where warehouse levels are positive, $X_n(t) > 0$.

The objective is to find such resource usage $r_n(t)$, $n=1,2,\ldots,N$ that satisfies constraints (6.22)-(6.24) while minimizing the following expected cost over the planning horizon T:

$$J = E\left[\int_0^T \sum_n h_n X_n(t)dt + \sum_n P(X_n(T)-D_n)\right] \rightarrow \min, \qquad (6.25)$$

where h_n is the inventory holding cost of one product of type n per time unit and

$$P(Z_n) = p_n^+ Z_n^+ + p_n^- Z_n^-, \qquad (6.26)$$

where $Z_n^+ = \max\{0, Z_n\}$, $Z_n^- = \max\{0, -Z_n\}$, p_n^+ and p_n^- are the costs of one product surplus and shortage respectively.

To simplify the analysis, we substitute (6.26) into the objective (6.25). Then, given probability density $\varphi(D_n)$ of the demand, we find:

$$J = \int_0^T \sum_n h_n X_n(t)dt +$$

$$\int_0^\infty \sum_n p_n^+ \max\{0, X_n(T) - D_n\}\varphi(D_n)dD_n + \int_0^\infty \sum_n p_n^- \max\{0, D_n - X_n(T)\}\varphi(D_n)dD_n$$

$$= \int_0^T \sum_n h_n X_n(t)dt$$

$$+ \sum_n p_n^+ \int_0^{X_n(T)} (X_n(T) - D_n)\varphi(D_n)dD_n + \sum_n p_n^- \int_{X_n(T)}^\infty (D_n - X_n(T))\varphi(D_n)dD_n. \quad (6.27)$$

The new objective (6.27) is subject to constraints (6.22) - (6.24), which together constitute a deterministic problem equivalent to the stochastic problem (6.22)-(6.25).

Analysis of the problem

To study the equivalent deterministic problem, we use co-state variables $\psi_n(t)$ satisfying the following co-state equation:

$$\dot{\psi}_n(t) = h_n \qquad (6.28)$$

with transversality (boundary) constraint:

$$\psi_n(T) =$$

$$\frac{\partial\left[\sum_n p_n^+ \int_0^{X_n(T)}(X_n(T)-D_n)\varphi(D_n)dD_n + \sum_n p_n^- \int_{X_n(T)}^{\infty}(D_n - X_n(T))\varphi(D_n)dD_n\right]}{\partial X_n(T)}$$

$$= -\int_0^{X_n(T)} p_n^+ \varphi(D_n)dD_n + \int_{X_n(T)}^{\infty} p_n^- \varphi(D_n)dD_n ,$$

that is,

$$\psi_n(T) = -p_n^+\Phi(X_n(T)) + p_n^-(1-\Phi(X_n(T))) . \qquad (6.29)$$

According to the maximum principle, the Hamiltonian is maximized by the optimal control variables $r_n(t)$:

$$H(t) = -\sum_n h_n X_n(t) + \sum_n \psi_n(t)k_n r_n(t) \to \max \qquad (6.30)$$

By rearranging only control-variable dependent terms of the Hamiltonian and introducing a new variable

$$y_n(t) = w_n r_n(t) \qquad (6.31)$$

we obtain:

$$H_u(t) = \sum_n \psi_n(t)\frac{k_n}{w_n} y_n(t) \to \max . \qquad (6.32)$$

s.t.

$$y_n(t) \ge 0 , n=1,2,\dots,N; \qquad (6.33)$$

$$\sum_n y_n(t) \le M . \qquad (6.34)$$

Since this term (6.32) is linear in $y_n(t)$, it can be readily verified that the optimal resource usage that maximizes the Hamiltonian is

$$y_n(t) = \begin{cases} M, \text{ if } \psi_n(t)\dfrac{k_n}{w_n} > \psi_{n'}(t)\dfrac{k_{n'}}{w_{n'}} \text{ for} \forall n' \ne n \text{ and } \psi_n(t)\dfrac{k_n}{w_n} > 0; \\[2mm] L\in[0,M], \text{if } \psi_n(t)\dfrac{k_n}{w_n} = \psi_{n'}(t)\dfrac{k_{n'}}{w_{n'}} > \psi_{n'}(t)\dfrac{k_{n'}}{w_{n'}} \text{ for} \forall n'' \ne n' \ne n \\[2mm] \qquad \text{and } \psi_n(t)\dfrac{k_n}{w_n} \ge 0 ; \\[2mm] 0, \text{ if } \psi_n(t) < 0. \end{cases} \qquad (6.35)$$

With respect to (6.31), conditions (6.35) imply that under the optimal solution the n-th product is not produced and thus, $r_n(t) = 0$ (no resource usage); or it is produced at a maximum rate of $r_n(t) = \dfrac{y_n(t)}{w_n} = \dfrac{M}{w_n}$ (full resource usage); or the resource is shared for simultaneous production of a number of products $r_n(t) \in [0, \frac{M}{w_n}]$ (singular resource usage).

Similar to the problem considered in the previous section we find that conditions (6.35) are the necessary and sufficient optimality condition.

Proposition 6.6. *Problem (6.22)-(6.24), (6.27) is a convex program, i.e., there is only one optimal value for the objective function.*
Proof: First of all, note that constraints (6.22)-(6.24) are linear. The objective function (6.27) consists of two terms. The first term is linear as well. The second term

$$R = \sum_n p_n^+ \int_0^{X_n(T)} (X_n(T) - D_n)\varphi(D_n)dD_n + \sum_n p_n^- \int_{X_n(T)}^\infty (D_n - X_n(T))\varphi(D_n)dD_n$$

is convex with respect to $X_n(T)$, because

$$\frac{\partial^2 R}{\partial X_n(T)^2} = (p_n^+ + p_n^-)\frac{\partial \Phi(X_n(T))}{\partial X_n(T)} \geq 0.$$

Thus, problem (6.22)-(6.24), (6.27) is convex.

Since the problem is convex, the maximum principle provides not only the necessary, but also the sufficient conditions of optimality. Therefore, all triplets $(r_n(t), X_n(t), \psi_n(t))$ that satisfy the primal (6.22)-(6.24), the dual (6.28)-(6.29), and (6.35) will minimize the objective function (6.27).

System-wide optimal solution

We next study the basic properties of the optimal solution. The first property is the so-called integrality property, which is due to the fact that the singular regime never exists on an optimal trajectory.

Proposition 6.7: *Given* $h_n\frac{k_n}{w_n} \neq h_{n'}\frac{k_{n'}}{w_{n'}}$ *and* $h_n \neq 0$, $n=1,2,...,N$, *there*

always exists an optimal solution, such that $r_n(t)$ *is equal to either* $\frac{M}{w_n}$,

or 0 at each measurable interval of time.
Proof: The optimality condition (6.35) implies that the singular regime is the only regime characterized by the control variable which may take values between 0 and $\frac{M}{w_n}$ at a measurable time interval, τ. According to (6.35), the singular regime may occur if either at least two gradients are

equal to one another, i.e., $\psi_n(t)\dfrac{k_n}{w_n} = \psi_{n'}(t)\dfrac{k_{n'}}{w_{n'}}$, or at least one gradient

equals zero, i.e., $\psi_n(t)\dfrac{k_n}{w_n} = 0$.

Given $h_n\dfrac{k_n}{w_n} \neq h_{n'}\dfrac{k_{n'}}{w_{n'}}$, assuming first that the singular regime condition

$\psi_n(t)\dfrac{k_n}{w_n} = \psi_{n'}(t)\dfrac{k_{n'}}{w_{n'}}$ holds over τ and differentiating this condition, we find:

$$\dot{\psi}_n(t)\frac{k_n}{w_n} = \dot{\psi}_{n'}(t)\frac{k_{n'}}{w_{n'}}.\tag{6.36}$$

By taking into account equation (6.28), we conclude that for equality (6.36) to hold, it is necessary that

$$h_n\frac{k_n}{w_n} = h_{n'}\frac{k_{n'}}{w_{n'}}.\tag{6.37}$$

which contradicts the conditions of the proposition.

Similarly, it is verified that for condition $\psi_n(t)\dfrac{k_n}{w_n} = 0$ to hold, it is nec-

essary that

$$\dot{\psi}(t) = 0,$$

which contradicts the co-state equation (6.28), $\dot{\psi}(t) = h \neq 0$.

Note that equality $h_n\dfrac{k_n}{w_n} = h_{n'}\dfrac{k_{n'}}{w_{n'}}$ implies that it is does not matter what product to produce first in terms of the objective function. We further eliminate this degraded case by assuming

$$h_n\frac{k_n}{w_n} \neq h_{n'}\frac{k_{n'}}{w_{n'}} \text{ and } h_n \neq 0 \text{, } n=1,2,\ldots,N.$$

The next two corollaries present two properties: the sequencing of the manufacturers, i.e., the order in which it is optimal for the manufacturers to begin producing and the non-preemption of the optimal solution.

Corollary 6.1. *Given product n' is switched on after product n, the following holds*

$$h_n\frac{k_n}{w_n} < h_{n'}\frac{k_{n'}}{w_{n'}}.$$

Proof: The proof immediately follows from optimality condition (6.35), Proposition 6.7 and the fact that the co-state variable is a continuous function, increasing in time as defined by equations (6.28).

Corollary 6.2. *The optimal schedule is non-preemptive, i.e., if product n has been switched on after product n', at a time,* t_n, *then the following will hold* $r_{n'}(t) = 0$ *for* $t_n < t \leq T$.

Proof: Using the same argument as in the proof of Corollary 6.1, linear functions $\psi_n(t)\dfrac{k_n}{w_n}$ and $\psi_{n'}(t)\dfrac{k_{n'}}{w_{n'}}$ can intersect only once at a switching point t_n. Therefore the optimality condition $\psi_n(t)\dfrac{k_n}{w_n} > \psi_{n'}(t)\dfrac{k_{n'}}{w_{n'}}$ cannot change after this point, i.e., product n' will never regain production.

Henceforth, without loss of generality, we assume that all products are ordered and numbered in increasing order of $h_n\dfrac{k_n}{w_n}$.

Propositions 6.6 and 6.7 along with Corollaries 6.1 and 6.2 reduce the continuous-time, dynamic control problem to discrete-time problems of sequencing product types and allocating jobs for processing them at a maximum rate without preemption. However, to solve the problem, we need to know either time points $t_1, t_2, ..., t_N$ at which the processing of products $1, 2, ..., N$ switches on, or optimal inventory order quantities $X_n(T)$. The following proposition states the timing property of the optimal solutions, which implies that unknowns $t_1, t_2, ..., t_N$ and $X_n(T)$ affect each other. This is to say that the optimal switching points and order quantities cannot be determined independently, rather they constitute a simultaneous solution of a system of non-linear equations.

Proposition 6.8. *Define time points* $0 \leq t_1, t_2, ..., t_N \leq T$ *and terminal inventories* $X_n(T) \geq 0$, *n=1,2,...,N to satisfy the following system of N algebraic non-linear equations*

$$\left(f(X_{n-1}(T)) - (T - t_n)h_{n-1}\right)\frac{k_{n-1}}{w_{n-1}} = \left(f(X_n(T)) - (T - t_n)h_n\right)\frac{k_n}{w_n}, \quad n=2,...,N \ (6.38)$$

$$(T - t_1)h_1 = f(X_1(T)), \tag{6.39}$$

where

$$f(X_n(T)) = \psi_n(T) = -p_n^+ \Phi(X_n(T)) + p_n^-(1 - \Phi(X_n(T))) \text{ and}$$

$$X_n(T) = k_n \frac{M}{w_n}(t_{n+1} - t_n), \ n=1,2,\ldots,N; \ t_{N+1} = T. \quad (6.40)$$

Then the optimal solution is given by:

$$r_n(t) = 0 \ \text{for} \ 0 \le t < t_n \ \text{and} \ t_{n+1} < t \le T;$$

$$r_n(t) = \frac{M}{w_n} \ \text{for} \ t_n \le t \le t_{n+1}, \ n=1,\ldots,N.$$

Proof: For the state (6.22) and co-state (6.28) equations, consider the following solution, which is determined by N switching points:

$$r_n(t) = 0 \ \text{for} \ 0 \le t < t_n \ \text{and} \ t_{n+1} < t \le T; \ r_n(t) = \frac{M}{w_n} \ \text{for} \ t_n \le t \le t_{n+1},$$

$$n=1,\ldots,N; \ X_n(T) = k_n \frac{M}{w_n}(t_{n+1} - t_n), \ n=1,2,\ldots,N; \ t_{N+1} = T.$$

$$\psi_n(t) = \psi(T) - h(t - t_n), \ t \ge t_n, \quad (6.41)$$

This solution satisfies the optimality conditions (6.35) if:

$$\left(\psi_{n-1}(T) - (T - t_n)h_{n-1}\right)\frac{k_{n-1}}{w_{n-1}} = \left(\psi_n(T)\right) - (T - t_n)h_n\right)\frac{k_n}{w_n}, \ n=2,\ldots,N. \ (6.42)$$

Next, by taking into account the boundary conditions (6.29), equations (6.41) and (6.42), we obtain the system (6.38)-(6.40) stated in the proposition. According to Proposition 6.6, a feasible solution to the state (6.6) and dual (6.28)-(6.29) equations, which satisfies the optimality conditions (6.35), is globally optimal. The feasibility of the constructed solution is ensured by the production horizon which is sufficiently long to produce all optimal inventory amounts $X_n(T)$, i.e., by $0 \le t_1$, $t_n \le T$ as stated in the proposition.

As the following corollary shows, solvable cases can be derived from Proposition 6.8 .

Corollary 6.3. *Let demands be characterized by the uniform distribution,*

$$\varphi(D_n) = \begin{cases} \dfrac{1}{d_n}, & \text{for } 0 \le D_n \le d_n; \\ 0, \text{otherwise.} \end{cases} \quad \text{and} \ \Phi(a) = \frac{a}{d_n}, \text{if } 0 \le t_1, \ t_N \le T,$$

if there exists a feasible solution of the system (6.38)-(6.40), then problem (6.22)-(6.27) is solvable in $0(2N^3)$ time.

Proof: Given that the demands are characterized by the uniform distribution, the non-linear equation (6.38) becomes linear, that is, (6.38)-(6.40) constitute the following system of $2N$ linear equation in N unknown switching points $0 \le t_1, t_2, \ldots, t_N \le T$ and N terminal inventories $X_n(T)$:

$$\left(-p_n^+ \frac{X_{n-1}(T)}{d_{n-1}} + p_n^-(1 - \frac{X_{n-1}(T)}{d_{n-1}}) - (T - t_n)h_{n-1}\right)\frac{k_{n-1}}{w_{n-1}} =$$
$$= \left(-p_n^+ \frac{X_n(T)}{d_n} + p_n^-(1 - \frac{X_n(T)}{d_n}) - (T - t_n)h_n\right)\frac{k_n}{w_n}$$
$$n=2,\ldots,N, \text{ (6.43)}$$

$$(T - t_1)h_1 = -p_1^+ \frac{X_1(T)}{d_1} + p_1^-(1 - \frac{X_1(T)}{d_1}),$$

$$X_n(T) = k_n \frac{M}{w_n}(t_{n+1} - t_n), n=1,2,\ldots,N; \ t_{N+1} = T.$$

Note, both Proposition 6.8 and Corollary 6.3 assume that a feasible solution $0 \le t_1, t_2, \ldots, t_N \le T$ of the system (6.38)-(6.40) exists, that is, the production horizon is sufficiently long. Although there is no exact *a priori* condition to check whether the production horizon is long enough and, thus, equations (6.38)-(6.40) define the globally optimal solution of problem (6.22)-(6.25), solving equations (6.43) can be considered as a polynomial-time verification of the solvability of the problem. In addition, the worst-case estimate of the production horizon is presented in the following proposition. The estimate is sufficient to ensure the solvability of the problem because it is based on the longest, possibly optimal, processing times.

Proposition 6.9. *Let* X_n^* *be determined by* $\Phi(X_n^*) = \dfrac{p_n^-}{p_n^+ + p_n^-}$. *If*

$T \ge \sum_n \dfrac{X_n^* w_n}{M}$, *then the optimal solution of problem (6.22)-(6.25) is determined by the system (6.38)-(6.40).*

Proof: According to the optimality conditions (6.35), product n is produced during an interval of time, if $\psi_n(t) \ge a \ge 0$ along this interval. With respect to equations (6.28) and (6.29) this implies that

$$\psi_n(T) = -p_n^+ \Phi(X_n(T)) + p_n^-(1 - \Phi(X_n(T))) \ge a$$

holds. Thus, $\Phi(X_n(T)) \le \dfrac{p_n^- - a}{p_n^+ + p_n^-}$ and the maximum terminal inventory is

determined by $a=0$ as $X_n(T) \le X_n^*$, where $\Phi(X_n^*) = \dfrac{p_n^-}{p_n^+ + p_n^-}$.

Since each product is produced at a rate of $\dfrac{M}{w_n}$ (see Proposition 6.8),

the maximum time required for processing all products (and, thus, ensuring a feasible solution $0 \le t_1, t_2, ..., t_N \le T$ of the system (6.38)-(6.40)) is

determined by $\sum_n \dfrac{X_n^* w_n}{M}$ as stated in the proposition.

Algorithm

Besides special solvable cases, such as the case with demands for products characterized by uniform probability distributions (see Corollary 6.3), the optimization problem (6.22)-(6.25) is reduced to a system of state and co-state equations. The system consists of N non-linear equations and N-linear equations in $2N$ unknowns (see Proposition 6.8), which, in general, are not analytically solvable. However, this system of $2N$ equations can be solved numerically to any desired precision by decomposing it into $2N$ recursively solvable equations. The algorithm presented below begins with the earliest possible time point for production to begin, $t_1 \ge 0$, and proceeds to improve the solution at each subsequent iteration. A lower bound for this starting point is determined in the following proposition.

Proposition 6.10. *Given that the production horizon is long enough for all products to be produced, i.e., $0 \le t_1,\ t_N \le T$. If $T \ge \dfrac{p_1^-}{h_1}$, then $t_1 \ge T - \dfrac{p_1^-}{h_1}$.*

Proof: Let us substitute condition (6.29) for $\psi_1(T)$ into (6.39). Then by rearranging terms in (6.38) we obtain

$$\Phi(X_1(T)) = \frac{p_1^- - (T - t_1)h_1}{p_1^+ + p_1^-}. \tag{6.44}$$

Since $\Phi(\cdot)$ is non-negative, we find the time point t_1 is feasible if the right-hand side of equation (6.44) is non-negative, i.e., $p_1^- - (T - t_1)h_1 \ge 0$, as stated in the proposition. Note, if, $T < \dfrac{p_1^-}{h_1}$, then the earliest feasible point to start the production is simply $t_1 \ge 0$.

Once the earliest switching point has been set at the initial step of the algorithm (as determined by Proposition 6.10, the unknown terminal quantity $X_1(T)$ can be found from the non-linear equation (6.39). The next switching point is then obtained from the linear equation (6.40). Consequently, the non-linear equation (6.38) can be solved for $n=2$ since it has only one unknown, $X_2(T)$. Given the unknown terminal quantity $X_2(T)$, we are then able to return to equations (6.40) and (6.28) to find subsequent switching time points and terminal inventories. If, however, an unknown is not feasible, the algorithm returns to the initial stage to correct the earliest switching point, t_1, and to resume solving consecutively all the equations again. Since t_1 is artificially set, 2N-1 equations are needed to calculate the remaining 2N-1 unknowns. The 2Nth equation, $X_N(T) = k_N \dfrac{M}{w_N}(T - t_N)$, which is Nth equation (6.40), is then used to verify the obtained solution. The process is terminated if either a feasible solution cannot be determined, i.e., the production horizon is not long enough to produce all the products, or all unknowns have been found and feasible. The latter fact indicates that the Nth equation (6.40) holds, i.e., an optimal solution has been computed with desired precision ξ.

The algorithm is summarized as follows.

INPUT: N; M; ξ ; $\varphi(\cdot), \Phi(\cdot)$; k_n, w_n, p_n^+, p_n^-, h_n, $n=1,\ldots,N$;

Step 1. Sort and renumber products in non-decreasing order of $h_n \dfrac{k_n}{w_n}$.

Set $g = T\sum_n h_n \dfrac{k_n}{w_n} + \max\{\sum_n p_n^+ \dfrac{k_n}{w_n}; \sum_n p_n^- \dfrac{k_n}{w_n}\}$, $\varepsilon_X = \max_n \dfrac{k_n}{w_n} M\varepsilon_t$ and

$\varepsilon_t = \dfrac{\xi}{Mg}$. Set $t_1 = t^1 = T - \dfrac{p_1^-}{h_1}$ if $T \ge \dfrac{p_1^-}{h_1}$, otherwise set $t_1 = t^1 = 0$. Set $t^2 = T$.

Step 2. Given t_1, solve equation (6.39) in $X_1(T)$ by a dichotomous search with accuracy ε_X. Set $n=1$.

Step 3. Given t_n and $X_n(T)$, solve equation (6.40) in t_{n+1}. Set $n=n+1$. If $t_n > T$, then go to Step 5.

Step 4. Given t_n, t_{n-1} and $X_{n-1}(T)$, solve equation (6.38) in $X_n(T)$ by a dichotomous search with accuracy ε_X. If there is no feasible solution for $X_n(T)$, then go to the next step, otherwise go to Step 7.

Step 5. If either

$$\frac{1}{p_n^+ + p_n^-}\left[((T-t_n)h_{n-1} - f(X_{n-1}(T))\frac{k_{n-1}w_n}{w_{n-1}k_n} - (T-t_n)h_n + p_n^-\right] > 1, \text{ or}$$

$$t_n > T,$$

then set $t^2 = t_1$ and $t_1 = \dfrac{t_1 + t^1}{2}$; otherwise set $t^1 = t_1$ and $t_1 = \dfrac{t_1 + t^2}{2}$.

Step 6. If $t^2 = t^1$, then *Stop*, the production horizon is too short and the problem is not solvable in polynomial-time. Otherwise set $n=1$ and go to Step 2.

Step 7. If $n=N$, then go to Step 9; otherwise go to Step 3.

Step 9. Verify equation (6.39). If either $\left|X_N(T) - k_N\dfrac{M}{w_N}(T-t_n)\right| \le \varepsilon_X$ or

$t_2 - t_1 \le \varepsilon_t$, then set the optimal solution as determined in Proposition 6.8 and Stop; otherwise go to the next step.

Step 10. If $X_N(T) > k_N\dfrac{M}{w_N}(T-t_n)$, then set $t^2 = t_1$ and $t_1 = \dfrac{t^2 + t^1}{2}$;

otherwise set $t^1 = t_1$ and $t_1 = \dfrac{t^2 + t^1}{2}$. Set $n=1$ and go to Step 2.

OUTPUT: Optimal solution $r_n(t)$ for $0 \le t \le T$, $n=1,2,\dots,N$.

The efficiency of the algorithm is due to the important properties of the system of equations (6.38), (6.39) and (6.40). To state these properties, we rearrange terms in equations (6.38)

$$\left[((T-t_n)h_{n-1} - f(X_{n-1}(T))\frac{k_{n-1}w_n}{w_{n-1}k_n} - (T-t_n)h_n + p_n^-\right] - (p_n^+ + p_n^-)\Phi(X_n(T)) = 0. (6.45)$$

and in equation (6.39)

$$p_1^- - (p_1^+ + p_1^-)\Phi(X_1(T)) - (T-t_1)h_1 = 0. \tag{6.46}$$

These equations can be efficiently solved as stated in the following proposition.

Proposition 6.11. Given t_n, $X_{n-1}(T)$ and

$$\lambda = ((T-t_n)h_{n-1} - f(X_{n-1}(T))\frac{k_{n-1}w_n}{w_{n-1}k_n} - (T-t_n)h_n + p_n^-, \ n>1,$$

function $F_n^1 = \lambda - \left(p_n^+ + p_n^-\right)\Phi(X_n(T))$ *is non-increasing in* $X_n(T)$. *Given*
t_1, *function* $F_1^1 = p_1^- - (T - t_1)h_1 - (p_1^+ + p_1^-)\Phi(X_1(T))$ *is non-increasing*
in $X_n(T)$.

Proof: Given t_n and $X_{n-1}(T)$, the proof is immediate, as function $\Phi(\cdot)$ is non-decreasing in its argument.

Proposition 6.11 shows that the functions induced by each of the equations (6.38)-(6.39) are monotone if these equations are resolved separately. However, this property (used in Steps 2 and 4 of the algorithm) is insufficient. To efficiently solve the problem, we need the function induced by the entire system (6.38)-(6.40) to be monotone in unknown t_1. Given this property, t_1, which is initially estimated by its lower bound, can then be effectively corrected until the optimal solution is found. It is due to this overall monotone property we are able to decompose the system of non-linear equations so as to solve them separately in a recursive manner. The following two propositions formalize this important property. Specifically, the property proven in Proposition 6.12 and used in Step 5 of the algorithm provides efficient verification of whether the production horizon is long enough for the problem to be solvable. Proposition 6.13 (used in Step 10), on the other hand, provides fast convergence of the algorithm to an optimal solution.

Proposition 6.12. *Given 2N equations (6.38)-(6.40) in 2N unknowns* t_n
and $X_n(T)$, *n=1,2,...,N. Let* F_n^2 *be a function which maps* $t_1 \rightarrow X_n(T)$,
then F_n^2, *n=1,2,...,N are non-decreasing in their argument.*

Proof: Let us consider $n=1$. Given t_1, $X_1(T)$ is determined by (6.44) derived from equation (6.39). Since the right-hand side of (6.44) is non-decreasing in t_1 and $\Phi(\cdot)$ is non-decreasing in its argument, F_1^2 is non-decreasing in its argument as well. This implies that if t_1 increases, $X_1(T)$ cannot decrease as stated in the proposition. Let us now set $n=2$ and t_1 increases. Then according to the corresponding linear equation (6.40), t_2 must increase even if $X_1(T)$ does not increase. By rearranging terms in the corresponding equation (6.38), i.e., for $n=2$, we obtain:

$$f(X_1(T))\frac{k_1}{w_1} + (T - t_2)\left(h_2\frac{k_2}{w_2} - h_1\frac{k_1}{w_1}\right) = f(X_2(T))\frac{k_2}{w_2}.$$

Note that function $f(X_n(T)) = -p_n^+\Phi(X_n(T)) + p_n^-(1 - \Phi(X_n(T)))$ is non-increasing in $X_n(T)$. Since t_2 increases and $X_1(T)$ either does not change or increases as well, $X_2(T)$ defined by the last equation cannot decrease if $h_2 \dfrac{k_2}{w_2} \geq h_1 \dfrac{k_1}{w_1}$. However, we have assumed that the products are sequenced in non-decreasing order of $h_n \dfrac{k_n}{w_n}$. Thus, F_2^2 is monotone as stated in the proposition. By setting $n=3,...,N$ and repeating the same arguments, we find that the remaining functions F_n^2 are monotone as well.

Corollary 6.4. *Given 2N equations (6.38)-(6.40) in 2N unknowns t_n and $X_n(T)$, $n=1,2,...,N$. Let F_n^3 be a function which maps $t_1 \to t_n$, then F_n^3, $n=1,2,...,N$ are non-decreasing in their argument.*

Proof: This property is obtained as a by-product in the proof of Proposition 6.12.

Proposition 6.13. *Given 2N equations (6.38)-(6.40) in 2N unknowns t_n and $X_n(T)$, $n=1,2,...,N$, then function $F_N^4 = X_N(T) - k_N \dfrac{M}{w_N}(T - t_N)$ is non-decreasing in t_1.*

Proof: This proof immediately follows from Proposition 6.12 and Corollary 6.4.

Based on Propositions 6.9 – 6.13, we are now ready to evaluate the complexity and accuracy of the algorithm.

Theorem 6.1. *If a feasible solution of the system (6.38)-(6.40) exists, then problem (6.22)-(6.25) is solvable with accuracy ξ in*

$$O\left(N \max\left\{\log N; \left(\log \frac{TMg}{\xi}\right)^2\right\}\right) \ time,$$

where $g = T\sum_n h_n \dfrac{k_n}{w_n} + \max\left\{\sum_n p_n^+ \dfrac{k_n}{w_n}; \sum_n p_n^- \dfrac{k_n}{w_n}\right\}$.

Proof: First note that the primal-dual algorithm constructs a solution based on Propositions 6.7-6.8 and Corollaries 6.1 - 6.2, thereby satisfying all optimality conditions derived from the maximum principle within the specified computational accuracy. Moreover, due to Proposition 6.6 the optimality conditions are not only necessary but also sufficient.

To prove the complexity and accuracy of the algorithm, we assess it step-by-step. Step 1 uses Corollary 6.1 and Proposition 6.10 to sort products and provide an initial value for t_1 in $O(N\log N)$ time. In Step 2, equation (6.39) is solved by a dichotomous search, which, according to Proposition 6.11, requires at most $\log \dfrac{X_n^M}{\varepsilon_X}$ operations, where $X_n^M = \min\{X_n^*; k_n \dfrac{M}{w_n} T\}$

is the maximum possible terminal inventory (see Proposition 6.9). Thus, the worst-case estimate of this step is $\log\left(\max_n \dfrac{k_n}{w_n} \dfrac{MT}{\varepsilon_X} \right)$. Step 3 is of $O(1)$ complexity. In Step 4, equation (6.38) is solved again by dichotomous search in $\log \dfrac{X_n^M}{\varepsilon_X}$ time. Step 5, which is based on Proposition 6.12, veri-

fies whether the found solution is feasible in $O(1)$ time. If it is not feasible, then, with the aid of Step 6, the algorithm returns to Step 2. Thus, Steps 2-6 can be repeated at most N times, resulting in $O\left(N\log\left(\max_n \dfrac{k_n}{w_n} \dfrac{MT}{\varepsilon_X} \right) \right)$

complexity. By taking into account that $\varepsilon_X \le \max_n \dfrac{k_n}{w_n} M\varepsilon_t$, the last esti-

mate simplifies to $O\left(N\log\left(\dfrac{T}{\varepsilon_t} \right) \right)$. Once a feasible solution has been

found, Step 9 verifies the accuracy of this solution in $O(1)$ time. If the accuracy is insufficient, t_1 is corrected at Step 10, according to Proposition 6.13, and the algorithm returns to Step 2. This overall dichotomous search

is carried out at most $\log \dfrac{T}{\varepsilon_t}$ times. Thus, by taking into account the com-

plexity of Step 1, we conclude that the worst-case estimate of the algorithm complexity is

$$O\left(N \max\left\{ \log N; \left(\log \dfrac{T}{\varepsilon_t} \right)^2 \right\} \right). \tag{6.47}$$

Finally, we estimate the relationship between the computational accuracy of the objective function value, ξ, and the accuracy of the dichotomous search, ε_t. By varying the objective function, we obtain

$$\xi \le \int_0^T \sum_n h_n \left| \Delta X_n(t) \right| dt + \sum_n p_n^+ \Phi(X_n(T)) \left| \Delta X_n(T) \right| - \sum_n p_n^- (1 - \Phi(X_n(T)) \left| \Delta X_n(T) \right|$$

that can be simplified to:

$$\xi \le T \sum_n h_n \max_t \left| \Delta X_n(t) \right| + \max \left\{ \sum_n p_n^+ \left| \Delta X_n(T) \right|; \sum_n p_n^- \left| \Delta X_n(T) \right| \right\} \quad (6.48)$$

Since $\max_t \left| \Delta X_n(t) \right| = \max \left| \Delta X_n(T) \right| \le M \dfrac{k_n}{w_n} \varepsilon_t$, condition (6.48) transforms

into the following form:

$$\xi \le M g \varepsilon_t, \quad (6.49)$$

where

$$g = T \sum_n h_n \frac{k_n}{w_n} + \max \left\{ \sum_n p_n^+ \frac{k_n}{w_n}; \sum_n p_n^- \frac{k_n}{w_n} \right\}.$$

Substitution of ε_t, computed in (6.49), into the complexity estimate (6.47), completes the proof.

To illustrate both analytical and numerical methods, we consider a simple two-product (two parallel manufacturers) supply chain.

Example 6.2.

Let $M=1$ resource units, $N=2$ and demands be characterized by the uniform distribution over horizon $T=10$ time units. The input data for such a system is presented in Table 6.2.

Table 6.2. Parameters of the production system

	d_n	h_n	w_n	k_n	p_n^+	p_n^-	$h_n \dfrac{k_n}{w_n}$
$n=1$	30.0	0.1	0.1	1.0	6.0	10.0	1
$n=2$	80.0	0.4	0.05	1.0	8.0	15.0	8

First, we note that $h_1 \dfrac{k_1}{w_1} < h_2 \dfrac{k_2}{w_2}$, that is, the condition of Corollary 6.1

is met and, therefore, we don't need to reorder and renumber the products. Since the uniform distribution characterizes the demands, the problem can be solved analytically. Namely, according to Corollary 6.3, the optimal solution must satisfy the system of equations (6.43) which, with respect to the data of Table 6.2, takes the following form:

$$
\left[-8\frac{X_1(T)}{30}+15\left(1-\frac{X_1(T)}{30}\right)-(10-t_2)0.1\right]10=
$$
$$
\left[-8\frac{X_2(T)}{80}+15\left(1-\frac{X_2(T)}{80}\right)-(10-t_2)0.4\right]20 \quad;
$$

$$
-6\frac{X_1(T)}{30}+10\left(1-\frac{X_1(T)}{30}\right)=(10-t_2)0.1;
$$

$$
X_1(T)=10(t_2-t_1)\;;\;\; X_2(T)=20(10-t_2)\,.
$$

Solving this system of $2N=4$ equations results in $t_1=5.84$ time units and $t_2=7.64$ time units, $X_1(10)=18$ product units and $X_2(10)=47.2$ product units. Next, according to Proposition 6.8, the optimal solution is
$r_1(t)=0$ for $0\le t<5.84$ and $7.64<t\le10$; $r_1(t)=10$ for $5.84\le t\le7.64$;
$$r_2(t)=0 \text{ for } 0\le t<7.64\,;\; r_2(t)=20 \text{ for } 7.64\le t\le10\,.$$
This solution is depicted in Figure 6.2.

Then the optimal value of the objective function is found from (6.27):

$$
J=\int_{5.84}^{7.64}0.1\cdot10(t-5.84)dt+0.1\cdot18(10-7.64)+\int_{7.64}^{10}0.4\cdot20(t-7.64)dt+
$$
$$
+\int_{0}^{18}6(18-D_1)\frac{1}{30}dD_1+\int_{0}^{47.2}8(47.2-D_2)\frac{1}{80}dD_2+\int_{18}^{30}10(D_1-18)\frac{1}{30}dD_1+\int_{47.2}^{80}15(D_2-47.2)\frac{1}{80}dD_2
$$
$$
=
$$
$$
\frac{(7.64-5.84)^2}{2}+1.8\cdot2.36+\frac{8(10-7.64)^2}{2}+\frac{18^2}{10}+\frac{47.2^2}{20}+\frac{(30-18)^2}{6}+\frac{3(80-47.2)^2}{32}=
$$
$$
=1.62+4.248+22.278+32.4+111.39+24+100.86=\$296.796.
$$

Next, we solve the problem numerically according to the algorithm, as if equations (6.43) have a general non-linear form. The accuracy of the optimal value of the objective function we require is $\xi=\$20$.

In the first step of the algorithm, we determine parameters $g=470$, $t_1=0$ and the maximum precision for calculating time points $\varepsilon_t=0.043$ and inventories $\varepsilon_x=0.85$. Consequently, $X_1(10)=16.875$ is found from (6.39) at Step 2. At Step 3, the algorithm calculates $t_2=1.69$ from (6.40) and sets $n=2$. $X_2(10)=38.4$ is determined from (6.38) in Step 4.

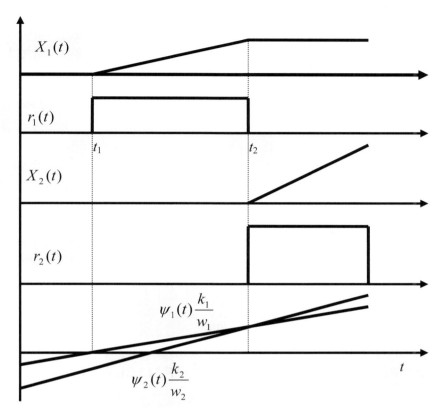

Figure 6.2. Optimal control of the horizontal supply chain with two producers sharing a limited resource

Since the obtained solution is feasible, the algorithm proceeds to Step 9. At Step 9, $X_2(10) = 38.4$ is compared to $k_2 \dfrac{M}{w_2}(T - t_2) = 166$. Since the difference between these two values exceeds $\varepsilon_x = 0.85$, Step 10 sets $t_1 = \dfrac{0 + 10}{2} = 5$ and returns the computation to Step 2. Then the second iteration is performed which results in $X_1(10) = 17.81$ (Step 2), $t_2 = 6.781$ (Step 3) and $X_2(10) = 45.92$ (Step 4). At Step 9, $X_2(10) = 45.92$ is compared to $k_2 \dfrac{M}{w_2}(T - t_2) = 64.38$. As a result, Step 10 sets $t_1 = \dfrac{5 + 10}{2} = 7.5$ and initiates a new iteration starting from Step 2. It is easy to verify that only eight iterations are required to obtain $t_1 = 5.856$ which meets $\varepsilon_t = 0.043$ and $X_1(10) = 17.97$; $t_2 = 7.653$; $X_2(10) = 46.94$.

Consequently, the value of the objective function (6.27) provided by the numerical algorithm is:

$$J' = \int_{5.856}^{7.653} 0.1 \cdot 10(t - 5.856)dt + 0.1 \cdot 17.97(10 - 7.653) + \int_{7.653}^{10} 0.4 \cdot 20(t - 7.653)dt +$$

$$+ \int_{0}^{17.97} 6(17.97 - D_1)\frac{1}{30}dD_1 + \int_{0}^{46.94} 8(46.94 - D_2)\frac{1}{80}dD_2 +$$

$$\int_{17.97}^{30} 10(D_1 - 17.97)\frac{1}{30}dD_1 + \int_{46.94}^{80} 15(D_2 - 46.94)\frac{1}{80}dD_2$$

$$=1.615+4.218+22.034+32.292+110.168+24.12+102.465=\$296.912.$$

Comparing this value with the optimal one obtained analytically, we conclude

$$J'-J=\$296.912-\$296.796=\$0.116,$$

which is better than the required accuracy $\xi = \$20$.

6.2.2 PRODUCTION CONTROL OF PARALLEL PRODUCERS WITH MAINTENANCE

As technology progresses, systems with shared resources become more complex. In order to obtain maximum availability and reliability, periodic maintenance is vital. The literature presents several methodologies for incorporating maintenance and system control policies in stochastic environments (e.g., Anderson 1981; Boukas and Liu 1999; Boukas and Haurie 1998) or in deterministic environments (e.g., Cho et al. 1993 and Maimon et al. 1998). In this section we follow the deterministic direction and deal with preventive maintenance. The maintenance service is outsourced to a firm which conducts a periodic check-up of a single resource shared by multiple producers (manufacturers). During maintenance, production processes are interrupted. Similar to the problem considered in the previous section, each manufacturer is responsible for producing a specific product. The products that the centralized supply chain is engaged with are non-substitutable.

Problem formulation

Consider a supply chain involving N parallel manufacturers producing N product-types to satisfy demand rate d_n, $n=1,2,...,N$. Since the manufacturers share a resource which is unreliable unless preventive maintenance is provided, production is periodically stopped for maintenance. Define t_s the time at which the production period starts; t_f the end of the maintenance period; P the production duration; and M the maintenance duration. We then have

$$t_f = t_s + P + M . \tag{6.50}$$

Assuming the system has reached the steady-state, then the cyclic behavior of the system can be described by the following differential equations:

$$\dot{X}_n(t) = A(t)u_n(t) - d_n, \ X_n(t_s) = X_n(t_f), \ n=1,2,...N , \tag{6.51}$$

where $X_n(t)$ is the surplus of product n at time t, if $X_n(t) \geq 0$, and the backlog, if $X_n(t) < 0$. $u_n(t)$ is the production rate and $A(t)$ is a periodic maintenance function defined as:

$$A(t) = \begin{cases} 1, \text{if } t_s \leq t < t_s + P; \\ 0, \text{if } t_s + P \leq t < t_s + P + M. \end{cases} \tag{6.52}$$

The production rate is a control variable, which is bounded by the maximum production rate U_n for product n:

$$0 \leq u_n(t) \leq U_n . \tag{6.53}$$

The production of each product utilizes a single resource with respect to the following normalized resource constraint

$$\sum_n \frac{u_n(t)}{U_n} \leq 1 . \tag{6.54}$$

In order to ensure that that the demand can be fulfilled at each production cycle, we also need that:

$$\sum_n \frac{d_n}{U_n} \leq \frac{P}{P+M} . \tag{6.55}$$

The objective is to find an optimal cyclic behavior $(u_n(t), X_n(t))$ of the manufacturers that satisfies constraints (6.51), (6.53)-(6.54) while minimizing the following piece-wise linear cost functional:

$$J = \int_{t_s}^{t_f} \sum_n \left[c_n^+ X_n^+(t) + c_n^- X_n^-(t) \right] dt , \tag{6.56}$$

Where

$$X_n^+(t) = \max\{X_n(t), 0\} , \ X_n^-(t) = \max\{-X_n(t), 0\} , \tag{6.57}$$

c_n^+ and c_n^- are the unit costs of storage (inventory) and backlog of product-type n, respectively.

We assume relatively large backlog costs are assigned to products that cause large inventory costs and vice versa as formalized below.

Assumption 6.1. *The inventory and backlog costs are agreeable, that is, if* $c_n^+ U_n > c_{n'}^+ U_{n'}$, *then* $c_n^- U_n > c_{n'}^- U_{n'}$ *and vice versa, for* $n, n' \in \Omega$, *where* $\Omega = \{1 \cdots N\}$.

Without losing generality, we also assume that if $c_n^+ U_n > c_{n'}^+ U_{n'}$ then $n > n'$, and if $n \neq n'$ then $c_n^+ U_n \neq c_{n'}^+ U_{n'}$, $n, n' \in \Omega$, where $\Omega = \{1 \cdots N\}$.

Analysis of the problem

Applying the maximum principle to problem (6.51)-(6.56), the Hamiltonian, is formulated as follows:

$$H = -\sum_n \left[c_n^+ X_n^+(t) + c_n^- X_n^-(t) \right] + \sum_n \psi_n(t)\left(u_n(t) - d_n \right). \qquad (6.58)$$

The co-state variables, $\psi_n(t)$, satisfy the following differential equations with the corresponding periodicity (boundary) condition:

$$\dot{\psi}_n(t) = \begin{cases} c_n^+, \text{if } X_n(t) > 0; \\ -c_n^-, \text{if } X_n(t) < 0; \\ a, a \in [-c_n^-, c_n^+], \text{if } X_n(t) = 0; \end{cases} \qquad \psi_n(t_s) = \psi_n(t_f). \qquad (6.59)$$

To determine the optimal production rate $u_n(t)$ when $A(t) \neq 0$, we consider the following four possible regimes, which are defined according to $U_n \psi_n(t)$.

Full Production regime FP: This regime appears if there is an n such that $U_n \psi_n(t) > 0$, and $U_n \psi_n(t) > U_{n'} \psi_{n'}(t)$, $\forall n' \neq n, n, n' \in \Omega$. In this regime, according to (6.58), we should have $u_n(t) = U_n$ and $u_{n'}(t) = 0$, $\forall n' \neq n, n, n' \in \Omega$. to maximize the Hamiltonian.

No Production regime NP: If $U_n \psi_n(t) < 0$, $\forall n \in \Omega$. In this regime we should have $u_n(t) = 0$, $\forall n \in \Omega$ to maximize the Hamiltonian.

Singular Production regime S-SP: This regime appears if there is $S \subset \Omega$, the rank of S (the rank of S is defined as the number of units in S and denoted $R(S)$) is greater than 1, and

$$U_n \psi_n(t) = U_{n'} \psi_{n'}(t) > 0, \forall n, n' \in S, \text{ and } U_n \psi_n(t) > U_m \psi_m(t), \forall n \in S, m \notin S.$$

In this regime there is a set of products S (the active set) for which the Hamiltonian gradients $U_n \psi_n(t) > 0$ are equal to each other and are greater than all the other gradients at an interval of time.

Singular Production regime Z-SP: This regime appears if there is a $Z \subset \Omega$ such that

$$U_n \psi_n(t) = U_{n'} \psi_{n'}(t) = 0, \forall n, n' \in Z, \text{ and } U_n \psi_n(t) > U_m \psi_m(t), \forall n \in Z, m \notin Z.$$

In this regime there is a set of products Z (the active set) for which the Hamiltonian gradients $U_n \psi_n(t) = 0$ and are greater than all the other gradients in an interval of time.

The optimal production rates under the singular production regimes are discussed in the following three propositions.

Proposition 6.14. *If there is an $n \in \Omega$ such that $U_n \psi_n(t) > 0$, then*

$$\sum_{m \in \Omega} \frac{u_m(t)}{U_m} = 1, \text{ and}$$

if $u_n(t) > 0$ then $U_n \psi_n(t) \geq U_{n'} \psi_{n'}(t)$ for all $n, n' \in \Omega$.

Proof: Since the optimal control maximizes the Hamiltonian (6.58), the first part of the proposition must hold, otherwise we could increase $u_n(t)$ to enlarge the Hamiltonian. To prove the second part of the proposition, assume there is an n' such that $U_n \psi_n(t) < U_{n'} \psi_{n'}(t)$. Also assume the portion of the resource allocated to part n is $\dfrac{u_n(t)}{U_n} = \alpha$. Then $\alpha U_n \psi_n(t) < \alpha U_{n'} \psi_{n'}(t)$ and if the same capacity were allocated to part n' instead of n, Hamiltonian H could be increased. But this violates the optimality assumption.

Proposition 6.15. *Let the S-SP regime with its active set S be in a time interval τ. Then the following hold for $t \in \tau$:*

$$X_{n^*}(t) \neq 0 \text{ and } u_{n^*}(t) = U_{n^*} \left(1 - \sum_{\substack{n \in S, \\ n \neq n^*}} \frac{d_n}{U_n} \right) \text{ for } n^* = \min_{n \in S} n;$$

$$u_n(t) = d_n, \ X_n(t) = 0 \text{ for all } n \neq n^*, \ n \in S;$$

$$u_n(t) = 0 \text{ for all } n \notin S.$$

Proof: According to the definition of the S-SP regime,

$$U_n \psi_n(t) = U_{n'} \psi_{n'}(t) > 0, \ t \in \tau \text{ for all } n, n' \in S, \tag{6.61}$$

$$U_n \psi_n(t) > U_l \psi_l(t), \ t \in \tau \text{ for all } n \in S, l \notin S. \tag{6.62}$$

By differentiating condition (6.61), we obtain:

$$U_n \dot{\psi}_n(t) = U_{n'} \dot{\psi}_{n'}(t), \ t \in \tau. \tag{6.63}$$

Considering Assumption 6.1 and the definition of $\psi_n(t)$ shown in (6.59), equation (6.63) can be met in only two cases.

Case 1: $X_n(t) = 0$ for all $n \in S$, and

Case 2: $X_{n^*}(t) \neq 0$, and $X_n(t) = 0$ for all $n \neq n^*$, $n \in S$ with $n^* = \min_{n \in S} n$

and

$$u_{n^*}(t) = U_{n^*}\left(1 - \sum_{\substack{n \in S, \\ n \neq n^*}} \frac{d_n}{U_n}\right) \tag{6.64}$$

If $X_n(t) = 0$ in a time interval for some $n \in S$, then differentiating $X_n(t) = 0$ and using state equation (6.51), we obtain:

$$u_n(t) = d_n. \tag{6.65}$$

But from (6.55) we have $1 - \sum_{n \in \Omega} \frac{d_n}{U_n} \geq 0$, thus

$$\frac{u_{n^*}(t)}{U_{n^*}} = 1 - \sum_{\substack{n \in S, \\ n \neq n^*}} \frac{d_n}{U_n} \geq \frac{d_{n^*}}{U_{n^*}}. \tag{6.66}$$

In case 1, $u_{n^*}(t) = d_{n^*}$. Thus the previous inequality implies that the Hamiltonian in Case 2 will be larger than the Hamiltonian in Case 1 and therefore Case 2 provides the optimal control. The maximization of the Hamiltonian also demands that $u_n(t) = 0$ for all $n \notin S$. From (6.65) we have $u_n(t) = d_n$, for all $n \neq n^*$, $n \in S$.

Proposition 6.16. *Let the Z-SP regime with its active set Z be in a time interval τ. Then $u_n(t) = d_n$, $X_n(t) = 0$ for all $n \in Z$ and $u_n(t) = 0$ for all $n \notin Z$, $t \in \tau$.*

Proof: Consider the Z-SP regime which by definition satisfies:

$$\psi_n(t) = 0, \ t \in \tau \text{ for all } n \in Z, \tag{6.67}$$

and

$$\psi_n(t) < 0, \ t \in \tau \text{ for all } n \notin Z.$$

First if $\psi_n(t) < 0$ to maximize the Hamiltonian we must have $u_n(t) = 0$. Next, by differentiating condition (6.67), we obtain:

$$\dot{\psi}_n(t) = \dot{\psi}_{n'}(t) = 0, \ t \in \tau \text{ for all } n, n' \in Z. \tag{6.68}$$

Using the same argument as in Proposition 6.15, we have:

$$X_n(t) = 0, \ u_n(t) = d_n, \ t \in \tau \text{ for all } n \in Z. \tag{6.69}$$

The next proposition shows that there must be a Z-SP regime with its active set $Z = \Omega$ in some time interval τ.

Proposition 6.17. Let $\displaystyle\sum_n \frac{d_n}{U_n} < \frac{P}{P+M}$. *Then during the production period P there must be a Z-SP regime with its active set $Z = \Omega$ in some time interval τ.*

Proof: We first notice that under the S-SP, Z-SP, and FP regimes $\displaystyle\sum_n \frac{u_n(t)}{U_n} = 1$. Also, based on the assumption of this proposition we have $\displaystyle\sum_n \frac{d_n}{U_n}(P+M) < P$. Therefore, during the production duration P, if we only use the S-SP, Z-SP, and FP regimes, we would have $\displaystyle\sum_n \frac{d_n}{U_n}(P+M) < \sum_n \frac{u_n(t)}{U_n}P$, which implies the production would exceed demand. This violates our cyclic production assumption. Accordingly there must be a time period $P_1 \subset P$, during which $\displaystyle\sum_n \frac{u_n(t)}{U_n} < 1$, and the only possible regimes during P_1 are Z-SP and NP. If $Z \neq \Omega$, either Z-SP or NP will result in some product(s) being not produced. That is, there exists some n such that $u_n(t)=0$, $t \in P_1$. We now argue that this cannot be the optimal solution.

For such n that $u_n(t)=0$, $t \in P_1$, we must have $\psi_n(t) < 0$ under Z-SP or NP regimes. If $X_n(t) < 0$, then $\dot{\psi}_n(t) < 0$ and thus product n will not be produced again. This contradicts the cyclic production assumption. If $X_n(t) > 0$, then we can certainly reduce the overall cost by doing the following. We first reduce the production in the period before P_1 so that $X_n(t_1) = 0$, where t_1 is the starting time of P_1. We then let $u_n(t)=d_n$, $t \in P_1$ maintain $X_n(t) = 0$, $t \in P_1$. Both will reduce the inventory cost. Thus we must have $u_n(t) \neq 0$, all $n \in \Omega$, $t \in P_1$. Therefore the only possible regime is Z-SP with $Z = \Omega$.

In the following we will establish the optimal production sequence, starting from Z-SP regime with $Z = \Omega$. First, Proposition 6.18 shows that the regime following the above Z-SP regime must be an S-SP regime with $S = \Omega$.

Proposition 6.18. *Let* τ_1 *and* τ_2 *be two consecutive time intervals,* τ_2 *following* τ_1. *If Z-SP regime is in* τ_1, *then* $u_n(t) > 0$ *for all* $n \in Z$, $t \in \tau_2$. *Further, if* $Z = \Omega$ *then there is an S-SP regime in* τ_2 *with* $S = \Omega$.

Proof: According to Proposition 6.16, $X_n(t) = 0$ and $\psi_n(t) = 0$ for $n \in Z$, $t \in \tau_1$. If $u_n(t) = 0$, $t \in \tau_2$, then from (6.51) and (6.59), we have $X_n(t) < 0$, $\dot{\psi}_n(t) < 0$, and $\psi_n(t) < 0$, $t \in \tau_2$. Therefore $\psi_n(t) < 0$ for $t > t_1$, where t_1 is the starting time of τ_2 and product n will never be produced again. This contradicts the assumption of the cyclic production requirement. If $Z = \Omega$, then $u_n(t) > 0$ for all $n \in \Omega$, $t \in \tau_2$. This can only happen if S-SP regime is in τ_2 with $S = \Omega$.

We now state the relationship between two consecutive S-SP regimes.

Proposition 6.19. *Let two S-SP regimes with their active sets* S_1 *and* S_2 *be in two consecutive time intervals* τ_1 *and* τ_2, τ_2 *following* τ_1 *and* $m = \min\limits_{n \in S_1} n$. *If* $X_m(t) > 0$, $t \in \tau_1$ *and* $m > n', \forall n' \in \Omega, n' \notin S_1$, *then* $S_1 = S_2 + m$.

Proof: If $n \in S_1$, $n > m$, then according to Proposition 6.15 we have $X_n(t) = 0$, $U_n \psi_n(t) = U_m \psi_m(t)$, $t \in \tau_1$. If $u_n(t) = 0$, $t \in \tau_2$, then $X_n(t) < 0$, $t \in \tau_2$. Further, since $n > m$, if $X_n(t) < 0$, $U_n \dot{\psi}_n(t) < U_m \dot{\psi}_m(t)$ (see (6.59) and Assumption 6.1). Therefore $U_n \psi_n(t) < U_m \psi_m(t)$ for all $t > t_1$, where t_1 is the starting time of τ_2. This ensures $u_n(t) = 0$ for all $t > t_1$ which contradicts the cyclic production requirement. Therefore $u_n(t) > 0$, $t \in \tau_2$. Thus $n \in S_2$.

We next show if $n \notin S_1$ then $n \notin S_2$. We first observe that by definition of an S-SP regime, $U_n \psi_n(t) > U_{n'} \psi_{n'}(t)$, $\forall n \in S_1$, $n' \notin S_1$, $t \in \tau_1$. Since $n > n'$ for $\forall n \in S_1, n' \notin S_1$ and $X_m(t) > 0$, $t \in \tau_1$ (assumptions of this proposition), we have $U_n \dot{\psi}_n(t) > 0$, $U_n \dot{\psi}_n(t) > U_{n'} \dot{\psi}_{n'}(t)$, $\forall n \in S_1, n' \notin S_1$ (see (6.59)). Therefore, $U_n \psi_n(t) > U_{n'} \psi_{n'}(t)$, $\forall n \in S_1$, $n' \notin S_1$, $t = t_1$, where t_1, as defined above, is the starting time of τ_2. Consequently, $n' \notin S_2$. Since $S_1 \neq S_2$, we must have $S_1 = S_2 + m$.

The above propositions show that there must be a Z-SP regime with $Z = \Omega$ (Proposition 6.17) followed immediately by an S-SP with $S = \Omega$ (Proposition 6.18). The possible regimes afterwards are S-SP regimes defined in Proposition 6.19. We now show that an FP regime must be the last regime before the maintenance period.

Proposition 6.20. *Let* τ_1 *and* τ_2 *be two consecutive time intervals,* τ_2 *following* τ_1. *Further, S-SP regime with its active set S is in* τ_1. *Then FP regime is in* τ_2 *if and only if R(S)=2. (Recall R(S) denotes the number of units in S.)*

Proof: If $R(S)>2$ there would exist $n_1 \in S$ and $n_2 \in S$ such that $n_1 > m$ and $n_2 > m$, where $m = \min_{n \in S} n$. If FP is in τ_2 then either $u_{n_1}(t) = 0$ or $u_{n_2}(t) = 0$, $t \in \tau_2$. But this contradicts the arguments established in the first part of Proposition 6.19.

If $R(S)=2$, there exists an $n \in S$, $n > m$. According to the argument in Proposition 6.19, the only possible regime in $t \in \tau_2$ is an FP regime.

It is easy to show that only the maintenance period can stop an FP regime. The above propositions established the optimal sequence of regimes between the Z-SP with $Z = \Omega$ and the maintenance period. It is summarized in the following proposition.

Proposition 6.21. *The optimal production regimes from the Z-SP regime to the maintenance period are the following:* $Z\text{-}SP \to S\text{-}SP_1 \to S\text{-}SP_2 \to ... S\text{-}SP_{N-1} \to FP \to Maintenance$, *where S-SP$_k$ is an S-SP regime with its active set being $S_k=\{k, k+1, ... , N\}$.*

A similar proposition will show that the optimal production regime after the maintenance period is the reverse of the sequence in Proposition 6.21 due to the agreeable cost coefficients (see Assumption 6.1): Maintenance $\to FP \to S\text{-}SP_{N-1}... \to S\text{-}SP_2 \to S\text{-}SP_1 \to Z\text{-}SP$.

Having determined the optimal control regime sequence, our next step is to determine t_n, the time instances at which the regimes change after Z-SP regime but before the maintenance, and t'_n, that after the maintenance as shown in Figure 6.3.

We further denote maintenance interval $[t_1^M, t_2^M]$, and time instance t_n^* at which inventory levels cross zero line, $n=1,2,...,N$. By integrating state equation (6.51), we immediately find:

$$\left(1 - \sum_{i=n+1}^{N-1} \frac{d_i}{U_i}\right) U_n(t_{n+1} - t_n) - d_n(t_n^* - t_n) = 0 , \; n=1,..,N, \; t_{N+1} = t_1^M ; \quad (6.70)$$

$$\left(1 - \sum_{i=n+1}^{N-1} \frac{d_i}{U_i}\right) U_n(t'_n - t'_{n+1}) - d_n(t'_n - t_n^*) = 0 , \; n=1,..,N, \; t'_{N+1} = t_2^M . \quad (6.71)$$

Integrating co-state equations (6.59) we will obtain another set of N equations:

$$\sum_{i=1}^{n-1} c_i^+ U_i(t_{i+1} - t_i) + c_n^+ U_n(t_n^* - t_n) = \sum_{i=1}^{n-1} c_i^- U_i(t_i' - t_{i+1}') + c_n^- U_n(t_n' - t_n^*), \; n=1,..,N.$$

(6.72)

$$t_{N+1} = t_1^M, \; t_{N+1}' = t_2^M.$$

The above $3N$ equations can then be used to determine the $3N$ unknown t_n, t_n', and t_n^*.

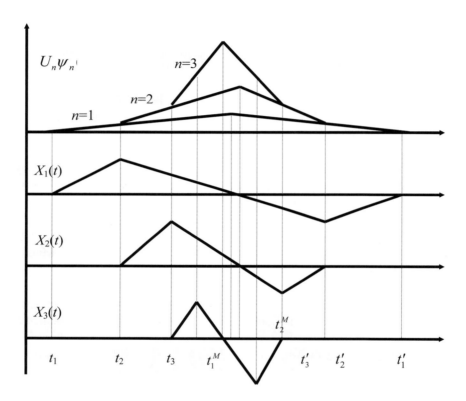

Figure 6.3. Optimal behavior of the state and co-state variables for $N=3$

Algorithm

Step 1: Sort products according to $c_n^+ U_n$ in ascending order.

Step 2: Find $3N$ switching points t_n, t_n', t_n^*, $n=1,..,N$ by solving $3N$ equations (6.70)-(6.72).

Step 3: Determine the optimal production rates in each regime according to Propositions 6.15 and 6.16.

Note that in the above algorithm the production is organized according to the weighted lowest production rate rule (WLPR), where the maximum production rate is weighted by the inventory or backlog costs. In contrast to most WLPR rules, which only allow one product with the lowest production rate to be produced at a time, this algorithm may assign a number of products to be produced concurrently, as there are multple manufacturers. Since the production rate is inversely proportional to the production time, the concurrent WLPR rule is consistent with the weighted longest processing time rule (WLPT) well-known in scheduling literature (Pinedo 1990). The complexity of the algorithm is determined by Step 2, which requires $O(N^3)$ time to solve.

6.3 SUPPLY CHAINS WITH RANDOM YIELD

In this section we consider a centralized vertical supply chain with a single producer and retailer. Similar to the problem considered in the previous section (6.2), since the retailer gains a fixed percentage from sales, control over the supply chain is not affected. The new feature is that a random production yield characterizes the manufacturer.

The stochastic production control in a product defect or failure-prone manufacturing environment is widely studied in literature devoted to real-time or on-line approaches (see, for example, the pioneering work of Kimemia (1982), Kimemia and Gershwin (1983), and Akella and Kumar (1986)). The optimal production rate $u(t)$, which minimizes the expected inventory holding and backlog costs, is usually a function of the inventory $X(t)$. To prove the optimality of the control, certain assumptions will have to be asserted, e.g., the observability of the inventory level and manufacturing states, and notably the Markovian supposition that stipulates that a continuous-time Markov chain describes the transition from an operational state to a breakdown state of the manufacturer.

Unfortunately, in certain manufacturing systems, the information about either manufacturing states or inventory levels may at best be imprecise, if not unobtainable. One example is the chip fabricating facility, where yield or production breakdowns are due to complex causes which are difficult to identify. The system, like many modern ones, could continue producing at the same rate even when there has been a malfunction, because it is the part inspection, at a much later production stage, that will eventually unveil the culprits.

It is also commonplace in some production systems that inventory levels are not continuously obtainable . This reality, in conjunction with the often ambiguous manufacturing states described above, warrants the exploration of an off-line control, which provides better system management when the above-mentioned information is lacking (Kogan and Lou 2005).

As with many other sections in this book, we assume here periodic inventory review and thus the problem under consideration can be viewed as one more extension of the classical newsvendor problem. This dynamic extension is due to the random yield. Accordingly, an optimal off-line control scheme is developed in this section for a production system with random yield and constant demand.

Many authors have considered random yields in various forms. Comprehensive literature reviews on stochastic manufacturing flow control and lot sizing with random yields or unreliable manufacturers can be found in Haurie (1995) as well as Yano and Lee (1995). In addition, Gerchak and Grosfeld-Nir (1998) and Wang and Gerchak (2000) consider make-to-order batch manufacturing with random yield. In these papers it is proven that the optimal policy is of the threshold control type—stop if and only if the stock is larger than some critical value. Gerchak and Grosfeld-Nir (1998) develop a computer program for solving the problem of binomial yields, while Wang and Gerchak (2000) study the critical value for different production cases.

The optimal control derived in this section is significantly different from the traditional threshold control expected under the Markovian assumption, which alternates between zero and the maximum production rate. Indeed, the production rate is not necessarily maximal when the expected inventory level is less than the critical value X^*. Nor is it necessarily zero when the inventory level is larger than X^*.

Problem formulation

Consider a single manufacturer, single part-type centralized production system with random yield characterized by a Wiener process. Similar to the Wiener-increment-based stochastic production models (Haurie 1995), the inventory level $X(t)$ is described by the following stochastic differential equation

$$dX(t) = \left(Pdt + \beta d\mu(t)\right)u(t) - Ddt, \qquad (6.73)$$

where $X(0)$ is a given deterministic initial inventory and $u(t)$ is the production rate,

$$0 \le u(t) \le U, \qquad (6.74)$$

P, $0 < P < 1$ ($U>D/P$), is the average yield - the proportion of the good parts produced; $\mu(t)$ is a Wiener process; $\overline{\beta}$ is the variability constant of the yield; $d\mu(t)$ is the Wiener increment; and D is the constant demand rate.

Similar to Shu and Perkins (2001) and Khmelnitsky and Caramanis (1998), we consider a quadratic inventory cost which is incurred when either $X(t)>0$ (inventory surplus), or $X(t)<0$ (shortage). The objective of the production control is to minimize the overall expected inventory cost:

$$J = E\left[\int_0^T X^2(t)dt\right] \tag{6.75}$$

subject to (6.73) and (6.74), where T is the planning horizon during which the state of the system can be evaluated.

To find the optimal production control, we introduce an equivalent deterministic formulation.

Proposition 6.22. *Problem (6.73) - (6.75) is equivalent to minimizing*

$$J = \int_0^T \left(\left[X(0) - Dt + P\int_0^t u(s)ds\right]^2 + \beta\int_0^t u^2(s)ds\right)dt, \tag{6.76}$$

s.t.

$$(6.74), \text{ where } \beta = \overline{\beta}^2.$$

Proof: Integrating equation (6.73) we have

$$X(t) = X(0) - Dt + \int_0^t Pu(s)ds + \int_0^t \overline{\beta}u(s)d\mu(s), \tag{6.77}$$

which leads to

$$X^2(t) = [X(0) - Dt]^2 + 2[X(0) - Dt]L(t) + [L(t)]^2, \tag{6.78}$$

where $L(t) = \int_0^t Pu(s)ds + \int_0^t \overline{\beta}u(s)d\mu(s)$. Using the fact that the expectation of the stochastic (Ito) integrals is zero, we obtain

$$E[X^2(t)] =$$

$$[X(0)-Dt]^2 + 2[X(0)-Dt]\int_0^t Pu(s)ds + E\left[\int_0^t Pu(s)ds + \int_0^t \overline{\beta}u(s)d\mu(s)\right]^2. \tag{6.79}$$

With respect to the Ito isometry, $E\left[\int_0^t A(\tau)dW(\tau)\right]^2 = \int_0^t E[A^2(\tau)]d\tau$ (Kloeden and Platen 1999), the last term in (6.79) can be rewritten as:

$$E\left[\int_0^t Pu(s)ds + \int_0^t \overline{\beta}u(s)d\mu(s)\right]^2 = E\left[\left(\int_0^t Pu(s)ds\right)^2 + 2\int_0^t Pu(s)ds\int_0^t \overline{\beta}u(s)d\mu(s) + \left(\int_0^t \overline{\beta}u(s)d\mu(s)\right)^2\right] =$$

$$= P^2\left[\int_0^t u(s)ds\right]^2 + \overline{\beta}^2\int_0^t u^2(s)ds.$$

Therefore we have

$$J = E\left[\int_0^T X^2(t)dt\right] = \int_0^T E[X^2(t)]dt =$$

$$\int_0^T ([X(0) - Dt]^2 + 2[X(0) - Dt]P\int_0^t u(s)ds +$$

$$P^2\left[\int_0^t u(s)ds\right]^2 + \overline{\beta}^2\int_0^t u^2(s)ds)dt .$$

Finally, by rearranging the terms in the last expression and using $\beta = \overline{\beta}^2$, we arrive at (6.76).   □

We use the maximum principle to solve the problem. Note that the objective function (6.76) is a summation of strictly convex functions. This implies that the problem is convex and has a unique optimal solution.

Since the objective function (6.76) contains integrals over independent variable t, it does not satisfy the canonical optimal control formulation needed for using the maximum principle. Hence we introduce the expected inventory, $X_E(t)$, which satisfies

$$\dot{X}_E(t) = Pu(t) - D, \ X_E(0) = X(0),$$ (6.80)

and the cumulative quadratic control, $Y(t)$, which satisfies

$$\dot{Y}(t) = u^2(t), \ Y(0) = 0 .$$ (6.81)

Then the objective function (6.76) takes the following form:

$$J = \int_0^T \left[X_E^2(t) + \beta Y(t)\right]dt \rightarrow \min .$$ (6.82)

Formulation (6.74), (6.80)-(6.82) is canonical. According to the maximum principle, the control $u(t)$ which maximizes the Hamiltonian $H(t)$ subject to constraint (6.74) is optimal for (6.80) - (6.82) and, thus, for the original problem. The Hamiltonian is defined as

$$H(t) = -X_E^2(t) - \beta Y(t) + \psi_X(t)(Pu(t) - D) + \psi_Y(t)u^2(t), \quad (6.83)$$

where the co-state variables $\psi_X(t)$ and $\psi_Y(t)$ satisfy the following co-state equations

$$\dot{\psi}_X(t) = 2X_E(t), \ \psi_X(T) = 0 ;$$ (6.84)

$$\dot{\psi}_Y(t) = \beta, \ \psi_Y(T) = 0 .$$ (6.85)

Analysis of the problem: two special cases

As delineated below, depending upon the level of the initial inventory $X(0)$, different optimal control formulations will have to be employed. The formulations are, unfortunately, rather involved, and their proofs convoluted. To make the results more comprehensible, we will start off by proving two special cases.

The first special case: $X(0) \geq DT$.

In this case, the initial inventory is large enough to meet the demand for the entire planning horizon T. Therefore the optimal policy, as one expects, is not to produce at all.

Proposition 6.23. *If* $X(0) \geq DT$, *then* $u(t) = 0$, $0 \leq t < T$ *is optimal.*

Proof: Since $X_E(0) \geq DT$ means $X_E(t) = X(0) + \int_0^t (Pu(\tau) - D)d\tau > 0$,

we have $\dot{\psi}_X(t) = 2X_E(t) > 0$, $0 \leq t < T$. But $\psi_X(T) = 0$, therefore $\psi_X(t) < 0$ and $\psi_Y(t) < 0$, $0 \leq t < T$ (see (6.84) and (6.85) respectively). Therefore, $u(t) = 0$, $0 \leq t < T$ maximizes (6.83) and is thus optimal.

The second special case: $X(0)$ is moderately large, but $X(0) < DT$.

As shown in Theorem 6.2 below, given two critical values, $X^* = -\dfrac{\beta D}{P^2}$ and $\hat{X}$ which can be evaluated through equations depending on the system and initial conditions, $\hat{X} > X^*$, we will have a three-phase control when $\hat{X} > X(0) \geq X^*$ (see Figure 6.4(a)). Initially the optimal production rate $u(t)$ is zero, and thus the average inventory level $X_E(t)$ decreases. This is the first phase, which is identical to the control in the preceding special case. At a time point t_ψ (a certain level of $X_E(t)$, $\hat{X} > X_E(t_\psi) > X^*$), the optimal $u(t)$ becomes positive but is still small enough so that $X_E(t)$ continues its decline. This is the second phase.

Finally, as soon as $X_E(t)$ reaches a critical value, X^* , (this time point is referred to as t_O), the optimal $u(t)$ becomes a constant, $u^*(t) = \dfrac{D}{P}$ and from that point on $X_E(t)$ and $u^*(t)$ will remain equal to X^* and $\dfrac{D}{P}$, respectively. This is the third phase during which the system enters the steady state. The optimal control when $X(0)$ is smaller than X^* is the mirror image of the described control (see Figure 6.4(b)) and therefore is not considered here. On the other hand, if $DT > X(0) \geq \hat{X}$, then the optimal control will include only the first two phases. Note, that the proofs of the equation for $\hat{X}$ and the existence of t_O when $\hat{X} > X(0) \geq X^*$, which utilize the asymptotic behaviors of the family of Bessel functions, are tedious and therefore excluded. To prove Theorem 6.2, we first need to establish the following proposition.

Proposition 6.24. *Assume functions* $\psi(t)$, $X(t)$ *and*

$$
u(t) = \begin{cases} 0, & \psi(t) < 0, \\ \dfrac{P\psi(t)}{2\beta(T-t)}, & \psi(t) \geq 0 \end{cases}
$$

satisfy $\dot{X}(t) = Pu(t) - D$ *and* $\dot{\psi}(t) = 2X(t)$ *for* $0 \leq t \leq T$, *where* $\beta > 0$, *P>0 and D>0 are constants. Furthermore, assume* $\psi(T) = 0$, $X(0) > \tilde{X} = -\dfrac{\beta D}{P^2}$ *and* $X(t') = \tilde{X}$ *for some* t', $0 \leq t' \leq T$ *and* $X(t) \neq \tilde{X}$, $0 \leq t < t'$. *Then*

$$
\psi(t) \leq -2\tilde{X}(T-t),\ 0 \leq t \leq t',\ and
$$
$$
X(t) = \tilde{X}\ and\ \psi(t) = -2\tilde{X}(T-t)\ for\ t' \leq t \leq T.
$$

Proof: We first show that $\psi(t) < -2\tilde{X}(T-t)$, $0 \leq t < t'$. Since $X(0) > \tilde{X}$, $X(t') = \tilde{X}$ and $X(t) \neq \tilde{X}$, $0 \leq t < t'$, we must have $X(t) > \tilde{X}$, $0 \leq t < t'$. Thus, there is a t'', $t'' < t'$, such that $\dot{X}(t) = Pu(t) - D < 0$ for $t'' \leq t < t'$.

Therefore $u(t) < \dfrac{D}{P}$, $t'' \leq t < t'$, which leads to

$$
\psi(t) < -2\tilde{X}(T-t)\ \text{for}\ t'' \leq t < t'. \tag{6.86}
$$

If $\psi(\bar{t}) > -2\tilde{X}(T-\bar{t})$ for some $\bar{t}$, $0 \leq \bar{t} < t''$, then because $\dot{\psi}(t) = 2X(t) > 2\tilde{X}$ for $0 \leq t < t'$, we would have $\psi(t'') > -2\tilde{X}(T-t'')$. But this contradicts (6.86). Therefore $\psi(t) < -2\tilde{X}(T-t)$, for $0 \leq t < t'$.

We now show that $\psi(t') = -2\tilde{X}(T-t')$. Assume the opposite were true, that is, $\psi(t') < -2\tilde{X}(T-t')$. Thus $u(t') < \dfrac{D}{P}$ and $\dot{X}(t') < 0$. Therefore there would exist a t''', $t' < t''' < T$ such that $X(t) < \tilde{X}$ for $t' < t \leq t'''$. Thus, $\dot{\psi}(t) = 2X(t) < 2\tilde{X}$ and $\psi(t) < -2\tilde{X}(T-t)$ for $t' < t \leq t'''$.

Furthermore, there would exist a t^*, $t' < t^* \leq T$, such that $X(t^*) = \tilde{X}$, otherwise $X(t) < \tilde{X}$ and, thus, $\dot{\psi}(t) = 2X(t) < 2\tilde{X}$ and $\psi(t) < -2\tilde{X}(T-t)$ for $t' < t \leq T$. This implies $\psi(T) < 0$, which contradicts the assumption that $\psi(T) = 0$. Since $X(t^*) = \widetilde{X}$ and $X(t''') < \widetilde{X}$, there would be a t_1, $t''' < t_1 \le t^*$ such that $X(t_1) = \widetilde{X}$ and $X(t) < \widetilde{X}$ for $t''' \le t < t_1$. Therefore $\dot{\psi}(t) = 2X(t) < 2\widetilde{X}$ and, thus, $\psi(t) < -2\widetilde{X}(T - t)$, $u(t) < \dfrac{D}{P}$, and finally $\dot{X}(t) < 0$, for $t''' < t < t_1$. Since $X(t''') < \widetilde{X}$, we would have $X(t_1) < \widetilde{X}$. But this contradicts the assumption that $X(t_1) = \widetilde{X}$. Therefore we must have $\psi(t') = -2\widetilde{X}(T - t')$.

We now show that $X(t) = \widetilde{X}$ and $\psi(t) = -2\widetilde{X}(T - t)$ for $t' \le t \le T$ by contradiction. Assume there existed some α_1 and α_2, $t' < \alpha_1 < \alpha_2 < T$ such that $X(t) = \widetilde{X}$ for $t' \le t \le \alpha_1$, and $X(t) \ne \widetilde{X}$ for $\alpha_1 < t \le \alpha_2$. This would mean that $\dot{X}(t) \ne 0$ at $t = \alpha_1$. But $X(t) = \widetilde{X}$ for $t' \le t \le \alpha_1$ and $\psi(t') = -2\widetilde{X}(T - t')$ should result in $\dot{\psi}(t) = 2\widetilde{X}$, $\psi(t) = -2\widetilde{X}(T - t)$, $u(t) = \dfrac{D}{P}$ and, thus, $\dot{X}(t) = 0$ for $t' \le t \le \alpha_1$ which contradicts $\dot{X}(t) \ne 0$ at $t = \alpha_1$. Therefore we must have $X(t) = \widetilde{X}$ and $\psi(t) = -2\widetilde{X}(T - t)$ for $t' \le t \le T$.

Theorem 6.2. Let $\hat{X} > X(0) \ge X^* = -\dfrac{\beta D}{P^2}$ and A, B, t_ψ, t_O satisfy the following equations:

$$AI_0\left(2\sqrt{CT}\right) + BK_0\left(2\sqrt{CT}\right) = \dfrac{2(X(0) - X^*)}{\sqrt{C}}, \tag{6.87}$$

$$AI_1\left(2\sqrt{C(T - t_O)}\right) + BK_1\left(2\sqrt{C(T - t_O)}\right) = 0, \tag{6.88}$$

$$AI_1\left(2\sqrt{C(T - t_\psi)}\right) + BK_1\left(2\sqrt{C(T - t_\psi)}\right) = 2X^*\sqrt{T - t_\psi}, \tag{6.89}$$

$$X^* = X(0) - Dt_\psi +$$

$$+\dfrac{P^2}{2\beta}\left[\dfrac{A}{\sqrt{C}}\left(I_0(2\sqrt{C(T - t_\psi)}) - I_0(2\sqrt{C(T - t_O)})\right) - \dfrac{B}{\sqrt{C}}\left(K_0(2\sqrt{C(T - t_\psi)}) - K_0(2\sqrt{C(T - t_O)})\right)\right] \tag{6.90}$$

$$\hat{X} = X^* + Dt_\psi \dfrac{I_0(2\sqrt{CT})}{I_0(2\sqrt{CT}) + I_0(\sqrt{C(T - t_\psi)}) - 2}, \tag{6.91}$$

where $C = \dfrac{P^2}{\beta}$.

Define

$$\psi_X(t) = \begin{cases} \sqrt{T-t} \cdot \left[AI_1\left(2\sqrt{C(T-t)}\right) + BK_1\left(2\sqrt{C(T-t)}\right) \right] - 2X^*(T-t), & 0 \le t < t_O \\ -2X^*(T-t), & t_O \le t \le T \end{cases}$$

(6.92)

where $I_n(z)$ is the Modified Bessel function of the first kind of order n and $K_n(z)$ is the Bessel function of the second kind of order n (Neumann function).

Then

$$u(t) = \begin{cases} 0, & 0 \le t < t_\psi, \\ \dfrac{P\psi_X(t)}{2\beta(T-t)}, & t_\psi \le t \le T \end{cases}$$

(6.93)

is optimal.

Proof: In order to show the optimality of $u(t)$, we need to prove that

(i) $\dot{\psi}_X(t) = 2X_E(t)$ and $\psi_X(T) = 0$,

(ii) $u(t)$ is feasible, and

(iii) $u(t)$ and $\psi_X(t)$ maximize the Hamiltonian (6.83).

First, $\psi_X(t)$, $0 \le t < t_O$ satisfies the following differential equation (Gradshteyn and Ryzhik 1980):

$$\ddot{\psi}_X(t) - P^2 \frac{\psi_X(t)}{\beta(T-t)} = -2D,$$

(6.94)

which can be rewritten as

$$\ddot{\psi}_X(t) = 2Pu(t) - 2D = 2\dot{X}_E(t).$$

(6.95)

One can also find from (6.87)-(6.92), that $\dot{\psi}_X(t)$ satisfies the following boundary condition

$$\dot{\psi}_X(0) = 2X_E(0).$$

(6.96)

Integrating both sides of (6.95) with respect to (6.96) shows that $\dot{\psi}_X(t) = 2X_E(t)$ for $0 \le t < t_O$. For $t_O \le t \le T$, substituting (6.92) into (6.93) leads to $u(t) = \dfrac{D}{P}$. Thus, $Pu(t) - D = \dot{X}_E(t) = 0$, which results in $X_E(t) = X^*$ for $t_O \le t \le T$. Differentiating (6.92) we show that $2X_E(t) = 2X^* = \dot{\psi}_X(t)$. Finally, it is easy to verify that $\psi_X(T) = 0$. Therefore (i) is proven.

Let us now show $u(t)$ is feasible, that is, $0 \le u(t) \le U$. First, it can be shown that $X_E(t_O) = X^*$ ((6.90) - (6.92)), $\psi_X(t_O) = -2X^*(T-t_O)$ ((6.86) and (6.92)), as well as, $\psi_X(t)$, $X_E(t)$ and $u(t)$ satisfy the remaining conditions of Proposition 6.24. According to that proposition,

$\psi_X(t) \le -2X^*(T-t)$ for $0 \le t \le t_0$. Thus, $\psi_X(t) \le -2X^*(T-t)$ and therefore $u(t) \le \dfrac{D}{P} < U$ for $0 \le t \le T$, which yields $\dot{X}_E(t) \le 0$, $0 \le t \le T$. Assume $X_E(0) > 0$ (in fact, this is ensured by the existence of t_ψ). Since $X_E(t)$ is non-increasing and $X^* < 0$, there must be a $t_X < t_0$, such that $X_E(t_X) = 0$. Therefore $X_E(t) \le 0$ and, thus, $\dot{\psi}_X(t) = 2X_E(t) \le 0$, $t_X \le t \le T$ and $\dot{\psi}_X(t) > 0$, $0 \le t < t_X$. Considering $\psi_X(T) = 0$, we have $\psi_X(t) \ge 0$, $t_X \le t \le T$. Thus $t_\psi < t_X$. Also $\psi_X(t) < 0$, $0 \le t < t_\psi$, and $\psi_X(t) \ge 0$, $t_\psi \le t \le T$. Taking (6.93) into account, we conclude that $0 \le u(t)$ for $0 \le t \le T$. Combining this with the fact $u(t) \le \dfrac{D}{P} < U$ for $0 \le t \le T$ that we have just proven, we conclude that $u(t)$ is feasible.

Finally, it is easy to observe, that $u(t)$ and $\psi_X(t)$ determined by (6.92) and (6.93) maximize the Hamiltonian (6.83).

Optimal control

The optimal control, when $X(0) \ge X^*$, is dependent upon the initial inventory $X(0)$ in the following manner.

Case 1: $X(0) \ge DT$, $u^*(t) = 0$, $0 \le t \le T$. This is the first special case in the last section. Only the first phase of the three-phase control is used.

Case 2: $DT > X(0) \ge \hat{X}$. The optimal control is defined as:

$$\psi_X(t) = A\sqrt{T-t} \cdot I_1\left(2\sqrt{C(T-t)}\right) + \frac{2D}{C}(T-t), \text{ for } 0 \le t \le T, \quad (6.97)$$

$$u^*(t) = \begin{cases} 0, & 0 \le t < t_\psi, \\ \dfrac{P\psi_X(t)}{2\beta(T-t)}, & t_\psi \le t \le T \end{cases} \quad (6.98)$$

and t_ψ is obtained by solving the following equation:

$$AI_1\left(2\sqrt{C(T-t_\psi)}\right) - 2X^*\sqrt{T-t_\psi} = 0, \quad (6.97)$$

where $A = \dfrac{2(\hat{X} - X^*)}{\sqrt{C}I_0(2\sqrt{CT})}$. Obviously, $t_\psi < T$ satisfies $\psi_X(t_\psi) = 0$.

This case has the first two phases described in Theorem 6.2: initially $u^*(t) = 0$ when $t < t_\psi$ and then $u^*(t)$ becomes positive, but still small enough so that the average inventory level $X_E(t)$ continues declining.

Since the initial inventory is relatively large, $X_E(t)$ will never reach the critical value X^* and thus the third phase of the control will not be entered.

Case 3: $\hat{X} > X(0) \geq X^*$. We have the second special case determined by Theorem 6.2 with a three-phase control, which is illustrated in Figure 6.4(a).

Note that the proof of Case 2 is very similar to that of Theorem 6.2 and thus omitted.

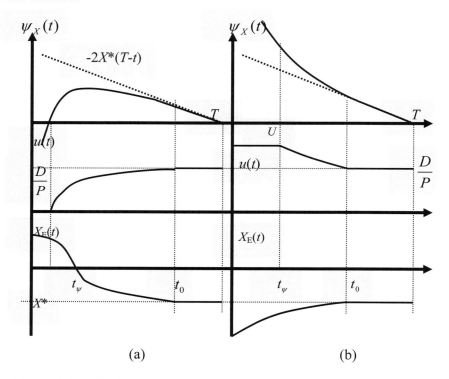

(a)   (b)

Figure 6.4. Optimal Behavior of the system for $X(0){>}X^*$ (a) and $X(0){<}X^*$ (b)

In summary, depending upon the initial inventory level, the optimal control may have up to three phases. In the first phase, the optimal production rate is either at its maximum or its minimum, as in the traditional threshold control. The optimal production rate in the second phase is determined by a set of complex non-linear equations containing Bessel functions. In the third phase, similar to the traditional threshold control, the system enters a steady state characterized by a constant production rate inversely proportional to the expected yield.

REFERENCES

Anderson MQ (1981) Monotone optimal preventive maintenance policies for stochastically failing equipment. *Naval Research Logistics Quarterly* 28(3):347-58.

Akella R, Kumar PR. (1986) Optimal control of production rate in a failure prone manufacturing system. *IEEE Transactions on Automatic Control*, AC-31(2): 116-126.

Boukas EK, Liu ZK (1999) Production and maintenance control for manufacturing system. *Proceedings of the 38th IEEE Conference on Decision and Control, IEEE* vol.1: 51-6 1999.

Boukas K, Haurie AR (1988) Planning production and preventive maintenance in a flexible manufacturing system: a stochastic control approach. *Proceedings of the 27th IEEE Conference on Decision and Control,* vol. 3: 2294-300.

Cho DI, Abad PL, Parlar M (1993) Optimal production and maintenance decisions when a system experiences age-dependent deterioration. *Optimal Control Applications & Methods* 14 (3):153-67, 1993.

Gerchak Y, Vickson R, Parlar M. (1988) Periodic review production models with variable yield and uncertain demand. *IIE Transactions* 20: 144-150.

Gradshteyn IS, Ryzhik IM (1980) *Tables of Integrals; Series and Products*, Academic Press, London.

Haurie A (1995) Time scale decomposition in production planning for unreliable flexible manufacturing system. *European Journal of Operational Research* 82: 339-358.

Khmelnitsky E, Caramanis M (1998) One-Manufacturer N-Part-Type Optimal Set-up Scheduling: Analytical Characterization of Switching Surfaces. *IEEE Transactions on Automatic Control* 43(11): 1584-1588.

Khouja M (1999) The Single Period (News-Vendor) Inventory Problem: A Literature Review and Suggestions for Future Research. *Omega* 27: 537-553.

Kimemia JG (1982) Hierarchical control of production in flexible manufacturing systems, Ph.D. dissertation, Mass. Inst. Technol., Cambridge, MA.

Kimemia JG, Gershwin SB (1983) An algorithm for the computer control of production in flexible manufacturing systems. *IEEE Transactions on Automatic Control* AC-15: 353-362.

Kloeden P, Platen E (1999) *Numerical Solution of Stochastic Differential Equations*, Springer.

Kogan K, Khmelnitsky E (1998). Tracking Demands in Optimal Control of Managerial Systems with Continuously-divisible, Doubly Constrained Resources. *Journal of Global Optimization* 13: 43-59.

Kogan K, Lou S (2003) Multi-stage newsboy problem: a dynamic model. *EJOR* 149(2): 2003, 448-458.

Kogan K,. Lou S (2005) Single Machine with Wiener Increment Random Yield: Optimal Off-Line Control. *IEEE Transactions on Automatic Control* 50(11): 1850-1854.

Kogan K, Lou S, Herbon A (2002) Optimal Control of a Resource-Sharing Multi-Processor with Periodic Maintenance. *IEEE Transactions on Automatic Control* 47(6): 1342-1346.

Liu L, Pu C, Schwan K, Walpole J (2000) InfoFilter: supporting quality of service for fresh information delivery. *New Generation Computing*, 18(4): 305-321, 2000.

Maimon O, Khmelnitsky E, Kogan K (1998) *Optimal Flow Control in Manufacturing Systems: Production Planning and Scheduling*, Kluwer Academic Publisher, Boston.

Shu C, Perkins JR (2001) Optimal PHP production of multiple part-types on a failure-prone manufacturer with quadratic warehouse costs. *IEEE Transactions on Automatic Control* AC-46: 541-549.

Silver EA, Pyke DF, Peterson R (1998). *Inventory Management and Production Planning and Scheduling*, John Wiley & Sons.

Pinedo M (1995) *Scheduling: Theory, Algorithms and Systems*, Prentice Hall, New Jersey.

Wang, Y, Gerchak Y. (2000). Input control in a batch production system with lead times, due dates and random yields. *European Journal of Operational Research* 126: 371-385.

Yano CA, Lee HL (1995) Lot sizing with random yields: A review. *Operations Research* 43(2): 311-334.

PART III
RISK AND SUPPLY CHAIN MANAGEMENT

7 RISK AND SUPPLY CHAINS

Risk results from the direct and indirect adverse consequences of outcomes and events that were not accounted for or that we were ill prepared for, and concerns their effects on individuals, firms or the society at large. It can result from many reasons both internally induced and occurring externally with their effects felt internally in firms or by the society at large (their externalities). In the former case, consequences are the result of failures or misjudgments while in the latter, consequences are the results of uncontrollable events or events we cannot prevent.

A definition of risk and risk management involves as a result a number of factors, each reflecting a need and a point of view of the parties involved in the supply chain. These are:

(1)   Consequences, individual (persons, firms) and collective (supply chains, markets).
(2)   Probabilities and their distribution, whether they are known or not, whether empirical or analytical and based on models or subjecttive.
(3)   Individual preferences and Market-Collective preferences, expressing a subjective valuation by a person or firm or organizationally or market defined—its price.
(4)   Sharing and transfer effects and active forms of risk prevention, expressing risk attitudes that seek to alter the risk probabilities and their consequence, individually or both.

These are relevant to a broad number of professions, each providing a different approach to the measurement, the valuation and the management of risk which *is motivated by real and psychological needs and the need to deal individually and collectively with problems that result from uncertainty and the adverse consequences they may induce and sustained in an often unequal manner between individuals, firms, a supply chain or the society at large.* For these reasons, risk and its management are applicable to many fields where uncertainty primes (for example, see Tapiero 2005a, 2005b). In supply chains, these factors conjure to create both a conceptual and technical challenge dealing with risk and its management.

7. 1 RISK IN SUPPLY CHAINS

Risk management in supply chains consists in using risk sharing, control and prevention and financial instruments to negate the effects of the supply chain risks and their money consequences (for related studies and applications see Anipundi 1993, Christopher 1992, Christopher and Tang 2006, Lee 2004, Tayur et al. 1998, Eeckouldt et al. 1995, Hallikas et al. 2004). For example, Operational Risks concerning the direct and indirect adverse consequences of outcomes and events resulting from operations and services that were not accounted for, that were ill managed or ill prepared for. These occur from many reasons, both induced internally and externally. In a former case, consequences are the result of failures in operations and services sustained by the parties individually or collectively due either to an exchange between the parties (in this case an endogenous risk) or due to some joint (external) risks the firms are confronted by. In the latter case it is the consequence of uncontrollable events the supply chain was not ready for or is unable to attend to. The effect of risk on the performance of supply chains can then be substantial arising due to many factors including the following.

- Exogenous (external) factors—factors that have nothing to do with what the supply chain firms do but due to some uncontrollable external events (a natural disaster, a war, a peace, etc.);
- It may be due to controllable events—endogenous, either because of human errors, mishaps of operating machines and procedures or due to the inherent conflicts that can occur when organization and persons in the supply chain may work at cross purpose. In such conditions, risk can be motivated, based on agents and firms' intentionality;
- It may be due to information asymmetry, leading adverse selection and moral hazard (as we shall see below and define) that can lead to an opportunistic behavior by one of the parties which have particular implications for the management of the supply chain;
- It may result from a lack of information, or the poor management of information and its exchange in the supply chain, such as forecasting—individually and collectively, the supply chain need, demands etc.;
- It may express a perception, where a risk attitude (by a party or a firm) may confer risk to events that need not be risky and vice versa. Risk attitude is then imbedded in a subjective perception of events that may be real or not;

- It may be the result of measurements—both due to the definition of its attributes or simply errors in their measurement.

For example, some agents may convey selective information regarding products or services and thereby enhance the appeal of these products and services. Some of this information may be truthful, but not necessarily! Truth-in-advertising and truth (transparency) in lending for example are important legislations passed to protect consumers just as truth in exchange and in transparency are needed to sustain a supply chain. In many cases however, it might be difficult to enforce. Further, usually firms are extremely sensitive to negative information or to the presumption that they have been misinformed. Such situations are due to an uneven distribution of information and power among the parties and induce risks we coin "Adverse Selection" and "Moral Hazard". Risk, information and information asymmetry are thus important issues supply chains are concerned with and are therefore the topic of essential interest and management.

Adverse Selection and "The Lemon Phenomenon"

Akerlof (1970) has pointed out that goods of different qualities may be uniformly priced when buyers cannot realize that there are quality differences. For example, one may buy a used car, not knowing its true state, and therefore the risk of such a decision may induce the customer to pay a price which would not reflect truly the value of the car and therefore misprice the car. In other words, when there is such an information asymmetry, valuation and prices are ill defined because of the mutual risks that exist due to the buyer and seller specific preferences and the latter having a better information. In such situations, informed sellers can resort to opportunistic behavior. Such situations are truly important. They can largely explain the desires of firms to seek "an environment" where they can trade and exchange in a truthful and collaborative manner. Some buyers might seek assurances and buy warranties to protect themselves against post-contract failures or to favor firms who possess service organizations (in particular when the products are complex or involve some up-to-date technologies). As a result, for transactions between producers and suppliers, the effects of uncertainty lead to dire needs to construct long-term and trustworthy relationships as well as a need for contractual engagements to assure that "the contracted intentions are also delivered". Such relationships may lead, of course, to the "birth" of a supply chain.

The Moral Hazard Problem

A characteristic that cannot be observed induces a risk to the non-informed. This risk is coined "Moral Hazard" (Holstrom, 1979, 1982; Hirschleifer and Riley 1979). For example, possibly, a supplier (or the provider of a service) may use such a fact to his advantage and not deliver the contracted amount. Of course, if we contract the delivery of a given level of quality and if the supplier does not knowingly maintain the terms of the contract that would be cheating. We can deal with such problems with various sorts of (risk-statistical) controls combined with incentive contracts which create an incentive not to cheat or lie. If a supplier were to supply poor quality and if it were detected, the supplier would then be penalized accordingly (according to the agreed terms of the contract or at least in his reputation and the probability that buyers will turn to alternative suppliers). If the supplier unknowingly provides products which are below the agreed contracted standard of quality, this may lead to a similar situation, but would result rather from the uncertainty the supplier has regarding his delivered quality. This would motivate the supplier to reduce the uncertainty regarding quality through various sorts of controls (e.g. through better process controls, outgoing quality assurance, assurances of various sorts and even service agreements) as will be discussed in the next chapter in far greater details. For such cases, it may be possible to share information regarding the quality produced and the nature of the production process (and use this as a signal to the buyer). For example, firms belonging to the same supply chain may be far more open to the transparency of their processes in order to convey a message of truthfulness. A supplier would let the buyer visit the manufacturing facilities as well as reveal procedures regarding the controls it uses, machining controls, the production process in general as well as the IT Technologies it has in place.

Examples of these risk prone problems are numerous. We outline a few. An over-insured logistic firm might handle carelessly materials it is responsible for; a warehouse may be burned or looted by its owner to collect insurance; a transporter may not feel sufficiently responsible for the goods shipped by a company to a demand point etc. As a result, it is necessary to manage the transporter and related relationship and thereby manage the risks implied in such relationships. Otherwise, there may be adverse consequences, leading, for example, to a greater probability of transports damage; leading to the "de-responsabilization" of parties or agents in the supply chain and inducing thereby a risk of moral hazard. It is for this reason that incentives, controls, performance indexation to and "on-the-job and co-responsibility" are so important and needed to minimize the risks of Moral hazard (whether these are tangible or intangible). For example, in decentralized supply chains, getting firms to assume fully their responsibilities may also be a means to care a little more for the supply chain overall performance and in everything they do that affects other firms' performance as well. A supplier who has a long term contract might not care to supply performing parts for the buyer who is locked-into such a relationship (contract). In such situations, co-dependence and collaboration provide some of means needed to mitigate these risks.

Throughout these examples, there are negative inducements to performances. To control or prevent these risks, it is necessary to proceed in a number of manners. The concern of a supply chain organizational design is a reflection of the need to construct relationships which do not induce counter-productive acts. For example, the demand for ever greater performance at lower risk has induced firms to define various organizational frameworks for supply chains which are altering the nature of doing business. GM and its suppliers are working closely together, albeit with many control and counter measures to assure managed exchanges. The same applies to almost any major manufacturer. It is increasingly believed that the reduction of the number of suppliers is leading to a sort of semi-integration and exchanges between producers and suppliers and therefore to some other risks arising due to potential moral hazard. Such exchanges are bi-directional relating to:

- Information which prevents faulty operations and services;
- Information needed for in-process services and operations;
- Information which induces collaborative exchange.

Steps that are important and can be followed include: Detecting signals of various forms and origins to reveal the supply chain agents' behaviors, rationality and performances; A greater understanding of the supply chain's intentions which can lead to a better design of the supply chain organization and its information systems; Managing and controlling the relationship between the supply chain's firms', their employees and workers. Earlier for example, we saw that information asymmetry can lead to opportunistic behavior such as cheating, lying and being counter productive, just because there may be an advantage to doing so without having to sustain the consequences of such a behavior. Developing an environment which is cooperative, honest, open and which leads to a frank exchange of information and optimal performances is thus a necessity to sustain a supply chain's existence. For additional references and applications in supply chains pertaining to issues of information and supply chains, the reader may consult as well Boone et al 2002, Cachon and Fisher 2000, Cheng and Wu 2005, Agrell et al. 2004, Aviv 2004, 2005). These problems are fundamentally important in designing and managing the contractual relationships that define the supply chain and its operations and applied in the many contexts in which supply chains operate (see also Agrawal and Shesadri 2000, Cohen and Agrawal 1999, Corbett 2001, Corbett and Groote 2000, Corbett and Tang 1998, Desiraju and Moorthy 1997, Lee et al. 1997a, 1997b).

Signaling and Screening

In conditions of information asymmetry, one of the parties may have an incentive to reveal some of the information it has. The seller of a product may have or have not an interest in making his product transparent to the buyer. He may do so in a number of ways, such as pricing it high and therefore conveying the message to the potential buyer that it is necessarily (at that price) a high quality product (but then, the seller may also be cheating!). The seller may also spend heavily on advertising the product, claiming that it is an outstanding product with special attributes and thereby inducing sales justified by the product quality (but then the seller my also be lying!). Claiming that the product is just "great" may be insufficient. Not all buyers are gullible. They require and look for signals that reveal the true properties of the product. Pricing, warranties, reputation and principles, advertising, are some of the means used by sellers to send signals. For example, the seller of a lemon with a warranty will eventually lose money. Similarly, a firm that wants to limit the entry of new competitors may signal that its costs are very low (and so if they decide to enter, they are likely to lose money in a price battle). Advertising heavily may be recuperated only through repeat purchase and therefore, over-advertising may be used as a signal that the over-advertised products are of good quality.

Uninformed parties, however, have an incentive to look for and obtain information. For example, shop and compare, search reliable suppliers etc. are instances of information seeking by uninformed parties. Such activities are called *screening*. A driver that has a poor accident record history is likely to pay a greater premium. If characteristics of customers are unobservable, firms can use self selection constraints as an aid in screening to reveal private information. For example, consider the phenomenon of rising wage profiles where workers get paid an increasing wage over their careers. An explanation may be that firms are interested in hiring workers who will stay for a long time. Especially if workers get training or experience which is valuable elsewhere this is a valid concern. Then they will pay workers below the market level initially so that only "loyal" workers will self select to work for the firm. Similar arguments may be used in selecting supply chain parties.

The classic example of "signaling" points out that high productivity individuals try to differentiate themselves from low productivity ones, by the amount of education they acquire. In other words, only the most productive workers invest in education. This is the case because the signaling cost to the productive workers is lower than to low productivity workers and therefore firms can differentiate between these two types of workers. By the same token, individual, parties, supply chains are very jealous of their reputation and the maintenance of their standards of services so that they can differentiate themselves from competitive alternatives.

7.2 RISK PRACTICE IN SUPPLY CHAINS

A search for "Supply Chains Risks" on internet will reveal a large number of interviews with practitioners, individual and academic contributions, consulting firms and papers that seek to bring attention to what supply chain managers are calling attention to—supply chains derived risks. For example, Chris D. Mahoney (UPS, October 2004, www.ism.ws/Pubs/ISMMag/100406.cfm) points out that "many companies have worked hard to streamline their supply chains. They've whittled down the field and built relationships with only the most competent suppliers. And many have gotten the desired result—supply chains that run like clockwork, reducing costs and bolstering customer service. But it turns out there's a downside—greater risk". These risks are also more complex, arising for many reasons transcending the traditional concern for operational (intra firm) and external (hazard) risks. Risks previously neglected have expanded because of supply chains entities dependencies, political, strategic and risks externalities, augmenting thereby the importance of their assessment and their management. For example, the unending drive for lean manufacturing, to reduce inventory, single sourcing of raw materials, or adopting just-in-time (JIT) manufacturing and delivery techniques, while cutting costs has also contributed to the size and adverse effects of supply chains risks. Greater attention and management to these risks is therefore needed both because of the potential catastrophic costs these risks imply and because the drive to expand and streamline into lean and cost-reducing supply chains has ignored these risks. Mahoney for example, raises the following questions:

> "If your main distribution center or plant sustained substantial damage, how much time would it take you to bounce back? How much inventory would you lose and what are the costs of recouping it? If inventory loss is sizeable, how rapidly can you adjust production lines and plans to accommodate new production goals? Can key suppliers ramp up swiftly? Or, if a product is de-emphasized, how will they handle the revenue loss? How much

revenue would your company stand to lose if order taking and filling were to come to a halt for a week, two weeks or a month? What are the legal and financial ramifications of being unable to satisfy contracts? How will your market share and brand be affected in the long-term? What sales and marketing initiatives will you need to adopt to handle customers, recoup revenues, and reclaim lost market share and goodwill?"

According to Mahoney, we require more risk management, more supply chain integration and stakeholder management, and network capacity. However the "answer is always part of the problem" and risk management in supply chains will need far more strategic and senior management involvement to provide directives for dealing with the following issues (Marsh's consulting Risk-Adjusted Supply Chain Practice www.marsh riskconsulting. com/st/):

- Do we fully understand the dependencies within our supply chain?
- Have we identified the weak links within our supply chain?
- Do we understand the risk that has been inadvertently built into our supply chain?
- Have we identified the supply chain risks that we might be able to mitigate, eliminate, or pass on to another supply chain member?
- Do we incorporate the element of risk when making strategic or tactical decisions about our supply chain?
- Is our supply chain nimble and flexible so that we can take advantage of both supply chain risks and opportunities?
- Have we fully captured our enterprise-wide risk profile?
- Do we know which supply chain risks may cause an adverse event that could cause a significant disruption to our supply chain?
- Do we have the necessary tools, skills, and resources to model our supply chain, including its risks and vulnerabilities, in order to understand the financial impact that various events and scenarios will have on our supply chain?
- Do we benchmark the activities that make up our supply chain?
- Have we identified — and do we monitor — key risk indicators of upstream or downstream activities that might result in a disruption in the supply chain?
- Have we fully integrated our business contingency plans and emergency response plans into our supply chain management initiatives?

Overwhelmingly, and as discussed earlier, supply chains are based on co-dependency and collaboration exchange between firms, each drawing a payoff whose risks it must also sustain and manage in as many ways as it may be able to measure and conjure. Collaboration for example, is a well trumpeted mechanism for maximizing payoffs while at the same time managing co-dependence risks that firms engaged in supply chain exchanges are dealing with. Collaboration is not always possible however, as presumed in our previous chapters (where Nash and Stackelberg solutions were used) for agreements that may be difficult to self-enforce. As a result co-dependence risks are strategic and can potentially be overwhelming for firms that operate in and out of supply chains.

By the same token, strategic focusing and outsourcing by firms while justified on some theoretical and economic grounds induce their own risks. These issues, specific to supply chains combined with the operational and external risks that supply chains are subject to, require that specific attention be directed to their measurement and to their management. Such measurement requires a greater understanding of firms' motivations' in entering supply chain relationships and the factors that determine their dependence risks (Bank 1996, Tapiero 2005a). However, the growth and realignment along supply chains of corporate entities in an era of global and strategically focused and market sensitive strategies is altering the conception of Corporate Risk in Supply Chains. Some of these risks are well known and well documented, including:

- Operational, and
- External-hazards risks
- Risks of globalization,
- Financial markets risks
- Strategic risks as well as
- Technological risk,
- Sustainability and risk externalities.

While risk exposure and risk management may use our abilities to deal ex-ante and ex-post with the adverse consequences of uncertainty (risk), the measurement and the valuation-pricing of risks remains a challenge (although great strides have been achieved in using financial instruments and real options, see Tapiero 2004 for example). If risk is money valued by some actor-agent, it need not be valued equally by the agents involved and collaborating in the supply chain, leading thereby to latent supply chain asymmetries with dire consequences to the management of risks as indicated earlier. There are, of course, non money measures such as measurements of variability (variance, semi variance, range etc. and other statistical and probability based measures). How valuable are these risks to firms? How valuable are they to the supply chain? And how does the market mechanism value-price these risks? It is through such a valuation and its price (the risk premium) that events assume a consequence defined as risk. For example, is the loss of capacity a risk measure? Is the cost of losing a client ...a risk measure? Is a demand's standard deviation a risk measure? These terms are risk-meaningful only to the extent that we are conscious of their effects and amplified by our ability to measure, value and price their consequences. Supply Chain Managers have thus an important role to assume in defining supply chain risks, in measuring and valuing these risks and "internalize them" in the costs and benefits calculations they are using to reach decisions and draw the essential attention that supply chains risks deserve. For related problems in supply chain see also Agrawal and Nahmias 1998, Akella et al 2002, Anupundi and Akella 1993, Harland et al 2003.

7.3 SUPPLY CHAIN RISKS AND MONEY

Risk is a consequence, expressing the explicit and latent objectives of the firm and the supply chain. In supply chains, as well in markets, the unit of exchange is essentially "money" and therefore, risk is ultimately measured, valued and managed by money. The concern of "Total" approaches in industrial management that seek to account for all potential risk effects—direct and indirect ones, are a departure from the traditional approaches, that recognize the significant performance effects of risks but do not always provide a quantification of their value nor recognize the derived (indirect) dependencies of such effects. Further, since we can only value what we can be aware of or can measure and inversely we can measure only what we value", attention has been directed to problems that are easier to identify rather than to the strategic problems that occur in supply chains. For example, quality measurements in industry have emphasized primarily non-quality because it can be measured, as a result, industrial managers have mostly been oblivious to "good quality" because they were hardly measured. By the same token, while the unknown demand for a product might be a source of (inventory) risk, it is not a risk. It is a measured risk when the consequence of such a demand uncertainty can be assessed in money terms. Similarly, if a party A is not responsible and does not pay for the costs it has inflicted to another party B in the supply chain, these costs will not be defined as a risk for that first party A. In this sense, risk measurement implies at the same time its risk definition and its valuation. Such an approach provides a far greater justification and incentive for performance measurement which becomes extremely important and in some cases may contribute to the growth of supply chains. For example, Barzel (1982), points out that "when two inputs have to be measured at two successive junctures, a rationale for an integrated (supply chain- our insertion) firm emerges". In this sense, measurement has its own error sources and commensurate risks which induce firms to network in order to work and operate in an environment where measurements and their risks are reduced.

Risk management in supply chains is, as a result, multi-faceted. It is based on both theory and practice. It is conceptual and technical, blending behavioral psychology, financial economics and decision making under uncertainty into a coherent whole that justify the selection of risky choices and manages their consequential risks. Its applications are also broadly distributed across many areas and fields of interest. The examples we shall treat in the next section are meant to highlight a number of approaches that address specific concerns in supply chains. In the next chapter we shall be concerned in particular with the management of quality in supply chains, recognizing the strategic aspects of risks in a supply chain. The approaches used are equally applicable to a broad variety of problems supply chains are confronted with.

In a supply chain, the management of risks is both active and reactive, requiring on the one hand that actions be taken to improve a valuable process and, on the other, preventing recurrent problems. Generally, problems occur for a number of reasons, enriched by the complexity supply chains induce. These include:

- Unforeseeable events we are ill prepared to cope with.
- Adversarial situations resulting from a conflict of interests between contract holders. Adversarial relationships combined with private information leading to situations when one might use information to the detriment of the other and thereby possibly resulting to opportunistic behavior.
- Information asymmetries inducing risks (moral hazard and adverse selection) that affect the supply chain parties and society at large.
- Oversimplification of the problems involved or their analysis (which is often the case when the problems of globalization are involved). Such oversimplification may lead to erroneous assessments of uncertain events and thereby can lead to the participants to be ill prepared for their consequences.
- Information is not available or improperly treated and analyzed. This has the effect of inducing uncertainty regarding factors that can be properly managed and, potentially leading to decisions that turn out to be wrong. Further, acting with no information breeds incompetence that contributes to the growth of risk.
- Poor organization and control of processes. For example, a process that does not search for information, does not evaluate outcomes

- and situations that can be costly. Similarly, a myopic approach with no controls of any sort, no estimation of severity and consequences and no long run evaluation of consequences can lead to costs that could have been avoided.
- Non adaptive procedures to changing events and circumstances. Decision makers oblivious to their environment, blindly and stubbornly following their own agenda are a guarantee for risk.

These problems recur in many areas and thus one can understand the universality and the importance of risk and its management.

The definitions of risk, risk measurement and risk management are closely related, one feeding the other to determine the proper-optimal levels of risk. Economists and Decision Scientists have attracted special attention to these problems. In this process a number of tools are used based on:

- Ex-ante risk management
- Ex-post risk management and
- Robust Design

Ex-ante risk management involves "before the fact" application of various tools such as: Risk sharing and transfer; Preventive controls; Preventive actions; Information seeking; Statistical analysis and forecasting; Design for reliability; Insurance etc..

"After the fact" or Ex-post risk management involves, by contrast, control audits and the design of flexible-reactive schemes that can deal with problems once they have occurred to limit their consequences. Option contracts for example, are used to mitigate the effects of adverse movements in stock prices (but not only as we saw earlier). With call options, in particular, the buyer of the option limits the downside risk to the option price alone while profits from price movements above the strike price are unlimited. For example, to manage the price of supplies and their delivery, a buyer may buy options contract (see Ritchken and Tapiero 1984).

Robust design, unlike ex-ante and ex-post risk management seeks to reduce risk by rendering a process insensitive (i.e. robust) to its adverse consequences. If a supplier fails due to an unforeseeable event, the effects of this event can be reduced by providing contingent supply opportunities. Technically, *risk management desirably alters the outcomes a party of the supply chain can be confronted with and reduce negative consequences to planned or economically tolerable levels.*

There are many ways to reach this goal, however each discipline devises the tools it can apply. For example, engineers and industrial managers apply reliability design and quality control techniques and TQM (Total Quality

Management) approaches to reduce the costs of poorly produced and poorly delivered products etc. as we shall see in the next chapter. Examples to these approaches in risk management in these and other fields abound. In supply chains, one may consult additionally AON 2005, Babich et al. 2004, Chopra 2004, Kleindorfer and Saad 2005, Nagurney et al. 2005, Parlar and Perry 1996, Rice and Caniato 2003, Shefy 2001 and Zsidisin et al 2001).

Controls such as audits, statistical quality and process controls (as will be seen in the next chapter) for example are applied ex-ante and ex-post. They are exercised in a number of ways in order to rectify processes and decisions taken after non-conforming events have been detected and specific problems have occurred. Auditing a firm in a supply chain, controlling a product performance over time etc. are simple examples of controls sought.

7.4   RISK VALUATION

Risk and uncertainty in supply chains are treated and valued in many disparate ways, conceptually and technically different. The approaches we use are subjective (personal) and objective (based on a market attitude and valuation-pricing). There is an extensive body of knowledge and numerous references that treat these problems. In supply chains, the treatment of risk and uncertainty assumes an added complexity however due to the strategic interactions (and game-like situations) that the parties of the supply chain are engaged in as we saw in previous chapters. We shall elaborate below on a number of cases to highlight some approaches we might use in analyzing "supply chain problems" when risk is an essential consideration to reckon with. In particular, we shall consider again Problem 2.1 of Chapter 2 and extend it in several directions emphasizing the introduction of uncertainty in the supplier and the retailer decisions. Second, we shall use a common VaR risk approach to an inventory problem in a supply chain (see also Jorion 2000, Alessandro et al. 2005, Tappiero 2000, 2005). A third problem considers an outsourcing problem while a final and fourth problem outlines an approach to (financially) market pricing a franchise contract. The problems are by no means exhaustive and are used instead to demonstrate a number of techniques adapted to our analysis of supply chain problems under risk and uncertainty. Prior to dealing with these problems, a brief review of valuation approaches is outlined.

Expected Utility and Risk Attitudes

Expected utility (EU) is a traditional way economists use to evaluate a random prospect or discrete payments made or received, described by "lottery" L or vector $(x_1,p_1; x_2,p_2; \ldots\ldots; x_n,p_n)$, meaning that the consequence x_1 is obtained with probability p_1, etc. Such a discrete lottery may designate for example a type of risk (investment, an industrial performance, a demand for a product, a machine's failure etc.). From the point of view of an individual endowed with a utility function $u(\cdot)$ and under $Eu(.)$ (the expected utility) it is worth $\sum_1^n p_i u(x_i)$, or: $Eu = \sum_1^n p_i u(x_i)$. In such an expression, the utility function imbeds the decision maker (investor, manager) preference for money. Further, in seminal and individual papers, Pratt and Arrow have shown that one version of the attitude toward risk can be recovered from the explicit specification of the utility function. Assume an individual endowed with a constant (non risky) asset C ("a bird in the hand") as well as with a potential and favorable prospect ("the bird in the bush"), a lottery $\tilde{x}$, i.e. with expected value with $E(\tilde{x}) > 0$. The risk attitude of a decision maker could then be characterized by that amount, priced p_a, the decision maker would be willing to *sell* the lottery or, in other words, the smallest *non random amount* he would be willing to *receive* to remove the risk faced (an amount often loosely called the "cash equivalent" of the lottery to the given individual decision maker). Using then the expected utility rule for reaching decision when faced with uncertainty (or the EU rule), we have a value equivalence between two different prospects, one certain and the other uncertain, or:

$$u(C + p_a) = E[u(C + \tilde{x})] \text{ or } p_a = u^{-1}(E[u(C + \tilde{x})]) - C \qquad (7.1)$$

where p_a denotes the risk premium. An alternative manner for expressing this premium would be to equate it to that amount of money the decision maker would be willing to pay for obtaining for sure the expected payoff of the lottery, namely, $p_a + CE = E(\tilde{x})$, where the certainty equivalent is a function of the prospects and, of course, the utility function, or $CE = \pi(C, \tilde{x})$ and thereby,

$$p_a = E(\tilde{x}) - \pi(C, \tilde{x}) \qquad (7.2)$$

Note that, if a decision maker seeks some risk protection (say by insurance), he (she) will accept to *pay* the risk premium defined by p_a. In this particular case, the decision maker displays a risk aversion. Note that to be accepted, the premium will be *at most*: $p_1(C, \tilde{x}) = -p_a = \pi(C, \tilde{x}) - E(\tilde{x})$. Finding a convenient analytical expression of $\pi(C, \tilde{x})$ seems thus an essential question to solve (albeit it is completely defined by the utility function and its inverse). Unfortunately, this is a very complicated function for discretionarily large values of $\tilde{x}$. Fortunately, Pratt showed that if $\tilde{x}$ displays finite variance and if its range is not too large with respect to C, then π can be satisfactorily approximated by the expression:

$$\pi \cong \frac{-u''[C + E(\tilde{x})]}{u'[C + E(\tilde{x})]} \cdot \frac{\sigma^2(\tilde{x})}{2} = \frac{-u''(W)}{u'(W)} \cdot \frac{\sigma^2(\tilde{x})}{2} = A(W) \cdot \frac{\sigma^2(\tilde{x})}{2}, \quad (7.3)$$

where

$$W = C + E(\tilde{x}), \; A(W) = -\frac{u''(W)}{u'(W)} = \frac{d}{dW} \log u'(W). \quad (7.4)$$

Where W is the expected wealth, which reduces obviously to C when $\tilde{x}$ is actuarially fair (i.e. has zero mean). This result can be proved easily by replacing the uncertain prospect $\tilde{x}$ with $\tilde{x} = E(\tilde{x}) + \varepsilon$, where ε is a zero mean random variable with finite variance $\sigma^2(\tilde{x})$. From the equation above, we see that any decision maker endowed with a concave utility function will have positive $A(W)$ and π, as $u'(W)>0$ for all possible values of W (by definition of a utility function) and as variances are always positive, the sign of π is opposite to that of $u''(W)$, i.e. π will be positive if and only if $u''(W)<0$, which characterizes a concave utility function. Thus, there exists a simple link between the complicated function in equation denoting π and the local approximation.

The function $A(W)$ is known as the *Arrow-Pratt coefficient of absolute risk aversion* (hereunder *ARA*). As our equations show, the larger $A(W)$, the smaller the *cash (certainty) equivalent* (or *selling price*) of any given lottery $\tilde{x}$. By definition therefore, a decision maker is said to be *locally risk averse* if and only if he (she) displays positive π and hence positive $A(W)$, *locally risk prone* (or risk lover) if she (he) displays negative π (and hence negative $A(W)$), locally risk neutral if he (she) displays null π (and hence null $A(\tilde{W})$). This corresponds respectively to a concave, convex, affine (or linear) utility function.

Risk can also appear as a multiplicative factor, for example when an amount a of the asset is invested with a risk favorable return $\tilde{g}$, the rest being held as cash C. Expected wealth (or the objective we are concerned with) is then equal to:

$$W = E(C + a.\tilde{g}). \quad (7.5)$$

Maximizing expected utility leads then to a *relative* risk premium $\pi^* \approx A^*(W).\sigma^2(g)/2$ by a similar computation. The relevant coefficient of *relative* local risk aversion is dimensionless (a percentage ratio, hereunder called *RRA*) is denoted here by $A^*(W)$, with:

$$A*(W) = W . \frac{-u''(W)}{u'(W)} = W . A(W) . \tag{7.6}$$

There are additional expressions and extensions of this approach, accounting for behavioral and psychological attitudes and profusely used in practice. We refer to the following references for further study however (see the bibliographical list at the end of this chapter). Instead, we shall focus on some of their implications and applications in problems that portent to supply chain management issues.

The Principle-Agent Problem in Supply Chains

Consider a supply chain consisting of two parties, operating in a co-dependent manner, with one party reaching a set of decisions and the other reaching another set of decisions, whose outcomes affect the parties' performance (Riordan 1984). Such situations can coexist with power and information asymmetries of one party over the other. Thus, while each of the parties may be free to reach some decisions, the consequences of these decisions depend on both parties' decisions with one of the party subjugated to the other in some manner or one party having some informational advantage over the other. Such situations lead to "Principal and Agents" problems. For example a supplier may be the "agent" for a firm—the producer who acts as a "principal", trusting the supplier to perform his job in the interest of the producer. In such a situation, the producer can use such an advantage to tailor his actions by taking into consideration the actions pursued by the agent—the supplier. Similar situations arise between a firm and its salesmen (some operating on commissions, quotas and difficult to manage). Salesmen effort allocations can be observed by the firm which uses such information to manage and motivate salesmen for greater efforts and profits. Similarly, in many situations, a firm belonging to a supply chain may provide incentives to motivate another firm to exchange with "the-agents" to be more sensitive to the firm's needs. In these situations, the actions taken by the latter firm may be observed only imperfectly. The principal agent problem consists then in determining the rules for sharing the outcomes resulting from say a supplier and a producer interactions. Asymmetry of information in such cases can lead of course to a situation of potential moral hazards and therefore to risks that firms in a supply chain seek to manage. To do so, there are several approaches, generally based on the design of an appropriate incentive system and audits (controls). Given the substantial risks this implies that such "incentive and control" systems are of great practical importance in the design and in the management of supply chains. In the next chapter, we shall focus our attention on such "systems" for the management of quality. Below, we shall merely elaborate on a simple utility quantification of the problem which parallels facets of the Stackelberg strategy used throughout this text.

A first approach to this problem which we consider here, coined *the first order approach* consists in the following. Let $\tilde{x}$ be a random variable which represents the gross return obtained by say, a producer--the principal. The principal-agent problem consists then in determining the amount transferred to the agent by the principal in order to compensate him for the efforts he has performed on behalf of the principal. To do so, assume that the agent utility is separable and is given by:

$$V(\tilde{y},a) = v(\tilde{y}) - w(a); \; v' > 0, v'' \le 0, w' > 0, w'' > 0 , \tag{7.7}$$

where v', v'', w', w'' are the first order and second order derivatives of the agent and the principal respectively and $\tilde{y}$ is the return of the agent, a random variable as well. In order to assure the agent's participation, it is necessary to give him at least an expected utility greater than the cost of the agent's effort, or:

$$EV(\tilde{y},a) \ge 0 \;\; or \;\; Ev(\tilde{y}) \ge w(a) . \tag{7.8}$$

In this case, the utility of the principal is:

$$u(\tilde{x} - \tilde{y}), u' > 0, u'' \le 0 . \tag{7.9}$$

Note that the problems we formulate depend then on the information distribution between the principal and the agent. Various assumptions pertaining to what one party knows of the other will lead of course to alternative problem formulations.

For simplicity, let the expected value of $\tilde{x}$ be x, agent's share y be a decision variable (replacing $\tilde{y}$), w be independent of a as well as the agent's effort a be observable to both parties and is proportional to the share y. That is, assume we are interested to determine a sharing rule $(x\text{-}y, y)$ and the corresponding proportional agent's effort. In this case, the problem of the producer-principal is formulated as follows.

$$\max_{y}[u(x - y)] \; s.t. \; v(y - w) \ge 0 . \tag{7.10}$$

By applying Kuhn-Tucker conditions for optimality to (7.10), the optimal solution is found to be:

$$\frac{u'(x - y)}{v'(y - w)} = \lambda , \tag{7.11}$$

which provides the sharing rule between the principal (receiving $x - y(x)$) and the agent (receiving $y(x)$) a function of their respective utilities. In other words, the parties' preferences expressed by their utility function, provide an expression for their mutual interest in maintaining an exchange.

Note that if we account for the fact that the agent's share y depends on the overall profit, i.e., $y=y(x)$, then by taking a second derivative of $u'(x-y(x)) = \lambda v'(y(x))$, with respect to x, we find an implied optimal relationship:

$$u''(x-y(x))(1-y'(x)) = \lambda v''(y(x)-w)y'(x). \qquad (7.12)$$

and therefore

$$\frac{u'(x-y(x))}{v'(y(x))} = \frac{u''(x-y(x))(1-y'(x))}{v''(y(x))(y'(x))}, \qquad (7.13)$$

where w is omitted to shorten the following expressions. As a result, we obtain an ordinary differential equation for the sharing rule between the principal and the agent, explicitly given by:

$$\frac{v''(y(x))/v'(y(x))}{u''(x-y(x))/u'(x-y(x))} = \frac{A_v(y(x))}{A_u(x-y(x))} = \frac{(1-y'(x))}{(y'(x))}, \qquad (7.14)$$

where $(A_v(y(x)), A_u(x-y(x)))$ are the Arrow-Pratt indices of risk aversions of the agent and principal ,

$$y'(x)(A_u(x-y(x)) + A_v(y(x))) = A_u(x-y(x)). \qquad (7.15)$$

For example, let the utilities of both principal and the agent are given by the exponential functions: $u(w) = 1 - e^{-\alpha w}$ and: $v(w) = 1 - e^{-\beta w}$. Then their derivatives are $u'(w) = \alpha e^{-\alpha w}$, $v'(w) = \beta e^{-\beta w}$, while the indices of risk aversion for both, are constant and given by (α, β) respectively. As a result, the exchange rule is simply defined by the following differential equation:

$$y'(x) = \frac{\alpha}{\alpha+\beta} \text{ and } y(x) = A + \frac{\alpha}{\alpha+\beta}x. \qquad (7.16)$$

The solution indicates then a "linear sharing rule", proportional to the indices of risk aversion of the principle and the agent. These results are important for explaining and dealing with a number of relationships that recur in supply chain contexts. Below we summarize a number of examples used by Dyane Reyniers (London School of Economics) in her classrooms.

Example 7.1. (The effort can be observed)

Assume a centralized firm that can observe the effort of the supply chain's parties with a direct relationship between their performance and the effort they provide. Let e be the effort and $P(e)$ be the resulting performance, a profit function. The party's cost is $C(e)$ while its reservation utility is u. There is a number of simple payment schemes that can motivate the party to provide the efficient amount of effort. These are: *Payments receive based on effort e*, and the party's return is $we+K$. The centralized problem is then to solve

$$Max \quad \pi = P(e) - (we + K) - C(e) \quad Subject\ to: \quad we + K - C(e) \geq u \,. \,(7.17)$$

The inequality constraint is called the "participation constraint" or the "individual rationality constraint". The centralized firm has no motivation to give more money to the party than his reservation utility. In this case, the effort selected by the firm will be at the level where marginal cost equals the marginal profit of effort, or: $P'(e^*) = C'(e^*)$. The party has to be encouraged to provide the optimal effort level however which leads to the *incentive compatibility constraint*. In other words, the party's net payoff should be maximized at the optimal effort or $w = C'(e^*)$. Thus, the party is paid according to his marginal disutility of effort and a lump sum K which satisfies the reservation utility.

Example 7.2. (Forcing contracts)

The central firm could propose to pay the party a lump sum L providing a reservation utility if the effort e^* is made, i.e. $L = u + C(e^*)$ and zero otherwise. Clearly, the participation and incentive compatibility constraints are satisfied under this simple payment scheme. This arrangement is called a forcing contrat because the party is forced to make the effort e^* (while above, the party is left to select his effort level).

Example 7.3. (Franchises)

Now assume that a franchisee can keep part of the profits of his effort in return for a certain payment to the principal—the franchiser. This can be interpreted as a franchise structure. To set the franchise fee, the franchiser proceeds as follows. First the franchisee maximizes $P(e) - C(e) - F$ and therefore chooses the same optimal effort as before such that: $P'(e^*) = C'(e^*)$. The principal-franchiser can charge a franchise fee which leaves the franchisee with his reservation utility: $F = P(e^*) - C(e^*) - u$. When the effort cannot be observed, the problem is more difficult. In this case, payment based on effort is not possible. If we choose to pay based on output, then the principal would choose a franchise structure. However, if the agent is risk averse, he will seek some payment to compensate the risk he is assuming. If the principal is risk neutral, he may be willing to assume the agent's risk and therefore the franchise solution will not be possible in its current form!

Utility Valuation and Market Pricing

"Utility", although an important tool to analyze decisions made under risk expresses an individual point of view, imbedded in the parameters that define the utility function. It does not express a market valuation of the prospect—its price. For practical purposes however, decisions in supply chain firms are valued and priced by "money" as set by the market. Advances in market pricing of financial assets and their derivatives, have contributed an important approach to determining asset prices in general using a plethora of financial products (options of various sort and broadly traded). These prices are risk attitude free, in the sense that they do not express the price a person or a firm is willing to pay for a given prospect based on a personal risk attitude but based on a broad (and liquid) market when such an exchange is being pursued. As a result, "price" is defined by an equilibrium in a large liquid financial market where assets are traded and their risk priced according to the risk attitudes of not one party, but the "multitudes" of parties (investors and speculators, risk averse, risk loving, risk neutral or not) engaged in such an exchange. Such prices are however relative, based on commonly observed prices of risk free assets or some other agreed on and observable assets. In such contexts, the price of an asset can be valued by the expected value of a distribution coined the Risk Neutral Distribution—RND. Such a distribution does not indicate a risk free attitude but is merely a belief regarding the market future prospects whose current price can be expressed in certain conditions (called Complete Markets) as an expectation (under the RND) discounted at the risk free rate. In this sense, the future expected value acts as a sort of cash and certainty equivalent for a future prospect currently valued at its on-going risk free rate.

The implication of this approach to supply chains is extremely important and provides grounds for fertile and extensive research that has practical applications to the management, the planning, the contracting and the organization of supply chains in general. Below, we shall merely point out to some basic relationships between the empirical distribution (or as determined by one or the other parties in the supply chain, or by the supply chain as a whole through collaborative forecasting) of the future prospect as defined by the financial market—the Risk Neutral Distribution, that underlies pricing of risky assets in complete markets. For further study, the reader is encouraged to consult the extensive literature in finance, in derivatives, in real options and in the many applications of financial market pricing to non-financial assets. Finally, we shall restrict ourselves to tractable problem examples.

The importance of a financial market approach to risk pricing is then in providing a price at which risk can be defined uniquely (in case, markets are complete) and exchanged (bought and sold). This is possible of course if an exchange between such investors and traders can be realized freely. When this is not the case, and risks are valued individually, implicitly due to individuals' or firms' attitudes to risk and their private information, a broad number of approaches is used to mitigate their effects and establish a subjective valuation and price for the underlying risk.

In supply chains, firms are not indifferent to risk and their corresponding subjective probability is thus different than the (market) RND. In fact, the RND would be adjusted upward (or downward) for all states in which money is more (or less) highly valued. Hence the higher the risk aversion, the more the RND and the subjective-empirical probability distributions differ. The risk aversion can thus be estimated from the joint observation of the two densities. Studies by Ait-Sahalia and Lo 1998, Anagnou et al. 2003, Bahra 1997 and others have contributed by pointing out that the index of absolute risk aversion $A(.)$, can be presented as a functional relationship involving empirical and the risk neutral distribution. That is, given any two, the other third might be determined (revealed):

$$A(.) = f(RND, \text{ Subjective Probability}). \qquad (7.18)$$

This was exploited by Jackwerth (1999, 2000) and Ait-Sahalia and Lo (1998, 2000) for extracting a measure of risk aversion in a standard dynamic exchange economy. Essentially, they assumed individuals would maximize the expected utility of future and uncertain prospects at a future date whose current price is known and measured by the (implied) risk neutral distribution. In this formulation, risk attitude (implied by the utility function) is a function of the risk neutral and the investor's subjective probability distributions. To determine their relationship, assume that markets are complete, that is there is a unique price for all assets that have the same returns and the same risks (thus there can be no arbitrage). Then, by definition of completeness, an asset price is necessarily equal to the discounted future price of the asset with expectation taken with respect to the risk neutral distribution and discounted at the risk free rate. A simple mathematical formulation that captures simultaneously an investor's risk attitude, his subjective-empirical probability and the market risk neutral distribution consists in maximizing the investor future expected utility subject to the current market pricing of the portfolio (using a risk neutral distribution). Letting for convenience the current market price (wealth) be W_0 and the future wealth at time T be W_T, we have the following problem:

$$\max_{W_T} E_P[u(W_T)], \text{ subject to } \frac{1}{(1+R_f)^T} E_{RN}[W_T] = W_0, \tag{7.19}$$

where R_f is the risk free market rate. Thus, to find a relationship between the private (subjective), subscript, P, and risk neutral, subscript RN, probability distributions we assume that the choice of the portfolio by a consensus investor is equivalent to choosing the wealth which will realize in each terminal state. Using the constraint Lagrange multiplier λ, the objective to optimize is:

$$\Phi = \int u(W_T) f_P(W_T) dW_T - \lambda \left(\frac{1}{(1+R_f)^T} \int W_T f_{RN}(W_T) dW_T - W_0 \right) \tag{7.20}$$

where $f_P(.)$ and $f_{RN}(.)$ are subjective (based on private information) and risk neutral distributions respectively. Applying the first order optimality condition to (7.20) and assuming that in equilibrium the consensus investor holds the market portfolio with return S, we find

$$u'(S) = \frac{\lambda}{(1+R_f)^T} \frac{f_{RN}(S)}{f_P(S)}. \tag{7.21}$$

In this case, for no arbitrage, that is for an economic equilibrium with a unique and known price, we have by differentiating condition (7.21):

$$u''(S) = \frac{\lambda}{(1+R_f)^T} \frac{f_{RN}'(S)f_P(S) - f_{RN}(S)f_P'(S)}{[f_P(S)]^2} \tag{7.22}$$

Combining (7.21) and (7.22), we obtain:

$$A_r(S) = -\frac{u''(S)}{u'(S)} = \frac{f_P'(S)}{f_P(S)} - \frac{f_{RN}'(S)}{f_{RN}(S)} \tag{7.23}$$

And therefore:

$$A_r(W_T) = \frac{\partial}{\partial W_T} \left(\ln f_P(W_T) - \ln f_{RN}(W_T) \right) = \frac{\partial}{\partial W_T} \left(\ln \left(\frac{f_P(W_T)}{f_{RN}(W_T)} \right) \right). \tag{7.24}$$

Consequently, we obtain that the index of risk aversion is given by a discrimination function of the subjective (private) and the risk neutral distributions, explicitly given by:

$$A(W_T) = \frac{\partial}{\partial W_T} g(W_T), \; g(W_T) = \ln \left\{ \frac{f_P(W_T)}{f_{RN}(W_T)} \right\}. \tag{7.25}$$

A broad number of techniques can then be used to estimate empirically the risk neutral distribution (by using market data on derivatives for example, and calculating the implied risk neutral distribution reflecting market prices). This establishes a specific relationship between the risk attitude and the subjective probability distribution. Note that for a risk neutral decision maker, the index of absolute risk aversion is null and thus, the ratio of the distributions $\dfrac{f_P(W_T)}{f_{RN}(W_T)}$ is constant. By the same token, for an exponential utility function, the index of absolute risk aversion is proportional to the future prospect and therefore, the function $g(W_T)$ is linear. In other words:

$$a + bW_T = \ln \frac{f_P(W_T)}{f_{RN}(W_T)} \tag{7.26}$$

and, therefore,

$$f_P(W_T) = e^{a+bW_T} f_{RN}(W_T), \tag{7.27}$$

where a and b are appropriately defined parameters. Of course, since the risk neutral probability distribution integral equals one, we have:

$$1 = \int f_P(W_T) dW_T = \int e^{a+bW_T} f_{RN}(W_T) dW_T$$

and

$$e^a = \frac{1}{\int e^{bW_T} f_{RN}(W_T) dW_T}. \tag{7.28}$$

Thus:

$$f_P(W_T) = \frac{e^{bW_T} f_{RN}(W_T)}{\int e^{bW_T} f_{RN}(W_T) dW_T}. \tag{7.29}$$

Since, also

$$\int e^{-a-bW_T} f_P(W_T) dW_T = \int f_{RN}(W_T) dW_T = 1 \tag{7.30}$$

we have:

$$\int e^{-bW_T} f_P(W_T) dW_T = e^a \tag{7.31}$$

and therefore,

$$f_{RN}(W_T) = \frac{e^{-bW_T} f_P(W_T)}{\int e^{-bW_T} f_P(W_T) dW_T}. \tag{7.32}$$

If the utility of an investor is assumed to be HARA (Hyperbolic Absolute Risk Aversion) with an index of absolute risk aversion $A(W_T)$, declining with wealth, given by:

$$u(W_T) = \frac{1-\gamma}{\gamma}\left\{\frac{aW_T}{1-\gamma}+b\right\}^{\gamma}, \; A(W_T) = \frac{a}{\left\{\frac{aW_T}{1-\gamma}+b\right\}} \tag{7.33a}$$

$$A(W_T) = \frac{a(1-\gamma)}{aW_T + b(1-\gamma)} = \frac{d}{dW_T} g(W_T), \tag{7.33b}$$

then,

$$f_P(W_T) = \left(aW_T + b(1-\gamma)\right)^{(1-\gamma)} f_{RN}(W_T). \tag{7.34}$$

In some problems it might be convenient to approximate the distributions by parametric known distributions. For example, say that both the subjective and the risk neutral distributions are normally distributed with means and variances given by: (μ_S, σ_S^2) and $(\mu_{RN}, \sigma_{RN}^2)$, then:

$$\frac{f_P(W_T)}{f_{RN}(W_T)} = \frac{\sigma_{RN}}{\sigma_P} \exp\left[-\frac{(W_T - \mu_P)^2}{2\sigma_P^2} + \frac{(W_T - \mu_{RN})^2}{2\sigma_{RN}^2}\right]. \tag{7.35}$$

And therefore, this implies that risk aversion is proportional to the prospect since:

$$A(W_T) = \frac{d}{dW_T}\left(\ln(\sigma_{RN} - \sigma_P) - \frac{(W_T - \mu_P)^2}{2\sigma_P^2} + \frac{(W_T - \mu_{RN})^2}{2\sigma_{RN}^2}\right) = W_T\left(\frac{\mu_P}{\sigma_P^2} - \frac{\mu_{RN}}{\sigma_{RN}^2}\right). \tag{7.36}$$

Thus, if $\dfrac{\mu_P}{\sigma_P^2} < \dfrac{\mu_{RN}}{\sigma_{RN}^2}$, risk aversion decreases linearly in wealth and vice

versa. Note that a risk aversion is then defined by the sign of $A(W_T)$, as defined in this artificial example. In this sense, a risk attitude can be revealed by observed behaviors (the financial market for example).

Higher order risk attitudes can be defined similarly. Explicitly, the index of prudence can be defined as well in terms of the subjective and risk neutral distributions logarithmic function $g(W_T)$ and given by:

$$A_P = \frac{g''(W_T)}{g'(W_T)} - g'(W_T), \; g''(W_T) = (A_P + A)A. \tag{7.37}$$

If an investor is risk averse, $A>0$ and, therefore, for $g''(W_T)>0$ (an increasing rate of discrimination rate with respect to wealth) implies prudence since $A_P>0$. However, if empirical analysis indicates $g''(W_T)<0$, then the prudence index is both negative and larger than the index of risk aversion. Extensive research, both theoretical and empirical has extended this approach and indicated a number of important results. For example, risk attitude is not only state varying but is time varying as well. Further, it clearly sets out the concept of risk attitude in terms of a distance between the subjective and the risk neutral (market) distributions. Again, in the case of the artificial normal example treated earlier, we have an index of prudence given by:

$$A_P = \frac{1}{A}\left(\frac{1}{\sigma^2_{RN}} - \frac{1}{\sigma^2_P}\right) - A \cdot \qquad (7.38)$$

In this equation, we clearly see the relationship between the risk neutral, the subjective and the risk attitude of the investor both with respect to the index of absolute risk aversion and the investor's prudence. This brief introduction to utility and risk valuation will be discussed subsequently using specific examples. Below, we consider a concept of preference based on quantile risk and profusely used as a measure of risk exposure. It is also used to manage risks in some financial firms and increasingly in operational and supply chain problems.

Preferences and Quantile Risks

In theory and in practice risk is measured and valued based on three essential approaches which we shall resume by: (1) Utility valuation, (2) Asymmetric (individual, whether "rational" or not) preferences based on regret and quantile risk specification (also coined as value at Risk) and (3) Market pricing of risk. The first approach, discussed previously, is based on the presumption that "persons" are not indifferent to the size of gains and losses. Rather, "persons" actions are motivated by an "underlying rationale", mostly specified by the expected utility approach. In the third approach, Market Pricing, risk is valued and priced by "market of risk" where investors and speculators interact and exchange current and future certain and uncertain prospects, till a state of equilibrium is reached at which these prospects have a defined price.

In other cases, "persons" actions are motivated by an underlying "behavioral and psychological" rationale. A typical rational is based on a concept of regret, expressing decision makers ex-post attitudes towards results they either did not expect or losses (the second approach). Such an approach is closely associated to the widely practiced quantile risk or Value at Risk as we shall see below. To express such subjective preferences, a number of approaches (based and often derived from expected utility arguments) are used and portending to represent persons' preferences. In some cases, as it is the case in finance and in related areas, risks and their value-price are measured relative to some well known and predictable state. For example, measure one set of uncertain returns relative to a sure one. In this context, a concept of certainty equivalence (if it can be measured or assessed and as seen earlier) is used to measure the premium a person would be willing to pay to do away altogether with the risk of a given prospect. In financial analysis, one often encounters the equivalent and relative price which we denoted by the risk free rate, providing the time value of an asset, which has no risk. The premium "a person" or "market" will pay for a given lottery or uncertain return in the present or in the future is then expressed relative to these commonly known values. In more general terms, an expression for the measurement and the pricing of risk (and therefore its management) needs to be specified in terms of data we can properly refer to and gather.

The quantile risk approach, unlike the utility or market price approach is based on "expectation and threshold" (rather than expectation and risk premium) sometimes used (Artzner et al. 1997, 1999, 2000, 2001; Embrecht 2000; Beckers 1996) to suggest an excess function defined in terms of a loss threshold K, and given by:

$$e(k) = E\left\{\tilde{L} - K \mid \tilde{L} - K\right\}. \tag{7.39}$$

In other words, this is the expected loss incurred beyond a (subjective) threshold "K", expressing thereby a risk exposure. For example, K might be a maximally allowable loss before a certain and unpleasant action is taken. This approach is in fact a complement to the celebrated and applied Value at Risk (VaR) quantile risk model which measures risk exposure by specifying the probability that the loss be greater than the threshold, or :

$$P\left\{\tilde{L} > K\right\} \leq 1 - \varsigma. \tag{7.40}$$

In this case, the threshold we have defined can be interpreted as a capital or a reserve set aside to meet such contingent losses. In this sense, this measure relates to money and implies as well a risk preference—based on a regret type preference. Explicitly, given ς, we have: $1 - F_L(K) \leq 1 - \varsigma$ or $K = F_L^{-1}(\varsigma)$. Thus given the quantile risk ς we may be willing to assume and given the loss function, the implied amount of money we have at risk (at the specified exposure probability) is K.

For these reasons, K is often denoted as the Value at Risk (VaR) and commonly defined as the expected loss from an adverse movement with specified probability over a specified period of time, expressing thereby a quantile risk measure of a risk exposure in money terms. The use of VaR can be justified as stated earlier by an ex-post disappointment decision model as we shall see and on the basis of which it can be applied to numerous supply chain decision making problems. In other words, it can be justified by a regret criterion which has inspired a number of approaches coined "regret-disappointments models" (Savage 1954; Loomes and Sugden 1986; are important references in this approach). According to Bell 1982, (see also Bell 1985, 1995, 1999) disappointment is a psychological reaction to an outcome that does not meet a decision maker's expectation. In particular, assume that the measurement of disappointment is assumed to be proportional to the difference between expectation and the outcome below the expectation. Elation, may occur when the outcome obtained is better than its expectation. A general treatment based on risk-value theory with respect to such an approach is suggested by Dyer and Jia 1997. Their approach uses a "value asymmetry" between "elation" and "regret-loss". Such an approach is based on the presumption that the "cost of reaching the wrong decision" may be, proportionately, greater than the payoff of having made the right decision. In other words, managers abhor losses, valuing them more than they value gains for having made the right decision. This explains a preference for flexibility more than justified by its expected payoff and of course contrasts classical Bayesian (ex ante) expected decision theory. For example, a performance below than expected can have disproportionate effects on value. The VaR approach provides an expression for these considerations consisting in specifying a time horizon T and the confidence level P_{VaR}, with VaR denoting the loss in value over the time horizon T that is exceeded with probability $(1 - P_{VaR})$. In other words, if ξ is a random variable denoting the amount of money made in a number of sales, then VaR is defined as that amount, where the probability of losing more than that amount is determined with the specified probability P_{VaR}. Explicitly, if we let $P_T(.)$ be the probability of gains and losses over the time period T, we then have:

$$1 - F_T(VaR) = P_{VaR}, \quad F_T(VaR) = \int_{-\infty}^{-VaR} P_T(\xi) d\xi . \tag{7.41}$$

Assuming the invertibility of cumulative distribution function, we have again

$$F_T(VaR) = 1 - P_{VaR} \quad \text{and} \quad VaR = F_T^{-1}(1 - P_{VaR}) . \tag{7.42}$$

Therefore the VaR is often used as the amount of money to set aside to meet such contingencies losses. In this sense (which is continuously questioned and criticized but used), the specification of (P_{VaR}, VaR) provides a constraint determining the risk exposure as well as the amount of money (our other resources) to set aside to meet such eventualities. The rationality of this approach will be considered in the next section within a simple target cost problem which we will generalize to a supply chain context.

7.5 SELECTED CASES AND PROBLEMS

To highlight some of the approaches elaborated in the previous section, we shall consider a number of problems where risks are measured in terms of money and managed in supply chains in terms of operational risk externalities.

Example 2.1 Revisited

We reconsider Example 2.1 (Chapter 2) consisting of one supplier, s, and one retailer r. As stated earlier, the supplier offers products at a known wholesale price w while the retailer buys q units of product which he sells at a set retail price $p=w+m$. In the pricing game considered in Example 2.1, the two firms maximize their profits; the demand is linear and downward sloping in price. Results were obtained under alternative assumptions regarding the relationships the supplier and the retailer were engaged in. Uncertainty in such problems arises in a number of manners. Of course, both the market demand (the quantity actually sold by the retailer) may be random, as well as the product price. Further, both the quantity sold and the price may depend (or not) on one another. Such uncertainty leads to inventories or shortages accumulating when market demand and the supply of a retailer are not synchronized. Further, price uncertainty can lead as well to performance risks, mitigated at times by using derivatives to hedge against price variations. While the retailer faces downstream "a market risk" (in demand and in price), he may also face risks "upstream" when dealing with the supplier, such as supply uncertainties, supply delays and uncertainty in these delays, supply prices etc. These risks, however, are mostly the result of the bilateral relationship between the supplier and the retailer and the contracts that bind them. In a supply chain, the problems that can be encountered are indeed very broad and therefore, we shall restrict ourselves to a sample few.

In our analysis, we shall assume first and for simplicity a supplier functioning in a purely competitive economy and pricing products sold at their marginal costs. That is, if S is the total quantity produced by the supplier at a cost $C(S)$ with a profit $wS - C(S)$, then the retailer price is $w = C'(S)$. In addition, we consider a retailer buying products at a wholesale price w, assumed fixed for simplicity. We also let π_0 be the current market price of this product and therefore, the retailer's realized profit margin at this time is $\pi_0 - w$. Since all order decisions are made at this time for selling at the next period, the profit margin of the retailer is a random variable which we denote by $\tilde{m}_1 = \tilde{\pi}_1 - w_1$ where $\tilde{\pi}_1$ is the market (random selling price) at the next period and w_1 is the current negotiated wholesale price. Let the order policy be the fixed quantity, $\overline{D}_1$ while the next period demand is a random variable $D_1(\tilde{\pi}_1)$, a function of prices only (subsequently, we shall consider a more general case, with a demand, a random function of random prices). Then, the retailer profit at the next period is:

$$P_r(1) = \begin{cases} (\tilde{\pi}_1 - w_1(1+R_f))\bar{D}_1 - Q(D_1(\tilde{\pi}_1) - \bar{D}_1) & \text{if } \bar{D}_1 \leq D_1(\tilde{\pi}_1) \\ \tilde{\pi}_1 D_1(\tilde{\pi}_1) - w_1(1+R_f)\bar{D}_1 - I(\bar{D}_1 - D_1(\tilde{\pi}_1)) & \text{if } \bar{D}_1 > D_1(\tilde{\pi}_1) \end{cases} \quad (7.43)$$

while the supplier profit is

$$P_S(1) = w_1 \bar{D}_1 \quad (7.44)$$

Note that when the order is smaller then the demand, $(D_1(\tilde{\pi}_1) \geq \bar{D}_1)$, a shortage or service cost $Q(D_1(\tilde{\pi}_1) - \bar{D}_1)$ is incurred while when the order is larger than the demand, $(\bar{D}_1 \geq D_1(\tilde{\pi}_1))$, then an inventory cost $I(\bar{D}_1 - D_1(\tilde{\pi}_1))$ is sustained. These costs are usually difficult to price. Therefore, in this section, we shall price these costs by using derivatives on the underlying product. Explicitly, say that the retailer buys initially y put options with a strike of h, with a current price of $P_{PUT}(\pi_0; h)$. Of course, to meet shortages, the retailer may buy on the spot market the quantity needed to meet his contracted demand or exercise the put options if this is economical. In this case, the quantity bought on the spot market is $z_P = (D_1(\tilde{\pi}_1) - \bar{D}_1) - y1_{\tilde{\pi}_1 \geq h}$ while the profit from such a transaction is the speculative profit only, or $yMax(h - \tilde{\pi}_1, 0)$. By the same token, when the demand is smaller than the order quantity, we have an end of period inventory equal to $(\bar{D}_1 - D_1(\tilde{\pi}_1))$. In such a case, if we were to sell initially x call options at a strike k, then the ending inventory after such a transaction is $z_C = (\bar{D}_1 - D_1(\tilde{\pi}_1)) - x1_{\tilde{\pi}_1 \leq k}$. To assure ourselves of a zero inventory, we require that: $0 = (\bar{D}_1 - D_1(\tilde{\pi}_1)) - x1_{\tilde{\pi}_1 \leq k}$, which can be obtained by buying a sufficient number of call options to meet any demand. For example, if the high price is $\pi^+_1 = Max(\tilde{\pi}_1)$, then the maximum inventory is $\bar{D}_1 - D_1(\pi^+_1)$ and therefore $X = \bar{D}_1 - D_1(\pi^+_1)$ will provide a guarantee that inventories will be sold at any strike (albeit at a loss, compensated by the premium collected initially) below the maximum price. By the same token, we can buy put options at the strike price $h = Max(\tilde{\pi}_1)$ in which case, when the demand is low, a profit will be made. The maximum shortage is then Y. In this case, we have the following retailer's profit:

$$P_r(1)=\begin{cases} \left(\tilde{\pi}_1-w_1(1+R_f)\right)\bar{D}_1-Y(1+R_f)P_{PUT}(\pi_0;h)+YMax(h-\tilde{\pi}_1,0) & if\ \bar{D}_1\le D_1(\tilde{\pi}_1) \\ \tilde{\pi}_1 D_1(\tilde{\pi}_1)-w_1(1+R_f)\bar{D}_1+X(1+R_f)C_{CALL}(\pi_0;h)-XMax(\tilde{\pi}_1-k,0) & if\ \bar{D}_1>D_1(\tilde{\pi}_1) \end{cases}$$

(7.45)

Of course, under risk neutral pricing, we have:

$$(1+R_f)P_{PUT}(\pi_0;h)=Max(h-\tilde{\pi}_1,0);\quad (1+R_f)C_{CALL}(\pi_0;h)=Max(\tilde{\pi}_1-k,0).(7.46)$$

And therefore, the retailer profit is:

$$P_r(1)=\begin{cases} \left(\tilde{\pi}_1-w_1(1+R_f)\right)\bar{D}_1 & if\ \bar{D}_1\le D_1(\tilde{\pi}_1) \\ \tilde{\pi}_1 D_1(\tilde{\pi}_1)-w_1(1+R_f)\bar{D}_1 & if\ \bar{D}_1>D_1(\tilde{\pi}_1) \end{cases}$$

(7.47)

Accordingly, under risk neutral pricing, the price of the ordering policy is:

$$P_r(0)=-w_1\bar{D}_1+\frac{1}{(1+R_f)}E_{RN}\left(P_r(1)\right)$$

(7.48)

Explicitly, it can be calculated as follows. First, since $\partial D_1(\pi_1)/\partial\pi_1<0$, we have,

$$\pi_1^-=D_1^{-1}(D_{Max});\quad \pi_1^+=D_1^{-1}(D_{Min}),\pi_1^*=D_1^{-1}(\bar{D}_1)$$

(7.49)

As a result:

$$P_r(0)=-w_1\bar{D}_1+\frac{1}{(1+R_f)}\left\{\bar{D}_1\int_{D_1(D_{Max})}^{D_1^{-1}(\bar{D}_1)}\pi_1 f_{RN}(\pi_1)d\pi_1+\int_{D_1^{-1}(\bar{D}_1)}^{D_1(D_{Min})}\pi_1 D_1(\pi_1)f_{RN}(\pi_1)d\pi_1\right\}$$

(7.50)

The optimal order policy is thus given by the maximization of $P_r(0)$. However, to do so, we require the underlying risk neutral distribution of prices which is not always easily available. Of course, in a binomial model, assuming that markets are complete, a solution, implied in our observation of spot, call and put prices, can be found easily. Explicitly, since, in a binomial complete market model we have:

$$\pi_0=\frac{1}{1+R_f}\left(p_{RN}\pi_1^++(1-p_{RN})\pi_1^-\right),\ or,\ p_{RN}=\frac{\pi_0(1+R_f)-\pi_1^-}{\pi_1^+-\pi_1^-},\ (7.51a)$$

$$P_{CALL}(\pi_0)=\frac{1}{1+R_f}p_{RN}(\pi_1^+-k)=\frac{1}{1+R_f}\frac{\left[\pi_0(1+R_f)-\pi_1^-\right](\pi_1^+-k)}{\pi_1^+-\pi_1^-},(7.51b)$$

$$P_{PUT}(\pi_0)=\frac{1}{1+R_f}(1-p_{RN})(h-\pi_1^-)=\frac{1}{1+R_f}\frac{\left(\pi_1^+-\pi_0(1+R_f)\right)(h-\pi_1^-)}{\pi_1^+-\pi_1^-}.(7.51c)$$

The last two equations can be solved simultaneously, providing thereby a solution for both the binomial process and the risk neutral (binomial) probability. Since there are two prices only, the demand implied by the complete market assumption is:

$$D_1^+ = D_1(\pi_1^+); \quad D_1^- = D_1(\pi_1^-); \quad D_1^+ < D_1^-. \tag{7.52}$$

And therefore:

$$P_r(1) = \begin{cases} (\pi_1^- - w_1(1 + R_f))\bar{D}_1 & \text{for } \bar{D}_1 \leq D_1(\pi_1^-) \\ \pi_1^+ D_1(\pi_1^+) - w_1(1 + R_f)\bar{D}_1 & \text{for } \bar{D}_1 > D_1(\pi_1^+) \end{cases} \tag{7.53}$$

$$P_r(0) = -w_1(1 + R_f)\bar{D}_1 + \frac{1}{1 + R_f}\left((1 - p_{RN})\pi_1^-\bar{D}_1 + p_{RN}\pi_1^+ D_1^+\right). \tag{7.54}$$

Note that the latter equation is not linear in the order quantity as this quantity is implicit as well in the calculations of the implied risk neutral probability. Using equation (7.50), a derivative with respect to the order quantity yields:

$$0 = -w_1 + \frac{1}{(1+R_f)}\left\{ \begin{array}{l} \displaystyle\int\limits_{D_1^{-1}(D_{Max})}^{D_1^{-1}(\bar{D}_1)} \pi_1 f_{RN}(\pi_1)d\pi_1 + \bar{D}_1 \frac{\partial D_1^{-1}(\bar{D}_1)}{\partial \bar{D}_1}\left[D_1^{-1}(\bar{D}_1)f_{RN}\left(D_1^{-1}(\bar{D}_1)\right)\right] \\[3ex] -\frac{\partial D_1^{-1}(\bar{D}_1)}{\partial \bar{D}_1} D_1^{-1}(\bar{D}_1)D_1\left(D_1^{-1}(\bar{D}_1)\right)f_{RN}\left(D_1^{-1}(\bar{D}_1)\right) \end{array} \right\} \tag{7.55}$$

Since $D_1\left(D_1^{-1}(\bar{D}_1)\right) = \bar{D}_1$, this equation is reduced to:

$$w_1(1 + R_f) = \int\limits_{D_1(D_{Max})}^{D_1^{-1}(\bar{D}_1)} \pi_1 f_{RN}(\pi_1)d\pi_1 = \int\limits_{\pi_1^-}^{\pi_1^*} \pi_1 f_{RN}(\pi_1)d\pi_1; \quad \bar{D}_1^* = D_1(\pi_1^*). \tag{7.56}$$

and therefore, if the demand is linear and given by $D_1(\pi) = a - b\pi$, then the optimal order quantity is $\bar{D}_1^* = a - b\pi^*$ a solution of the above equation. For example, say that the risk neutral distribution is a uniform distribution in the time interval $\left[x_1^-, x_1^+\right]$, then we have:

$$w_1(1 + R_f) = \int\limits_{\pi_1^-}^{\pi_1^*} \frac{\pi}{\pi_1^+ - \pi_1^-}d\pi = \frac{1}{2}\frac{\pi_1^{*2} - \pi_1^{-2}}{\pi_1^+ - \pi_1^-} \tag{7.57}$$

and therefore,

$$\pi_1^* = \sqrt{\left(\pi_1^-\right)^2 + 2w_1(1 + R_f)\left(\pi_1^+ - \pi_1^-\right)}. \tag{7.58}$$

Finally,

$$\bar{D}_1^* = D_1\left(\sqrt{\left(\pi_1^-\right)^2 + 2w_1(1 + R_f)\left(\pi_1^+ - \pi_1^-\right)} \right). \qquad (7.59)$$

More generally, calculating the implied risk neutral distribution in terms of derivatives will be more precise, albeit numerically based.

When the demand is a random function of random prices as well, a similar approach can be used by considering a stochastic constraint, summarizing the potential demand realizations in terms of an order and a portfolio of derivatives contracts that replicate (that is can meet) the demand at all prices and in all situations.

The implication of our approach is that a retailer acting on a risk neutral probability (i.e., in complete markets) is subject to the same laws of finance, presuming that "without assuming risks", there are no profits. In other words, a retailer, acting as an intermediary between, say, the supplier and end market customers, "must have some informational advantage" or some other advantages that will allow him to make (arbitrage) profits. Of course in a practical setting, the retailer and the supplier base their analyses on both observations of market behavior and their instructed beliefs in regard to the future states of potential prices. Such beliefs recur both with respect to the demand and price uncertainty, which are combined in retailers risk attitudes in determining an optimal order policy. The risk neutral probability imbeds these beliefs (assuming their expression in the price of derivatives) while the relationship between demand and price, has made it possible to remain within the simple complete markets framework that has allowed our calculations. A generalization to more complex situations can be considered as well.

From the analysis in the previous section, we clearly saw that our results depend on knowing the risk neutral probability. In practice however, retailers and suppliers possess private information which can lead them to believe that they have in fact an informational advantage. In this case, while market prices are what they are, a retailer for example, may think that the market errs in the specification of the risk neutral distribution. In other words, say that the retailer has a private information regarding the future demand and therefore an information regarding prices (assuming that the demand is indeed a function of prices), or a direct information regarding the future prices. Let $f_P(.)$ be a private probability estimate of the future prices and let the retailer utility function (expressing his risk attitudes) be $u(.)$. In this case, the retailer private utility and information would lead him to maximize the expected next period profit subject to the current prices. An explicit relationship, between these (see also Jackwerth

1999, 2000, Ait-Sahalia .and Lo 1998, 2000) is given by (as seen and proved earlier):

$$A_r(.) = \frac{d}{d(.)} \left\{ \ln\left(\frac{f_P(.)}{f_{RN}(.)}\right) \right\}, \quad \exp\left(\int A_r(z)dz\right) = \frac{f_P(.)}{f_{RN}(.)}. \quad (7.60)$$

This relationship states that a decision maker's index of risk aversion $A_r(.)$, expressing his personal risk attitude, combined with his subjective assessment of future states (prices for example) determines the risk neutral distribution of these future states. Thus, given any two of these terms, the third can be calculated. Therefore, introducing in our optimality equation the implied risk neutral distribution we have:

$$w_1(1+R_f) = \int_{\pi_1^-}^{\pi_1^*} \pi_1 \exp\left(-\int_{\pi_1^-}^{\pi_1} A_r(z)dz\right) f_P(\pi_1)d\pi_1; \quad \bar{D}_1^* = D_1(\pi_1^*). \quad (7.61)$$

For example, if the retailer has an exponential utility function whose index of absolute risk aversion is α, then integration of (7.61)) yields:

$$\xi + \alpha\pi = \ln\frac{f_P(\pi)}{f_{RN}(\pi)} \quad \text{or} \quad f_{RN}(\pi) = e^{-\xi - \alpha\pi} f_P(\pi). \quad (7.62)$$

where ξ is an integration constant defined by the risk neutral distribution:

$$1 = \int_0^\infty f_{RN}(\pi)d\pi = e^{-\xi} \int_0^\infty e^{-\alpha\pi} f_P(\pi)d\pi = e^{-\xi} L(\alpha). \quad (7.63)$$

In equation (7.63), α is the Laplace Transform of the retailer subjective estimate of the future prices (i.e. his private information). As a result, $e^\xi = L(\alpha)$ and finally,

$$f_{RN}(\pi) = \frac{e^{-\alpha\pi} f_P(\pi)}{L(\alpha)}. \quad (7.64)$$

As a result, we have:

$$w_1(1+R_f) = \int_{\pi_1^-}^{\pi_1^*} \pi_1 \frac{e^{-\alpha\pi_1} f_P(\pi_1)}{L(\alpha)} d\pi_1; \quad \bar{D}_1^* = D_1(\pi_1^*). \quad (7.65)$$

Again, let the private information of the retailer indicate a price distribution which is uniform, that is:

$$w_1(1+R_f) = \frac{1}{L(\alpha)\left[\pi_1^+ - \pi_1^-\right]} \int_{\pi_1^-}^{\pi_1^*} \pi_1 e^{-\alpha\pi_1} d\pi_1; \quad \bar{D}_1^* = D_1(\pi_1^*), \quad (7.66)$$

or

$$w_1(1+R_f) = \frac{1}{L(\alpha)\left[\pi_1^+ - \pi_1^-\right]} \left\{ e^{-\alpha\pi_1^-}\left(-\frac{\pi_1^-}{\alpha} - \frac{1}{\alpha^2}\right) - e^{-\alpha\pi_1^*}\left(\frac{\pi_1^*}{\alpha}\frac{1}{\alpha^2}\right) \right\}. \quad (7.67)$$

Therefore, it is a solution of:

$$e^{-\alpha\pi_1^*}\left(\frac{\pi_1^*}{\alpha}+\frac{1}{\alpha^2}\right)=e^{-\alpha\pi_1^-}\left(\frac{\pi_1^-}{\alpha}+\frac{1}{\alpha^2}\right)-w_1(1+R_f)L(\alpha)\left[\pi_1^+-\pi_1^-\right]; \quad \overline{D}_1^*=D_1\left(\pi_1^*\right),(7.68)$$

where:

$$L(\alpha)=\frac{1}{\left[\pi_1^+-\pi_1^-\right]}\int_{\pi_1^-}^{\pi_1^+}e^{-\alpha x}dx=\frac{e^{-\alpha\pi_1^-}-e^{-\alpha\pi_1^+}}{\left[\pi_1^+-\pi_1^-\right]}. \quad (7.69)$$

Consequently,

$$e^{-\alpha\pi_1^*}\left(\frac{\pi_1^*}{\alpha}\frac{1}{\alpha^2}\right)=e^{-\alpha\pi_1^-}\left(\frac{\pi_1^-}{\alpha}+\frac{1}{\alpha^2}\right)-w_1(1+R_f)\left(e^{-\alpha\pi_1^-}-e^{-\alpha\pi_1^+}\right); \quad \overline{D}_1^*=D_1\left(\pi_1^*\right).(7.70)$$

In this situation as well, the price estimates of the derivatives by the retailer are:

$$\pi_0=\frac{1}{1+R_f}\int_{\pi_1^-}^{\pi_1^+}\pi\frac{e^{-\alpha\pi}f_P(\pi)}{L(\alpha)}d\pi. \quad (7.71)$$

$$C_0(\pi_0;k)=\frac{1}{1+R_f}\int_{\pi_1^-}^{\pi_1^+}Max(\pi-k,0)\frac{e^{-\alpha\pi}f_P(\pi)}{L(\alpha)}d\pi, \quad (7.72)$$

$$P_0(\pi_0;h)=\frac{1}{1+R_f}\int_{\pi_1^-}^{\pi_1^+}Max(h-\pi)\frac{e^{-\alpha\pi}f_P(\pi)}{L(\alpha)}d\pi. \quad (7.73)$$

Value at Risk, Safety is First and Target Costing

In this illustration, we shall provide an application of Value at Risk as outlined in the previous section. Assume that in a supply chain an individual firm target costs Q a part for a product to be assembled by a supply chain and let a random variable, z, be the realized cost with a known probability distribution function. The actual development cost is a function of the firm's operational strategy and investments. To finance the production cost, an amount equal to the target cost is borrowed at the bank at the rate r. The following objective is then defined, consisting of the costs of over or under meeting the target cost. Explicitly, if $z>Q$, then the firm (a supplier) is penalized at a rate of $\alpha>r$ while if the supplier cost is below the target, $z<Q$, then the resulting cost of such a deviation is penalized at a rate of $\beta<\alpha$. As a result, the following (asymmetric) objective is defined

$$Min \ \Phi_Q=rQ+\alpha E\left[z-Q|z\geq Q\right]+\beta E\left[z-Q|z\leq Q\right]; \quad \beta<\alpha, \ \alpha>r.(7.74)$$

Let the cumulative probability distribution of the cost be $F_C(.)$, then the expected cost is:

$$Min \quad \Phi_Q = rQ + \alpha \int_Q^\infty (z - Q) dF_C(z) + \beta \int_0^Q (z - Q) dF_C(z); \quad \beta > r \cdot \quad (7.75)$$

The least objective target cost is thus found by setting to zero the first derivative of the objective function above, or:

$$r - \alpha \left[1 - F_C(Q)\right] - \beta F_C(Q) = 0 \cdot$$

Thus, an optimal target cost is

$$\xi = \frac{\alpha - r}{\alpha - \beta} = F_C(Q), \quad r < \alpha, \quad \beta < \alpha, \quad (7.76)$$

or, in other words, the target cost is given by

$$Q^* = F_C^{-1}(\xi) = F_C^{-1}\left(\frac{\alpha - r}{\alpha - \beta}\right), \quad r < \alpha, \quad \beta < \alpha, \quad (7.77)$$

where Q^* is the optimal value.

Of course, if development costs are a function of their effort (or some other variable of interest) and the costs are charged when performing above or below the target cost, then a firm's objective consists in minimizeing the following:

$$Min \quad \Phi_u = h(u) + \alpha \int_Q^\infty z dF_C(z|u) + \beta \int_0^Q z dF_C(z|u) \cdot \quad (7.78)$$

For example, let the cost probability distribution have a Weibull probability distribution defined by $f(z) = ab(u)z^{a-1}e^{-b(u)z^a}, \tau \geq 0, a > 0$, then the cumulative probability distribution is $F(z) = 1 - e^{-b(u)z^a}$ with mean and variance

$$E(z) = b(u)^a \Gamma\left(\frac{1}{a} + 1\right), \quad var(z) = b(u)^{2a}\left[\Gamma\left(\frac{2}{a} + 1\right) - \Gamma^2\left(\frac{1}{a} + 1\right),\right].(7.79)$$

Thus, the optimal target cost is in this case is:

$$Q^* = F_C^{-1}\left(\ln\frac{\alpha - r}{\alpha - \beta}\right) = \left\{\ln\left(\frac{\beta - r}{\alpha - r}\right)^{\frac{1}{b(u)}}\right\}^{\frac{1}{a}}, \quad r < \alpha, \quad \beta < \alpha, \cdot \quad (7.80)$$

The Target Costing problem defined above can be generalized further to the firm outsourcing parts to multiple suppliers. In this case, a simple formulation of the problem faced by the "central firm" is given by

$$Min \quad \Phi_Q = r\sum_{k=1}^n Q_k + \alpha \sum_{k=1}^n \int_{Q_k}^\infty (z - Q_k) dF_{C,k}(z) + \beta \sum_{k=1}^n \int_0^{Q_k} (z - Q_k) dF_{C,k}(z); \quad \beta > r \cdot (7.81)$$

while each supplying firm, seeks to minimize the following

$$Min \quad \Phi_u = h(u_k) + \alpha \int_{Q_k}^{\infty} z dF_{C,k}(z|u_k) + \beta \int_{0}^{Q_k} z dF_{C,k}(z|u_k). \quad (7.82)$$

These equations define a game between the supply chain manager and the individual supplying firms. In this game, the "central firm" determines the target cost for each firm, while the individual firms optimize with respect to the efforts furnished. Evidently, for each firm k, we have the following:

$$r - \alpha\left(1 - F_{C,k}(Q_k)\right) - \beta F_{C,k}(Q_k) = 0; \quad \beta > r. \quad (7.83)$$

Or

$$F_{C,k}(Q_k) = \frac{r - \alpha}{\beta - \alpha}, \quad (7.84)$$

while optimization of the effort by the individual firm is defined as stated earlier.

7.6 COLLABORATION, RISKS AND SUPPLY CHAINS

Collaboration in supply chains assumes a growing importance due to the profit that results from economies of scale, in technology, in production, in market power, in introducing entry barriers and thereby reducing some of the associated risks for firms. At the same time however, internal risks such as lock-in contracts, risk sharing, risk transfer, size risk etc, have to be dealt with. The risks sustained are of course a function of the contractual, behavioral and collaboration attitudes in use. For example, vertical integration or hierarchical control; subcontractors-contractual relationships; franchises; joint ventures and partnerships; strategic alliances; reciprocity agreements etc. all have benefits and risks. These risks derive mostly form the supply chain leadership rules and incentives (inducing power asymmetries) and by information asymmetry. For example, when two parties engage in a contractual relationship which is costly to break apart, or lock-in contracts, there may be risks for one or the other party or both. For this reason, the profit of collaboration by reducing the number of suppliers and building trustworthy relationships, engaging in long term supplies and exchange, locking oneself in dependence of any kind (such as joint technology, Intranets, joint planning, technology sharing etc.) is also a "two edge sword". In this sense collaboration in supply chains is not a "free lunch". Celebrated cases are of course outsourcing and franchises.

Outsourcing and Risks

Outsourcing (as discussed in Chapters 2 and 4) is essentially defined as the transfer of previously in-house activities to a third party (see also, Gattorna 1988; La Londe and Cooper 1989 for additional review of this problem). In such a transfer, economies of scale may be reached while fixed cost investments can be reduced rendering the outsourcing more agile-flexible. At the same time, there may be opportunity risks based on the search for self interest such as information asymmetries as stated above. The questions firms struggle with prior to outsourcing are therefore both complex and numerous. Should a firm strive to maintain its capacity or turn to an external (and therefore hardly controllable) supplier? Will a firm's technological positioning (and therefore its knowledge base in the future) be reduced? What are the firm's strategic options and contingent plans? These and other questions are important risk problems to contend with. For example, an essential motivation when outsourcing inventory, arises from economies of scale, risk and focus. These motivations presume that economic advantages arise from collaboration and exchange between firms, leading to a firm restructuring its organization to deal with its external supply chain. A typical example would in practice be to focus on a JIT (Just in Time) manufacturing strategy while outsource the management of inventories to a carefully selected supplier (although outsourcing and JIT might not be correlated). Such a practice can lead to numerous problems however. Specifically, when several firms act on the same market and outsource to a common supplier, they may augment significantly the demand volatility faced by the supplier (and thereby augment costs). Such risk considerations are therefore essential and to be accounted for when reaching the decision to outsource inventories.

Thus, inventory outsourcing involves not only reduced costs and the potential to focus on core competencies, but also risks. The two main risks (ex-ante and ex-post) in this case include the outsourcing of critical inventory activities, namely, risks assumed at the inventory and order stage and risks assumed ex-post once uncertainty in demands is revealed and supplies received. These risks are of course dependent on different factors such as supply delays and the preferential supplier-firm relationship. As a result, inventory outsourcing can be conceived in numerous ways, based on model relationships, which involve wholly or partly, arm's length contractual and conflicting partnerships. From a supplier's point of view the concern to maintain firms-clients that have outsourced as well as minimizing the costs of managing inventories are the prime objective. See also Baghana and Cohen 1998, Janssen and Kok 1999, Ritchken and Tapiero 1986,

Tapiero and Grando 2006, Van Donk and van der Vaart 2005, Tsay et al. 1998.

To manage outsourcing risks, a number of approaches is suggested in the literature. For example, essential factors to reckon with in reaching the decision to outsource require that we understand the specific competitive advantage the firm has, and recognize the firm's resource heterogeneity, the effects of imperfect mobility and its internal alignment. In this context, a firm to manage risks and seek one or several suppliers ought to: (i) Retain the resources responsible for competitive advantage; (ii) Avoid monopolistic or oligopolistic supply markets and (iii) Manage the risk of post-contractual dependency. In implementing the decision to outsource, negotiations relating to supply prices, supply security and assurances, back up and alternative supplies are the issues a firm will be confronted with. Should the firm have one or more suppliers? To what extend can a firm depend on its suppliers? Can contracts negotiated between two firms be reciprocal, in a manner that one will depend reciprocally on the other! What are the penalties for non conformance to contract terms? These are a sample of the many questions one may raise that can have risk implications. For this reason, in car manufacturing supply chains in Japan, several suppliers are used, emphasizing the independent development of parts, integrated into a whole at the Car manufacturer. As a result, outsourcing and external supply relationships are extremely varied with different types of supplier relationship; with different costs and rewards associated in each relationship. They are also varied with relationships designed to meet the supply chain specific needs and spanning "arm's length"- contractual, conflictual, limited or full partnership that may be fixed or varying over time.

As seen earlier, each relationship entails its own risks of supplying faulty material and products, information asymmetry and power risks (of moral hazard, adverse selection). In such an environment, the risk management of suppliers and outsourcing depends far more on organizational and properly conceived contracts than just technical analysis, albeit such an analysis is important as we shall see below through examples and in the next chapter as well. For example, single sourcing versus multiple sourcing can compound the supplies variation of firms, long term and locked in contracts can lead one firm to be totally dependent on the other (although long term contracts are considered important for sustaining a supply chain). Of course, a mutual commitment, a shift form a conflictual to a collaborative based on trade-offs and sharing, maximizing mutual understanding and an exchange of information leading to trustworthy and credible commitments are basic ingredients in outsourcing, supplier and supply chain relationships. These problems are of the utmost importance requiring a strategic approach to risk. For simplicity, we often reduce these problems to a treatable format as will be shown in a specific case below.

Franchises

Franchises are an old and broadly practiced economic arrangement, originating in the Middle Ages (X and XII the Centuries) where landed lords granted territorial rights to cultivate land by some in their local population. It expanded dramatically at the beginning of the century in both the US and Europe. The French Cotton firm (Lainiere de Roubaix, Laine Penguin) seeking to sell its textile expanded into 350 franchisees in less than ten years while in the US, Antitrust Laws of 1929 led US firms to the creation of distribution franchises by US car manufacturers. The expansion of franchises, mostly in services, has been since then spectacular, accounting for a substantial percentage of service and logistics activity. In France for example, there were 34 Franchisers in 1970 compared to 600 in 1990 and, of course, this number has expanded since the European Union integration. Franchises are an approach to collaboration between a franchiser—the firm, and franchisees, contracted for the purpose of exploiting a particular concept or advantage provided by the franchiser. It is mostly an economic agreement based on an exchange between parties made for profit, with each of the parties expecting to draw some advantage from the agreement. This general principle underlies franchise contracts, outsourcing agreements, joint partnerships etc. Franchises in particular, are essentially a contract between two legally independent firms establishing a long-term relationship where the franchiser grants to the franchisee the right to use the franchiser's trademark, the use of a specific (potentially patented) technology etc., In exchange, the franchisee pays a lump sum fee and annual royalties at an agreed percentage of sales.

A franchise may involve several other provisions as well as options that each of the parties may grant to the other. For example, risk sharing, exclusive territories with optional agreement appended to these agreements, promotional efforts sharing, buy-back provisions (Marvel 1982, Rey 1992, Rey and Tirole 1986, Tirole 1988, Mathewson and Winter 1986, Klein and Saft 1885). These contractual relationships are broadly used. Over one third of all retail sales in the US occur through a franchise system. For example, in many cases, production may be centralized while distribution may be franchised (e.g. car selling, some food and department stores, fast food, clothing trademarks etc.). In some cases as well, image and advertising is centralized but production is decentralized, franchised to companies focused in manufacturing (as it is increasingly the case).

The economic rationale for franchises arises due to the very high set up costs in selling as well as to problem of managing complex and diffused distribution systems. Thus, a franchisor may construct a franchising system where franchisees would invest parts, if not all, of the required local investment. Typically, such an agreement is made for definite or indefinite periods of time, which the owner of a protected trademark grants to franchisees, for some consideration, the right to operate under this trademark for the purpose of producing or distributing a product or service (Caves and Murphy 1976). Because the value of such assets is defined by their use, these contracts involve difficult contractual relations. Franchisee fees assume then many variations such as royalties, or commission, resale price maintenance, exclusive territories, exclusive dealing as well exclusivity relationships of various sorts with reciprocal agreements for the conduct of mutual services. The study of franchises involves as a result many issues such as resource constraints (thus the franchise will grant access to financial capital, market expertise and managerial talent of franchisee); incentive issues where the franchise system provides strong incentive for both parties to perform well; and of course an economy of scale where the franchiser assumes responsibility for economic activities where economies of scale can be realized.

Traditionally, an expected utility framework based on the parties' utilities for money is used to value franchise contracts (see, for example, Blair and Kaserman 1982, Caves and Murphy 1976, Mathewson and Winter 1986, Rubin 1978, Rubin and Carter, 1990). Such an approach is subjective however expressing the value that each of the parties draws from the agreement based on valuations that are no easily revealed. For example, each of the parties may calculate the discounted utility of gains and losses, summarized in a "flow of funds", over a relevant planning horizon. And on the basis of appropriate assumptions regarding the policies and managerial procedures adopted, a pricing "objective" is determined (see Kaufman and Dant 2001, Lafontaine 1992, Kaufman and Lafontaine 1994, Sen 1993). This price is not the market price for the franchise agreement and does not convey the true discount rate (which is both time and risk sensitive).

In addition franchising risk is imbedded in both the franchise contract and the ex-post controls applied to manage the franchisee-franchisor relationships once the contract is signed. For example, a typical franchise contract consists of a lump sum payment which may or may not be refundable and involves optional choices just as the relationship maintained over at least a certain length of time, at which the franchise can be renegotiated (as a way to commit the franchisee to entrepreneurial activity and safeguard from misuse of the franchise). Similarly, an advantage (or disadvantage) can be gotten through a tax on current inputs, such as selling current input at prices larger than the franchisor's marginal cost. For some contracts the franchisor supply parts of the fixed operating costs (when he leases land that he owns) combined (or not) with provisions to recapture the franchise (which alters the franchisee utility). Thus, even with the most stringent contract, franchises are subject to many risks. Risks of "milking" the franchise; asymmetry risks (in power and in information) and other risks resulting in sub-performing franchisees can harm the franchise brand as a whole. We will next consider a number of simple problems to highlight only some of these issues.

Below we shall consider two problems demonstrating alternative approaches to dealing with risk in both outsourcing and in franchises. The former is a straightforward expected minimization problem, while the latter provides an approach to pricing the franchise.

Outsourcing Inventory in a Supply Chain

We consider first a problem of "inventory outsourcing" (Tapiero and Grando 2006) with the supplier a leader, having full information of the outsourcing firm's demand distributions and parameters. This leads as we saw earlier to a Stackelberg game meaning that one of the parties in the game is a leader, aware of the other party—the follower, his motivations and his decisions. When dealing with an independent demand of the parties, the supplier benefits from (statistical) risk aggregation. On the other hand, if parties demands are dependent, this may lead to an unwieldy situation which requires that a risk management policy be adopted by the supplier (such as building an aggregate inventory as well as buying call options for further supplies, as shown in Ritchken and Tapiero 1986 and highlighted in the revised Example 2.1 in this chapter).

Say that we have a number of individual firms j, $j=1,2,\ldots$ managing inventories independently and ordering the quantities R_j inducing inventory and shortage costs given by c_{1j} and c_{2j} respectively, where $\tilde{D}_j$ is the individual firm demand for these quantities. The total incurred cost for each firm, j, is random and is defined by

$$\tilde{C}_j = p_m R_j + c_{1j} \left(R_j - \tilde{D}_j \right)^+ + c_{2j} \left(R_j - \tilde{D}_j \right)^- . \qquad (7.85)$$

Note that $p_m R_j$ is the value of materials to be ordered while the latter are cost items measured at the end of the period. Further, we use the notation:

$$(x)^+ = \begin{cases} x & \text{if } x > 0 \\ 0 & \text{if } x \le 0 \end{cases}, (x)^- = \begin{cases} -x & \text{if } x < 0 \\ 0 & \text{if } x \ge 0 \end{cases}, \begin{array}{l} x = (x)^+ - (x)^- \\ |x| = (x)^+ + (x)^- \end{array}. \quad (7.86)$$

The total expected first two moments of the costs are thus

$$E(\tilde{C}_j) = p_m R_j + c_{1j} E(R_j - \tilde{D}_j)^+ + c_{2j} E(\tilde{D}_j - R_j)^+, \quad (7.87)$$

$$\text{var}(\tilde{C}_j) = c_{1j}^2 \left\{ E\left[(R_j - \tilde{D}_j)^+\right]^2 - \left[E(R_j - \tilde{D}_j)^+\right]^2 \right\}$$

$$+ c_{2j}^2 \left\{ E\left[(R_j - \tilde{D}_j)^-\right]^2 - \left[E(R_j - \tilde{D}_j)^-\right]^2 \right\} - 2c_{1j}c_{2j} E(\tilde{D}_j - R_j)^+ E(R_j - \tilde{D}_j)^-$$

,(7.88)

where p_m is the current market price of buying the good (a part that might be needed in a production process). A risk neutral optimal ordering policy based on expected costs minimization can be found by minimizing $\hat{C}_j = E(\tilde{C}_j)$ above where $F_j(\tilde{D}_j)$ is the cumulative function of the jth firm demand. For such a firm, the optimal order policy is given by the first necessary conditions for optimality $\partial \hat{C}_j / \partial R_j = 0$, which leads to an optimal quantile risk specification for the inventory policy. Namely, we have:

$$F_j(R_j^*) = \alpha_j \text{ or } 1 - F_j(R_j^*) = 1 - \alpha_j = (c_{1j} + p_m)/(c_{1j} + c_{2j}). \quad (7.89)$$

This expression defines, at the least inventory cost, the shortage risk sustained by the outsourcing firm. As a result, the optimal ordering policy of an inventory managing firm is $R_j^* = F_j^{-1}(\alpha_j)$. Due to focusing and economies of scale, the supplier may acquire goods at a lower price $p_s \le p_m$, which he may use to set a selling price p_{ms}, $p_s \le p_{ms} \le p_m$ to outsourcing firms, lower than the market price. The supplier's holding and shortage costs are assumed given by parameters (c_{1s}, c_{2s}). Using the same model, the supplier adopts an optimal order policy given by

$$F_N(R^*) = \alpha_S \text{ where } 1 - \alpha_s = (c_{1s} + p_{ms})/(c_{1s} + c_{2s}), \quad (7.90)$$

where $F_N(.)$ is the cumulative distribution function of the aggregate demand $\sum_{j=1}^{n} \tilde{D}_j$ by all firms, assumed to be normal. The mean and variance are given respectively by

$$\mu = E\left(\sum_{j=1}^{n} \tilde{D}_j\right); \quad \sigma^2 = \text{var}\left(\sum_{j=1}^{n} \tilde{D}_j\right) = \sum_{j=1}^{n} \text{var}\left(\tilde{D}_j\right) + \sum_{j\neq i}^{n}\sum_{i=1}^{n} \text{cov}(\tilde{D}_i, \tilde{D}_j).(7.91)$$

When demands are independent, we have:

$$\sigma^2 = \sum_{j=1}^{n} \text{var}\left(\tilde{D}_j\right). \tag{7.92}$$

When demands are dependent, the demand variance faced by the supplier can be much greater (or smaller, depending on demand correlations).

For our special purpose case, assume that firms outsource to the supplier (and therefore do not hold inventories). If firm j is supplied $V_j \leq \tilde{D}_j$, a shortage $\left(\tilde{D}_j - V_j\right)^+ = Max\left(\tilde{D}_j - V_j, 0\right)$ is incurred with shortage risk $1 - F_j\left(V_j\right)$. To assume a risk of shortage smaller than the risk sustained by self-managing inventory, the outsourcing firm would require that

$$1 - F_j\left(V_j\right) \leq \frac{c_{1j} + p_m}{c_{1j} + c_{2j}} \quad \text{or} \quad \frac{c_{2j} - p_m}{c_{1j} + c_{2j}} \leq F_j\left(V_j\right). \tag{7.93}$$

As a result the least supply of an outsourcing firm requires (protected by an appropriate outsourcing contract)

$$V_j \geq F_j^{-1}\left(\frac{c_{2j} - p_m}{c_{1j} + c_{2j}}\right). \tag{7.94}$$

By outsourcing inventories, firms may thus profit not only by an expected cost reduction but by reducing inventory risks as well.

Valuation, Pricing and Franchises with A Binomial Process

For exposition purposes, we consider a one period binomial process representing a simplified franchise exchange agreement (for extensions and additional developments, see Tapiero, 2007). The terms of exchange consist in the transfer of a lump sum from the franchisee to the franchisor and a royalty payment. Furthermore, we also assume that the franchiser guarantees the franchisee by providing a buy back option that the franchisee can exercise at any time at a set price. In this sense, the franchisee has a Put option defined by the terms of the franchise contract defined by both the profitability of the franchise and the terms set by the franchisor. This particular characteristic is used to price the franchise price by replicating it to an equivalent Put option traded (if a market can be found for such trades) in some financial markets. Thus, the question we address is: what is the market price for the franchisee's investment, or equivalently, what is the franchisee's risk premium when investing in the franchise, as required by the franchisor? Further, what are the terms that the franchiser can provide in such a franchise contract? Unlike game theoretic approaches to such problems, based on the parties' interest and information, our approach is based on the existence of complete markets and both parties use such markets to price the terms of the franchise contract. In this sense, there is no conflict, but a price equilibrium that the franchiser and the franchisee use in determining the terms they would accept and at what price.

Say that a franchisee initial investment is K, part of which, βK, $0 < \beta < 1$, is transferred to the franchiser as a lump sum for the right to exercise under the franchiser banner. The net starting investment of the franchise is, thus, $(1-\beta)K$. In addition, assume that the franchisee pays out to the franchiser a royalty at a proportional rate $0 \le \alpha < 1$ to the profit made in the period. The price of the franchisee a period later is then either $(1-\beta)K(1+h)$, $h > 0$ or $(1-\beta)K(1-\ell)$, $\ell > 0$, where (h, ℓ) are the rates of return in case of economic success or economic failure. The transfer amounts to the franchiser are therefore $\alpha(1-\beta)K(1+h)$, $h > 0$ or $\alpha(1-\beta)K(1-\ell)$, $\ell > 0$, with their complement values remaining as part of the franchisee's income. If markets are complete (in a financial sense), and assuming an implied risk neutral probability for such markets, the current price equals the discounted future price at the risk free rate R_f, or

$$(1-\beta)K = \frac{1}{1+R_f}\left[p_{RN}(1-\beta)K(1+h) + (1-p_{RN})(1-\beta)K(1-\ell)\right].(7.95)$$

And therefore, the implied risk neutral probability is

$$p_{RN} = \frac{R_f + \ell}{h + \ell}, \quad R_f < h \cdot \tag{7.96}$$

Given the terms of exchange between the franchisee and the franchiser, each of the parties will be faced with the following cash flows (where the first term is the franchisee and the second the franchiser and as explained below).

At time $t=0$, $(-K, \beta K)$, while at time $t=1$:

$$\begin{cases} (1-\alpha)(1-\beta)K(1+\xi) + K(1-\beta)\big(\rho + \max\big(\mu(\xi)-\rho,0\big)\big) \\ K\beta(1+r) + \alpha(1-\beta)K(1+\xi) - K(1-\beta)\max\big(\rho-\mu(\xi),0\big) \end{cases} . \tag{7.97}$$

If the franchisee exits the franchise at time $t=1$, we have:

$$\begin{cases} Q + \Phi \\ -Q - \Phi \end{cases}$$

where

$$\Phi = Max\left((1-\alpha)(1-\beta)K(1+\xi)-Q,0\right). \tag{7.98}$$

Here ξ denotes the binomial states, namely whether the franchise was highly profitable or less, r denotes the rate of return the franchiser collects on the lump sum payment, while $\left(K-(1-\alpha)(1-\beta)K(1+\xi)\right)$ denotes the franchisee return for the period and therefore a rate of return for the period is given by

$$\mu(\xi) = \frac{\left(1-(1-\alpha)(1-\beta)(1+\xi)\right)}{(1-\beta)} - 1. \tag{7.99}$$

Of course, if the franchisee is assured a least rate of return equal to ρ by the franchise, then we have a rate of return, $K(1-\beta)\max\left(\mu(\xi),\rho\right)$. Since, $\max\left(\mu(\xi),\rho\right) = \rho + \max\left(\mu(\xi)-\rho,0\right)$, the franchiser is responsible only for the complement in case the franchisee does not reach the guaranteed return, or $\max\left(\rho-\mu(\xi),0\right)$. We have, as a result, the cash flow indicated in our equation (7.97) above.

By the same token, if the franchiser provides an exit price to the franchisee as a function of the franchisee's investment, then when the franchisee exercises this option (in fact, a perpetual American option), then at the exit time, the franchisee collects the maximum of $(1-\alpha)(1-\beta)K(1+\xi)$ and the exit price, denoted by, say, Q. In other words, the franchiser supplies the franchisee with the following option

$$Max\left((1-\alpha)(1-\beta)K(1+\xi),Q\right) = Q + \Phi,$$

where

$$\Phi = Max\left((1-\alpha)(1-\beta)K(1+\xi)-Q,0\right). \tag{7.100}$$

This option is evidently a cost to the franchiser, as stated above since it involves a transfer of funds to the franchisee. As a result, at time $t=1$, the franchisee and the franchiser collect (or pay out) $P_f(1)$ and $P_F(1)$ respectively and by risk neutral pricing

$$P_f(0) = \frac{1}{1+R_f}E_{RN}\left(P_f(1)\right); \; P_F(0) = \frac{1}{1+R_f}E_{RN}\left(P_F(1)\right), \tag{7.101}$$

where

$$P_f(0) = K; \; P_F(0) = -K\beta. \tag{7.102}$$

In other words, the present value to the franchisee equals in a complete market his investment, while the franchiser collecting the lump sum, $K\beta$, is receiving such a payment to meet future obligations to the franchisee. If, $\mu(h) - \rho > 0$, then for the franchisee we have

$$E_{RN}\left(P_f(1)\right) = (1-\alpha)(1-\beta)K\left(p_{RN}(h+\ell)+(1-\ell)\right) +$$
$$+ K(1-\beta)\left(\rho + p_{RN}(\mu(h)-\rho)\right) + p_{RN}\left((1-\alpha)(1-\beta)K(1+h)\right) + (1-p_{RN})Q \tag{7.103}$$

while for the franchiser

$$E_{RN}\left(P_F(1)\right) = K\beta(1+r) + \alpha(1-\beta)K\left(p_{RN}(h+\ell)+(1-\ell)\right)$$
$$- K(1-\beta)(1-p_{RN})\left(\rho-\mu(\ell)\right) - (1-p_{RN})\left(Q-(1-\alpha)(1-\beta)K(1-\ell)\right) \tag{7.104}$$

Note that the last term in the equation above corresponds to the money exchange in case the franchisee chooses to exit the franchise agreement. As a result, we obtain the following system of equations expressed in exchange terms of the franchise,

$$K\left(1+R_f\right) = (1-\alpha)(1-\beta)K\left(p_{RN}(h+\ell)+(1-\ell)\right) +$$
$$+ K(1-\beta)\left(\rho + p_{RN}(\mu(h)-\rho)\right) + p_{RN}\left((1-\alpha)(1-\beta)K(1+h)\right) + (1-p_{RN})Q$$
$$-K\beta\left(1+R_f\right) = K\beta(1+r) + \alpha(1-\beta)K\left(p_{RN}(h+\ell)+(1-\ell)\right) \tag{7.105}$$
$$- K(1-\beta)(1-p_{RN})\left(\rho-\mu(\ell)\right) - (1-p_{RN})\left(Q-(1-\alpha)(1-\beta)K(1-\ell)\right)$$

Following some elementary manipulations, we have

$$\frac{(1+R_f)}{(1-\beta)} - (1-\alpha)(1-\ell) - \rho = p_{RN}\left((1-\alpha)(1+2h+\ell)+\mu(h)-\rho\right) + \frac{(1-p_{RN})Q}{(1-\beta)K} \tag{7.106}$$
$$\frac{\beta(R_f-r)}{(1-\beta)} + \alpha(1-\ell) = p_{RN}\alpha(h+\ell) + (1-p_{RN})\left[\frac{Q}{(1-\beta)K}+\rho-\mu(\ell)-(1-\alpha)(1-\ell)\right]$$

Using the risk neutral pricing probabilities calculated earlier, we have for the franchisee and the franchiser:

$$\frac{(1+R_f)}{(1-\beta)} - (1-\alpha)(1-\ell) - \rho = \left(\frac{R_f+\ell}{h+\ell}\right)\left((1-\alpha)(1+2h+\ell)+\mu(h)-\rho\right) + \left(\frac{h-R_f}{h+\ell}\right)\frac{Q}{(1-\beta)K} \tag{7.107}$$

$$\frac{\beta(R_f-r)}{(1-\beta)} + \alpha(1-\ell) = \alpha\left(R_f+\ell\right) + \left(\frac{h-R_f}{h+\ell}\right)\left[\frac{Q}{(1-\beta)K}+\rho-\mu(\ell)-(1-\alpha)(1-\ell)\right] \tag{7.108}$$

Consequently, for a fixed sharing agreement, the lump sum transfer payment is found by equating these two equations and solving them. Alternatively, for a fixed royalty contract, the lump sum payment can be determined by equating these equations. We can then calculate the resulting ration Q/K, expressing the proportion of the franchisee investment which is guaranteed by the franchiser. A numerical analysis of these equations is considered below emphasizing the substitution between the problem's parameters. Of course, given one parameter, the other can be calculated. Such an approach can be used in various other manners. For example, the potential returns of the franchise may be determined by the efforts of franchisees and the franchiser (for example through greater advertising), altering thereby the potential returns, $(h, -\ell)$, and their risk neutral probabilities and the actions that ensue these implied probabilities. Of course, greater investment in such returns will increase the price of the franchise. While, milking the franchise, will dim its prospects and reduce its price. If the franchiser and the franchisee are mutually aware of each other preferences and the implications of their acts and policies, a game might follow, priced also by the market as a function of their resulting strategies. In such cases, distrust and non-collaborative behaviors can result in large losses by both parties. The consequences of such gains and losses can be assessed using the framework we have outlined, appropriately expanded to be time sensitive (i.e., in a multi-period context) and more specific in terms of the return processes unfolding as the franchisee and the franchiser adopt their respective policies. Such situations and games may be topics for additional research, albeit the approach followed in such research would conceptually be the same as that pursued here.

Example 7.5

For the following parameters: $h=0.3$, $l=0.15$, $R_f=0.05$, $r=0.12$, $\alpha=0.08$, $\rho=0.16$ we computed in equation (7.98), Q/K, the ratio of the exit strike to the franchisee investment, as a function of β, the proportion paid upfront to the franchiser. The results are shown in Figure 7.1 below.

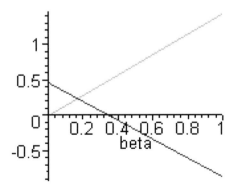

Figure 7.1. Equations (7.107) and (7.108) of Q/K as a function of β

The intersection of the two equations provides the simultaneous solution in β, which is found with Maple to be 0.1693159796. In other words, if a franchisee were to invest \$100,000 for his acquiring and operating a franchise, then the upfront payment to the franchiser would be \$16,931. In this sense, the terms the franchisee would be confronted will be the down payment of 16,931 to the franchiser and a transfer of 8% of all future earnings.

REFERENCES

Agrawal V, Seshadri S (2000). Risk intermediation in supply chains. *IIE Transactions* 32: 819-831.

Agrell PJ, Lindroth R, Norrman A (2004) Risk, information and incentives in telecom supply chain. *International Journal of Production Economics* 90 (1): 1-16.

Ait–Sahalia Y, Lo A (1998) Nonparametric estimation of state-price densities implicit in financial asset prices. *Journal of Finance* 53: 499-548.

Ait-Sahalia Y, Lo AW (2000) Nonparametric risk management and implied risk aversion., *Journal of Econometrics* 94: 9-51.

Akella R, Araman VF, Kleinknect J (2002) B2B markets: procurement and supplier risk management in business. In: Geunes J, Pardalos PM, Romeijn HE (Eds.) *Supply Chain Management—Applications and Algorithms,* Kluwer, pp. 33-66.

Akerlof G (1970) The Market for Lemons: Quality Uncertainty and the Market Mechanism. *Quarterly Journal of Economics* 84: 488-500.

Brun A, Caridi_M, Fahmy-Salama K, Ravelli I (2006) Value and risk assessment of supply chain management improvement projects. *Int. J. Production Economics* 99: 186-201.

Anagnou I (2003) The Relation between Implied and Realized Probability Density Functions, Working Paper,, University of Warwick.

Anupindi R (1993) *Supply Management Under Uncertainty.* Ph.D. Thesis, Graduate School of Industrial Administration, Carnegie Mellon University.

Anupindi R, Akella R (1993) Diversification under supply uncertainty. Management Science 39(8): 944-963.

Aviv Y (2004) Collaborative forecasting and its impact on supply chain performance. In: Simchi-Levi D, Wu D, Zhen Z (Eds.), *Handbook of Quantitative Supply Chain Analysis*, Kluwer Publisher, Dordrecht.

Aviv Y (2005) On the benefits of collaborative forecasting partnerships between retailers and manufacturers. Working Paper, Olin School of Management, Washington University.

AON (2005) Protecting Supply Chains Against Political Risks, available from (www.AON.com).

Artzner P, Delbaen F, Eberand JM, Heath D (1997) Thinking coherently, *RISK*, 10: 68-71.

Artzner P, Delbaen F, Eber JM, Heath D (1999) Coherent risk measures. *Mathematical Finance* 9: 203-228.

Artzner P, Delbaen F, Eber JM, Heath D (2000) Risk management and capital allocation with coherent measures of risk, Available from www. math.ethz.ch/finance.

Artzner P, Delbaen F, Eber JM, Heath D, Ku H (2001). Coherent multi-period risk adjusted values, Available from www.math.ethz.ch/finance.

Babich V, Burnetas A, Ritchken P (2004) Competition and diversification effects in supply chains with supplier default risk. Working Paper, Department of Industrial and Operations Engineering, University of Michigan.

Bagahana MP, Cohen M (1998) The stabilizing effect of inventory in supply chains. *Operations Research* 46: 572-583.

Bahra B (1997) Implied Risk-Neutral Probability Density Functions from Option Prices: Theory and Application', Working Paper, Bank of England.

Barzel Y (1982) Measurement cost and the organization of markets, *Journal of Law and Economics* 25: 27-47.

Beckers S (1996) A Survey of Risk Measurement Theory and Practice, in *Handbook of Risk Management and Analysis,* Alexander C (ed).

Bell DE (1982) Regret in decision making under uncertainty. *Operations Research* 30: 961-981.

Bell DE (1985) Disappointment in decision making under uncertainty., *Operation Research* 33: 1-27.

Bell DE (1995) Risk, return and utility. *Management Science* 41: 23-30.

Blair RD, Kaserman DL (1982) Optimal franchising, *Southern Economic Journal* 49(2): 494-505.

Boone Y, Ganeshan R, Stenger A (2002) The benefits of information sharing in a supply chain: An exploratory simulation study. In: Geunes J, Pardalos P, Romeijn E (Eds.) *Supply Chain Management Models, Applications and Research Directions.* Kluwer Publishers, Dordrecht.

Cachon G, Fisher M (2000) Supply chain inventory management and the value of shared information. *Management Science* 46: 1032–1048.

Caves RE, Murphy WE (1976) Franchising firms, markets and intangible assets. *Southern Economic Journal,* 42: 572-586.

Cheng TCE, Wu YN (2005) The impact of information sharing in a two-level supply chain with multiple retailers. *Journal of the Operational Research Society* 56: 1159-1165.

Chopra S, Sodhi M (2004) Avoiding supply chain breakdown. *Sloan Management Re*view 46(1): 53-62.

Christopher M (1992) *Logistics and Supply Chain Management*. Pitman, London.

Tang CS (2006) Perspectives in supply chain risk management. *Int. J. Production Economics* 103: 451-488.

Cohen MA, Agrawal N (1999) An analytical comparison of long and short term contracts. *IIE Transactions* 31: 783-796.

Corbett C (2001) Stochastic inventory systems in a supply chain with asymmetric information: Cycle stocks, safety stocks, and consignment Stocks. *Operations Research* 49: 487–500.

Corbett C, de Groote X. (2000) A supplier's optimal quantity discount policy under asymmetric information. *Management Science* 46: 444–450.

Corbett C, Tang CS (1998) Designing supply contracts: Contract type and information asymmetric information. In: Tayur et al. (Eds.), *Quantitative Models for Supply Chain Management*. Kluwer Publisher, Dordrecht.

Desiraju R, Moorthy S (1997) Managing a distribution channel under asymmetric information with performance requirements. *Management Science* 43: 1628-44.

Dyer JS, Jia J (1997) Relative Risk-Valuemodel. *Euro. J. of Operations Research* 103: 170-185.

Eeckhoudt L, Gollier C, Schlesinger H (1995). The risk-averse (and prudent) newsboy. *Management Science* 41(5): 786-794.

Embrechts P (Ed.) (2000) *Extremes and Integrated Risk Management*. Risk Books, London.

Gattorna J (Ed.) (1998) *Strategic Supply Chain Alignment*, Chap. 27, Gower, Aldershot.

Harland C, Brencheley H, Walker H (2003) Risk in supply network. *Journal of Purchasing and Supply Management* 9(2): 51-62.

Hallikas J, Karvonen I, Pulkkinen U, Virolainen VM, Tuominen M (2004). Risk management processes in supplier networks. *International Journal of Production Economics* 90(1): 47-58.

Hirschleifer J, Riley JG (1979) The Analysis of Uncertainty and Information: An Expository Survey. *Journal of Economic Literature* 17: 1375-1421.

Holmstrom B (1979) Moral hazard and observability. *Bell J. of Economics* 10(1): 74-91.

Holmstrom B (1982) Moral hazard in teams. *Bell Journal of Economics.* 13(2): 324-40.

Jackwerth JC (1999) Option implied risk neutral distributions and implied binomial trees: a literature review. *Journal of Derivatives* 7: 66-82.

Jackwerth JC (2000) Recovering risk aversion from option prices and realized returns. *The Review of Financial Studies* 13(2): 433-451.

Janssen F, de Kok T (1999) A two-supplier inventory model. *International Journal of Production Economics* 59: 395-403.

Jorion P (2000) VaR: *The New Benchmark for Managing Financial Risk.* McGraw Hill, New York.

Kaufmann PJ, Dant RP (2001) The pricing off franchise rights, *Journal of Retailing* 77: 537-545.

Kaufman PJ, Lafontaine F (1994) Costs of Control: The source of economic rents for McDonald's franchises. *The Journal of Law and economics* 37(2): 413-453.

Klein B, Saft LF (1985) The law and economics of franchise tying contracts. *Journal of Law and Economics* 345-349.

Kleindorfer P, Saad G (2005) Managing disruption risks in supply chains. *Production and Operations Management* 14: 53-68.

Lafontaine F (1992) Contract theory and franchising: some empirical results. *Rand Journal of Economics* 23(2) 263-283.

La Londe B, Cooper M (1989) *Partnership in providing customer service: a third-party perspective*, Council of Logistics Management, Oak Brook, IL.

Lee H (2004) The triple—a supply chain. *Harvard Business Review*, 102–112.

Lee HL, Padmanabhan V, Whang S (1997a). Information distortion in a supply chain: The bullwhip effect. *Management Science* 43: 546–548.

Lee HL, Padmanabhan V, Whang S (1997b) The bullwhip effect in supply chains. *Sloan Management Review* 38: 93–102.

Loomes G, Sugden R (1986) Disappointment and Dynamic Consistency in Choice Under Uncertainty. *Review of Economic Studies* 53: 271-282.

Mathewson GF, Winter RA (1986) The economics of franchise contracts. *Journal of Law and Economic* 28: 503-526.

Marvel H (1982) Exclusive dealing. *Journal of Law and Economics* 25: 1-26.

Munier B, Tapiero CS (2008) Risk Attitudes, *Encyclopedia of Quantitative Risk Assessment and Analysis,* Wiley, (Forthcoming).

Nagurney A, Curz J, Dong J, Zhang D (2005) Supply chain networks, electronic commerce, and supply side and demand side risk. *European Journal of Operational Research* 164: 120–142.

Parlar M, Perry D (1996) Inventory models of future supply uncertainty with single and multiple suppliers. *Naval Research Logistics* 43: 191–210.

Rey P (1992) The economics of franchising, ENSAE Paper, February, Paris.

Rey P, Tirole J (1986) The logic of vertical restraints, *American Economic Review* 76: 921-939.

Rice B, Caniato F (2003) Supply chain response to terrorism: Creating resilient and secure supply chains. Supply Chain Response to Terrorism Project Interim Report, MIT Center for Transportation and Logistics, MIT, Massachusetts.

Riordan M (1984). Uncertainty, asymmetric information and bilateral contracts. *Review of Economic Studies* 51: 83-93.

Ritchken P, Tapiero CS (1986) Contingent Claim Contracts and Inventory Control. *Operations Research* 34: 864-870.

Rubin PH (1978) The theory of the firm and the structure of the franchise contract. *Journal of Law and Economics* 21: 223-233.

Rubin PA, Carter JR (1990) Joint optimality in buyer–supplier negoti-ations. *Journal of Purchasing and Materials Management* 26(1): 54-68.

Savage LJ (1954) The *Foundations of Statistics*, Wiley.

Sen KC (1995) The use of initial fees and royalties sin business format franchising. *Managerial and Decision Economics* 14(2) 175-190.

Sheffi Y (2001) Supply chain management under the threat of international terrorism. *International Journal of Logistics Management* 12(2): 1-11.

Tapiero CS (2000) Ex-Post Inventory Control. *International Journal of Production Research* 38(6): 1397-1406.

Tapiero CS (2005) Value at Risk and Inventory Control. *European Journal of Operations Research* 163(3): 769-775.

Tapiero CS (2005a) Risk Management, John Wiley *Encyclopedia on Actuarial and Ris*k Management, Wiley, New York-London.

Tapiero CS (2005b) Risk *and Financial Management: Mathematical and Computational Methods*, Wiley, London and New York.

Tapiero CS (2006) Consumers Risk and Quality Control in a Collaborative Supply Chains. *European Journal of Operations research*, (available on line, October 18).

Tapiero CS (2007) Market Pricing Franchise Contracts, Working Paper, Polytechnic University, New York.

Tapiero CS, Grando A (2006) Supplies Risk and Inventory Outsourcing, *Production Planning and Control* 17(5): 534-539.

Tayur S, Ganeshan R, Magazine M (1998) *Quantitative Models for Supply Chain Management*. Kluwer Publisher, Dordrecht.

Tirole J (1988) *The Theory of Industrial Organization.* The MIT Press, Cambridge, MA.

Tsay AA, Nahmias S, Agrawal N (1998) Modeling supply chain contracts: a review. In: Tayur, Ganeshan, Magazine (Eds.) *Quantitative Models for Supply Chain Management.* Kluwer Academic Press, Norwell, MA, pp. 299–336.

Van Donk DP, van der Vaart T (2005) A case of shared resources, uncertainty and supply chain integration in the process industry. *International Journal of Production Economics* 96: 97 108.

Zsidisin G, Panelli A, Upton R (2001) Purchasing organization involvement in risk assessments, contingency plans, and risk management: An exploratory study. *Supply Chain Management: An International Journal,* 5(2): 187-197.

8 QUALITY AND SUPPLY CHAIN MANAGEMENT

Managing (non) quality and the risk consequences have generally assumed that the underlying uncertainty faced by firms, individually and collectively, is neutral! In other words, uncertainty and risk are not motivated while issues relating to information, information and power and parties' intentionalities are mostly neglected. Supply chains however are beset by multiple parties interacting with broadly varying motivations, information and power asymmetries. In addition, there may be conflicting and competitive objectives as well as environments (in the form of governments, other supply chains and interest groups) that render the management of quality in supply chains far more strategic. This raises many problems that are specific to supply chains and require particular attention. In figure 8.1 some techniques and a number of factors are pointed out, summarizing a number of concerns that will be considered in this chapter.

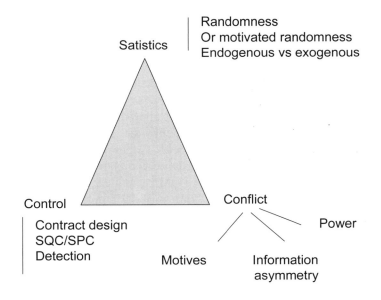

Figure 8.1. Quality related techniques and factors

Three elements include: statistics, conflict and control. Statistics deals with the uncertainties we face and how they are defined empirically-quantitatively. For example, is uncertainty originating in (unmotivated) randomness, is randomness motivated (as it would be if it were to depend on other parties actions and intentions). Is randomness the product of external and uncontrollable hazard or it is endogenous, resulting from our actions and the incentive we apply in managing quality and the supply chain? Conflict refers to the mutual behavioral and organizational relationships that evolve in a supply chain. If there is an information asymmetry or if there is a power asymmetry, and if there are separate and potentially non-identical objectives, one of the parties may resort to an opportunistic behavior. In this sense, the questions stating to what extent are the supply chain parties independent or dependent, which party is a leader and which is a follower are important ones. To control quality and the quality of the relationship in a supply chain, it means to control what the parties of the supply chain do, both ex-ante and ex-post. For example, are parties complying in meeting the contractual agreements they have agreed to; is the quality delivered, the quality agreed on between the parties etc.

The implications of such questions and of the control of quality in supply chains are of course inherent in the assumptions we are willing to make regarding the parties implied, their behavior and the characteristics of the underlying supply chain processes. For example, are there quality incentive contracts to manage the quality of products transferred from one firm to another? What are the implications to any of the parties of a non-quality originating in any specific firm of the supply chain? How is quality controlled across the supply chain and what are the pre-posterior controls (contract design) that allow both a monitoring-control and the choice of actions (which have, of course, consequences for the responsible supply parties)? These problems require technical approaches such as game theory, random payoff games and other approaches that are far more sensitive to the types of problems we have to deal with. Generally, in managing quality in supply chains, three essential approaches can be used. First, the plethora of human based approaches based on TQM (which are not considered here), Contracts negotiations and the economics of such contracts and finally, a strategic approach to the management of quality and its control (based on the endogenous uncertainty that arises due to the game parties engage in). We shall consider only some of these approaches, demonstrated profusely through examples.

8.1 QUALITY AND CONTRACTS

A contract is a bilateral binding agreement by which agreed upon exchange terms between two parties are used as substitutes to market mechanisms. This may involve contracts regarding work practices, payments and salary scales and a set of clauses intended to protect each of the parties against possible non-compliance by one of the parties bound by the contract. The essential advantage resulting from a contract is to protect both parties, reduce the uncertainty they may face and thereby stabilize their respective operating environments. For example, a supplier who enters into a contractual relationship with a specific producer may secure a certain level of sales which brings both profits and stability to its operational plans (as it is the case in supply chains). A producer could assure (through inspection sampling) that special care be given by the supplier to materials and parts. Pre-contract negotiations, which vary from situation to situation, provide an opportunity to clarify future terms of exchange and provide protection for each of the parties once the contract is signed. A poorly designed contract may be disastrous for the supplier and the producer alike, since ex-post-contract disagreements can lead to litigations, which are usually very costly and therefore important source of risks for the proper supply chain operations. For example, if delivery of quality products is not specifically stated as special clauses, suppliers may be tempted to supply sub-standard products. Similarly, union-management negotiations, over generous terms for one of the parties can lead to an environment which will induce non-quality by one of the parties taking advantage of situations as they arise and potentially cheating or non-conforming to the terms of the employment contract. For example, in the beginning of the industrial revolution, overly harsh working conditions induced workers to sabotage their machines by putting their sabot (wooden shoes) in the machine requiring thereby both direct and statistical controls. By the same token, overly protective measures for work, or pay scales based on piece work only can have adverse effects on inventory accumulation and on the production of quality produced (if they are not sensitive to the quality of work as well).

When there is an information asymmetry, a party may take advantage of special situations, in contradiction with the terms of the contract negotiated (in letter or in spirit). Such behaviors include opportunistic behavior, cheating, hiding facts, interpreting falsely or to one's own interest certain outcomes and situations. In other words, one of the parties (or both) may resort to opportunistic behavior. For example, when the cost of inspecting quality is large and there is an information asymmetry, a supplier can be tempted to supply a sub-standard quality (in contrast to the agreed upon and negotiated quality supply contract). To reduce such risk (usually called moral hazard as discussed in Chapter 7), once a contract has been signed the producer must devise a strategy which will provide an incentive to meet the terms of the contract and sufficient protection in case of supplier default. In these circumstances, inspection helps detect sub-standard quality. For example, if a part is tested and found defective, a rebate (negotiated at the time the quality contract was signed) can be paid by the supplier which in effect reduces the price of parts to the producer on the one hand and provide an incentive to the supplier to perform as agreed upon by the terms of the contract. Foreseeing such situations and providing rules for sharing the costs of non-quality are extremely important in determining the actual quality delivered as well as for instituting controls by both the supplier and the producer alike. To highlight some of the economic issues associated to contracts we shall consider specific problems.

For demonstration purposes, say that a contract consists of a quality Q for a product and a transfer payment p. The supplier has a private knowledge regarding Q which is given by z, providing a statistical information regarding Q and probability density function $F(.)$ while the buyer provides information which is given by y with a probability density function $G(.)$. The supplier and the buyer profits are given with respect to sales $R(Q,y)$ and production cost $C(Q,y)$ as:

$$\pi_B = R(Q, y) - p$$
$$\pi_S = p - C(Q, y)$$

The price p is therefore a transfer between the parties. If these were collaborating, then the sum of their profits would be:

$$\pi_B + \pi_S = R(Q, y) - C(Q, y)$$

Then, a quality contract is a pair of valued function $\{Q(y,z), P(y,z)\}$ which depends on the information available to each. When quality Q is a function of the private information available to both parties and the (transfer) price is a function of this information, an efficient contract can be defined by the following: "*A contract is said to be efficient if it maximizes the profit of one agent given some level of profit for the other*".

$$Max \ \pi_B \ \text{Subject to:} \ \pi_S \geq \overline{\pi}_S$$
$$Max \ \pi_S \ \text{Subject to:} \ \pi_B \geq \overline{\pi}_B$$

If the parties in the supply chain collaborate, then the centralized profit is given y: $\pi(Q, y, z) = R(Q, y) - C(Q, z)$. Mathematically, this means that for a strongly convex function of profits, we require:

$$\frac{\partial \pi}{\partial Q} = \frac{\partial R}{\partial Q} - \frac{\partial C}{\partial Q} = 0, \quad \frac{\partial^2 R}{\partial Q^2} - \frac{\partial^2 C}{\partial Q^2} < 0,$$

which provides the quality level that maximizes a centralized profit. Therefore the maximum profit, $\pi^*(Q, y, z)$, is defined as well in terms of the contractual terms agreed on by the parties. When the parties have their own private information (in which case there is an information asymmetry), the problem is slightly more complex, as indicated earlier.

We consider next the case of information asymmetry, in which case we assume that:

- Buyer observes y but not z;
- Seller observes z but not y.

In such circumstances *a contract is said to be* incentive compatible *in the following condition. Say that the buyer observes y and reveals y*.* Then, the contract is incentive compatible for the buyer if:

$$E\{R(X(y,z), y) - P(y,z)|y\} > E\{R(X(y,z), y) - P(y,z)|y'\}$$

For contract to be strongly incentive compatible it requires that truthfulness be always optimal even after knowing the private information of the opposing agent. For example, Supplier will be truthful even if it already knows the sampling plan of the producer. Incentive compatibility thus requires that an agent will always find it optimal to reveal private information before knowing the private information of the opposing agent. To highlight these problems we shall consider the following problems.

- Discuss three forms of producers-supplier relationships in a supply chain: (i) Conflictual, (ii) Contractual and (iii) Partnering. What are the advantages and the disadvantages of these organizations? Formulate the corresponding optimization problems.
- Consider a conflicting relationship between a producer and his supplier. What would be the effects of an information asymmetry between the two and what would be the effects of sharing information. Discuss two imagined situations involving, in the first case, information regarding product quality and, in the second, information regarding demand requirements for some parts.
- What are the effects of the dependence of a producer on a unique supplier? Contrast the advantages and disadvantages of a single versus a multiple suppliers outsourcing policy (to do so, use the concept of Stackelberg games considered earlier in the book).

Example: Types of quality-quantity contracts

Information asymmetry beset the many types of quality contracts that can be defined, including for example: *Requirement contract* that grants quantity discretion to the buyer within pre-negotiated limits, but allows quantity to move outside those limits with the mutual consent of both buyer and seller. *Output contracts* are the converse, giving limited quantity discretion to the seller. Finally, *quantity contracts* are intermediate between both parties; pre-negotiating a definite transaction quantity but permits parties to mutually agree to deviate from that quantity

8.2 MUTUAL SAMPLING

Supply chains are as stated earlier organizational frameworks based on exchange and dependence between firms, each with its own objectives and motivations and drawing a payoff whose risks it must also sustain and manage. Collaboration is not always possible however, for agreements may be difficult to self-enforce and as a result dependence risks of various sort may lead some firms to take advantage of their position in the supply chain network either because of power or information asymmetries. Further, profit from collaboration must also be justified for parties which will be involved together if the supply chain collaborates in fact. The traditional control of quality however assumes mostly that the underlying uncertainty faced by firms, individually and collectively, is neutral. In other words, the risk consequences measured by non-conforming quality are not motivated and therefore, the traditional approach to quality and its control has ignored the strategic and competitive effects of managing quality in an environment where firms act for their self interest (for exceptions see Reyniers and Tapiero 1995a, 1995b).

The implications of such an environment to the control of quality in supply chains are of course inherent in the assumptions we are willing to make regarding the supply chain organization on the one hand and the quality contract engaging the parties on the other. For example, are there incentives to deliver conforming quality between the parties? What are the risk and economics consequences of delivering poor quality? Is quality controlled across the supply chain and what are the pre-posterior controls that allow both a monitoring-control and a choice of actions by the parties. Typically, in supply chains, uncertainty arises not only due to the uncertainties in the underlying processes producing quality but also on the motivations and preferences that each of the parties has. In this sense, in addition to statistical uncertainty, the management of quality may include strategic uncertainty arising due to conflicts latent between the supply chain firms. As a result, in such an environment, games of strategies and the control ex-ante and ex-post of quality might lead to quality control strategies that are "mixed", with both strategic (threats and menaces) and statistical (information and assurance based) considerations. In this section, we consider the control of quality contracts from a number of perspectives, emphasizing both competition and collaboration.

Consider a contract to deliver parts or products of "acceptable" quality defined by the contract on the one hand and economic consequences for each of the parties in case the terms of the contract are not met. The essential advantage resulting from a contract is therefore to protect both parties, reduce the uncertainty they may face and thereby stabilize their respective operating environments. In a producer-supplier environment, a producer could assure (through inspection sampling) that special care be given by the supplier to materials and parts. Pre-contract negotiations, which vary from situation to situation, provide an opportunity to clarify future terms of exchange and provide protection for each of the parties once the contract is signed. For example, if delivery of quality products is not specifically stated in special clauses, suppliers may be tempted to supply sub-standard products.

To resolve some of these issues such problems may lead to, we assume that money and risk define the parties objectives. To focus our attention we consider some examples and calculate the risks and the control associated to specific supply chain organizations. Although the problems we formulate can be analyzed analytically in a very limited number of cases, numerical calculations can be reached with relative ease. To keep matters tractable however, some simplifications are made.

In a lone-firm framework, control-sample selection consists in minimizing a consumer risk (or a type II risk $\beta_c(n,k)$ in a Neymann-Pearson statistical framework) which consists in accepting a lot which is "not conforming" subject to a Producer risks $\alpha_C(n,k)$ (or type I error) which consists in rejecting a "good lot". These risks will be explained further below and are usually and explicitly defined in terms of control inspection parameters, for example (n,k) where n is a sample size on the basis of which a decision (based on the result of such an inspection compared to parameter k) is reached. This can be formulated as follows:

$$\underset{(n\geq 0, 0\leq k\leq n)}{Min}\left[\beta_c(n,k)\right] \text{ Subject to: } \alpha_C(n,k)\leq\overline{\alpha}_C \qquad (8.1)$$

That is, minimizing the type II errors (of say accepting a bad lot) subject to a type I error (of say, rejecting a good lot). The parameter $\overline{\alpha}_C$ is usually specified by the parties, representing the risk it is willing to assume. In a producer-supplier environment, both the statistical risks of the supplier and the producer are to be considered and the economic consequences, negotiated, resulting from a game that both parties engage in. We consider such a game by considering a number of situations co-existing in supply chains. These examples highlight the approach we use. We shall begin with some essential assumptions needed to obtain analytical results however.

8.2.1 THE RISK NEUTRAL GAME

Consider the strategic quality control game between a producer and a supplier engaged in an exchange with outcomes defined by the bimatrix random payoff game defined by $\left[\tilde{A}, \tilde{B} \right]$ below. The strategies that each of the parties can choose consist in selecting a quality control (sampling) strategy for product assurance and supply controls. Such strategies assume many forms, although we shall focus our attention on the selection of elementary sampling strategies (for example, apply a specific sampling strategy, or do nothing). The consequences of such choices by the parties in the supply chain (for example, a supplier and a producer) are statistical, denoting by the entries in the random payoff matrix, where ~ denotes a random variable.

$$\left[\tilde{A}, \tilde{B} \right] = \begin{bmatrix} \tilde{a}_{00} & \tilde{a}_{01}; & \tilde{b}_{00} & \tilde{b}_{01} \\ \tilde{a}_{10} & \tilde{a}_{11}; & \tilde{b}_{10} & \tilde{b}_{11} \end{bmatrix}. \tag{8.2}$$

In such a game, there are two essential considerations faced by the supply chains parties—economic and risk, which are imbedded in the bi-matrix random payoff entries. For example, say that the sampling strategies that each of the parties can follow are: Use a binomial control sample $\left(n_j, k_j \right), j = 0,1$ or do nothing. Here, the index j=0 denotes for example a supplier while the index j=1 denotes a downstream producer. Of course, generally, we can consider a finite set of alternative control strategies that each of the parties can pursue. In this sense, sampling control by the producer acts both as a quality control and as a "threat" to the producer, expressing "lack of trust" in the supplier's quality. As commonly practiced in sampling control we let $\left(\theta_1, \theta_2 \right)$ to be the proportions of parts defectives in a lot where $\theta_1 < \theta_2$ denotes a conforming lot (also called the AQL in statistical quality control) and the latter proportion, θ_2,denoting a non conforming lot (also called LTFD in statistical quality control). In such a case, the "probability risks" associated to each of these sampling strategies, coined type I and type II errors in a Neymann-Pearson statistical control framework are for the supplier and the producer defined by (Wethehill 1977; Tapiero 1996):

$$1 - P\left(j \le k_P \middle| n_P, \theta_1\right) = \alpha_P; \quad 1 - P\left(j \le k_S \middle| n_S, \theta_1\right) = \alpha_S$$
$$P\left(j \le k_P \middle| n_P, \theta_2\right) = \beta_P; \quad P\left(j \le k_S \middle| n_S, \theta_2\right) = \beta_S;$$

(8.3)

In this approach, the parameters $\left(\theta_1, \theta_2\right)$ are negotiated contract quality terms which we assume given while the statistical control strategies $\left(n_j, k_j\right)$, $j = 0,1$ can be parameters defined by each of the parties together with their decision to apply such controls or not. Further, the production technology used by the supplier is assumed to be by its propensity to produce confirming lots, given by:

$$\tilde{\theta} = \begin{cases} \theta_1 & \text{with probability } 1-v \\ \theta_2 & \text{with probability } v \end{cases}.$$

(8.4)

In this sense, a supplier can both improve his process reliability by decreasing v (but of course, production might be costlier) or augment the amount of quality inspection controls and apply more stringent control rules. Given these risks and the parties strategic behavior, the economic consequences, are necessarily random, expressed as a function of the sampling results and the uncertain consequences due to the facts that the process of producing non-quality is also random (since non conforming lots are produced in a random manner that the parties seek to control). For demonstration purposes, assume that the following costs are defined for both the producer and the supplier: I_j denote sampling inspection costs; E_j denote consumers' costs borne by the party in case a bad lot is accepted; D_j denotes the cost if both parties sample while the second party (producer) detects the non-conforming lot; And finally, C_j, denotes the cost if a good lot is rejected. In this case, the bi-matrix random payoff game between the producer and the supplier on the basis of which we shall pursue our analyses. Note that in this formulation, we have a random costs matrix, a function of the risks probabilities each of the parties assumes and a function of the organizational process (in this case, a single supplier and a single producer):

$$
[\tilde{A}] = \begin{bmatrix}
I_p + \begin{cases} \begin{array}{lll} E_p & wp & \beta_p\beta_s v \\ D_p & wp & (1-\beta_p)\beta_s v \\ C_p & wp & \alpha_p(1-v) \\ 0 & wp & 1-''''' \end{array} \\ \begin{array}{lll} E_p & wp & \beta_s v \\ 0 & wp & 1-\beta_s v \end{array} \end{cases} & I_p + \begin{cases} \begin{array}{lll} E_p & wp & \beta_p v \\ D_p & wp & (1-\beta_p)v \\ C_p & wp & \alpha_p(1-v) \\ 0 & wp & 1-''''' \end{array} \\ \begin{array}{lll} E_p & wp & \beta_s v \\ 0 & wp & 1-\beta_s v \end{array} \end{cases}
\end{bmatrix} \quad (8.5)
$$

$$
[\tilde{B}] = \begin{bmatrix}
I_S + \begin{cases} \begin{array}{lll} E_s & wp & \beta_p\beta_s v \\ D_s & wp & (1-\beta_p)\beta_s v \\ C_s & wp & \alpha_S(1-v) \\ 0 & wp & 1-''''' \end{array} \end{cases} & \begin{array}{lll} E_s & wp & \beta_p v \\ D_s & wp & (1-\beta_p)v \\ \\ 0 & wp & 1-''''' \end{array} \\
I_S + \begin{cases} \begin{array}{lll} E_s & wp & \beta_s v \\ \\ C_s & wp & \alpha_S(1-v) \\ 0 & wp & 1-''''' \end{array} \end{cases} & \begin{cases} \begin{array}{lll} E_s & wp & v \\ \\ 0 & wp & 1-v \end{array} \end{cases}
\end{bmatrix} . \quad (8.6)
$$

While these economic costs are self explanatory, we shall briefly discuss them. Assume that both the supplier and the producer apply a statistical control procedure and consider the first entry in the producer bi-matrix game. The cost C_p is the cost incurred if the producer rejects a good lot received from the supplier and produced by the supplier with a technology whose characteristic is defined by the probability parameter $(1-v)$. Since the risk probability of such an event in case the producer applies his statistical control sample is α_p, the probability of such an event is $\alpha_p(1-v)$. By the same token D_p is the cost that the producer assumes if he rejects a bad lot (with risk probability $1-\beta_p$) produced by the supplier with probability v. To do so however, the supplier must have accepted such a bad lot which he would with probability β_S. As a result we obtain the appropriate entry in the producer payoff (costs) matrix. This cost may also be shared or passed on back to the supplier, as specified by the contract between these parties when drafted. Consider next the cost E_p which the producer sustains because of his accepting a bad lot passed on to consumers, who, unavoidably will detect its non-conforming quality. The risk probability of such a cost would necessarily be $\beta_P\beta_S v$. Further, a commensurate cost would be passed on to the supplier such that the total end-customer cost is $E_P + E_S$. A similar interpretation is associated to each of the terms in the bi-matrix game.

The strategic quality control random payoff (costs) game can then provide some insights on the amount of controls parties will exercise. To resolve the problems associated with the solution of this random payoff game, we shall maintain the Neymann-Pearson risk framework and associate type I and type II risks to each strategy the parties adopt and explicitly given below. First define by $P_{I,S}(i, j)$, the probability of the supplier accepting a good lot when applying a strategy i and the producer applying strategy j and let $P_{II,S}(i, j)$, be the probability that the supplier accepts a bad lot, although it is good and each of the parties follows sampling control strategies i and j. Let (x, y), $0 \le x \le 1$, $0 \le y \le 1$, be the probabilities that the producer and the supplier sample, then the risk probabilities assumed by the parties are in expectation given for the supplier by:

$$P_{I,S} = (1-\alpha_S)(1-v)xy + (1-\alpha_S)(1-v)(1-x)y = (1-\alpha_S)(1-v)y$$
$$P_{II,S} = (1-y)v + xy\beta_S v + (1-x)y\beta_S v = (1-y)v + y\beta_S v \tag{8.7}$$

And for the producer (who receives lots from the supplier), given by:

$$P_{I,P} = x(1-\alpha_p)(1-v) + (1-x)(1-v)$$
$$P_{II,P} = \beta_p\beta_S xyv + x(1-y)\beta_p v + (1-x)y\beta_S v + (1-x)(1-y)v = \tag{8.8}$$
$$= v(x\beta_p + 1 - x)(\beta_S y + 1 - y)$$

Note that in the first case, when calculating the probability of accepting a good lot, if a lot is properly produced, the prior actions taken by the supplier are not relevant. Therefore the probability of accepting a good lot is essentially determined by the probability that it has been manufactured properly. While in the latter case, the probability is based on the strategies adopted by the supplier and the producer, based on sample results. Now, say that we impose (based on negotiations and agreements between the parties) the following expected acceptable risk parameters (A_S, A_P), consisting in the probability of rejecting a good lot for both the supplier and the producer. That is:

$$1 - P_{I,S} \le A_S \quad \text{and} \quad 1 - P_{I,P} \le A_P. \tag{8.9}$$

By the same token, we define the risk parameters (B_S, B_P) such that:

$$P_{II,S} \leq B_S \text{ and } P_{II,P} \leq B_P \tag{8.10}$$

Equations (8.7)-(8.10), thus provide a set of risk constraints which will be helpful to determine a solution to our strategic collaborative and competitive quality control games, faced by the supplier and the producer. We shall consider first a number of results, providing some theoretical insights on the effects of strategic games on sampling control (and in fact contracts controls) in supply chains. First we consider the risk neutral game, in which only expected costs are minimized. Subsequently, we shall consider a collaborative and risk control game and provide an alternative approach to obtaining collaborative controls in supply chains. For simplicity, some of our results (when they are based on straightforward analysis of the underlying games) are summarized by propositions. First, say as stated above, that the supplier and the producer are risk neutral. In this case, the expected costs for the producer and the supplier are:

$$\left[\hat{A}\right] = \begin{bmatrix} I_p + E_p \beta_p \beta_S v + D_p (1-\beta_p) \beta_S v + C_p \alpha_p (1-v) & I_p + E_p \beta_p v + D_p (1-\beta_p) v + C_p \alpha_p (1-v) \\ E_p \beta_S v & E_p v \end{bmatrix} \tag{8.11}$$

$$\left[\hat{B}\right] = \begin{bmatrix} I_S + E_S \beta_p \beta_S v + D_S (1-\beta_p) \beta_S v + C_S \alpha_S (1-v) & E_S \beta_p v + D_S (1-\beta_p) v \\ I_S + E_S \beta_S v + C_S \alpha_S (1-v) & E_S v \end{bmatrix} \tag{8.12}$$

These two matrices, define a 2-persons non-zero sum game whose solution can be found by an application of the well known Nash equilibrium (Nash 1950; Moulin 1995). The following sampling strategies result which we summarize in the proposition below proved in the appendix.

Proposition 8.1. *Define* $d_k = D_k / E_k$; $c_k = C_k / E_k$; $i_k = I_k / E_k$, *then the supplier and the producer Nash equilibrium sampling policies are defined by:*

$$x = \begin{cases} 1 & if & v \leq \dfrac{i_S + c_S \alpha_S}{(1-\beta_S) + c_S \alpha_S} \\ 0 & if & v \geq \dfrac{i_S + c_S \alpha_S}{\left(d_S(1-\beta_p) - \beta_p\right)(1-\beta_S) + c_S \alpha_S} \\ x^* & otherwise \end{cases} \quad (8.13)$$

$$x^* = \frac{v(1-\beta_p)(1-d_p) - i_p + c_p \alpha_p(1-v)}{v(1-\beta_S)(1-\beta_p)(1-d_p)}$$

and

$$y = \begin{cases} 1 & if & v \geq \dfrac{i_p + c_p \alpha_p}{(1-d_p)(1-\beta_p)\beta_S + c_p \alpha_p} \\ 0 & if & v \leq \dfrac{i_p + c_p \alpha_p}{(1-d_p)(1-\beta_p) + c_p \alpha_p} \\ y^* & otherwise \end{cases} \quad (8.14)$$

$$y^* = \frac{v(1-\beta_S) - i_S + c_S \alpha_S(1-v)}{v(1-\beta_S)(1-\beta_p)(1-d_S)}$$

Proof: The proof is a straightforward application of Nash equilibrium to non-zero sum games.

In this solution a number of insights results. First, note that the greater the production technology reliability the smaller the incentive to sample and vice versa. In this sense production technology and statistical sampling controls are substitutes. If the propensity to produce non conforming lots is larger than $\dfrac{i_p + c_p \alpha_p}{(1-d_p)(1-\beta_p)\beta_S + c_p \alpha_p}$, then the supplier will fully sample while the producer will sample fully only if that same propensity is smaller than $\dfrac{i_S + c_S \alpha_S}{(1-\beta_S) + c_S \alpha_S}$. This is the case, because the producer will presume that it would be in the best interest of the supplier to fully sample (and therefore there would be no need for him to do so as well). By the same token, if the propensity to produce non conforming units is smaller than $\dfrac{i_p + c_p \alpha_p}{(1-d_p)(1-\beta_p) + c_p \alpha_p}$, then the supplier presuming that his tech-

nology is reliable, will not sample at all. Interestingly, when the production technology is unreliable with

$$v \le \frac{i_S + c_S \alpha_S}{(1-\beta_S) + c_S \alpha_S} \tag{8.15}$$

then the producer will sample fully. For all other regions, there will be partial sampling as indicated in the proposition. The value for each of the parties in such a situation is given from equation (8.2) by:

$$
\begin{aligned}
V_P(x,y) &= \hat{a}_{00} xy + \hat{a}_{01} x(1-y) + \hat{a}_{10}(1-x)y + \hat{a}_{11}(1-x)(1-y) \\
V_S(x,y) &= \hat{b}_{00} xy + \hat{b}_{01} x(1-y) + \hat{b}_{10}(1-x)y + \hat{b}_{11}(1-x)(1-y)
\end{aligned}
\tag{8.16}
$$

Thus, for an interior solution we have (as calculated explicitly in Proposition 8.1) the following probabilities of sampling:

$$y^* = \frac{\hat{a}_{11} - \hat{a}_{01}}{\hat{a}_{00} - \hat{a}_{10} + \hat{a}_{11} - \hat{a}_{01}}, \quad x^* = \frac{\hat{b}_{11} - \hat{b}_{10}}{\hat{b}_{00} - \hat{b}_{10} + \hat{b}_{11} - \hat{b}_{01}}, \tag{8.17}$$

which leads to the following Nash values:

$$V_P^N(x^*, y^*) = \frac{\hat{a}_{11}\hat{a}_{00} - \hat{a}_{10}\hat{a}_{01}}{\hat{a}_{00} - \hat{a}_{10} + \hat{a}_{11} - \hat{a}_{01}} , \text{ or}$$

$$V_P^N(x^*, y^*) = E_p v \frac{i_P + c_P \alpha_P (1-v)}{v(1-\beta_P)(1-d_P) - c_P \alpha_P (1-v)} \tag{8.18}$$

and

$$V_S^N(x^*, y^*) = \frac{\hat{b}_{00}\hat{b}_{11} - \hat{b}_{10}\hat{b}_{01}}{\hat{b}_{00} - \hat{b}_{10} + \hat{b}_{11} - \hat{b}_{01}} , \text{ or}$$

$$V_S^N(x^*, y^*) = E_S \frac{i_S + c_S \alpha_S (1-v)}{1 - \beta_S} \tag{8.19}$$

Of course all cases $(x,y=0,1)$ ought to be analyzed as well, corresponding to all the situations we have indicated in our proposition. We can also see from (8.18) and (8.19) the effects of the ex-post (customers) quality costs on both the supplier and the producer alike. The larger these costs the larger the costs for the producer. While, for the supplier, it seems

that the Nash equilibria costs given by: $V_S^N(x^*, y^*) = \dfrac{I_S + C_S \alpha_S (1-v)}{1 - \beta_S}$ are

only functions of the amount of inspection carried and the expected cost of rejecting good lots, augmented by $1/(1-\beta_S)$ which is the inverse of the probability of rejecting a good lot. For example, for the following parameters: $\nu = 0.1$, $\theta_2 = 0.3$, $\theta_1 = 0.01$ with the following specified risks $\alpha_p = 0.10, \beta_p = 0.05$; $\alpha_S = 0.05, \beta_S = 0.05$ arising from the choice of sampling techniques of the supplier and the producer and the following costs parameters for the producer and the supplier: $E_p = 30$, $D_p = 10$, $I_p = 0.75, C_p = 2$; $E_S = 20$, $D_S = 4, I_S = 0.5$, $C_S = 4$ we find an interior solution to sampling by both the producer and the supplier, which is given by: $x^* = 0.8448$, $y^* = 0.6259$.

8.2.2 CENTRALIZED CONTROL AND COLLABORATION

When the supplier and the producer collaborate by setting up a centralized control over the chain to minimize the overall supply chain cost, the resulting system-wide cost is:

$$
V_P^C(x,y) + V_S^C(x,y) = \left(\hat{a}_{00} + \hat{b}_{00}\right)xy + \left(\hat{a}_{01} + \hat{b}_{01}\right)x(1-y) +
$$
$$
+\left(\hat{a}_{10} + \hat{b}_{10}\right)(1-x)y + \left(\hat{a}_{11} + \hat{b}_{11}\right)(1-x)(1-y)
$$
(8.20)

The Hessian matrix of function (8.20) is indefinite. Therefore, the sampling solution in such a case is a corner solution, in which case, the supplier will always fully sample or not, and similarly for the producer. Consequently, both the supplier and the producer disregard their own costs and risks with 4 potential solutions to be compared:

$$
V_P^C(1,1) + V_S^C(1,1);\ V_P^C(1,0) + V_S^C(1,0);\ V_P^C(0,1) + V_S^C(0,1);\ V_P^C(0,0) + V_S^C(0,0). (8.21)
$$

This approach however is neither interesting nor practical because it negates the existence of the risks that both the supplier and the producer seek to manage. Thus, in a collaborative framework, both the expected costs for the parties and the risks implied by both the producer and the supplier are to be accounted for. In this case, an appropriate formulation of the random payoff game, in terms of expected costs and the controlling Neymann-Pearson constraints (8.7)-(8.8) are given by:

$$\underset{0\le x\le 1, 0\le y\le 1}{Min} \quad V_P^{CR}(x,y)+V_S^{CR}(x,y)$$

Subject to:

$$1-P_{I,S} \le A_S \quad \text{and} \quad 1-P_{I,P} \le A_P$$

$$P_{II,S} \le B_S \quad \text{and} \quad P_{II,P} \le B_P$$

(8.22)

This is a straightforward nonlinear optimization problem whose solution can be reached by standard numerical methods. The disadvantage of this formulation is that it still assumes full collaboration or vertical integration of the supply chain, which is rarely possible and ignores individual costs and costs transfers between the parties.

Alternatively, we can obtain a collaborative binary as well as interior solutions that are sensitive to both the risk constraints and individual costs of the supplier and producer by assuming that the producer's propensity to control quality is proportional to that of the supplier, denoted for convenience by $x=ky$. With such an assumption a number of possibilities are neglected and can be verified separately. These possibilities include the following six sampling strategies:

$$(x,1),(x,0),(1,y),(0,y),(1,0),(0,1)$$

(8.23)

For example, for a sampling strategy $(x,0)$, we have the following (using equations (8.7), (8.8) and the objectives stated above):

$$1-P_{I,S}=1, \quad P_{I,P}=x(1-\alpha_p)(1-v)+(1-x)(1-v)\le A_p$$

$$P_{II,S}=v, \quad P_{I,P}=x(1-\alpha_p)(1-v)+(1-x)(1-v)\le A_P, \quad P_{II,P}=v(x\beta_p+1-x)\le B_p$$

(8.24)

and therefore the risk constraints are reduced to:

$$\frac{1-\dfrac{B_p}{v}}{(1-\beta_p)}\le x\le \frac{A_p-v}{(1-v)\alpha_p},$$

(8.25)

while the joint objective of the collaborating supply chain is:

$$V_P^{CR}(x,y)+V_S^{CR}(x,y)=\hat{a}_{11}+\hat{b}_{11}+(\hat{a}_{01}-\hat{a}_{11}+\hat{b}_{01}-\hat{b}_{11})x .$$ (8.26)

A solution is then necessarily determined by the risk constraints. Namely, $x=0$ if $\hat{a}_{01}+\hat{b}_{01}>\hat{a}_{11}+\hat{b}_{11}$, violating the risk constraint (8.25) and therefore $x\ne 0$ necessarily. When the inequality is reversed, we obtain also a sampling program determined by the upper constraint imposed by the type I risk of the producer. As a result:

$$
x = \begin{cases} \dfrac{1 - \dfrac{B_p}{v}}{(1 - \beta_p)} & \text{if } \hat{a}_{01} + \hat{b}_{01} \leq \hat{a}_{11} + \hat{b}_{11} \\[4mm] \dfrac{A_P - v}{(1 - v)\alpha_p} & \text{if } \hat{a}_{01} + \hat{b}_{01} > \hat{a}_{11} + \hat{b}_{11} \end{cases} \tag{8.27}
$$

Similarly, other constraints can be treated. For convenience, consider interior solutions, by letting $x = ky$. The collaborative objectives of the supplier and the producer are then convex and therefore a global solution can be found, as summarized by the following proposition.

Proposition 8.2. *Let risk constraints not to be binding and define*

$$
\xi = \frac{2\hat{a}_{00} + 2\hat{b}_{00} - \hat{a}_{01} - \hat{a}_{10} - \hat{b}_{01} - \hat{b}_{10}}{\hat{a}_{11} + \hat{b}_{11} - \hat{a}_{10} - \hat{b}_{10}} \tag{8.28}
$$

If $k > \dfrac{1}{1 + \xi}$ *then (8.23) has a unique interior optimal solution.*

Proof: The proof is straightforward by verifying the second order optimality condition along with binary constraints.

The advantage of this collaborative approach is that once a solution for sampling is determined in terms of the parameter k, we can employ k for fine tuning the supply chain to prevent violations of the risk constraints. Specifically, substituting $x = ky$ into the objective function (8.20) we have the collaborative cost:

$$
y^{**} = \frac{(\hat{a}_{11} + \hat{b}_{11})(1 + k) - (\hat{a}_{01} + \hat{b}_{01})k - \hat{a}_{10} - \hat{b}_{10}}{2k(\hat{a}_{00} - \hat{a}_{01} + \hat{a}_{11} - \hat{a}_{10} + \hat{b}_{00} - \hat{b}_{01} + \hat{b}_{11} - \hat{b}_{10})}, \, x^{**} = ky^{*}, \tag{8.29}
$$

while the collaborative cost is:

$$
\begin{aligned}
V^{CR} &= ky^2(\hat{a}_{00} - \hat{a}_{01} + \hat{a}_{11} - \hat{a}_{10} + \hat{b}_{00} - \hat{b}_{01} + \hat{b}_{11} - \hat{b}_{10}) \\
&\quad - y((\hat{a}_{11} + \hat{b}_{11})(1 + k) - (\hat{a}_{01} + \hat{b}_{01})k - \hat{a}_{10} - \hat{b}_{10}) + \hat{a}_{11} + \hat{b}_{11}
\end{aligned} \tag{8.30}
$$

To obtain feasible solution, satisfying the producer and the supplier risk constraints, we thus solve the following problem:

Find $0 \leq k$ such that:

$$
1 - P_{I,S}(k) \leq A_S \quad \text{and} \quad 1 - P_{I,P}(k) \leq A_P \tag{8.31}
$$

$$
P_{II,S}(k) \leq B_S \quad \text{and} \quad P_{II,P}(k) \leq B_P
$$

Explicitly, this is given by:

Find $0 \le k$ such that:

$$1-(1-\alpha_S)(1-v)y \le A_S$$
$$x(1-\alpha_p)(1-v)+(1-x)(1-v) \le A_P \qquad (8.32)$$
$$(1-y)v+y\beta_s v \le B_S$$
$$v(x\beta_p +1-x)(\beta_s y+1-y) \le B_P$$

If the risk constraints are not binding, then there is a non empty interval defined by (k_1,k_2) where k turns out to be a potentially negotiating parameter, *defining both the economic costs sustained by the producer and the supplier and the type I and type II risks, (both a function of k).* This is illustrated in Figure 8.2 below. Explicitly, assume the following parameters

$$E_p = 30, D_p = 10, I_p = 0.75,$$
$$C_p = 2,\ E_S = 20,\ D_S = 4, I_S = 0.5, C_S = 4$$

with risk parameters $\alpha_S = 0.1, \alpha_p = 0.05, \beta_P = 0.1, \beta_S = 0.1, v = 0.055$ (a function of the sampling plans adopted). For these system parameters we see the effects of parameters k on the costs sustained by each of the parties. Clearly, the sum of the Nash equilibrium costs for both parties is much larger than the sum of collaborative costs. In addition, we see also that collaborative costs are increasing in k as stated in Proposition 8.3 below.

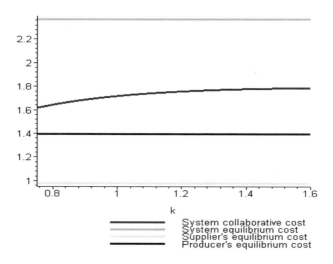

Figure 8.2. System collaborative and equilibrium costs

Proposition 8.3. *If the probability of non conforming production lots satisfies the condition below:*

$$v \le \frac{I_p}{\left(E_p + E_S\right)} + \frac{\left(D_p + D_S\right)\left(1 - \beta_p\right)v}{\left(E_p + E_S\right)} + \frac{C_p \alpha_p \left(1 - v\right)}{\left(E_p + E_S\right)}$$

then the smaller the collaboration parameter k the lower the collaborative supply chain cost.

Proof: The proof is obtained by differentiating the cost function (8.30) with respect to k resulting in the condition $\hat{a}_{11} + \hat{b}_{11} \le \hat{a}_{01} + \hat{b}_{01}$, which requires such a result to be positive.

The implication of this proposition is that the party with larger inspection costs will reduce the amount of inspection and thus the associated cost while transferring some of the inspection effort and cost to the other party. This is observed in Figures 8.3 and 8.4 below. Specifically, the supplier's cost (Figure 8.3) and inspection effort (Figure 8.4) decrease while the producer's cost and inspection effort increase as k increases. At the intersection point of the two cost curves in Figure 8.3 both parties incur identical costs. At this point $k<1$, pointing out to unequal inspection efforts exercised by the parties (see Figure 8.4). Furthermore, at this point the parties attain equal individual costs, but do not minimize the system-wide cost of the collaborative supply chain (see Figure 8.2).

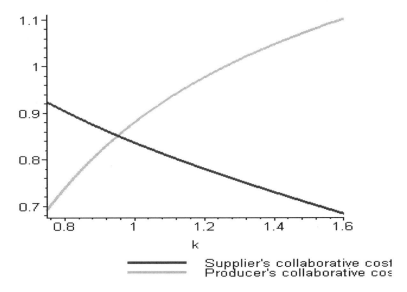

Figure 8.3. Supplier's and Producer's collaborative costs

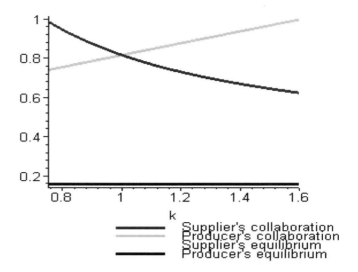

Figure 8.4. Collaborative versus equilibrium costs

Finally, Figure 8.5 outlines the effects of parameter k on the risk constraints. For the parameters selected, we note that the maximum errors tolerated by the producer and the supplier are as defined in the figure. The conclusion to be drawn from such a numerical analysis confirms the intuition that having the supplier augment the control of quality (meaning a smaller k), relative to that of the producer, will result in lower risks for both the producer and the supplier. In this sense, the conventional wisdom that sampling upstream the supply chain is efficient is verified here as well. Further, as stated earlier, the parameter k, is shown to be a parameter where both costs and risks substitution can be determined.

This problem has taught us on the one hand that in a competitive state, there may be an interior solution to the inspection game as stated in Proposition 8.1. The decision to control or not for the supplier is then a function of the underlying process reliability. For the producer, a reliable process may require as well some inspection (of course, we are not considering in this case the extreme case of zero default production). The propensity to inspect by the producer is then merely a result of the parties' motivations and the mutual distrust implied in the Nash solution.

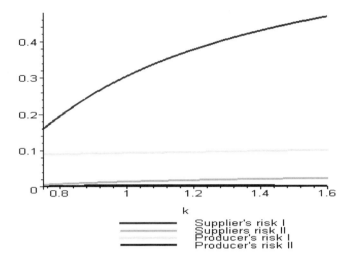

Figure 8.5. Supplier's and producer's risks

A similar analysis under a collaborative framework turns out to be trivial, with a solution to the inspection game turning out to be an "all or nothing" solution for one or both parties. Of course such a solution is not realistic nor does it confirm the observed behavior of industrial firms operating in a supply chain. This is the case because we have neglected the risk effects that are particularly important in the control of quality. In this sense, the risk neutral problem has limited interest while the random-payoff inspection game is difficult to resolve in a practical sense. A potential approach entertained was to assume a risk attitude (Munier and Tapiero 2007) by both the producer and the supplier and thereby transform the competitive game into a deterministic non-zero sum game. Such an approach is not appropriate however as it introduces risk attitude parameters that are only implicit in decision makers' actions rather than known explicitly. Further, when studying the risk attitude sensitive problem we also reached the same conclusion for a collaborative supply chain, neglecting again the implied risk constraints that underlie the decision to control quality or not. For these reaons, following an approach set by Tapiero (2005a-2005b), the random payoff strategic quality control game was transformed into a Neymann-Pearson risk constraints game. In other words, while maintaining the risk neutral valuation of economic costs, the approach has appended to parties decision processes the risk qualifications (type I and type II risks in the Neymann-Pearson statistical framework). Explicitly, since parties strategies are defined in terms of both the choice of sampling plans and the randomized strategies applied in selecting these plans, we have introduced a concept of "expected" type I and type II risks to be sustained by both the producer and the supplier. Such an approach leads to a broad number of potential equilibria, when combined with sampling plans selections. Further, in a collaborative framework, the model assessed will also lead to results that might not be practical due to producer's and suppliers' focus on sepecfic parameters and selecting the relationships that they ought to maintain one with respect to the other. In this sense, assuming that there is an interior solution to the game where such a relationship is maintained, defined by parameter k, we demonstrated (Proposition 8.2) that in the collaborative game there can be an interior solution to the sampling random payoff game meeting the parties risk constraints that can be used to select an optimal sampling strategy by each of the parties on the one hand and selecting the compatible optimal sampling plan on the other.

8.3 YIELD AND CONTROL

Consider next a game between a supplier of parts and a manufacturer, who uses these parts in its production process. We focus our attention on a supplier of parts, whose manufacturing yield, p, expresses the propensity to produce a part which conforms to acceptable standards of manufacture and whose manufacturing cost is $C(p)$, $\partial C/\partial p > 0$, $\partial^2 C/\partial p^2 > 0, C(1) = \infty$. The yield can be improved either by investments in technologies enhancing the process reliability or by investing in preventive measures taken to improve the quality delivered by the supplier and measuring the actual proportion of faulty products delivered. In this sense, a supplier's strategy is defined by selecting the yield determined by investments in the manufacturing process and its investment in control measures, exercised once the part has been manufactured. Of course, practically, the yield may itself be random. In this case, we are still confronted with a random payoff game, which is more difficult to solve and decision parameters will be based on the statistical properties of the yield. Such an issue as well as the asymmetric knowledge between the supplier and the customer regarding this parameter is a topic for further research however.

A customer's strategy consists in either accepting the supplier's deliveries without any inspection, or in submitting the supplier's parts to control inspection (conformance tests), assuring that parts conform to the agreed contract. Nonconformance or delivery of substandard parts, entailing costs which are defined by the contract terms between the supplier and the customer-manufacturer. To represent such a situation, we shall construct again a non-zero sum two persons game with random payoffs which we simplify and analyze. This analysis will clarify the effects of the contract terms between the supplier and the manufacturer-customer on the inspection-control policies of both parties and the manufacturing yield, which is selected by the supplier. We assume that the supplier preventive policy consists in the inspection-control of a part, at a cost of c_i, prior to its sale to the manufacturer-customer (who pays a price π for the part and who may test it faultlessly at a cost of c_b). Furthermore, the sale of a part is defined by a contract which protects the customer against the delivery (whether on purpose or not) of defective parts by the supplier, on the condition that parts are inspected and detected by the customer. Such a contract is thus defined by a schedule (π, T) denoting both the part price and the cost incurred by the supplier when a defective part is detected. When the customer accepts a part, it is then the customer's responsibility. That is, the customer only (who in turn sells the part as one of the elements in a complex product, which it may produce or assemble), will incur warranty and post sales defective costs. The supplier's "transfer cost", incurred when a product is detected by the customer, provides as we shall see, an incentive for the supplier to provide good parts on the one hand (either through inspection or through the selection of an appropriate manufacturing technology resulting in the yield p) and for the customer to inspect incoming products on the other. In this section we shall consider the effects of the contract parameters on the equilibria of a non zero sum game random payoff game played by the supplier and the customer and on their behavior.

In this section, unlike previous ones, we consider explicitly the nonlinear costs of production as a function of the yield and determine the substitution effects between the yield and the inspection policy of the supplier. To assess some of the relationships between the customer and the supplier inspection-control policies, the manufacturing yield and the parameters of the two parties contract; a sensitivity analysis is performed. Existence of a number of equilibria for the game is shown to depend on the production cost function $C(p)$ and on the contract terms of the supplier-customer.

Thus, we will be able to show that a production technology (resulting in the yield p) and the negotiated contract agreements generate various yield offerings as well as a number of potential inspection-control policies. Furthermore, using the framework, we note that inspection-control is also strategic, recognizing conflict and information asymmetry between contractual parties as essential problems to reckon with in the management of quality. Finally, we are also addressing the problem of sampling design by both a customer and a supplier and discuss the substitution effects between preventive (based on inspection and detection of defective parts prior to their entering the production process) and the application of high yield (albeit costly) manufacturing technologies.

This approach to supply chains management thus recognizes that uncertainty for both the supplier and the customer, can be generated endogenously, i.e. it is determined by the acts that each of the parties, bound by a contract and conflicting objectives, will follow. Inspection (control) is thus required and determined according to the assumptions made regarding the game participants, their motivations and their behavior and not only the uncertainty with respect to the process. In other words, these controls are determined by the solution of a game, which recognizes the realistic conflict between the supplier and the customer and the uncertainty such conflict generates. We shall also clarify the idea of equilibrium as a mechanism to generate yield and inspection-control policies (rather than optimality of some function, which expresses an individual point of view). These particular facets of our problem are more in tune with the practical setting and the environment within which suppliers and manufacturers-customers operate.

8.3.1 THE SUPPLY QUALITY AND CONTROL GAME

Assume that a contract for the delivery of materials or parts has been negotiated and signed by a supplier and a producer. Suppose that this contract stipulates a price π and a transfer cost T by the supplier (if a part is delivered and found defective by the manufacturer-customer). Further, let p be the yield and assume that once a unit is inspected by the supplier, and found defective, then all subsequent units are inspected until a unit is demonstrably non defective (that is, the supplier uses a corrective sampling technique). Let ϕ be the manufacturer-customer-selling price of the part (to some end customer). Then if the producer inspects it with probability y and the supplier with probability x, the bi-matrix (random payoff) game results, as summarized in Table 8.1. Note that in this game, the manufacturer-customer has two alternatives: inspect or not the incoming part. If we denote the randomized strategy of the customer by $0 \le y \le 1$, then this defines also the customer's "strategic" sampling (an outcome in the game solutions). The supplier has however, two decisions to reach one regarding the manufacturing yield and the other regarding the amount of out-going parts inspection. Both are defined over a continuum of probabilities (x, p) where x is the randomized strategy based on the two alternatives, "inspect versus do not inspect". We assume that the supplier uses the games' value to determine the yield. The cost of consistently producing a good part (equivalent to a "zero-defects" technology) is infinite, $C(1) = \infty$. Thus, we presume that the cost of producing only defective parts is not acceptable and the only remaining possibility is for the yield to be some probability $0 < p < 1$ defined by the game's random payoffs. Of course, both the "yield" and the "control-inspection" strategies of the supplier are dependent, for the choice of one affects the other. For example, if p is very large (close to one), then it is possible that a 0% inspection policy may be optimal.

Table 8.1. The (Supplier, Producer) Payoff Matrix

Sampling Policies	Test	No test
Test	$\begin{pmatrix} \pi - [C(p)+c_i] & w.p.\ p \\ \pi - [C(p)+c_i](1+\tilde{k}_1) & w.p.\ 1-p \end{pmatrix}$; $\phi - \pi - c_b$	$\begin{pmatrix} \pi - [C(p)+c_i] & w.p.\ p \\ \pi - [C(p)+c_i](1+\tilde{k}_1) & w.p.\ 1-p \end{pmatrix}$; $\phi - \pi$
No Test	$\begin{pmatrix} \pi - C(p) & w.p.\ p \\ \pi - C(p) - [C(p)+c_i](\tilde{k}_1) - T & w.p.\ 1-p \end{pmatrix}$; $\begin{pmatrix} \phi - \pi - c_b & w.p.\ p \\ \phi - \pi - c_b + T & w.p.\ 1-p \end{pmatrix}$	$\pi - C(p)$; $-\pi$

The bi-matrix game entries are evident. For example, say that the supplier produces a defective part (with probability $1-p$) and inspects it as well (with probability x). Then, whether the customer inspects the part or not, the expected payoff is equal to the unit price π less the production cost $C(p)$ and its inspection cost c_i. Since the part is defective, it must be replaced by another part which is also inspected until the part manufactured is found to be non-defective. Since p is the manufacturing yield, the number of parts manufactured until one is non-defective is a random variable $\tilde{k}_1$ whose probability distribution is geometric and given by $p(1-p)^j$, $j = 1,2,...$ whose mean is $(1-p)/p$. Thus, although only one unit is sold at a price of π, the expected number of parts manufactured and tested in this case equals $E(1+\tilde{k}_1)=1/p$. The customer's corresponding payoff is equal to $\phi - \pi - c_b$ in case of inspection and $\phi - \pi$ in case of no inspection by the customer. If the supplier does not inspect an outgoing defective part and if the customer detects this part for sure through inspection, then the supplier's payoff is reduced by T, the contracted amount transferred to the customer which is in fact a price break for the sale of defective parts. If the customer does not inspect the incoming part, he will incur a total loss of $-\pi$. Of course, we do not consider in this special case, the post sales failure costs such as warranty and related costs which can be substantial, sustained by the customer who is using the part in the assembly of the manufacturing of some other more complex products.

Using the defined game, we can formulate the following proposition for the supplier control-inspection policy.

Proposition 8.4. *For a risk neutral, maximizing expected payoff supplier, using a yield p, it is never optimal to fully sample. Further,*

(i) *If $T \leq C(p) + c_i / (1 - p)$ it is never optimal to sample, i.e.*
$$x^* = 0.$$

(ii) *If $T \geq C(p) + c_i / (1 - p)$ it is optimal to sample, i.e. $x^* > 0$.*

(iii) *For a risk neutral manufacturer-customer, the supplier inspection probability under condition (ii) above is given by:*
$$x^* = 1 - \frac{c_b}{\phi + T(1 - p)}$$

Proof: For each of the entries of the supplier's game, we consider the expected utility game, where the utility function $u(.)$, $u'(.) > 0$, $u''(.) \leq 0$,

$$\hat{u}_{11} = pu(\pi - [C(p) + c_i]) + (1 - p)Eu(\pi - C(p) - [C(p) + c_i](1 + \tilde{k}_1)), \ \hat{u}_{12} = \hat{u}_{11}$$

$$\hat{u}_{21} = pu(\pi - C(p)) + (1 - p)Eu(\pi - C(p) - [C(p) + c_i](\tilde{k}_1) - T), \ \hat{u}_{22} = u(\pi - C(p)),$$

which is, of course, reduced to simple expectation for the risk neutral supplier.

Note that $\hat{u}_{22} > \hat{u}_{11} (= \hat{u}_{12})$. Further, $\hat{u}_{21} > \hat{u}_{11}$ if $T \leq C(p) + c_i / (1 - p)$. Therefore the strategy to never sample by the supplier is always optimal as stated in (i) in the proposition. When this is not the case, both the full sampling and the no sampling strategies are not dominating and therefore the only remaining possibility is for a mixture of these alternatives, which leads to a sampling probability of $0 < x^* < 1$ as stated in (ii) above. Further, these two cases cover the proposition's statement that it is never optimal to fully sample.

When the game solution is a randomized strategy, it is well known to be:

$$x^* = \frac{v_{21} - v_{22}}{v_{12} - v_{11} + v_{21} - v_{22}},$$

where v_{ij}, $i = 1,2; \ j = 1,2$ are the expected values corresponding to the entries in the bimatrix games.

$$v_{11} = \phi - \pi - c_b$$
$$v_{12} = \phi - \pi$$
$$v_{21} = \phi - \pi - c_b + T(1-p)$$
$$v_{22} = -\pi$$

In this case, we have:

$$x^* = \frac{\phi - \pi - c_b + T(1-p) + \pi}{\phi - c_b + T(1-p) + \pi + \phi - \pi - c_b + T(1-p) + \pi},$$

which is reduced to the proposition statement.

The implication of this proposition is that if the contracted transfer cost T, is smaller than the expected cost of production plus the expected cost of sampling a good unit, then it is not optimal to test. However, when it is larger, it is optimal to incur the inspection cost to avoid such a payment. In this case, the inspection probability is expressed in terms of the contract parameters and the manufacturer-customer inspection cost. Note that the larger the inspection cost c_b, the smaller the amount of control-inspection to be carried (since the manufacturer-customer will tend to exercise an inspection policy less often). Further the larger the selling price and the larger the transfer cost, the more the supplier will sample. In this sense, the terms of the contract determine the amount of sampling to be carried by the supplier. We turn next to the manufacturer-customer and prove the following proposition.

Proposition 8.5. *The manufacturer-customer has no dominating no sampling strategy. Further,*
   If $T \le C(p) + c_i / (1-p)$, the supplier -customer optimal policy is to sample fully
   If $T \ge C(p) + c_i / (1-p)$ the optimal sampling strategy is to sample partly with probability given by:

$$0 < y^* = \frac{1}{1 + \dfrac{p(T(1-p) - c_i - (1-p)C(p))}{(1 - p(1-p))c_i + (1-p)C(p)}} < 1$$

Proof: Consider again the following entries

$$v_{11} = \phi - \pi - c_b$$
$$v_{12} = \phi - \pi$$
$$v_{21} = \phi - \pi - c_b + T(1-p)$$
$$v_{22} = -\pi$$

Note that $v_{11} < v_{12}$ while $v_{21} > v_{22}$ which rules out a dominating sampling strategy by the manufacturer-customer. As a result, the sampling strategy is a randomized one, which is given by solving for y,

$$\underset{y}{Max}\left\{v_{11}xy+v_{12}x(1-y)+v_{21}(1-x)y+v_{22}(1-x)(1-y)\right\}.$$

As we saw in the previous proposition, two situations arise. First, $x=0$, which occurs when $T\le C(p)+c_i/(1-p)$. In this case,

$$\underset{y}{Max}\left\{v_{21}y+v_{22}(1-y)\right\}=\underset{y}{Max}\,v_{22}+y\left\{v_{21}-v_{22}\right\}$$

Since $v_{21}>v_{22}$, the optimal sampling policy is full sampling, that is $y^*=1$. Now consider the case, $T\ge C(p)+c_i/(1-p)$ where the supplier policy is to sample. In this case, we have:

$$\underset{0<y<1}{Max}\left\{v_{11}xy+v_{12}x(1-y)+v_{21}(1-x)y+v_{22}(1-x)(1-y)\right\}$$

$$\underset{0<x<1}{Max}\left\{\hat{u}_{11}xy+\hat{u}_{12}x(1-y)+\hat{u}_{21}(1-x)y+\hat{u}_{22}(1-x)(1-y)\right\}$$

and

$$0=\left\{v_{11}x-v_{12}x+v_{21}(1-x)-v_{22}(1-x)\right\}$$
$$0=\left\{\hat{u}_{11}y+\hat{u}_{12}(1-y)-\hat{u}_{21}y-\hat{u}_{22}(1-y)\right\}.$$

This leads to:

$$0=v_{21}-v_{22}+\left\{v_{11}-v_{12}-v_{21}+v_{22}\right\}x$$
$$0=\hat{u}_{12}-\hat{u}_{22}\left\{\hat{u}_{11}-\hat{u}_{12}-\hat{u}_{21}+\hat{u}_{22}\right\}y.$$

Therefore,

$$y^*=\frac{\hat{u}_{22}-\hat{u}_{12}}{\left\{\hat{u}_{11}-\hat{u}_{12}-\hat{u}_{21}+\hat{u}_{22}\right\}}$$

and inserting the relevant terms of the bimatrix, we have:

$$\hat{u}_{11}=pu(\pi-\left[C(p)+c_i\right])+(1-p)Eu(\pi-C(p)-\left[C(p)+c_i\right](1+\tilde{k}_1)),\ \hat{u}_{12}=\hat{u}_{11}$$

$$\hat{u}_{21}=pu(\pi-C(p))+(1-p)Eu(\pi-C(p)-\left[C(p)+c_i\right](\tilde{k}_1)-T),\ \hat{u}_{22}=u(\pi-C(p))$$

$$y^*=\frac{\dfrac{(1-p(1-p))c_i+(1-p)C(p)}{p}}{\left\{T(1-p)-c_i-(1-p)C(p)\right\}+\dfrac{(1-p(1-p))c_i+(1-p)C(p)}{p}}.$$

Or

$$0<y^*=\frac{1}{1+\dfrac{p\left(T(1-p)-c_i-(1-p)C(p)\right)}{(1-p(1-p))c_i+(1-p)C(p)}}<1$$

and

$$0<y^*=\frac{1}{1+\dfrac{\left((T-C(p))(1-p)-c_i\right)}{(1-p(1-p))c_i+(1-p)C(p)}}<1,$$

which is the randomized strategy when $T \geq C(p) + c_i / (1 - p)$, as sated in the proposition.

The implications of this proposition are again demonstrating the dependence of the manufacturer-customer sampling policy of the yield p, its cost function and, of course, the contract parameters (see Nash 1950).

These propositions confirm partially Deming's "conventional wisdom" that it is often optimal to fully (or not at all) sample (see Burke et al 1993). Explicitly, Deming, incorporating costs, has developed a decision rule, which would recommend a 100% inspection, no inspection or turning to acceptance sampling. This was translated to a teaching game highlighting the importance of Deming's argument against traditional acceptance sampling techniques. It is noteworthy, however, that even though Deming did not use a game theoretic approach, he pointed out to the use of a "randomized sampling strategy" which, of course has been suggested here and which generalizes traditional sampling techniques. Our result points out however, that it is "often optimal" to sample *for strategic reasons* as indicated by the solution of the Nash game above. In this sense, sampling for the control of quality has a strategic effect, which has an importance previously neglected.

In summary for the more usual case $T \geq C(p) + c_i / (1 - p)$, there is one unique equilibrium in mixed strategies while when $T \leq C(p) + c_i / (1 - p)$, there is also a unique equilibrium consisting of no sampling by the supplier and full sampling by the manufacturer-customer.

8.3.2 OPTIMAL YIELD

To determine the optimal yield, the supplier will then consider the value of the game when the Nash game above has the solution expounded by Propositions 8.4 and 8.5. In this case, the optimal yield is found by solving and comparing the value of the game under the following two (equilibria) optimization problems.

$$\underset{0 < p < 1}{Max} \left\{ \hat{u}_{11} x^* y^* + \hat{u}_{12} x^* (1 - y^*) + \hat{u}_{21} (1 - x^*) y^* + \hat{u}_{22} (1 - x^*)(1 - y^*) \right\}$$

s.t.

$$T \geq C(p) + c_i / (1 - p),$$

which is equivalent to:

$$\underset{0 < p < 1}{Max} \frac{\hat{u}_{11} \hat{u}_{22} - \hat{u}_{12} \hat{u}_{21}}{\hat{u}_{11} + \hat{u}_{22} - \hat{u}_{12} - \hat{u}_{21}} = \underset{0 < p < 1}{Max} \frac{\hat{u}_{11} (\hat{u}_{22} - \hat{u}_{21})}{\hat{u}_{22} - \hat{u}_{21}} = \underset{0 < p < 1}{Max} \hat{u}_{11}$$

s.t.

$$T \geq C(p) + c_i / (1-p),$$

where

$$\hat{u}_{11} = \pi - \left(\frac{1}{p}\right)C(p) - \left(\frac{1 - p(1-p)}{p}\right)[c_i].$$

As a result, if there is an interior solution, we have:

$$c_i = \left(\frac{pC'(p) - C(p)}{1 - p^2}\right).$$

Note that, this requires as well that $C'(p) > C(p)/p$. For example, if $C(p) = A/(1-p)$, then,

$$\frac{c_i}{A} = \left(\frac{2p-1}{(1+p)(1-p)^3}\right) \text{ with } p > \frac{1}{2}.$$

Alternatively, under the second equilibrium, we have:

$$\underset{0<p<1}{Max} \{\hat{u}_{21}\} = \underset{0<p<1}{Max} \ \pi - \left(\frac{1 - p(1-p)}{p}\right)C(p) - [c_i]\frac{(1-p)^2}{p} + T(1-p)$$

s.t.

$$T \leq C(p) + c_i / (1-p).$$

Let $\left(p_1^*, p_2^*\right)$ be the optimal solutions under both equilibria, then, obviously, the supplier will adopt the solution leading to the largest expected payoff of the game. By changing the assumptions regarding the relative power each of the parties has, we will obtain, of course, other solutions. These are discussed below.

Each of the solutions considered here can be altered by an appropriate selection of contract parameters which can lead to a pre-posterior game analysis evaluated in terms of (p, T) (see also Reyniers and Tapiero 1995a for the analysis of contracts). If the supplier and the producer do not cooperate (and thus the Nash equilibrium solutions defined here are appropriate), it is possible to create an incentive for the supplier to supply quality parts by the selection of contract parameters.

A sensitivity analysis of some of these solutions follows. For convenience, we consider only the case with interior solutions and study the effects of T on the propensity to sample. Obviously, the larger T, the less the manufacturer-customer will sample since $\frac{\partial y}{\partial T} < 0$.

Further,

$$\frac{\partial x}{\partial T} = \frac{c_b}{(1-p)(\phi + T)^2} > 0.$$

Similar relationships can be found by treating other parameters. The implications are that increases in T provide an incentive for the supplier to sample while for the customer to sample less.

Stackelberg equilibrium

When either the customer or the supplier is a leader and the other a follower, we define a Stackelberg game (Stackelberg 1952). For example, say that the supplier is a leader and the customer is a follower. Then, for a given (x, p), the customer problem is:

$$\underset{y}{Max} \ V(y, x; p) = \left[\phi(1-x)(1-p) + T(1-x)(1-p) - c_b\right]y$$

and therefore the customer sampling policy is either to inspect all of the time (y=1) or none at all (y=0). Of course:

$$y = \begin{cases} 1 & \text{if } \phi(1-x)(1-p) + T(1-x)(1-p) \geq c_b \\ 0 & \text{if } \phi(1-x)(1-p) + T(1-x)(1-p) < c_b \end{cases}$$

The supplier's problem consists then in selecting a strategy (x,p) based on the customer's response $y(x,p)$ given above. Namely,

$$\underset{x, p}{Max} \ U(y; x; p) = \pi - C(p) - c_i x - \left[\left(C(p) + c_i\right)\frac{(1-p)^2}{p}\right](x + y(1-x)) - T(1-x)(1-p)y$$

s.t.

$$y(x, p) = 1, 0 \ ,$$

as sated above.

In this case, we note that the sampling decision is always an all or nothing sampling policy. For example, say that $y=1$, then the supplier turns to full sampling if $T(1-p) \geq c_i$, otherwise the supplier will not sample at all. However, if the customer does not sample, then we note that the supplier does not sample either, since

$$\underset{x, p}{Max} \ U(y; x; p) = \pi - C(p) - x\left(c_i + \left[\left(C(p) + c_i\right)\frac{(1-p)^2}{p}\right]\right)$$

has always a solution $x = 0$. Of course, the supplier yield will then be minimal (and therefore the quality will be the worst possible). In this sense, when one of the parties has power over the other the quality will be low (as it is the case in Stackelberg games but which does not hold true in Nash conflict games).

Centralized problem

In industrial situations, it is common that cooperative solutions are sought. In this case (if we do not consider for simplicity the distribution of spoils resulting from cooperation), the problem faced by the supplier and the manufacturer customer alike is given by:

$$V(y,x,p)+U(y,x,p)=\phi-C(p)-c_rx-c_sy-\phi(1-x)(1-p)(1-y)-\left[(C(p)+c_i)\frac{(1-p)^2}{p}\right](x+y(1-x))$$

which we maximize with respect to (y,x,p). Of course, $x=y=0$ and therefore the optimum yield is found by a solution of:

$$V(0;0;\ p) + U(0;0;\ p) = -C(p) + \phi p$$

and therefore, the optimal yield is: $\dfrac{\partial C(p)}{\partial p} = \phi$ which expresses the classical relationship between the marginal cost and the marginal revenue for the optimal yield. In this sense, cooperation will lead to the highest yield while an asymmetric power relationship as the one stated above will lead to the least yield.

In conclusion, we note that producers' and suppliers' inspections are, as we discussed, function of the industrial contract in effect between a supplier and a customer. This provides a wide range of interpretations and potential approaches for selecting a quality inspection policy. This section has shown that there is a clearly important relationship between the terms of a contract and the acceptance sampling policy. There are, of course, many facets to this problem, which could be considered and have not been considered in sufficient depth. For example, risk aversion, more complex contracts and the design of yield delivery contracts have not been considered. Nevertheless, these are topics for further research. The basic presumption of this section is that once supplier-customer contracts are negotiated and signed, there may be problems when enforcing these contracts. As a result, some controls are needed to ensure that contracts are carried out as agreed on. The approach is based on solving the post-contract game between the supplier and the producer where the resultant inspection and quality supplies equilibrium policies are given by the randomized strategies available to each of the parties.

8.4 RISK IN A COLLABORATIVE SUPPLY CHAIN

We consider next the supply chain organizational structures and their associated rules of leaderships. We also use the statistical Neyman-Pearson quantile risk framework for hypothesis testing (and quality control), as done earlier. Based on such risks we shall construct a variety of control programs that respond to the specific needs and the specific organizational structure of a supply chain. To demonstrate the usefulness of this approach, a number of problems are also solved. To keep matters tractable however, some simplifications are made.

8.4.1 A NEYMANN-PEARSON FRAMEWORK FOR RISK CONTROL

For simplicity and exposition purposes, assume that lots of size N are delivered by a supplier to a buyer (a producer of finished products), parts of which are sampled and tested. To assure contract compliance, both the supplier and the buyer can use a number of sampling programs, each with stringency tests of various degrees (spanning the no sampling case and thereby accepting the lot as is, to the full sampling case and thereby inspecting the whole lot) and assuming no risks. Let $j = 1,...M \leq N$ be the M alternative sampling-control programs used by the client-buyer and $i = 1,...N$ be the alternative sampling-control programs used by the provider-supplier. Correspondingly, we denote by $\left(\alpha_{p,i}, \beta_{p,i}\right)$; $\left(\alpha_{S,j}, \beta_{S,j}\right)$, the producers and consumers risks for the producer and the supplier respectively. These are the probabilities of rejecting a good lot and accepting a bad one by a producer (indexed p) and a supplier (indexed S), under each specific and alternative statistical sample selected. These risks are summarized in the matrix below.

$$\begin{pmatrix} \left(\left(\alpha_{p,1}, \beta_{p,1}\right); \left(\alpha_{S,1}, \beta_{S,1}\right)\right) & \cdots & \cdots & \left(\left(\alpha_{p,1}, \beta_{p,1}\right); \left(\alpha_{S,m}, \beta_{S,M}\right)\right) \\ \left(\left(\alpha_{p,2}, \beta_{p,2}\right); \left(\alpha_{S,1}, \beta_{S,1}\right)\right) & & & \left(\left(\alpha_{p,2}, \beta_{p,2}\right); \left(\alpha_{S,m}, \beta_{S,M}\right)\right) \\ & \cdots & & \cdots \\ \left(\left(\alpha_{p,N}, \beta_{p,N}\right); \left(\alpha_{S,1}, \beta_{S,1}\right)\right) & \cdots & \cdots & \left(\left(\alpha_{p,N}, \beta_{p,N}\right); \left(\alpha_{S,m}, \beta_{S,M}\right)\right) \end{pmatrix} \quad (8.33)$$

The selection of a control program can be unique and randomized, reflecting strategic considerations such as signals by a producer to indicate that they control their suppliers and vice versa for suppliers to indicate that they are careful to deliver acceptable quality items. These controls and their outcomes may also be negotiated and agreed on in contractual agreements to include penalties and incentives based on the control-sample outcomes. In this sense, associated to the risk specifications of equation (8.33), there may be as well a bi-matrix of costs summarizing the expected and derived costs implied by the parties control strategies. For simplicity and brevity, this section will consider only a specification of type I risks and the collaborative minimization of type II risks, in the spirit of the traditional Neyman-Pearson theory.

Explicitly, assume for simplicity binomial sampling distributions with parameters $\left(n_{p,i}, c_{p,i}\right)$ for the producer and $\left(n_{S,j}, c_{S,j}\right)$ for the supplier where the indices i and j denote a set of finite and alternative sampling plans available to the producer and the supplier respectively. Let AQL be a contracted proportion of acceptable defectives (or the Acceptable Quality Limit) and $LTFD$ be a contracted proportion of unacceptable defectives in a lot (or the Lowest Tolerance Fraction Defectives). Then the risks sustained by the producer (buyer) and by the supplier, when each selects sampling plans $\left(n_{p,i}, c_{p,i}\right)$ and $\left(n_{S,j}, c_{S,j}\right)$ are respectively (see also Wetherhill, 1977, Tapiero, 1996):

$$\alpha_{k,i} = 1 - \sum_{\ell=0}^{c_{k,i}} \binom{n_{k,i}}{\ell}(AQL)^{\ell}(1-AQL)^{n_{k,i}-\ell} \; ; \; \beta_{k,i} = \sum_{\ell=0}^{c_{k,i}} \binom{n_{k,i}}{\ell}(LTFD)^{\ell}(1-LTFD)^{n_{k,i}-\ell}.$$

(8.34)

where $k = p, S$. For example, if the supplier fully samples (i.e. $j=N$) and attends to all non conforming units, then $\alpha_{S,N} = 1$, $\beta_{S,N} = 0$. If the buyer knew for sure that this were the case, he would use always a costless no-inspection alternative. Similarly, say that the supplier accepts a bad lot (the supplier'S consumer risk). The buyer-consumer risk will in this case be determined by the stringency of controls used by the supplier. If the buyer-producer also accepts this defective lot, the probability corresponding to the producer and the supplier sampling strategies defines bi-matrices with entries: $\left[\alpha_{p,i}(1-\overline{A}_{S}); \alpha_{S,j}\right]$ and $\left[\beta_{p,i}\overline{B}_{S}; \beta_{S,j}\right]$ for type I (producer) and type II (consumer) risks. In these entries, $(\overline{A}_{S}, \overline{B}_{S})$ denote the average supplier control risks, assumed known (or contracted) by the producer. These risks will be altered, of course, as a function of the mutual relationships established between the supplier and the producer. If the supplier assumes responsibility for a consumer's risk only if it is detected by the producer, then the supplier and the producer consumer risks will rather be $\left[\beta_{p,i}\overline{B}_{S}; \beta_{S,j}(1-\overline{B}_{p})\right]$ instead of $\beta_{S,j}$, as stated in the type II risk bi-matrix above. Note that $(\overline{A}_p, \overline{B}_p)$ denote the average producer and consumers risks of the buyer-producer. Other cases may be considered as well, based on the exchange of information between the supplier and the producer. For example, if the supplier reports to the producer his choice of control techniques, then the risk bi-matrices for both, will be instead $\left[\alpha_{p,i}(1-\alpha_{S,j}); \alpha_{S,j}\right]$ and $\left[\beta_{p,i}\beta_{S,j}; \beta_{S,j}(1-\overline{B}_p)\right]$. In other words, the organization structure of the supply chain and the information-controls exchange combined with the "various degrees and forms" of collaboration (or none at all) will determine both the control programs applied and the risks sustained by the supply chain parties. Each of theses cases can be treated separately, although the approach we use here is essentially the same.

Assume that average risks sustained by the supply chain parties are agreed on (or contracted) and let each of the parties selects a control program in randomized manner over the following risk bi-matrices $\left[\alpha_{p,i}(1-\overline{A}_S); \alpha_{S,j}\right]$ and $\left[\beta_{p,i}\overline{B}_S; \beta_{S,j}\right]$. Let x_i be the probability that the producer selects a control strategy i while y_j is the probability that the supplier selects control strategy j. The average risks for the supplier are then: $\sum_{j=1}^{M} y_j \alpha_{S,j} = \overline{\alpha}_S$; $\sum_{j=1}^{M} y_j \beta_{S,j} = \overline{\beta}_S$ where $(\overline{\alpha}_S, \overline{\beta}_S)$ is the average type I risk associated to a selection of sampling plans using the randomized sampling strategies y_j, $j = 1, 2, ... M$ used by the supplier. It is not, of course, the actual average risks sustained by the supplier, since such risks will depend on the action followed by producer as well. In this case, we use capital letters to denote the actual type I and II risks sustained. In this special case, $\overline{A}_S = \overline{\alpha}_S, \overline{A}_p = \overline{\alpha}_p(1-\overline{A}_S)$ and $\overline{B}_S = \overline{\beta}_S, \overline{B}_p = \overline{\beta}_p \overline{B}_S$ where $(\overline{A}_p, \overline{B}_p)$ are the corresponding average risks of the producer with sampling specific average risks $(\overline{\alpha}_p, \overline{\beta}_p)$ defined as randomized sampling strategies $\sum_{i=1}^{N} x_i \alpha_{p,i} = \overline{\alpha}_p, \sum_{i=1}^{N} x_i \beta_{p,i} = \overline{\beta}_p$. Note that in such notations, the sampling-control risk problems faced by both the producer and the supplier are then given by minimizing the consumers (type II) risks subject to some constraints on their producers (type I) risks, explicitly stated as follows.

$$\underset{\substack{(n_{p,i},c_{p,i}),0\le x_i\le1,\sum_{i=1}^{N}x_i=1,\ i=1,2,...N}}{Min} \quad \overline{B}_p=\overline{\beta}_p,\overline{B}_S=\overline{\beta}_p\overline{\beta}_S \quad \text{Subject to: } \overline{A}_p \le A_{PC}$$

$$\underset{\substack{(n_{S,j},c_{S,j}),0\le y_j\le1,\sum_{j=1}^{M}y_j=1,\ j=1,2,...M}}{Min} \quad \overline{B}_S=\overline{\beta}_S \quad \text{Subject to: } \overline{A}_S \le A_{SC}$$

(8.35)

where risk minimization is reached with respect to the available alternative sampling plans and their randomization (namely, selecting a number of sampling plans through a randomization rule to be found by the solution of the game). Here, $\left(A_{PC},A_{SC}\right)$ stands for specific parameters while $\overline{A}_S = \overline{\alpha}_S, \overline{A}_p = \overline{\alpha}_p(1-\overline{\alpha}_S)$. The solution of the constrained game (8.35) subject to risks (8.34) determines therefore an adaptation of the Neyman-Pearson lemma to a supplier-producer situation, which can be solved according to the available information we have regarding alternative sampling plans and assumptions on the behavioral relationships that exist between the supplier and the producer. For example, assuming power (leader-led) relationships and collaborative strategies that both the producer and the supplier will adopt, a number of games might be developed. Explicitly, if the producer is a leader in a Stackleberg game, fully informed of the supplier objectives, then the sampling-control selection problem is defined by:

$$\underset{\substack{(n_{p,i},c_{p,i}),0\le x_i\le1,\sum_{i=1}^{N}x_i=1,\ i=1,2,...N}}{Min} \quad \overline{B}_p=\overline{\beta}_p,\overline{B}_S=\overline{\beta}_p\overline{\beta}_S$$

Subject to: (8.36)

$$\underset{\substack{(n_{S,j},c_{S,j}),0\le y_j\le1,\sum_{j=1}^{M}y_j=1,\ j=1,2,...M}}{Min} \quad \overline{B}_S=\overline{\beta}_S$$

where the type I risks, dropped out of equation (8.36) are implied as in equation (8.35). When it is the supplier who leads and the producer is led, then producer risk is minimized first and the supplier uses this information to minimize his risks. Further, if both the supplier and the producer collaborate in controlling risks, then the problem they face can be stated as a weighted (Pareto optimal) solution to the game (8.35). In this case, we presume that there is a parameter $0 \le \lambda \le 1$ expressing the negotiating of each of the parties, such that:

$$\underset{\substack{\left(n_{p,i},c_{p,i}\right),0\leq x_i\leq1,\sum_{i=1}^{N}x_i=1,\ i=1,2,...N;\\ \left(n_{S,j},c_{S,j}\right),0\leq y_j\leq1,\sum_{j=1}^{M}y_j=1,\ j=1,2,...M}}{Min}\quad \left\{\lambda\overline{B}_p+(1-\lambda)\overline{B}_S\right\}\qquad(8.37)$$

subject to both the producer and the supplier type I risks constraints is minimized. Alternatively, we may consider other objectives such as economic and sampling costs as well as the costs associated with the type I and type II risks of both the producer and the supplier. If we consider the type II risk bi-matrices $\left[\beta_{p,i}\overline{B}_S;\beta_{S,j}(1-\overline{B}_p)\right]$ and $\left[\beta_{p,i}\beta_{S,j};\beta_{S,j}(1-\overline{B}_p)\right]$ instead, then the sampling-control problem's formulations in (8.35)-(8.37) remain the same with average risks respectively defined instead by:

$$\overline{B}_S=\overline{\beta}_S(1-\overline{B}_p)=\frac{\overline{\beta}_S}{1+\overline{\beta}_p\overline{\beta}_S},\quad \overline{B}_p=\overline{\beta}_p\overline{B}_S=\overline{\beta}_p\overline{\beta}_S(1-\overline{B}_p)=\frac{\overline{\beta}_p\overline{\beta}_S}{1+\overline{\beta}_p\overline{\beta}_S}$$

$$(8.38)$$

and

$$\overline{B}_S=\overline{\beta}_S(1-\overline{B}_p)=\overline{\beta}_S(1-\overline{\beta}_p\overline{\beta}_S),\quad \overline{B}_p=\overline{\beta}_p\overline{\beta}_S.\qquad(8.39)$$

Evidently, other situations arise, a function of the information available to each of the parties and the exchange they engage in and the behavioral assumptions made regarding the potential collaboration and/or conflict that exists between the supplier and the producer. To obtain tractable results and for demonstration purposes we restrict ourselves to simple solutions for a supplier and a producer, each considering two alternative control programs. Essential results are then summarized and discussed. Subsequently, special cases and numerical examples are treated to highlight both the implications and the applicability of the approach. Below, we begin with non-collaborating supplier and producer to subsequently compare to the effects of collaboration.

Proposition 8.6. *Let* $\left(\alpha_{p,i},\beta_{p,i}\right),i=1,2$ *and* $\left(\alpha_{S,j},\beta_{S,j}\right),j=1,2$ *be type I and II and risks of a producer and a supplier engaged in mutual (and conflicting) binomial sampling-controls as in equation (8.34) with* $\beta_{p,2}<\beta_{p,1}$ *and* $\beta_{S,2}<\beta_{S,1}$ *. Then if type I risks are satisfied by both strategies, the optimal sampling-control is a pure strategy where both the supplier and the producer adopt intensive control strategies (with type II risks* $\left(\beta_{p,2},\beta_{S,2}\right)$*). If type I risks constraints are binding then the supplier* *and the producer can turn to randomized sampling strategies given by the solution of:*

$$\bar{\alpha}_p = A_{PC}/(1 - A_{SC}) \quad \text{and} \quad \bar{\alpha}_S = A_{SC} \tag{8.40}$$

While average type II risks minimized by each of the parties are explicitly given by:

$$\underset{0 < c_{S,j} < n_{S,j}, j=1,2}{\text{Min}} \quad \bar{B}_S = \left(y\beta_{S,1} + (1-y)\beta_{S,2} \right)$$

$$\underset{0 < c_{p,i} < n_{p,i}, i=1,2}{\text{Min}} \quad \bar{B}_p = \left(x\beta_{p,1} + (1-x)\beta_{p,2} \right)\left(y\beta_{S,1} + (1-y)\beta_{S,2} \right) \tag{8.41}$$

where $(n_{S,j}, c_{S,j}), j = 1,2$ *and* $(n_{p,i}, c_{p,i}), i = 1,2$ *are two known sampling plans available to the supplier and to the producer, while* $(\alpha_{S,k}, \beta_{S,k}), k = 1,2$ *and* $(\alpha_{p,k}, \beta_{p,k}), k = 1,2$ *are the types I and II risks associated with each of these sampling plans by the supplier and the producer. Then the optimal randomized strategies for selecting one or the other sampling plans are:*

$$y^* = \frac{A_{SC} - \alpha_{S,2}}{\alpha_{S,1} - \alpha_{S,2}}; \quad x^* = \frac{A_{PC}/(1 - A_{SC}) - \alpha_{p,2}}{\alpha_{p,1} - \alpha_{p,2}}. \tag{8.42}$$

Proof: The proof is straightforward since in the bi-matrix $\left[\beta_{p,2}\bar{B}_S; \beta_{S,2} \right] = \left[\beta_{p,2}\beta_{S,2}; \beta_{S,2} \right]$, intensive sampling by both the supplier and the producer are dominating all other strategies. This observation might be practically misleading because it ignores the costs associated with sampling and of course all other risk costs. Of course, if the type I risks are set to their maximal values, then:

$$\bar{A}_p = \bar{\alpha}_p(1 - \bar{\alpha}_S) = A_{PC} \quad \text{or} \quad \bar{\alpha}_p = A_{PC}/(1 - A_{SC}) \quad \text{and} \quad \bar{A}_S = \bar{\alpha}_S = A_{SC},$$

which provides a system of equations in the randomizing parameters (x,y), or using equation (8.34), we have (with $0 \leq y_1 = y, y_2 = 1 - y, 0 \leq x_1 = x, x_2 = 1 - x$):

$$\sum_{j=1}^{2} y_j \left(1 - \sum_{\ell=0}^{c_{S,j}} \binom{n_{S,j}}{\ell}(AQL)^{\ell}(1 - AQL)^{n_{S,j}-\ell} \right) = A_{SC},$$

and

$$\left(\sum_{i=1}^{2} x_i \left(1 - \sum_{\ell=0}^{c_{p,i}} \binom{n_{p,i}}{\ell}(AQL)^{\ell}(1 - AQL)^{n_{p,i}-\ell} \right) \right) * \left(1 - \sum_{j=1}^{2} y_j \left(1 - \sum_{\ell=0}^{c_{S,j}} \binom{m_j}{\ell}(AQL)^{\ell}(1 - AQL)^{n_{S,j}-\ell} \right) \right) = A_{PC}$$

Given a solution for (x,y) in terms of the sampling-control parameters, the type II risks of the supplier, $\bar{B}_S = y\beta_{S,1} + (1-y)\beta_{S,2}$ is minimized with respect to $(c_{S,j}, n_{S,j})$ while the risk of the producer $\sum_{j=1}^{2}\sum_{i=1}^{2} x_i y_j \beta_{p,i} \beta_{S,j}$ is minimized with respect to $(c_{p,i}, n_{p,i})$. This proposition remains valid when we use instead (8.38) and (8.39). Of course, the solution to this problem requires that we apply numerical techniques to select the appropriate sampling control parameters. Non-cooperation implies therefore that the firm uses as much as possible sampling-controls and do not randomize sampling strategies (unless type I risk constraints are violated, as stated in the proposition above). When type I risks constraints are binding, then

$$y^* = \frac{A_{SC} - \alpha_{S,2}}{\alpha_{S,1} - \alpha_{S,2}}; \quad x^* = \frac{A_{PC}/(1-A_{SC}) - \alpha_{p,2}}{\alpha_{p,1} - \alpha_{p,2}}$$

and therefore the sampling control problem is reduced to a nonlinear optimization problem stated in the proposition and explicitly given by:

$$\underset{0<c_{S,j}<n_{S,j}, j=1,2}{Min} \quad \bar{B}_S = \beta_{S,2} + \left(\beta_{S,1} - \beta_{S,2}\right)\left(\frac{A_{SC} - \alpha_{S,2}}{\alpha_{S,1} - \alpha_{S,2}}\right)$$

$$\underset{0<c_{p,i}<n_{p,i}, i=1,2}{Min} \quad \bar{B}_p = \left(\beta_{p,2} + \left(\beta_{p,1} - \beta_{p,2}\right)\left(\frac{A_{PC}/(1-A_{SC}) - \alpha_{p,2}}{\alpha_{p,1} - \alpha_{p,2}}\right)\right)\left[\beta_{S,2} + \left(\beta_{S,1} - \beta_{S,2}\right)\left(\frac{A_{SC} - \alpha_{S,2}}{\alpha_{S,1} - \alpha_{S,2}}\right)\right]$$

Producers and suppliers can reduce control costs if they collaborate. In this case, equation (8.37) is resolved subject to the type I risk constraints. Of course, if these risks are binding, then (8.37) is reduced to:

$$\underset{0<c_{p,i}<n_{p,i}, i=1,2;\, 0<c_{S,j}<n_{S,j}, i,j=1,2}{Min} \quad \lambda\left[\frac{1-\lambda}{\lambda} + \left(\beta_{p,2} + x\left(\beta_{p,1} - \beta_{p,2}\right)\right)\right]\left(\beta_{S,2} + y\left(\beta_{S,1} - \beta_{S,2}\right)\right) \quad (8.43)$$

or

$$\underset{0<c_{p,i}<n_{p,i}, i=1,2;\, 0<c_{S,j}<n_{S,j}, i,j=1,2}{Min} \quad \Gamma$$

$$\Gamma = \lambda\left[\frac{1-\lambda}{\lambda} + \left(\beta_{p,2} + \left[\frac{A_{PC}/(1-A_{SC}) - \alpha_{p,2}}{\alpha_{p,1} - \alpha_{p,2}}\right]\left(\beta_{p,1} - \beta_{p,2}\right)\right)\right] * \left(\beta_{S,2} + \left[\frac{A_{SC} - \alpha_{S,2}}{\alpha_{S,1} - \alpha_{S,2}}\right]\left(\beta_{S,1} - \beta_{S,2}\right)\right)$$

$$(8.44)$$

Examples will elaborate both the usefulness of this approach as well as deviations from a complete collaboration between the supplier and the producer. To keep our calculations simple, some simplifications are made.

8.4.2 SPECIAL CASES AND EXTENSIONS

Example 8.1.

If there is only one firm (say the supplier), then the problem is reduced to the standard quality assurance approach with randomized sampling plans which uses Neyman-Pearson theory. In this case, we have:

$$Min\ \overline{\beta}_S = \sum_{j=1}^{M} y_j \beta_{S,j}\ \text{Subject to}: \sum_{j=1}^{M} y_j \alpha_{S,j} \le \overline{\alpha}_S,\ \sum_{j=1}^{M} y_j = 1,\ y_j \ge 0. (8.45)$$

In this simple problem, we note the potential for randomizing inspection strategies in costs reduction so that in effect inspection controls assume a strategic perspective. Such an idea is also pointed out by Deming (see Burke et al. 1993) who claims intuitively that "one either samples fully or not". When firms compete, the solution is for maximal sampling as stated here (although in practice, cost considerations will imply that randomized controls can be optimal).

Example 8.2.

We consider theoretically and numerically problem (8.41) when both—the supplier and the producer use two strategies: no sampling and sampling m and n (for the supplier and the producer respectively). If the parties do not sample, the probabilities of rejecting a good lot (the producer risk) is null for both while the probability of accepting a bad lot is 1, or: $\alpha_{p,1} = 0$; $\alpha_{S,1} = 0$, and $\beta_{p,1} = 1$; $\beta_{S,1} = 1$. This special situation results in the following risks:

$$\begin{pmatrix} & \{y, m = 0\} & \{1 - y, m > 0\} \\ \{x, n = 0\} & (0,0); (1,1) & (0, \alpha_S); (\beta_S, \beta_S) \\ \{1 - x, n > 0\} & (\alpha_p, 0); (\beta_p, 1) & (\alpha_p(1 - \alpha_S), \alpha_S); (\beta_p \beta_S, \beta_S) \end{pmatrix}. (8.46)$$

If type I risks are binding and the producer and the suppliers are not collaborating, we have the following randomizing parameters:

$$y^* = 1 - \frac{A_{SC}}{\alpha_S}; \quad x^* = 1 - \frac{A_{PC}/(1 - A_{SC})}{\alpha_p}. \tag{8.47}$$

And the sampling-control problem is reduced to:

$$(a) \quad \underset{0<m}{Min} \ \overline{B}_S = \beta_S + (1 - \beta_S) \left[1 - \frac{A_{SC}}{\alpha_S} \right] \ ;$$

$$(b) \quad \underset{0<n}{Min} \ \overline{B}_p = \overline{B}_S^* \left(\beta_p \beta_S^* + (1 - \beta_p \beta_S^*) \left[1 - \frac{A_{PC}/(1 - A_{SC})}{\alpha_p} \right] \right)$$

(8.48)

where starred variables are optimal values resulting from the supplier risk minimization. In case of collaboration, we have instead:

$$\underset{0<c<n;0<d<m}{Min} \ \lambda \left[\frac{1-\lambda}{\lambda} + \left(\beta_p + \left[1 - \frac{A_{PC}/(1 - A_{SC})}{\alpha_p} \right] (1 - \beta_p) \right) \right] * \left(\beta_S + \left[1 - \frac{A_{SC}}{\alpha_{S,2}} \right] (1 - \beta_S) \right) \ (8.49)$$

Next, assume that there is no collaboration. In this case, for parties minimizing type II risks, a non-zero sum game with a pure (costliest) strategy at $x = 0$ and $y = 0$ is reached with type II risks $(\beta_p \beta_S, \beta_S)$ for the producer and the supplier respectively. The supplier's sampling program selection will consist then in minimizing the type II risk subject to a type I constraint. By the same token, the type II risk minimization by the producer subject to the type I risk is $Min \ \beta_p \beta_S$ subject to the producer type I constraint which is a function of the supplier's assumed risks. If the producer is informed of the control procedure set in place by the supplier, then, of course, such information can be used to reduce the amount of sampling (and therefore costs) by the producer. If both the supplier and the producer collaborate ex-ante by an exchange of information regarding the quality strategies (but maintain their independence by selecting in a game-like manner the strategies to sample or not, resulting in the pure dominant strategy), then the amount of sampling to be performed on the same lot by both parties will be necessarily reduced. This can be verified by minimizing the producer risk with respect to the amount of sampling performed by both the supplier and the producer as well as selecting the jointly optimal critical test parameter. In other words, the optimal sampling program for both the supplier and the producer would be (once we insert the sampling distributions): $Min \ \beta_p \beta_S$ subject to type I constraints for the producer and the supplier. Of course, if collaboration between the supplier and the producer is complete, there might be a randomized strategy for sampling (in which case, either or both the producer and supplier may prefer not to sample), reducing thereby the amount of sampling. The problem to be minimized is then given by simplifying equation (8.37) which is reduced to: minimizing $\beta_S \{ \lambda \beta_p + (1 - \lambda) \}$ (since $\overline{B}_p = \beta_p \beta_S$

and $\bar{B}_S = \beta_S$). If both the supplier and the producer do not sample with probabilities (y, x), then the resulting type II risks are $\beta_S = \left(y + (1-y)\beta_{S,2}\right)$ and $\beta_p = \left(x + (1-x)\beta_{p,2}\right)$ and therefore, problem (8.37) is reduced to:

$$\underset{\substack{(n_{p,2},c_{p,2}),0\leq x\leq 1 \\ (n_{S,2},c_{S,2}),0\leq y\leq 1}}{\text{Min}} \quad \left(y + (1-y)\beta_{S,2}\right)\left[\lambda\left(x + (1-x)\beta_{p,2}\right) + (1-\lambda)\right]$$

Subject to: (8.50)

$$(1-y)\alpha_{S,2} \leq A_{SC}, \quad (1-x)\alpha_{p,2}\left(1 - (1-y)\alpha_{S,2}\right) \leq A_{PC}$$

with $\left(A_{PC}, A_{SC}\right)$ specified type I risk constraints (as defined in equation (8.35)). If there is an interior solution, it is easy to show that optimal sampling by the supplier and the producer is given by the marginal effect of a sample increment on type II risks, equaling the odds of not sampling, or $-\partial\beta_S / \partial n_{S,2} = y/(1-y)$, $-\partial\beta_p / \partial n_{p,2} = x/(1-x)$. Interestingly, if the type I constraints are binding, we will have then optimal sampling lot sizes $\left(n_{S,2}, n_{p,2}\right)$ which are given by:

$$-\frac{\partial\beta_S}{\partial n_{S,2}} = \frac{y}{1-y} = \frac{\alpha_{S,2}}{A_{SC}} - 1; \quad -\frac{\partial\beta_p}{\partial n_{p,2}} = \frac{x}{1-x} = \frac{\alpha_{p,2}(1-A_{SC})}{A_{PC}} - 1 \text{ (8.51)}$$

which are sets of equations, each a function of one variable with:

$$\alpha_{S,2} = 1 - (1-AQL)^{n_{S,2}-1}\left[1 - (n_{S,2}-1)AQL\right]; \quad \beta_{S,2} = (1-LTFD)^{n_{S,2}-1}\left[1 - (n_{S,2}-1)LTFD\right]$$
(8.52)

$$\alpha_{p,2} = 1 - (1-AQL)^{n_{p,2}-1}\left[1 - (n_{p,2}-1)AQL\right]; \quad \beta_{p,2} = (1-LTFD)^{n_{p,2}-1}\left[1 - (n_{p,2}-1)LTFD\right]$$
(8.53)

When we use risk bi-matrices such as (8.38) and (8.39), we obtain different results, expressing the interdependencies of risk presumed by the producer and supplier interdependent organization and risk transfer agreements. For example, when the relationship between the producer and supplier is altered, with the supplier fully responsible for any detected non conforming lot by the producer, the amount of control exercised by each will necessarily reflect this relationship. Explicitly, set the risk bi matrix

$$\left[\beta_{p,i}\bar{B}_S; \beta_{S,j}(1-\bar{B}_p)\right], \quad i = 1, 2; j = 1, 2,$$

where $\beta_{p,1} = 1, \beta_{S,1} = 1$, $\beta_{p,2} < 1, \beta_{S,2} < 1$, the average type II risks are then calculated as follows. Using the bimatrix:

$$\begin{bmatrix} \left[\overline{B}_S ; (1-\overline{B}_p) \right] & \left[\overline{B}_S ; \beta_{S,2}(1-\overline{B}_p) \right] \\ \left[\beta_{p,2}\overline{B}_S ; (1-\overline{B}_p) \right] & \left[\beta_{p,2}\overline{B}_S ; \beta_{S,2}(1-\overline{B}_p) \right] \end{bmatrix}$$

we have:

$$\overline{B}_p = x\overline{B}_S + (1-x)\beta_{p,2}\overline{B}_S = \overline{\beta}_p \overline{B}_S$$

$$\overline{B}_S = y(1-\overline{B}_p) + (1-y)\beta_{S,2}(1-\overline{B}_p) = \overline{\beta}_S(1-\overline{B}_p)$$

which is reduced to:

$$\overline{B}_S = \frac{\overline{\beta}_S}{1 + \overline{\beta}_p \overline{\beta}_S}, \quad \overline{B}_p = \frac{\overline{\beta}_p \overline{\beta}_S}{1 + \overline{\beta}_p \overline{\beta}_S} \tag{8.54}$$

with $\overline{\beta}_S = \left(y + (1-y)\beta_{S,2} \right)$, $\overline{\beta}_p = \left(x + (1-x)\beta_{p,2} \right)$, as stated above, while type I risks remain as in equation (8.50). Again, consider a collaborative solution that minimizes (8.37) (with (8.54) inserted into (8.37)). Namely, we minimize the objective:

$$\lambda \overline{B}_p + (1-\lambda)\overline{B}_S = \frac{\overline{\beta}_S}{1 + \overline{\beta}_p \overline{\beta}_S}\left(\lambda \overline{\beta}_p + (1-\lambda) \right),$$

explicitly specified by:

$$\underset{0<c_{S,2}<n_{S,2}; 0<c_{p,2}<n_{p,2}; 0<(x,y)<1}{Min} \quad \frac{\overline{\beta}_S}{1 + \overline{\beta}_p \overline{\beta}_S}\left(1 - \lambda(1-\overline{B}_p) \right) \tag{8.55}$$

with $(\overline{\beta}_p, \overline{\beta}_S)$ as stated above. Of course, minimization of (8.55) is a function of the "sharing" parameter λ and optimal sampling-control parameters that provide a feasible interior solution (i.e. a solution that satisfies the type I risks constraints). If type I risks are binding, then, of course, (x, y) are given by (8.47). Numerical examples to these effects will be considered subsequently.

Example 8.3. Multi-echelon and assembly supply chains

Consider a supplier who supplies a producer who in turn supplies another producer. The production process is thus a series assembly process supply chain, each producer with a risk attitude reflected by the consumer and producer risks assumed. Thus, letting the first supplier be indexed "1", we have:

$$A^{(1)} = \sum_{j=1}^{M_1} y_j^{(1)} \alpha_j^{(1)} \le A_{SC}^{(1)}, \quad B^{(1)} = \sum_{j=1}^{M_1} y_j^{(1)} \beta_j^{(1)}, \tag{8.56}$$

while for subsequent producers-suppliers we have recursive equations for type I and II risks explicitly given by:

$$A^{(k+1)} = \sum_{j=1}^{M_{k+1}} y_j^{(k+1)} \alpha_j^{(k+1)} \left(1 - A^{(k)}\right) \le A_{SC}^{(k+1)}, \quad B^{(k+1)} = \sum_{j=1}^{M_{k+1}} y_j^{(k+1)} \beta_j^{(k+1)} B^{(k)}. \quad (8.57)$$

In other words, for the second firm, the type I risk $A^{(2)}$ equals the probability that the first firm has not committed a type I error (and rejected a good lot with probability $\left(1 - A^{(1)}\right)$) times the probability that it commits such an error, as stated in equation (8.56) for the first supplier-firm. This results therefore in $A^{(2)} = \sum_{j=1}^{M_2} y_j^{(2)} \alpha_j^{(2)} \left(1 - A^{(1)}\right)$. Similarly, the type II error that a second firm commits is equal to the probability that the first firm (the supplier) has committed such an error (with probability $B^{(1)}$) times the probability that it commits such an error under all M_2 available sampling strategies, each selected with probability $y_j^{(2)}$, or $B^{(2)} = \sum_{j=1}^{M_2} y_j^{(2)} \beta_j^{(2)} B^{(2)}$, as stated in equation (8.57). Of course, in this equation both $\left(\alpha_j^{(k+1)}, \beta_j^{(k+1)}\right)$ are given in terms of the sampling-control parameters specified in equation (8.34).

For an assembly process, we can proceed similarly in two manners. Either the producer samples production ex-ante or ex-post (i.e. finished product). In the latter case, when the producer has several suppliers, we have:

$$\sum_{j=1} y_j^{\ell} \alpha_j^{\ell} = A^{\ell}; \quad \sum_{j=1} y_j^{\ell} \beta_j^{\ell} = B^{\ell}; \quad \ell = 1,2,..... \quad (8.58)$$

$$\sum_{i=1}^{n} x_i \alpha_i^{(A)} \left(\sum_{\ell} (1 - A^{(\ell)})\right) = A^{(A)}; \quad \sum_{i=1}^{n} x_i \beta_i^{(A)} \left(\sum_{\ell} B^{(\ell)}\right) = B^{(A)}. \quad (8.59)$$

Finally, if each supplier is tested individually, then we are in the specific case treated and summarized by our proposition. That is:

$$\sum_{j=1} y_j^{\ell} \alpha_j^{(\ell)} = A^{(\ell)}; \quad \sum_{j=1} y_j^{\ell} \beta_j^{(\ell)} = B^{(\ell)} . \quad (8.60)$$

$$\sum_{i=1}^{n} x_i^{\ell} \alpha_i^{(A,\ell)} (1 - A^{(\ell)}) = A^{(A,\ell)}; \quad \sum_{i=1}^{n} x_i^{\ell} \beta_i^{(A,\ell)} B^{(\ell)} = B^{(A,\ell)}; \quad (8.61)$$

while for ex-post assembly, we have the following consumers and producers risks:

$$A^{(A)} = \sum_{i=1}^{n} x_i^A \alpha_i^{(A)} \prod_{\ell=1} (1 - A^{(\ell)}) A^{(A,\ell)} \qquad B^{(A)} = \sum_{i=1}^{n} x_i^A \beta_i^{(A)} \prod_{\ell=1} B^{(A,\ell)} \quad (8.62)$$

From these expressions we clearly see (due to the mutliplicative effects of the risks borne by downstream firms of the supply chain) the important risk effects sustained by an assembler-producer when he uses multiple suppliers. Such an observation can therefore be used to justify the fact that a growth of assembly technologies in manufacturing necessarily implies a need for more reliable and responsible suppliers (and therefore, industrial organizations that are based on supply chains).

We conclude by providing a number of numerical examples.

Example 8.4. Curtailed sampling

For simplicity, we shall consider in this numerical example a randomized curtailed sampling technique, consisting in applying a curtailed sample in probability or doing nothing. When the parties do nothing, nothing is detected, while when the sample is tested, a non-conforming unit is detected in probability according to the stringency of the tests applied (the sample size). Curtailed sampling thus, specifies that the first time that a non defective unit is detected, then the lot is rejected. Say that AQL is an acceptable quality limit and let $LTFD$ be the lowest tolerance fraction defectives. Thus, the probability of accepting a bad lot (when the control test is applied) equals the probability that all units sampled are accepted, in other words $\beta_p = (1 - LTFD)^n$ where n is the producer sample size. By the same token, the probability of rejecting a good lot (the producer's risk) is equal to the probability of not accepting a good lot, or $\alpha_p = 1 - (1 - AQL)^n$. Similar results are obtained for the supplier who applies also a randomized curtailed sampling technique with a sample size m. Let (x, y) be the probabilities that the producer and the suppliers do not sample. Then, we have by equation (8.54), the following average type II risks:

$$\overline{\beta}_S = \left(y + (1 - y)(1 - LTFD)^m \right) \text{ and } \overline{\beta}_p = \left(x + (1 - x)(1 - LTFD)^n \right)$$

For type I risks, we have:

$$(1-y)\left[1-(1-AQL)^n\right] \le A_{SC},$$

$$(1-x)\left[1-(1-AQL)^n\right]\left(1-(1-y)\left[1-(1-AQL)^m\right]\right) \le A_{PC}$$

If the mean type I risk constraints are binding, the two equations above are equalities and can therefore be solved for the probabilities of not sampling at all, or:

$$(1-x) = \frac{A_{PC}}{\left(\left[1-(1-AQL)^n\right] - A_{SC}\left[1-(1-AQL)^m\right]\right)},$$

$$(1-y) = \frac{A_{SC}}{\left[1-(1-AQL)^n\right]}$$

Explicitly, let $AQL = 0.05$, $LTFD = 0.15$ and $A_{SC} = 0.10$, $A_{PC} = 0.08$. Thus if type I risks are binding, we have probabilities of sampling which a function of the sample size only:

$$1-y = \frac{0.10}{\left[1-(0.95)^n\right]}, \quad 1-x = \frac{0.08}{\left(\left[1-(0.95)^n\right] - 0.10\left[1-(0.95)^m\right]\right)}$$

Of course, for the supplier, the probability of sampling is the same regardless of the producer sample size (as shown in Table 8.3 where computations for alternative sample sizes selected by the supplier and the producer are summarized). However, the sample probability of the producer is always dependent on the sample size of the supplier. The higher the supplier's sample size, the smaller the probability of sampling by the producer. This relationship expresses therefore a sensitivity of the producer to controls made upstream by the supplier. By the same token, given the sampling probabilities, the average type II risks for the supplier and the producer are calculated by:

$$\bar{\beta}_S = \left((0.85)^m + \left[1 - \frac{0.10}{\left[1-(0.95)^n\right]}\right]\left[1-(0.85)^m\right]\right),$$

$$\bar{\beta}_P = \left((0.85)^n + \left[1 - \frac{0.08}{\left(\left[1-(0.95)^n\right] - 0.10\left[1-(0.95)^m\right]\right)}\right]\left[1-(0.85)^n\right]\right)$$

Note again that these risks are also a function of the sample sizes only. Using equation (8.54), or $\overline{B}_S = \overline{\beta}_S / (1 + \overline{\beta}_p \overline{\beta}_S)$, $\overline{B}_p = \overline{\beta}_p \overline{B}_S$, we find that the average risk sustained by the supplier and the producer is a function of the firms' sample size. Of course, if the supplier and the producer cooperate, we can consider a weighted sum of these risks which can be minimized with respect to the sample size to be applied by each firm. A likely result would be to sample more upstream and less downstream by the producer. If we set an $AQL=0.05$, and an $LTFD=0.15$ and assume that average type I risks for the supplier and the producer are bounded by $A_{SC} = 0.15$, $A_{PC} = 0.20$, then for sample sizes (5,10,15,20), selected by the supplier and the producer alike, we obtain the results in Table 8.3. Explicitly, if the producer and the supplier choose a sample size of 15 units each, then the probabilities of not sampling by the producer and the supplier are .56 and .72 respectively while the average type II risks for each are .308 and .514. The average type II risk, equally shared by the producer and the supplier, in case they cooperate, would be .411. In this table, we see that the average "shared type II risk" decreases when the supplier increases the sample size. However, when the supplier maintains a fixed sample size (say $m=10$), then the average shared type II risk increases when the producer increases the sample size. When the producer maintains a fixed sample size and the supplier augments the sample size, the shared type II risk always decline. This observation explains the common practice to augment the amount of quality control upstream at the expense of downstream quality control. Additional observations drawn from Table 8.3 indicate that the no sampling probability of the supplier and the producer increase as a function of their sample size, although the producer is more sensitive to the supplier sample size than the supplier to the producer sample size, as indicated by the equations for y and x given above.

When type I risks are not binding, we obtain the results stated in Table 8.2. In this case, we minimize equation (8.55), with

$$\overline{\beta}_S = \left(y + (1-y)(1-LTFD)^m \right) \text{ and } \overline{\beta}_p = \left(x + (1-x)(1-LTFD)^n \right)$$

with respect to $0 \le (x, y) \le 1$ and (n,m) and then calculate the resultant type I risks. In our analysis, we see that we sample more (since the probabilities x and y are smaller than in the case treated in Table 8.3) and that the type II risks are significantly reduced, albeit the resulting type I errors are significantly increased. For example, if the producer uses a sample size of 10 and the supplier a sample size of 15, then the shared type

II risk is equal to .1134 compared to .424 as indicated in Table 8.3. The type I risks are equal to .240 and .401 however, compared to .15 and .20 which we used as type I constraints in Table 8.3. In this sense, the intricate relationship between a producer and a his supplier as well as the risk specifications for type I and II risks for each combined with probabilities of doing nothing lead to complex relationships that can provide a broad number of potential control combinations. Finding sample sizes (n,m) and randomization parameters (x,y) for the producer and the supplier that meet risk constraints on both type I and II risks may thus require extensive analysis and in some cases extensive sampling by both parties. The sampling can, however, be significantly reduced if in fact, both the producer and the supplier turn to collaboration. A more extensive analysis would, in this case, assess the risks economic implications and proceed to their economic cost minimization.

Concluding, we note that Neyman-Pearson theory in statistics can be adapted to deal with sample control-inspection problems in supply chains. For simplicity, we have considered some applications and examples, although the approach used is quite general. Applications including economic sampling, conflicts, negotiations and contracts design in supply chains could be considered as well. We have focused attention on an extended application of Neyman-Pearson theory to risk control in a multi-agent environment when agents may collaborate or not (as it is the case in supply chains). We have also used a number of examples to highlight the approach and its applicability. Further research is needed both from a game theoretic perspective, emphasizing repeated and random payoffs game as well as from an economic valuation perspectives (emphasizing the economic valuation of collaboration, truthful sharing of information and sampling-control mechanism instituted to assure that agents act ex post as they have contracted to act ex-ante). Finally, the discussion was essentially based on the presumption that there is a strategic value to sampling-control which ought to be considered in designing cooperative partnerships. This is the case since, information and power asymmetries can lead to opportunistic behaviour while statistical controls can mitigate the adverse effects of such asymmetries.

This is coherent with the modern practice of quality management which has gone beyond the mere application of statistical tools but at the same time has maintained these tools as an essential facet of the management of quality and its control.

Table 8.2. Type I and type II Risks when Type I risks are not binding

m n	5	10	15	20
5: Type II	.267	.1308	.060	.027
Type I (P,S)	.175, .226	.135, .226	.10, .22	.081, .226
Type II (P,S)	.1644,.370	.080, .18	.037, .08	.0169, .038
10: Type II	.224	.1134	.05	.023
Type I (P,S)	.31, .401	.240, .401	.186, .401	.0143, .401
Type II (P,S)	.08, .408	.037, .189	.017. .085	.007, .038
15: Type II	.232	.105	.047	.021
Type I (P,S)	.415. .536	.32, .536	.24, .536	.192, .536
Type II (P,S)	.03, .427	.01, .19	.007..08	.003, .038
20: Type II	.226	.101	.0452	.020
Type I (P,S)	.496, .641	.384, .641	.29, .641	.223, .641
Type II (P,S)	.0169, .436	.0075, .195	.003, .08	.001, .038

Table 8.3. Type I and type II Risks when Type I risks are binding

m n	5	10	15	20
5:	Not feasible	Not feasible	Not feasible	Not feasible
10: Type II	.428	.390	.371	.362
(x,y)	.4555, .6261	.4136, .6261	.3764, .6261	.344 .6261
Type II (P,S)	.308, .547	.27, .51	.247 .495	.232, .49
15: Type II	.449	.424	.411	.405
(x,y)	.6022, .720	.580, .720	.5615, .720	.545, .720
Type II (P,S)	.349, .549	.323, .524	.308, .514	.299, .512
20: Type II	.459	.439	.429	.425
(x,y)	.67007, .766	.65559, .766	.643, .766	.63, .766
Type II (P,S)	.372, .52	.352, .526	.34, .51	.334, .516

REFERENCES

Burke R, Davis DR, F.C. Kaminsky (1993) The (k_1,k_2) game. Quality Progress 26: 49-53.

Corbett C, Tang C (1999) Designing supply contracts: contract type and information asymmetry in Quantitative Models for Supply Chain Management. Tayur S, Ganeshan R, Magazine M, eds., Kluwer Academic Publishers, Norwell, MA.

Eppen GD, Hurst EG Jr. (1974) Optimal location of inspection stations in a multistage production process. *Management Science* 20: 1194-2000.

Hurst EG Jr. (1974) Imperfect inspection in a multistage production process. *Management Science* 20: 378-384.

Lee HL, Tagaras G (1992) Economic acceptance sampling plans in complex multi-stage production systems. *International Journal of Production Research* 30: 2615-2632.

Lim W (2001) Producer-supplier contracts with incomplete information. *Management Science* 47(5): 709-715.

Moulin H (1995) Cooperative Microeconomics: A Game-Theoretic Introduction. Princeton University Press. Priceton, New Jersey.

Mukhopadhyay SK, Kouvelis P (1997) A differential game theoretic model for duopolistic competition on design quality. *Operations Research* 45: 886-893.

Nash F (1950) Equilibrium points in N-person games, *Proceedings of the National Academy of Sciences* 36:48-9.

Raz T (1986) A survey of models for allocating inspection effort in multistage production systems. *Journal of Quality Technology* 18(4): 239-247.

Reyniers DJ (1992) Supplier-Customer interaction in Quality Control. *Annals of Operations Research* 34: 307-330.

Reyniers DJ, Tapiero CS (1995a) The delivery and control of quality in supplier-producer contracts. *Management Science* 41: 1581-1589.

Reyniers DJ, Tapiero CS (1995b) Contract design and the control of quality in a conflictual environment. *Euro J. of Operational Research* 82: 373-382.

Shubik M (2002) Game theory and operations research: some musings 50 years later. *Operations Research* 50: 192-196.

Stackleberg von HV (1934) *Marktform and Gleichgweicht*, Vienna, Springer Verlag.

Stackleberg HV (1952) *The Theory of the Market Economy*, Translated by Peacock AT, London, William Hodge and Co.

Starbird SA (1994) The effect of acceptance sampling and risk aversion on the quality delivered by suppliers. *Journal of the Operational Research Society* 45: 309-320.

Stuart HW Jr. (2001) Cooperative games and business strategy. In *Game Theory and Business Applications,* Chatterjee K, Samuelson WF, editors. Kluwer Academic Publishers.

Tagaras G, Lee HL (1996) Economic models for vendor evaluation with quality cost analysis. *Management Science* 42: 1531-1543.

Tapiero CS (1995) Acceptance sampling in a producer-supplier conflicting environment: Risk neutral case. *Applied Stochastic Models and Data Analysis* 11: 3-12.

Tapiero CS (1996) *The Management of Quality and Its Control*, Chapman and Hall, London.

Tapiero CS (2001) Yield and and Control in a Supplier-Customer Relationship. *International Journal of Production Research* 39: 1505-1515.

Tapiero CS (2004) *Risk and Financial Management: Mathematical and Computational Concepts,* Wiley, London.

Tapiero CS (2005a) Environmental Quality Control and Environmental Games, *working paper.*

Tapiero CS (2005b) Modeling Environmental Queue Control: A Game Model, *Stochastic Environmental Research and Risk Assessment.*

Tsay A, Nahmias S, Agrawal N (1998) Modeling supply chain contracts: A review, in Tayur S, Magazine M, Ganeshan R. (eds) *Quantitative Models of Supply Chain Management*, Kluwer International Series.

Von Neumann J, Morgenstern O (1944) *Theory of Games and Economic Behavior.* Princeton University Press.

Wetherhill GB (1977) *Sampling Inspection and Quality Control,* Chapman and Hall, New York.

APPENDIX: OPTIMALITY CONDITIONS IN SINGLE- AND TWO-PLAYER DYNAMIC GAMES

A1.1 DYNAMIC PROBLEMS

System dynamics

Consider a dynamic system characterized by an object whose coordinates or states can be changed in time by exercising an action or control over a planning horizon $T\text{-}t_0$. Let such a change of state, n, $n=1,..,N$ be described by a set of differential equations

$$\frac{dx_n(t)}{dt} = f_n(x_1(t),x_2(t),..,x_N(t),u_1(t),u_2(t),..,u_M(t),t) \text{ , } n=1,..,N, t_0 \leq t \leq T \text{ , (A1.1)}$$

where t – time; t_0 – initial time point; $x_n(t)$ is a state variable; $u_m(t)$ is a control (or decision) variable; and functions $f_n(.)$ describe internal properties of the object and account for external effects. To simplify the presentation, we may use a vector form and omit t wherever the time-dependence is obvious,

$$\frac{dx}{dt} = f(x,u,t). \tag{A1.2}$$

Equations (A1.1) and (A1.2) assume continuity of the state variables $x^T=(x_1,..,x_N)$ and piecewise-continuous control functions $u^T=(u_1,..,u_m)$, where superscript T of a vector stands for its transpose. This, however, is not always the case in real-life. If state equations involve infinite jumps, the derivatives in (A1.1) and (A1.2) are replaced with differentials. Furthermore, dynamic processes have some limitations in real-life which can be formalized by a number of constraints.

Boundary state constraints

There may be an initial boundary constraint of state n
$$x_n(t_0)=x_0. \tag{A1.3}$$
If planning horizon T is finite, then a terminal boundary constraint can be imposed
$$x_n(T)=x_T, \tag{A1.4}$$

Or, if the dynamic process has a periodic character of length T-t_0 so that the terminal states of a period are identical to the initial states of the period, then

$$x(t_0) = x(T). \tag{A1.5}$$

Control constraints

Control is rarely arbitrary. Let $U(t)$ be a given set of possible controls from R^M, $t_0 \leq t \leq T$. Then the control constraints are described as

$$u(t) \in U(t), \; t_0 \leq t \leq T. \tag{A1.6}$$

State constraints

Let $G(t)$ be a given set of possible states from R^N. Then the state constraints are described as

$$x(t) \in G(t), \; t_0 \leq t \leq T. \tag{A1.7}$$

In addition, one can encounter constraints which combine different types of constraint. For example, state and control constraints can be mixed. The control is normally exercised to achieve a certain goal which we refer to as the objective function, J. The problem of choosing the best control for a dynamic system is further referred to as an optimal control problem.

The objective function

Let T be fixed, $L(x,u,t)$ and $R(x(0),x(T))$ be given cost functions. Note that if either the initial or terminal state is fixed, that is, if either constraint (A1.3) or (A1.4) is imposed, then we have $R(x_0,x(T))$ and $R(x(0),x_T)$ respectively. Consequently, the objective is to minimize an integral measure of the system's behavior along the planning horizon as well as the cost associated with its initial and/or terminal state.

$$J = \int_{t_0}^{T} L(x(t),u(t),t)dt + R(x(0),x(T)). \tag{A1.8}$$

If the planning horizon is not fixed, then the objective function is

$$J = \int_{t_0}^{T} L(x(t),u(t),t)dt + R(x(0),x(T)) + S(T\text{-}t_0), \tag{A1.9}$$

where $S(T$-$t_0)$ is the cost associated with the length of the planning horizon. If a state equation involves a stochastic process, then expectation, E, is typically added to the objective function if the goal is to minimize the expected cost.

Thus the optimal control problem is to find an admissible control so that

$$J(x_0, x_T, u, x, t_0, T) \to \inf . \qquad (A1.10)$$

An admissible control in terms of the imposed constraints which provide (A1.10) is referred to as optimal control, u^*. To study optimal control problems, we will now assume that all functions defining the system dynamics, constraints and objective function are continuous and piecewise continuously differentiable in x and u. All these functions can have a finite number of jumps in t. Moreover, we assume that function $L(x,u,t)$ is convex in control u for minimization problems. In addition, we further distinguish between two types of solutions. One is an open-loop optimal control $u^*=u^*(t)$ which is determined as a function of time and, in terms of the object state, depends only on the boundary state value x_0 (x_T). The other, an optimal solution found as a function of state history, $u^*=u^*(t, x(\tau)|0 \le \tau \le t)$, is referred to as a closed-loop solution. In a special, memoryless case of $u^*=u^*(x(t), t)$, the solution is referred to as a feedback control. If the optimal control problem is deterministic, it is often possible to find one form of solution, for example, an open-loop solution, and then transform it into the other, i.e., the closed-loop solution. This, however, is rarely possible for stochastic problems. Moreover, in stochastic problems an update (feedback) on the object state $x(t)$ may not be available at each point of time.

Example A1.1

Consider a manufacturer who continuously produces a single (aggregate) product type in response to a demand, $d(t)$. If cumulative production exceeds the cumulative demand, then excessive inventories are stored in a warehouse, otherwise the shortages are backlogged. The goal is to minimize both inventory holding and backlog costs. The described problem is a well-known optimal production control problem (see, for example, Kogan and Khmelnitsky, 2000). The state equation for this problem takes the following form

$$\dot{x}(t) = u(t) - d(t),$$

where $x(t)$ is the inventory level at time t (surplus, if $x(t)>0$ and backlog if $x(t)<0$); $u(t)$ is the production rate (the number of products per time unit) at t; and $d(t)$ is the demand rate at time t.

Note that the demand can be an exogenous function of time, or endogenous, depending, for example, on the inventory on hand, $d=d(x(t),t)$. Then the state differential equation is

$$\dot{x}(t) = u(t) - d(x(t),t).$$

Another possibility is that the demand, $d(x(t),t)$ is a stochastic process with jumps or instantaneous disturbances. The classical example of the latter type of demand is described by the Wiener process, $w(t)$. In such a case the state differential equation is a stochastic Ito differential equation (see, for example, Tapiero, 1988)

$$dx(t) = (u(t) - a(t))dt + \sigma(x,t)dw,$$

where $a(t)$ is a deterministic component of the demand; dw – Weiner increment, a stochastic component of the demand; and $\sigma(x,t)$ is the variability of the demand.

Assuming that the production rate cannot be negative and that the production capacity U, the maximum number of products which can be produced per time unit, is bounded, the control constraint is

$$0 \leq u(t) \leq U.$$

The initial boundary condition (A1.3) implies that in such a case the initial inventory level which we have in the warehouse, if x_0, is positive. Otherwise it is the backlog we need to take into account when planning production.

A typical example of a state constraint in production control problems is the maximum backlog, x^B, the manufacturer can afford without losing sales, i.e.,

$$x(t) \geq x^B.$$

If the demand process d is stochastic, then the state constraint is typically imposed on the probability P of the backlog,

$$P[x(t)<0]=1-\alpha,$$

or alternatively

$$P[x(t) \geq 0] = \alpha,$$

where α is the service level.

Let $t_0=0$ and inventory associated costs be leaner with product unit holding cost per time unit, h^+ equal to the unit backlog cost h^- per time unit, $h^+= h^-=h$. Then the goal of the optimal production control is to find $u(t)$, $0 \leq t \leq T$, which minimizes the following objective function subject to the described constraints:

$$J = \int_0^T h|x(t)|dt.$$

If the production process is stochastic, then expectation, E, is added

$$J = E \int_0^T h|x(t)|dt.$$

A1.2 DYNAMIC PROGRAMMING

Consider the following deterministic optimal control problem

$$J(t_0, x_0, u) = \int_{t_0}^{T} L(x(t), u(t), t)dt + R(x(T)) \to \inf , \qquad (A1.11)$$

s.t.

$$\frac{dx}{dt} = f(x, u, t), \ t_0 \leq t \leq T , \ x(t_0) = x_0, \qquad (A1.12)$$

$$u(t) \in U(t), \ x(t) \in G(t), \ t_0 \leq t \leq T . \qquad (A1.13)$$

Discrete-time dynamic programming

To obtain an approximate solution of the problems, select mesh points

$$t_0 < t_1 < \ldots t_{K-1} < t_K = T.$$

Then, using Euler's scheme, the differential equations can be replaced with difference equations and the integral with summations:

$$J_0(x^0, \mathbf{u}^0) = \sum_{i=0}^{K-1} L(x^i, u^i, t_i)(t_{i+1} - t_i) + R(x^K) \to \inf , \qquad (A1.14)$$

s.t.

$$x^{i+1} = x^i + f(x^i, u^i, t_i)(t_{i+1} - t_i), \ i=0,1,..,K-1, \ x^0 = x_0, \qquad (A1.15)$$

$$\mathbf{u}^0 = (u^0, u^1, ..., u^{K-1}), \ u^i \in U(t_i), \ x^i \in G(t_i), \ i=0,1,..,K-1. \qquad (A1.16)$$

Denoting

$$L^i = L(x^i, u^i, t_i)(t_{i+1} - t_i), \ f^i(x^i, u^i) = x^i + f(x^i, u^i, t_i)(t_{i+1} - t_i),$$

$$G^i = G(t_i) \text{ and } U^i = U(t_i),$$

equations (A1.14)-(A1.16) simplify to

$$J_0(x^0, \mathbf{u}^0) = \sum_{i=0}^{K-1} L^i(x^i, u^i) + R(x^K) \to \inf , \qquad (A1.17)$$

s.t.

$$x^{i+1} = f^i(x^i, u^i), \ i=0,1,..,K-1, \ x^0 = x_0, \qquad (A1.18)$$

$$\mathbf{u}^0 = (u^0, u^1, ..., u^{K-1}), \ u^i \in U^i, \ x^i \in G^i, \ i=0,1,..,K-1. \qquad (A1.19)$$

The objective function (A1.17) determines the overall cost incurred along the planning horizon when only the initial inventory level is known, $x^0 = x_0$. Similarly, if x^1 is known, then the new problem would be

$$J_1(x^1, \mathbf{u}^1) = \sum_{i=1}^{K-1} L^i(x^i, u^i) + R(x^K) \to \inf ,$$

s.t.

$$x^{i+1} = f^i(x^i, u^i), \ i=1,..,K-1, \ x^1 \text{ - fixed,}$$

$$\mathbf{u}^1 = (u^1,..,u^{K-1}),\ u^i \in U^i,\ x^i \in G^i,\ i=1,..,K-1.$$

To present the dynamic programming approach, let us introduce an auxiliary optimization problem

$$J_k(x^k,\mathbf{u}^k) = \sum_{i=k}^{K-1} L^i(x^i,u^i) + R(x^K) \to \inf, \qquad (A1.20)$$

s.t.

$$x^{i+1} = f^i(x^i,u^i),\ i=k,..,K-1, \qquad (A1.21)$$

$$\mathbf{u}^k = (u^k,..,u^{K-1}),\ u^i \in U^i, x^i \in G^i,\ i=k,..,K-1, \qquad (A1.22)$$

where x^k and integer k are fixed. When $k=0$, we evidently obtain initial formulation (A1.17)-(A1.19). Denote the set of all possible (in terms of condition (A1.22)) controls $\mathbf{u}^k$ by $V^k(x^k)$, so that the corresponding trajectory $\mathbf{x}^k = (x^k, x^{k+1},..,x^{K-1})$ satisfies state equations (A1.21) and state constraints $x^i \in G^i$, $i=k,..,K-1$. Correspondingly, denote the set of all admissible controls only at stage k, for which $\mathbf{u}^k \in V^k(x^k)$, by $U^k(x^k)$. It is easy to observe that

$$J_0(x^0,\mathbf{u}^0) = L^0(x^0,u^0) + J_1(x^1,\mathbf{u}^1),$$

and more generally,

$$J_k(X,\mathbf{u}^k) = L^k(x^k,u^k) + J_{k+1}(x^{k+1},\mathbf{u}^{k+1}).$$

Given $\inf_{V^{k+1}(x^{k+1})} J_{k+1}(x^{k+1},\mathbf{u}^{k+1})$ as a function of x^{k+1}, then backward optimization of $J_k(X,\mathbf{u}^k)$ leads to the following optimality conditions.

Let us introduce a new value function,

$$B_k(x^k) = \inf_{V^k(x^k)} J_k(x^k,\mathbf{u}^k),\ k=0,1,..,K-1, \qquad (A1.23)$$

which is referred to as the Bellman (cost-to-go) function. Then the principle of optimality is determined by the following recursive dynamic programming equations

$$B_k(x^k) = \inf_{U^k(x^k)} \left\{ L^k(x^k,u^k) + B_{k+1}(x^{k+1}) \right\},\ k=0,1,..,K-1, \qquad (A1.24)$$

$$B_K(x^K) = R(x^K).$$

With respect to the state equation $x^{i+1} = f^i(x^i,u^i)$, condition (A1.24) can be presented in a more convenient form for optimization

$$B_k(x^k) = \inf_{U^k(x^k)} \left\{ L^k(x^k,u^k) + B_{k+1}(f(x^k,u^k)) \right\},\ k=0,1,..,K-1.$$

This principle implies that if an optimal solution, $\mathbf{u}^{*0} = (u^{*0},u^{*1},..,u^{*K-1})$, exists (and thus "inf" can be replaced with "min"), then it can be found backward in time by first solving in u^{K-1},

$$B_{K-1}(x^{K-1}) = \min_{U^{k-1}(x^{k-1})} \left\{ L^{K-1}(x^{K-1},u^{K-1}) + R(x^K) \right\},$$

which, with respect to $x^K = f^{K-1}(x^{K-1}, u^{K-1})$ (see (A1.21)) is

$$B_{K-1}(x^{K-1}) = \min_{U^{K-1}(x^{k-1})} \left\{ L^{K-1}(x^{K-1}, u^{K-1}) + R(f^{K-1}(x^{K-1}, u^{K-1})) \right\}.$$

This allows us to find $u*^{K-1} = u*^{K-1}(x^{K-1})$. Next using $x^{K-1} = f^{K-2}(x^{K-2}, u^{K-2})$, we solve

$$B_{K-2}(x^{K-2}) = \min_{U^{k-2}(x^{K-2})} \left\{ L^{K-2}(x^{K-2}, u^{K-2}) + B_{K-1}(f^{K-2}(x^{K-2}, u^{K-2})) \right\}$$

to find $u*^{K-2} = u*^{K-2}(x^{K-2})$. We continue this way until $B_0(x_0)$ and accordingly the optimal feedback control is found for all stages, $u*^k = u*^k(x^k)$, $k=0$, $1,..,K-1$, $x^0 = x_0$.

Continuous-time dynamic programming

Consider problem (A1.11)-(A1.13) and let the auxiliary problem be

$$J(t,x,u) = \int_t^T L(x(\tau), u(\tau), \tau) d\tau + R(x(T)) \to \inf,$$

s.t.

$$\frac{dx}{d\tau} = f(x,u,\tau), \ t \le \tau \le T, \ x(\tau) = x,$$

$$u(\tau) \in U(\tau), \ x(\tau) \in G(\tau), \ t \le \tau \le T.$$

Denote the Bellman function as

$$B(x_1(t),..,x_N(t),t) = \inf_{V(x,t)} J(t, x_1(t),..,x_N(t), u_1(t),..,u_M(t)), \quad (A1.25)$$

where V(x,t) is the set of all controls $u(\tau) \in U(\tau)$, $t \le \tau \le T$ so that the corresponding trajectory $x(\tau) = x(\tau, u)$, $t \le \tau \le T$ satisfies state equations $\frac{dx}{d\tau} = f(x,u,\tau)$ *and state constraints $x(\tau) \in G(\tau)$, $t \le \tau \le T$. Correspondingly, we denote the set of all admissible controls at time t by U(x,t).*

Construct the Hamiltonian as

$$H\left(x, \frac{\partial B(x,t)}{\partial x}, u, t\right) = L(x,u,t) + \left(\frac{\partial B(x,t)}{\partial x}\right)^T f_n(x,u,t), \quad (A1.26)$$

where the Lagrange multipliers $\left(\dfrac{\partial B}{\partial x}\right)^T = (\dfrac{\partial B}{\partial x_1},..,\dfrac{\partial B}{\partial x_N})$ *are frequently referred to as co-state or adjoint variables.* $\dfrac{\partial B}{\partial x_n}$ *presents the shadow price of state n, i.e., the gain in the objective function value which can be obtained by reducing state n by one more unit at time t.*

Then the optimal control, $u^* = u^*(x, \frac{\partial B(x,t)}{\partial x}, t) \in V(x,t)$ minimizes the

Hamiltonian

$$\min_{U(x,t)} \left\{ H(x, \frac{\partial B(x,t)}{\partial x_1}, u, t) \right\}, \tag{A1.27}$$

and satisfies the Hamiltonian-Jacobi-Bellman equation

$$H(x, \frac{\partial B(x,t)}{\partial x_1}, u^*, t) = -\frac{\partial B(x,t)}{\partial t}, \tag{A1.28}$$

$$B(x, T) = R(x).$$

Consequently, the solution approach is to first find

$$u^* = u^*(x, \frac{\partial B(x,t)}{\partial x}, t) \in V(x,t) \tag{A1.29}$$

from (A1.27). Then substitute it into the Hamiltonian-Jacobi-Bellman equation (A1.28), which, if solved analytically, provides $B(x,t)$. Next, differentiating $B(x,t)$ with respect to x and substituting it into (A1.29) we obtain $u^* = u^*(x, t)$.

Example A1.2

Consider a transportation problem. Let x_0 be the amount of products to be transported within a limited time T. The transportation rate is not bounded, but increasingly expensive as it incurs quadratic cost $u^2(t)$ at each time point t. If by time T, any products $x(T)$ are still undelivered, an increasingly high penalty $ax^2(T)$ is levied. Thus, we encounter the following minimization problem

$$J = \int_0^T u^2(t)dt + ax^2(T) \to \min,$$

$$\dot{x}(t) = u(t), x(0) = x_0,$$

where x_0 and $a > 0$ are given constants and $u(t) \in R^1$. With respect to (A1.26) the Hamiltonian is

$$H(x, \frac{\partial B(x,t)}{\partial x}, u, t) = L(x, u, t) + \frac{\partial B(x,t)}{\partial x_n} f(u, x, t)$$

and thus,

$$H(x, \frac{\partial B(x,t)}{\partial x}, u, t) = u^2 + \frac{\partial B(x,t)}{\partial x} u.$$

Then the optimal control, $u^* = u^*(x, \frac{\partial B(x,t)}{\partial x}, t)$ is achieved by minimizing the Hamiltonian. That is, differentiating the Hamiltonian with respect to u we find

$$u = -\frac{1}{2} \frac{\partial B(x,t)}{\partial x}.$$

Note that the Hamiltonian is concave and consequently the first order optimality condition is not only necessary but also sufficient. Substituting this into the Hamiltonian-Jacobi-Bellman equation (A1.28)

$$H\left(x, \frac{\partial B(x,t)}{\partial x}, u, t\right) = -\frac{\partial B(x,t)}{\partial t},$$

we obtain

$$-\frac{1}{4}\left(\frac{\partial B(x,t)}{\partial x}\right)^2 + \frac{\partial B(x,t)}{\partial t} = 0 \text{ and } B(x,T) = ax^2.$$

We search for solution $B(x,t)$ of this partial differential equation in the form of a polynomial

$$B(x,t) = A_0(t) + A_1(t)x + A_2(t)x^2.$$

Substituting this expression into the derived partial differential equation along with the boundary condition, we have

$$\dot{A}_0 + \dot{A}_1 x + \dot{A}_2 x^2 - \frac{(A_1 + 2A_2 x)^2}{4} = 0, \qquad 0 \le t \le T$$

$$A_0(T) + A_1(T)x + A_2(T)x^2 = ax^2,$$

which, by comparing terms of the same power, leads to the following system of ordinary differential equations:

$$\dot{A}_0 + \frac{A_1^2}{4} = 0, \ \dot{A}_1 - A_1 A_2 = 0, \ \dot{A}_2 - A_2^2 = 0, \ 0 \le t \le T,$$

$$A_0(T) = 0, \ A_1(T) = 0, \ A_2(T) = a.$$

Thus, we find

$$A_0(t) \equiv A_1(t) \equiv 0, \ A_2(t) = \frac{a}{1 - a(t - T)}$$

and therefore the Bellman function and the co-state variable are

$$B(x,t) = \frac{ax^2}{1 - a(t-T)} \text{ and } \frac{\partial B(x,t)}{\partial x} = \frac{2ax}{1 - a(t-T)}.$$

Finally, the optimal feedback policy is

$$u^*(x,t) = -\frac{1}{2} \frac{\partial B(x,t)}{\partial x} = \frac{-ax}{1 - a(t-T)}.$$

The relationship between discrete and continuous-time dynamic programming

To see the relationship between discrete and continuous-time dynamic programming formulations, we apply the following heuristic considerations. Let $t_{i+1} - t_i = \Delta t$, $i=0,1,..,$ $K-1$, $t_i=t$, $x^i=x(t)$, $V^*(x^k)=V(x,t)$, $U^*(x^k)=U(x,t)$ and $B_k(x^k)= B(x,t)$. Then recalling that

$$L^i = L(x^i,u^i,t_i)(t^{i+1} - t^i),$$

and letting $\Delta t \rightarrow 0+$, we obtain

$$B(x,t) = \inf_{U(x,t)} \{L(x,u)dt + B(x + dx,t + dt))\}, \qquad (A1.30)$$

$$B(x, T)=R(x), \qquad (A1.31)$$

where

$$B(x,t) = \inf_{V(x,t)} J(t,x,u).$$

Assuming that the Bellman function is differentiable in time and state, and employing Taylor serious approximation of $B(x + dx,t + dt)$ with the terms only of the first order of Δt,

$$B(x + dx,t + dt) = B(x,t) + \left(\frac{\partial B(x,t)}{\partial x}\right)^T dx + \frac{\partial B(x,t)}{\partial t} dt, \quad (A1.32)$$

we obtain from (A1.30):

$$\inf_{U(x,t)} \left\{ L(x,u,t) + \left(\frac{\partial B(x,t)}{\partial x}\right)^T \frac{dx}{dt} + \frac{\partial B(x,t)}{\partial t} \right\} = 0. \qquad (A1.33)$$

Recalling that $\dfrac{dx}{dt} = f(x,u,t)$, we have from (A1.33), the continuous-time optimality conditions

$$\inf_{U(x,t)} \left\{ L(x,u,t) + \left(\frac{\partial B(x,t)}{\partial x}\right)^T f(x,u,t) + \frac{\partial B(x,t)}{\partial t} \right\} = 0. \text{ (A1.34)}$$

$$B(x, T)=R(x).$$

Denote the Hamiltonian, H as,

$$H(x,\frac{\partial B(x,u)}{\partial x},u,t) = L(x,u,t) + \left(\frac{\partial B(x,t)}{\partial x}\right)^T f(x,u,t). \text{ (A1.35)}$$

Then, if an optimal control policy exists, condition (A1.34) results in

$$\min_{U(x,t)} \left\{ H(x,\frac{\partial B(x,t)}{\partial x},u,t) \right\} = -\frac{\partial B(x,t)}{\partial t}. \qquad (A1.36)$$

Let the optimal control, which minimizes (A1.36) be $u^*=u^*(x,\dfrac{\partial B(x,t)}{\partial x}$,
$t)$. Then, substituting it into (A1.35), we find the Hamiltonian-Jacobi-Bellman partial differential equation,

$$L(x,u^*,t)+\left(\frac{\partial B(x,t)}{\partial x}\right)^T f(x,u^*,t) = -\frac{\partial B(x,t)}{\partial t}, \qquad (A1.37)$$

or equivalently

$$H(x,\frac{\partial B(x,t)}{\partial x},u^*) = -\frac{\partial B(x,t)}{\partial t}, \; B(x,\,T)=R(x). \qquad (A1.38)$$

Note that this result is identical to that stated in the previous section. It, however, is not always correct. If the state variables have instantaneous changes, as for example, with the Weiner process, then we would need a more precise analysis. Specifically, if $B(x,t)$ is twice differentiable with respect to x, we will have more terms in the first order of Δt in the Taylor expansion (A1.32)

$$B(x+dx,t) = B(x,t)+\left(\frac{\partial B(x,t)}{\partial x}\right)^T dx+\frac{1}{2}(dx)^T \frac{\partial^2 B(x,t)}{\partial x^2}dx+\frac{\partial B(x,t)}{\partial t}dt .(A1.39)$$

Then substituting (A1.39) into (A1.30) we have

$$\inf_{U(x,t)}\left\{L(x,u,t)dt +\left(\frac{\partial B(x,t)}{\partial x}\right)^T dx+\frac{1}{2}(dx)^T \frac{\partial^2 B(x,t)}{\partial x^2}dx+\frac{\partial B(x,t)}{\partial t}dt\right\}=0,(A1.40)$$

where

$$\frac{\partial^2 B}{\partial x^2} = \begin{vmatrix} \dfrac{\partial^2 B}{\partial x_1^2} & \cdots & \dfrac{\partial^2 B}{\partial x_1 \partial x_N} \\ \dfrac{\partial^2 B}{\partial x_N \partial x_1} & \cdots & \dfrac{\partial^2 B}{\partial x_n^2} \end{vmatrix}$$

and the partial differential equation for a dynamic system with states which may have instantaneous changes,

$$L(x,u^*,t)dt +\left(\frac{\partial B(x,t)}{\partial x}\right)^T dx+\frac{1}{2}(dx)^T \frac{\partial^2 B(x,t)}{\partial x^2}dx+\frac{\partial B(x,t)}{\partial t}dt = 0 .(A1.41)$$

A1.3 STOCHASTIC DYNAMIC PROGRAMMING

Consider first the discrete-time problem (A1.17)-(A1.19) which involves stochastic variable w^i

$$J_0(x^0, \mathbf{u}^0) = E\{\sum_{i=0}^{K-1} L^i(x^i, u^i) + R(x^K)\} \to \inf,$$

s.t.

$$x^{i+1} = f^i(x^i, u^i, w^i), \; i=0,1,...,K-1, \; x^0=x_0,$$

$$\mathbf{u}^0=(u^0, u^1,...,u^{K-1}), \; u^i \in U^i, \; x^i \in G^i, \; i=0,1,...,K-1.$$

Using the same definition of the Bellman (optimal cost-to-go) function at time t_k with starting state value x_k,

$$B_k(x^k) = \inf_{V^k(x^k)} J_k(x^k, \mathbf{u}^k), \; k=0,1,...,K-1,$$

the principle of optimality (A1.24) straightforwardly transforms into the following recursive dynamic programming equations

$$B_k(x^k) = \inf_{U^k(x^k)} \{E[L^k(x^k, u^k) + B_{k+1}(x^{k+1})]\}, \; k=0,1,...,K-1, \quad (A1.42)$$

$$B_K(x^K)=R(x^K).$$

Similar to the deterministic dynamic programming, these equations are solved backward, starting from $k=K-1$ with $B_K(x^K)=R(x^K)= R(f^{K-1}(x^{K-1}, u^{K-1}, w^{K-1}))$,

$$B_{K-1}(x^{K-1}) = \min_{U^{K-1}(x^{K-1})} \{E[L^{K-1}(x^{K-1}, u^{K-1}) + R(f^{K-1}(x^{K-1}, u^{K-1}, w^{K-1}))]\}$$

and continuing until all controls are found.

Consequently, the continuous-time formulation is obtained for problem (A1.11)-(A1.13) with stochastic disturbances $w(t)$:

$$J(t_0, x_0, u) = E[\int_{t_0}^{T} L(x(t), u(t), t)dt + R(x(T))] \to \inf,$$

s.t.

$$dx = df(x, u, w, \tau), \; t_0 \le t \le T, \; x(t_0)=x_0,$$

$$u(t) \in U(t), \; x(t) \in G(t), \; t_0 \le t \le T$$

and auxiliary problem

$$J(t, x, u) = E[\int_{t}^{T} L(x(\tau), u(\tau), \tau)d\tau + R(x(T))] \to \inf,$$

s.t.

$$dx = df(x, u, w, \tau), \; t \le \tau \le T, \; x(\tau)=x,$$

$$u(\tau) \in U(\tau), \; x(\tau) \in G(\tau), \; t \le \tau \le T.$$

Specifically, using the same definition

$$B(x, t) = \inf_{V(x,t)} J(x, u, t) \qquad (A1.43)$$

the dynamic programming optimality conditions (A1.30)-(A1.31) are

$$B(x,t) = \inf_{U(x,t)} E\{L(x,u)dt + B(x+dx,t+dt))\}, \qquad (A1.44)$$

$$B(x, T)=R(x), \qquad (A1.45)$$

where, similar to the previous analysis, the Taylor serious approximation can be used for $B(x+dx,t+dt))$. Taking into account (A1.40) found for a process with instantaneous changes in state, we have

$$\inf_{U(x,t)} E\left\{L(x,u,t)dt + \left(\frac{\partial B(x,t)}{\partial x}\right)^T dx + \frac{1}{2}(dx)^T \frac{\partial^2 B(x,t)}{\partial x^2} dx + \frac{\partial B(x,t)}{\partial t} dt\right\} = 0. (A1.46)$$

Example A1.3

Consider the following stochastic optimization problem

$$J(t_0, x_0, u) = E\int_0^T L(x(t), u(t), t)dt \rightarrow \inf,$$

s.t.

$$dx(t) = f(u(t))dt + \sigma(u)dw, \ 0 \le t \le T, \ x(t_0)=x_0,$$

$$u(t) \in U,$$

where dw – the Weiner increment with $E[dw]=0$ and $VAR[dw]=dt$.

For our example with the only state variable, condition (A1.46) takes the following form

$$\inf_{u \in U} E\{L(x,u,t)dt + \frac{\partial B(x,t)}{\partial x}dx + \frac{1}{2}\frac{\partial^2 B(x,t)}{\partial x^2}(dx)^2 + \frac{\partial B(x,t)}{\partial t}dt\} = 0.$$

There are only two terms in the last expressions which involve uncertainty: dx and $(dx)^2$. With respect to the state equation $dx(t) = f(u(t))dt + \sigma(u)dw$,

$$E[dx] = E[f(u)dt + \sigma(u)dw] = f(u)dt.$$

Assuming that $(dt)^2$ is very small and using $E[(dw)^2]= VAR[dw]=dt$, we have

$$E[(dx)^2] = E[(f(u)dt + \sigma(u)dw)^2] = \sigma^2(u)dt$$

Thus, (A1.46) simplifies to

$$\inf_{u \in U}\{L(x,u,t)dt + \frac{\partial B(x,t)}{\partial x}f(u)dt + \frac{1}{2}\sigma^2\frac{\partial^2 B(x,t)}{\partial x^2}dt + \frac{\partial B(x,t)}{\partial t}dt\} = 0.$$

That is, the optimal control $u^*=u^*(x)$ is determined by

$$\inf_{u \in U}\{L(x,u,t) + \frac{\partial B(x,t)}{\partial x}f(u) + \frac{1}{2}\sigma^2\frac{\partial^2 B(x,t)}{\partial x^2}\} \qquad (A1.47)$$

and thus the Hamiltonian-Jacobi-Bellman equation is

$$L(x,u^*,t) + \frac{\partial B}{\partial x}f(u^*) + \frac{1}{2}\sigma^2(u^*)\frac{\partial^2 B}{\partial x^2} + \frac{\partial B}{\partial t} = 0, \ B(x, T)=0. \ (A1.48)$$

If σ does not depend on u, then denoting

$$\inf_{u \in U} \{ L(x,u,t) + \frac{\partial B}{\partial x} f(u) \} = H(x, \frac{\partial B(x,t)}{\partial x}, u^*)$$

we have the following partial differential equation

$$H(x, \frac{\partial B(x,t)}{\partial x}, u^*) + \frac{1}{2}\sigma^2 \frac{\partial^2 B(x,t)}{\partial x^2} + \frac{\partial B(x,t)}{\partial t} = 0.$$

A1.4   THE MAXIMUM PRINCIPLE

As shown in the previous sections, dynamic programming in continuous-time problems involves the use of the Hamiltonian for deriving a partial differential equation for the Bellman value function, which, in turn, identifies the corresponding feedback policy. Although the dynamic programming approach is a powerful tool for developing optimal feedback policies, partial differential equations are not easy to solve, especially when the problem is featured with control and state constraints. In such a case, approximate (discrete-time) solutions can be found for continuous-time problems.

On the other hand, the use of the Hamiltonian could be extended to study optimal behavior of the dynamic system under an open-loop control with respect to the state and co-state differential equations rather than solving the partial differential equation for a feedback control. An approach which focuses on optimizing the Hamiltonian is referred to as the maximum principle.

The maximum principle provides a set of the necessary optimality conditions for identifying whether a solution is optimal or not. When reduced to solving a two-point boundary value system of ordinary differential equations, the maximum principle provides us with an additional chance to solve the problem.

Consider the following dynamic problem

$$J = \int_0^T L(x,u,t)dt + R(x(T)) \rightarrow \min \qquad (A1.49)$$

$$\frac{dx}{dt} = f(x,u,t)), \ 0 \le t \le T , \qquad (A1.50)$$

$$g_k(u,t) \le 0 , \ k=1,\ldots,K, \qquad (A1.51)$$

$$x(0)=x_0. \qquad (A1.52)$$

According to the maximum principle, the optimal control is achieved by maximizing for each time point t, the Hamiltonian

$$H(x,\psi,u) = \psi^T(t)f(x,u,t) - L(x,u,t), \qquad (A1.53)$$

as a function of controls $u_m(t)$, $m=1,..,M$, where $\psi^T(t) = (\psi_1(t), ..., \psi_n(t))$ are continuous co-state variables which satisfy the following co-state differential equations

$$\dot{\psi} = -\frac{\partial H(x,\psi,u)}{\partial x}. \qquad (A1.54)$$

The correspondence between the co-state variable of the dynamic programming and of the maximum principle is $\dfrac{\partial B}{\partial x} = -\psi$ (see, for example, Basar and Olsder, 1999).

The boundary condition for (A1.54) is

$$\psi(T) = -\frac{\partial R(x(T))}{\partial x(T)}. \qquad (A1.55)$$

Note that the Hamiltonian is not required to be always differentiable in x. At a point, $\hat{x}_n$, where the derivative does not exist, the co-state equation changes to involve the sub-gradient of the Hamiltonian

$$\dot{\psi}_n(t) \in \partial H(\hat{x}_n) = [a,b], \qquad (A1.56)$$

where

$$a = \lim_{x_n \to \hat{x}_n, x_n < \hat{x}_n} -\frac{H(x_n) - H(\hat{x}_n)}{x_n - \hat{x}_n} \text{ and } b = \lim_{x_n \to \hat{x}_n, x_n > \hat{x}_n} -\frac{H(x_n) - H(\hat{x}_n)}{x_n - \hat{x}_n}. \quad (A1.57)$$

Other boundary constraints

Instead of initial boundary constraint (A1.52), the problem may have terminal constraints

$$x(T) = x_T. \qquad (A1.58)$$

and correspondingly a payment for initial state $\breve{R}(x(0))$ in place of $R(x(T))$. Then the boundary condition (A1.55) is replaced with

$$\psi(0) = \frac{\partial \breve{R}(x(0))}{\partial x(0)}. \qquad (A1.59)$$

If the dynamic process has a periodic nature so that the terminal states of a period, T, are identical to the initial states,

$$x(0) = x(T). \qquad (A1.60)$$

then the boundary constraint (A1.52) is replaced with

$$\psi(0) = \psi(T). \qquad (A1.61)$$

State constraints

Let a constraint on state n be imposed as
$$l_n(x_n(t)) \leq 0, \ 0 \leq t \leq T. \tag{A1.62}$$
Then the co-state variable $\psi_n(t)$ may have jumps which are presented by measures $d\mu_n(t) \geq 0$. As a result, the co-state equation takes the following form

$$d\psi_n = -\frac{\partial H(\psi, x, u)}{\partial x_n} dt + \frac{\partial l_n(x_n)}{\partial x_n} d\mu_n \tag{A1.63}$$

and a complementary slackness condition is added so that $d\mu_n(t) = 0$, when the corresponding state constraint is not active, $x_n(t) < 0$,

$$\int_0^T l_n(x_n(t)) d\mu_n = 0. \tag{A1.64}$$

The objective function with variable planning horizon

If the planning horizon is not fixed (a decision variable) as defined in (A1.9) and the objective function is

$$J = \int_0^T L(x(t), u(t), t) dt + R(x(T)) + S(T), \tag{A1.65}$$

where $S(T)$ is the cost associated with the length of the planning horizon, then the only change in the optimality conditions is that they include an additional constraint

$$H = \frac{\partial S(T)}{\partial T}, \ 0 \leq t \leq T. \tag{A1.66}$$

Example A1.4

Consider again the problem from Example A1.2:

$$J = \int_0^T u^2(t) dt + a x^2(T) \to \min,$$

$$\dot{x}(t) = u(t), \ x(0) = x_0,$$

where x^0 and $a > 0$ are given constants and $u(t) \in R^1$. Now, instead of dynamic programming-based minimization in Example A1.2, we apply the maximum principle, where the optimal control is found by maximizing the Hamiltonian

$$H = -u^2(t) + \psi(t) u(t).$$

The co-state equation is

$$\dot{\psi} = -\frac{\partial H}{\partial x} = 0$$

with the boundary condition

$$\psi(T) = -2ax(T).$$

Thus we conclude that $\psi(t) = -2ax(T)$, $0 \le t \le T$. Since there are no constraints imposed on control, to find the optimal control we simply differentiate the Hamiltonian,

$$\frac{\partial H}{\partial u} = -2u + \psi = 0,$$

which results in $u = \dfrac{\psi}{2} = -ax(T)$. Note that the Hamiltonian is concave and thus the first order optimality condition is not only necessary but also sufficient. Integrating the state differential equation we have

$$x(T) = x_0 - \int_0^T ax(T)dt = x_0 - ax(T)T.$$

That is, $x(T) = \dfrac{x_0}{1 + aT}$, which along with $u = -ax(T)$ results in

$$u*(t) = -\frac{ax^0}{1 + aT}, \quad 0 \le t \le T.$$

Thus we have found an open-loop optimal control. Next we transform the control into a closed loop form. For this equivalent representation, we first express x_0 from the open-loop equation

$$x^0 = -\frac{u(1 + aT)}{a}.$$

Consequently, we substitute x_0 with the last expression in the state equation, $x(t) = x_0 + \int_0^t u d\tau$, which results in

$$u = \frac{x - x^0}{t} = \frac{x}{t} + \frac{u(1 + aT)}{ta},$$

that is, $u*(x,t) = \dfrac{-ax}{1 - a(t - T)}$, which is identical to the optimal feedback control found with dynamic programming in Example A1.2. Moreover, recalling that $u = \dfrac{\psi}{2}$, we obtain the co-state variable

$$\psi = -\frac{2ax}{1 - a(t - T)}.$$

This value differs only in sign from that found with dynamic programming in Example A1.2,

$$\frac{\partial B}{\partial x} = \frac{2ax}{1 - a(t - T)}.$$

The difference in signs, $\dfrac{\partial B}{\partial x} = -\psi$ of the two shadow prices is due to the fact that in contrast to the maximum principle, the corresponding Hamiltonian of the dynamic programming formulation that we employed is intended for minimization.

A1.5   NON-COOPERATIVE DYNAMIC GAMES

Game theory is concerned with situations involving conflict and cooperation between the players. Until now we assumed that there was a single object, controlled by a single decision maker (player) who was willing to optimize the object's behavior with respect to some objective. However, in real life, a number of decision makers who are not necessarily ready to cooperate may control several objects. In this section we discuss the optimality conditions in dynamic non-cooperative games which are a natural extension of the principles of optimality discussed in previous sections. Dynamic games arise when players gain some dynamic information throughout the decision process and use it over time. Our focus is on two important concepts – Nash and Stackelberg equilibriums – for dealing respectively with simultaneous and sequential decision-making among multiple players.

Similar to the static games, to find Nash equilibrium, we would need the optimality conditions of each player to be met simultaneously. This implies solving jointly the number of optimal control problems equal to the number of players. Accordingly, to determine a Stackelberg equilibrium, we need to find first the best response function of the follower (solve the follower's control problem) and then optimize the leader's behavior subject to the best follower's response. Since optimal solution of any single player control problem is either open-loop or closed-loop, the simultaneous optimization in a multi-player game is either open- or closed-loop. As discussed earlier, the maximum principle enables us to study an open-loop solution, while the dynamic programming allows us to search for a feedback solution. Thus, utilization of the two standard techniques of optimal control theory, the maximum principle and dynamic programming leads to open-loop and feedback equilibrium solutions respectively. This may be seen, for example, in discrete-time stochastic problems involving two players A and B.

$$J_0^A(x^0,\mathbf{u}_A^0,\mathbf{u}_B^0) = E\{\sum_{i=0}^{K-1} L_A^i(x^i,u_A^i,u_B^i) + R_A(x^K)\} \to \inf,$$

$$J_0^B(x^0,\mathbf{u}_A^0,\mathbf{u}_B^0) = E\{\sum_{i=0}^{K-1} L_B^i(x^i,u_A^i,u_B^i) + R_B(x^K)\} \to \inf,$$

s.t.

$$x^{i+1} = f^i(x^i,u_A^i,u_B^i,w^i), \ i=0,1,..,K-1, \ x^0{=}x_0,$$

$$\mathbf{u}_A^0{=}(u_A^0,u_A^1,..,u_A^{K-1}), \ u_A^i \in U_A^i, \ U_A^i \subseteq R^{AM}, \ U_A^i \subseteq R^{BM}, \ i=0,1,..,K-1,$$

$$\mathbf{u}_B^0{=}(u_B^0,u_B^1,..,u_B^{K-1}), \ u_B^i \in U_B^i, \ i=0,1,..,K-1.$$

The set of strategies $\{u_A^{*k}{=}u_A^*(x^k), u_B^{*k}{=}u_B^*(x^k), k{=}0,1,..,K{-}1\}$ provides a feedback Nash equilibrium solution if there exist functions with starting state value x_k at time t_k,

$$B_k^A(x^k) = \inf_{u_A^i \in U_A^i, i=k,..,K-1} J_k^A(x^k,\mathbf{u}_A^k,\mathbf{u}_B^{*k}),$$

$$B_k^B(x^k) = \inf_{u_B^i \in U_B^i, i=k,..,K-1} J_k^B(x^k,\mathbf{u}_A^{*k},\mathbf{u}_B^k), \ k{=}0,1,..,K{-}1,$$

such that the recursive dynamic programming equations

$$B_k^A(x^k) = \inf_{U_A^k}\{E[L_A^k(x^k,u_A^k,u_B^{*k}) + B_{k+1}^A(f^k(x^k,u_A^k,u_B^{*k},w^k)]\}, \ k{=}0,1,..,K{-}1,$$

$$B_k^B(x^k) = \inf_{U_B^k}\{E[L_B^k(x^k,u_A^{*k},u_B^k) + B_{k+1}^B(f^k(x^k,u_A^{*k},u_B^k,w^k))]\}, \ k{=}0,1,..,K{-}1,$$

$$B_K^A(x^K){=}R_A(x^K), \ B_K^B(x^K){=}R_B(x^K)$$

are met.

On the other hand, with the aid of the maximum principle for the continuous-time deterministic problems involving two players A and B with control variables $u_A^T{=}(u_{A1},..,u_{AM})$ and $u_B^T{=}(u_{B1},..,u_{BM})$ we can deal with the following problem for each of the players

$$J^A = \int_0^T L_A(x,u_A,u_B,t)dt + R_A(x(T)) \to \min$$

$$J^B = \int_0^T L_B(x,u_A,u_B,t)dt + R_B(x(T)) \to \min$$

$$\dot{x}(t) = f(x,u_A,u_B,t), \ 0 \le t \le T,$$

$$x(0){=}x_0,$$

$$g_k^A(u_A,t)) \le 0, \ g_i^B(u_B,t)) \le 0, \ k{=}1,\ldots,K_A, \ i{=}1,\ldots,K_B.$$

The set of strategies $\{u^*_A=u_A^*(t),\ u^*_B=u_B^*(t),\ 0 \le t \le T \}$ provides an open-loop Nash equilibrium solution if the following conditions hold.

The optimal control maximizes for each time point t, the Hamiltonian of player A

$$H^A(x,\psi_A,u_A,u_B^*) = \psi_A^T(t)f(x,u_A,u_B^*,t) - L_A(x,u_A,u_B^*,t)$$

with respect to u_A, $g^A_k(u_A,t)) \le 0$, k=1,...,K_A and the Hamiltonian of player B

$$H^B(x,\psi_B,u_A^*,u_B) = \psi_B^T(t)f(x,u_A^*,u_B,t) - L_B(x,u_A^*,u_B,t)$$

with respect to u_B, $g^B_i(u_{B,}t) \le 0$, i=1,...,K_B.

The co-state variables, $\psi_A(t)$ and $\psi_B(t)$, satisfy the following co-state differential equations

$$\dot{\psi}_A = -\frac{\partial H^A}{\partial x} \quad \text{and} \quad \dot{\psi}_B = -\frac{\partial H^B}{\partial x}$$

with boundary conditions

$$\psi_A(T) = -\frac{\partial R_A}{\partial x(T)} \quad \text{and} \quad \psi_B(T) = -\frac{\partial R_B}{\partial x(T)}.$$

Example A1.5

Consider a modification of the problem from Example A1.2. Let there be a retailer B who orders products from a supplier or manufacturer A. In response, the supplier, either produces or orders products from a distributor. The difference between the cumulative production rate of supplier A and the cumulative order rate of the retailer B constitutes the supplier's warehouse inventory level, $x(t)$:

$$\dot{x}(t) = \beta u_A(t) - u_B(t),\, x(0)=x^0>0,$$

where $0<\beta<1$ is the production efficiency coefficient.

The supplier's objective is to maximize payments for the retailer's orders, $u_B^2(t)$ and to minimize production cost, $u_A^2(t)$, both over the planning horizon. The cost of leftovers is minimized by the end of the horizon, $ax^2(T)$.

$$J_A = \int_0^T (u_A^2(t) - u_B^2(t))dt + ax^2(T) \rightarrow \min,$$

The retailer, on the other hand, naturally wants to pay less for his orders, but is interested in having the supplier produce more products and thus spend more on production. The reason for this is if the supplier produces more, he is likely to have leftovers which the retailer can utilize to ensure high level customer service.

$$J_B = -\int_0^T (u_A^2(t) - u_B^2(t))dt - ax^2(T) \to \min$$

This is an example of the classical zero-sum game, $J_B + J_A = 0$, which implies $Ha = -H_b$ and $\psi_A = -\psi_B$.

Let $(1 - \beta^2)aT > 1$. To find an open-loop Nash equilibrium, we first construct the Hamiltonians:

$$H_B = u_A^2(t) - u_B^2(t) + \psi_B(t)(\beta u_A(t) - u_B(t)),$$
$$H_A = -u_A^2(t) + u_B^2(t) + \psi_A(t)(\beta u_A(t) - u_B(t)).$$

Note that since the Hamiltonians are concave, the first-order optimality condition is not only necessary but also sufficient. The co-state equations are

$$\dot{\psi}_A = -\frac{\partial H_A}{\partial x} = 0 \text{ and } \dot{\psi}_B = -\frac{\partial H_B}{\partial x} = 0$$

with the boundary conditions

$$\psi_A(T) = -2ax(T) \text{ and } \psi_B(T) = 2ax(T).$$

Thus we conclude that $\psi_A(t) = -2ax(T)$ and $\psi_B(t) = 2ax(T)$, $0 \le t \le T$. Since there are no constraints imposed on control, to find optimal control we simply differentiate the Hamiltonians,

$$\frac{\partial H_A}{\partial u_A} = -2u_A + \beta\psi_A = 0 \text{ and } \frac{\partial H_B}{\partial u_B} = -2u_B - \psi_B = 0,$$

which results in $u_A = \dfrac{\beta\psi_A}{2} = -a\beta x(T)$ and $u_B = -\dfrac{\psi_B}{2} = -ax(T)$. Integrating the state differential equation we have

$$x(T) = x^0 - \int_0^T \beta^2 ax(T)dt + \int_0^T ax(T)dt = x^0 + (1 - \beta^2)ax(T)T.$$

That is, $x(T) = \dfrac{x^0}{1 - (1 - \beta^2)aT}$, which is along with $\dfrac{u_A}{\beta} = u_B = -ax(T)$

results in

$$u_A * (t) = -\frac{\beta a x^0}{1 - (1 - \beta^2)aT}, \ u_B * (t) = -\frac{a x^0}{1 - (1 - \beta^2)aT}, \ 0 \le t \le T.$$

As a result, we have found an open-loop optimal control. Next we transform the control into a closed-loop form. To do this, we first express x^0 from the open-loop equation

$$x^0 = -\frac{u_B(1 - (1 - \beta^2)aT)}{a}.$$

Consequently, we substitute x^0 with the last expression and $u_A = \beta u_B$ in the state equation, $x(t) = x^0 + \int_0^t (\beta u_A(t) - u_B(t)) dt$ and, hence, $x = x^0 + \beta^2 u_B t - u_B t$ which results in

$$u_B *(x,t) = -\frac{ax(t)}{1 + a(1 - \beta^2)(t - T)},$$

and since, $u_A = \beta u_B$,

$$u_A *(x,t) = -\frac{\beta a x(t)}{1 + a(1 - \beta^2)(t - T)}.$$

Another way to find a feedback Nash equilibrium is to employ corresponding continuous-time dynamic programming conditions. Specifically, using the same definition of the Hamiltonian as in the maximum principle, we have for the Hamiltonian-Jacobi-Bellman equations:

$$\max_{u_A} H^A(x, \psi_A, u_A, u_B^*, t) = -\frac{\partial B^A(x,t)}{\partial t},$$

$$g^A_k(u_A, t) \le 0, \ k=1,\ldots,K_A$$

$$\max_{u_B} H^B(x, \psi_B, u_A^*, u_B, t) = -\frac{\partial B^B(x,t)}{\partial t},$$

$$g^B_i(u_B, t) \le 0, \ i=1,\ldots,K_B$$

$$B^A(x, T) = -R_A(x) \text{ and } B^B(x, T) = R_B(x),$$

where $\psi_A = \dfrac{\partial B^A(x,t)}{\partial x}$ and $\psi_B = \dfrac{\partial B^B(x,t)}{\partial x}$.

Example A1.6

Once again we resolve here Example A1.5 to find a feedback a Nash equilibrium with dynamic programming rather than the maximum principle. As in Example A1.5, from the same Hamiltonians

$$H_A = -u_A^2(t) + u_B^2(t) + \psi_A(t)(\beta u_A(t) - u_B(t)),$$
$$H_B = u_A^2(t) - u_B^2(t) + \psi_B(t)(\beta u_A(t) - u_B(t)),$$

we have $u_A = \dfrac{\beta \psi_A}{2}$ and $u_B = -\dfrac{\psi_B}{2}$. Substituting this into the Hamiltonian-Jacobi-Bellman equations

$$H^A = -\frac{\partial B^A(x,t)}{\partial t} \text{ and } H^B = -\frac{\partial B^B(x,t)}{\partial t}$$

and taking into account that $\psi_A = \dfrac{\partial B^A}{\partial x}$, $\psi_B = \dfrac{\partial B^B}{\partial x}$, we obtain

$$\frac{1}{4}\left(\frac{\beta \partial B^A}{\partial x}\right)^2 + \frac{1}{4}\left(\frac{\partial B^B}{\partial x}\right)^2 + \frac{\partial B^A}{\partial x}\frac{\partial B^B}{2\partial x} = -\frac{\partial B^A(x,t)}{\partial t},$$

$$\frac{1}{4}\left(\frac{\beta \partial B^A}{\partial x}\right)^2 + \frac{1}{4}\left(\frac{\partial B^B}{\partial x}\right)^2 + \frac{\partial B^B}{\partial x}\frac{\beta^2 \partial B^A}{2\partial x} = -\frac{\partial B^B(x,t)}{\partial t}.$$

Recalling that for a zero-sum game, $\psi_A = -\psi_B$, and $B^B(x,t)=-B^A(x,t)$, we find a single partial differential equation

$$\frac{1}{4}\left(\frac{\partial B^A}{\partial x}\right)^2 (\beta^2 - 1) = -\frac{\partial B^A(x,t)}{\partial t}, \quad B^A(x,\,T)=-ax^2(T).$$

We search for solution $B(x,t)$ of this partial differential equation in the form of a quadratic function

$$B(x,t)= A(t)x^2.$$

Substituting this expression into the derived partial differential equation along with the boundary condition, we have

$$(\beta^2 - 1)x^2 A^2 + \dot{A}x^2 = 0$$
$$A(T)x^2 = -ax^2$$

which leads to the following ordinary differential equation

$$(\beta^2 - 1)A^2 + \dot{A} = 0, \; A(T)=-a.$$

Thus, we find

$$A(t) = \frac{a}{a(\beta^2 - 1)(t - T) - 1}$$

and, therefore, the Bellman function and the co-state variable for player A are

$$B^A(x,t) = \frac{ax^2}{a(\beta^2 - 1)(t - T) - 1} \text{ and } \psi^A = \frac{\partial B^A(x,t)}{\partial x} = \frac{2ax}{a(\beta^2 - 1)(t - T) - 1}.$$

Finally, the optimal feedback policies are

$$u_A^*(x,t) = \frac{\beta \psi_A}{2} = -\frac{a\beta x(t)}{a(1 - \beta^2)(t - T) + 1},$$

$$u_B^*(x,t) = -\frac{\psi_B}{2} = \frac{\psi_A}{2} = -\frac{ax(t)}{a(1 - \beta^2)(t - T) + 1}.$$

This result is identical to that from Example 1.5 which was derived from an open-loop Nash equilibrium.

REFERENCES

Basar T, Olsder GL (1982). *Dynamic Noncooperative Game Theory*. Academic Press, London.

Kogan K, Khmelnitsky E (2000). *Scheduling*: *Control-based Theory and Polynomial-Time algorithms*, Kluwer Academic Publisher, Boston.

Tapiero CS (1988) *Applied Stochastic Models and Control in Management*, North-Holland, Amsterdam-New York.

INDEX

Early Titles in the
INTERNATIONAL SERIES IN
OPERATIONS RESEARCH & MANAGEMENT SCIENCE
Frederick S. Hillier, Series Editor, *Stanford University*

Early Titles in the
INTERNATIONAL SERIES IN
OPERATIONS RESEARCH & MANAGEMENT SCIENCE
(Continued)

Early Titles in the
INTERNATIONAL SERIES IN
OPERATIONS RESEARCH & MANAGEMENT SCIENCE
(Continued)

Reveliotis/ *REAL-TIME MANAGEMENT OF RESOURCE ALLOCATIONS SYSTEMS: A Discrete Event Systems Approach*
Kall & Mayer/ *STOCHASTIC LINEAR PROGRAMMING: Models, Theory, and Computation*

* A list of the more recent publications in the series is at the front of the book *

Printed in the United States of America